Anthropogenic Pollution of Aquatic Ecosystems

AF841799

Donat-P. Häder · E. Walter Helbling ·
Virginia E. Villafañe
Editors

Anthropogenic Pollution of Aquatic Ecosystems

Springer

Editors
Donat-P. Häder
University of Erlangen-Nürnberg
Erlangen, Germany

E. Walter Helbling
Estación de Fotobiología Playa Unión
Rawson, Argentina

Virginia E. Villafañe
Estación de Fotobiología Playa Unión
Rawson, Argentina

ISBN 978-3-030-75604-8 ISBN 978-3-030-75602-4 (eBook)
https://doi.org/10.1007/978-3-030-75602-4

© The Editor(s) (if applicable) and The Author(s), under exclusive license to Springer Nature Switzerland AG 2021
This work is subject to copyright. All rights are solely and exclusively licensed by the Publisher, whether the whole or part of the material is concerned, specifically the rights of translation, reprinting, reuse of illustrations, recitation, broadcasting, reproduction on microfilms or in any other physical way, and transmission or information storage and retrieval, electronic adaptation, computer software, or by similar or dissimilar methodology now known or hereafter developed.
The use of general descriptive names, registered names, trademarks, service marks, etc. in this publication does not imply, even in the absence of a specific statement, that such names are exempt from the relevant protective laws and regulations and therefore free for general use.
The publisher, the authors and the editors are safe to assume that the advice and information in this book are believed to be true and accurate at the date of publication. Neither the publisher nor the authors or the editors give a warranty, expressed or implied, with respect to the material contained herein or for any errors or omissions that may have been made. The publisher remains neutral with regard to jurisdictional claims in published maps and institutional affiliations.

This Springer imprint is published by the registered company Springer Nature Switzerland AG
The registered company address is: Gewerbestrasse 11, 6330 Cham, Switzerland

Preface

Aquatic ecosystems exceed terrestrial ones in size and rival them in productivity. In addition to offering food to a rapidly growing human population, they render a wide array of services. The oceans take up most of the anthropogenically produced heat and a large share of the emitted trace gases such as CO_2 from human activities. The productivity of the oceans is affected by a plethora of environmental stress factors such as rising temperatures, ocean acidification and deoxygenation. Of growing importance is the increasing pollution of aquatic habitats. Because of the enormous size of the oceans, humans have always used them to get rid of their wastes. With the explosion of the human race, this behavior is becoming a burning problem because in addition to biodegradable material, toxic and long-lived wastes are being dumped into aquatic ecosystems. Terrestrial runoff transports fertilizers into coastal ecosystems which increase eutrophication. In addition, pesticides, drugs and pharmaceuticals accumulate in the water. The growing industry and lacking or inadequate cleaning strategies result in heavy metals concentrating in the rivers which transport these toxic materials into the oceans. Accidental and deliberate oil spills are found both in coastal areas and in open ocean waters affecting the biota from microorganisms to vertebrates. After their invention, plastic materials were celebrated as cheap, non-toxic and versatile blessings for human life. But, the accidental or deliberate disposal in the landscape, coastal zones and the oceans has accumulated the debris to enormous waste piles. Solar UV radiation and mechanical forces of wind and waves result in the fragmentation of larger wastes into mico- and nanoplastics which are a threat for most marine organisms which ingest the particles which subsequently bioconcentrate in the food web. All these growing polluting materials threaten our race, and fast and efficient solutions have to be developed to stop the attacks on human health.

Erlangen, Germany — Donat-P. Häder
Rawson, Argentina — E. Walter Helbling
Rawson, Argentina — Virginia E. Villafañe
Spring 2021

Contents

Contributors

Anthony Andrady Department of Chemical and Biomolecular Engineering, North Carolina State University, Raleigh, NC, USA

Marina Arcagni Centro Atómico Bariloche, Comisión Nacional de Energía Atómica, Bariloche, Argentina

Azizullah Azizullah Department of Biology, The University of Haripur, Haripur, Pakistan

Anastazia T. Banaszak Integrative Reef Conservation Research Laboratory, Unidad Académica de Sistemas Arrecifales, Instituto de Ciencias Del Mar Y Limnología, Universidad Nacional Autónoma de México, Puerto Morelos, Mexico

José Bonomi Baruffi Phycology Laboratory, Botany Department, Federal University of Santa Catarina, Florianópolis, Santa Catarina, Brazil;
Center of Physical and Mathematical Sciences, Federal University of Santa Catarina, Florianópolis, Santa Catarina, Brazil

Eduardo Bastos Phycology Laboratory, Botany Department, Federal University of Santa Catarina, Florianópolis, Santa Catarina, Brazil

Göran Björk Dept. Marine Sciences, University of Gothenburg, Gothenburg, Sweden

Marco J. Cabrerizo Centro de Investigación Mariña da Universidade de Vigo (CIM-UVigo), Vigo, Spain;
Departamento de Ecología y Biología Animal, Universidade de Vigo, Facultad de Ciencias del Mar, Vigo, Spain

Verena A. Campodonico Centro de Investigaciones en Ciencias de la Tierra, Consejo Nacional de Investigaciones Científicas y Técnicas y Universidad Nacional de Córdoba, Córdoba, Argentina

Giulia Burle Costa Phycology Laboratory, Botany Department, Federal University of Santa Catarina, Florianópolis, Santa Catarina, Brazil;
Center of Physical and Mathematical Sciences, Federal University of Santa Catarina, Florianópolis, Santa Catarina, Brazil

Enrique Crespo Centro para el Estudio de Sistemas Marinos (CESIMAR)—Centro Nacional Patagónico—CONICET, Puerto Madryn, Chubut, Argentina

Lineu del Ciampo Program in Health and Environment, University of Joinville Region Univille, Joinville, SC, Brazil

Pedro J. Depetris Academia Nacional de Ciencias, Córdoba, Argentina

Giovanna Destri Phycology Laboratory, Botany Department, Federal University of Santa Catarina, Florianópolis, Santa Catarina, Brazil;
Center of Physical and Mathematical Sciences, Federal University of Santa Catarina, Florianópolis, Santa Catarina, Brazil

Cristian Durante Centro para el Estudio de Sistemas Marinos (CESIMAR)—Centro Nacional Patagónico—CONICET, Puerto Madryn, Chubut, Argentina

Gilmar S. Erzinger Postgraduate Program in Health and Environment, University of the Region of Joinville—UNIVILLE, Joinville, Santa Catarina, Brazil

Mauricio Faleschini Laboratorio de Oceanografía Química y Contaminación de Aguas, CESIMAR, CCT CONICET-CENPAT, Puerto Madryn, Chubut, Argentina

Ana Fazio Centro Para El Estudio de Sistemas Marinos (CESIMAR-CONICET, CCT CONICET-CENPAT), Puerto Madryn, Chubut, Argentina

Alessandra Larissa Fonseca Center of Physical and Mathematical Sciences, Federal University of Santa Catarina, Florianópolis, Santa Catarina, Brazil;
Marine Biogeochemistry Laboratory, Oceanography Department, Federal University of Santa Catarina, Florianópolis, Santa Catarina, Brazil

Monique Fröhlich Program in Health and Environment, University of Joinville Region Univille, Joinville, SC, Brazil

Kunshan Gao State Key Laboratory of Marine Environmental Science and College of Ocean and Earth Sciences, Xiamen University, Xiamen, China

Raúl González Escuela Superior de Ciencias Marinas, Universidad Nacional del Comahue and CONICET (CIM Almirante Storni), San Antonio Oeste, Río Negro, Argentina

Donat-P. Häder Department of Biology, Friedrich-Alexander Universität, Erlangen, Germany

E. Walter Helbling Estación de Fotobiología Playa Unión, Chubut and Consejo Nacional de Investigaciones Científicas y Técnicas (CONICET), Rawson, Chubut, Argentina;
Consejo Nacional de Investigaciones Científicas y Técnicas (CONICET), Buenos Aires, Argentina

Paulo Antunes Horta Phycology Laboratory, Botany Department, Federal University of Santa Catarina, Florianópolis, Santa Catarina, Brazil;
Center of Physical and Mathematical Sciences, Federal University of Santa Catarina, Florianópolis, Santa Catarina, Brazil

Peng Jin School of Environmental Science and Engineering, Guangzhou University, Guangzhou, China

Romina Juncos Centro Atómico Bariloche, Comisión Nacional de Energía Atómica, Bariloche, Argentina

Sarzamin Khan Department of Chemistry, University of Swabi, Swabi, Pakistan

Neha Kumari Laboratory of Photobiology and Molecular Microbiology, Centre of Advanced Study in Botany, Institute of Science, Banaras Hindu University, Varanasi, India

Carla Keite Machado Program in Health and Environment, University of Joinville Region Univille, Joinville, SC, Brazil

Sonal Mishra Laboratory of Photobiology and Molecular Microbiology, Centre of Advanced Study in Botany, Institute of Science, Banaras Hindu University, Varanasi, India

Maite Narvarte Escuela Superior de Ciencias Marinas, Universidad Nacional del Comahue and CONICET (CIM Almirante Storni), San Antonio Oeste, Río Negro, Argentina

Luis F. H. Niencheski Instituto de Oceanografia, Fundação Universidade Federal do Rio Grande, Rio Grande, RS, Brazil

Marina L. Nievas El Makte Centro Para El Estudio de Sistemas Marinos (CESIMAR-CONICET, CCT CONICET-CENPAT), Puerto Madryn, Chubut, Argentina;
Facultad Regional Chubut, Universidad Tecnológica Nacional, Puerto Madryn, Chubut, Argentina

Abha Pandey Laboratory of Photobiology and Molecular Microbiology, Centre of Advanced Study in Botany, Institute of Science, Banaras Hindu University, Varanasi, India

Andrea I. Pasquini Centro de Investigaciones en Ciencias de la Tierra, Consejo Nacional de Investigaciones Científicas y Técnicas y Universidad Nacional de Córdoba, Córdoba, Argentina

Rosana Polifroni Centro Para El Estudio de Sistemas Marinos (CESIMAR-CONICET, CCT CONICET-CENPAT), Puerto Madryn, Chubut, Argentina

Sabeela Rehman Department of Chemistry, University of Swabi, Swabi, Pakistan

Lyllyan Santos Rocha Phycology Laboratory, Botany Department, Federal University of Santa Catarina, Florianópolis, Santa Catarina, Brazil;
Center of Physical and Mathematical Sciences, Federal University of Santa Catarina, Florianópolis, Santa Catarina, Brazil

Leonardo Rubi Rörig Phycology Laboratory, Botany Department, Federal University of Santa Catarina, Florianópolis, Santa Catarina, Brazil;
Center of Physical and Mathematical Sciences, Federal University of Santa Catarina, Florianópolis, Santa Catarina, Brazil

Juan Seco Pon Instituto de Investigaciones Marinas y Costeras, IIMyC—Universidad Nacional de Mar del Plata—CONICET, Mar del Plata, Argentina

Marcela A. Sepúlveda Centro Para El Estudio de Sistemas Marinos (CESIMAR-CONICET, CCT CONICET-CENPAT), Puerto Madryn, Chubut, Argentina

Vidya Singh Laboratory of Photobiology and Molecular Microbiology, Centre of Advanced Study in Botany, Institute of Science, Banaras Hindu University, Varanasi, India

Rajeshwar P. Sinha Laboratory of Photobiology and Molecular Microbiology, Centre of Advanced Study in Botany, Institute of Science, Banaras Hindu University, Varanasi, India

Sebastian M. Strauch Postgraduate Program in Health and Environment, University of the Region of Joinville—UNIVILLE, Joinville, Santa Catarina, Brazil

Nadia Taimur Department of Botanical and Environmental Sciences, Kohat University of Science and Technology, Kohat, Pakistan

Américo I. Torres Laboratorio de Oceanografía Química y Contaminación de Aguas, CESIMAR, CCT CONICET-CENPAT, Puerto Madryn, Chubut, Argentina

Virginia E. Villafañe Estación de Fotobiología Playa Unión, Chubut and Consejo Nacional de Investigaciones Científicas y Técnicas (CONICET), Rawson, Argentina;
Consejo Nacional de Investigaciones Científicas y Técnicas (CONICET), Buenos Aires, Argentina

Juan I. Vizzo Estación de Fotobiología Playa Unión, Rawson, Chubut, Argentina; Consejo Nacional de Investigaciones Científicas y Técnicas (CONICET), Buenos Aires, Argentina

Sten-Åke Wängberg Dept. Marine Sciences, University of Gothenburg, Gothenburg, Sweden

Liping Zhu Center for Advanced Low-Dimension Materials, State Key Laboratory for Modification of Chemical Fibers and Polymer Materials, Donghua University, Shanghai, China

Chapter 1
Introduction

Donat-P. Häder, E. Walter Helbling, and Virginia E. Villafañe

Abstract Aquatic ecosystems cover more than 70% of our planet. They produce about the same amount of biomass as all terrestrial ecosystems taken together, and they absorb about 90% of the anthropogenically produced heat. They take up a large amount of the emitted CO_2 and transport it to the deep ocean, thus curbing global climate change. Aquatic producers and consumers are under the effect of multiple stressors such as rising temperature, acidification and UV radiation. Pollution is a major stress factor affecting all aquatic ecosystems including fresh, coastal and open ocean waters. The main pollutants are oil spills, persistant organic pollutants, fertilizers and pharmaceuticals which reach the hydrosphere by terrestrial runoff. Toxic chemicals and heavy metals have long been dumped into the oceans affecting marine life such as invertebrates and vertebrates. A more recent advent is the growing pollution by microplastics.

Keywords Aquatic productivity · Ocean acidification · Solar UV radiation · Climate change · Ocean warming · Interactive effects · Pollutants · Stress factors

Even though water is the second most abundant molecule in the universe, the Earth is the only planet in the solar system to hold a considerable amount of water. About 70.7% are covered by water which corresponds to 361.2 million sq. km. Most of the water is concentrated in the oceans with a mean depth of 3617 m. Less than 1% of the available water represents liquid freshwater plus between 1.75 and 2% frozen in glaciers, ice and snow (Khyade and Swaminathan 2016). The aquatic ecosystems fix about 50–60 Pg carbon per year in the form of CO_2 from the atmosphere. This amount is equal to the uptake by all terrestrial ecosystems combined (Falkowski

---

D.-P. Häder (✉)
Department of Biology, Friedrich-Alexander Universität, 91054 Erlangen, Germany
e-mail: donat@dphaeder.de

E. W. Helbling · V. E. Villafañe
Estación de Fotobiología Playa Unión, Chubut and Consejo Nacional de Investigaciones Científicas y Técnicas (CONICET), Casilla de Correos No 15, 9103 Rawson, Argentina

© The Author(s), under exclusive license to Springer Nature Switzerland AG 2021
D.-P. Häder et al. (eds.), *Anthropogenic Pollution of Aquatic Ecosystems*,
https://doi.org/10.1007/978-3-030-75602-4_1

2013) even though the standing crop of aquatic ecosystems is only about 1% of that of their terrestrial counterparts.

About 10% of the organic biomass produced in the process of photosynthesis sinks out of the photic zone into the deep sea in the form of dead organisms and fecal pellets which can be detected as marine snow (Laurenceau-Cornec et al. 2015). In addition, these pellets contain mineral forms of N and P as well as Ca, Mg and K. 99.5% of this organic material is mineralized by bacteria, and about 0.5% reaches the deep sea sediment. This removal of atmospheric carbon dioxide is called biological pump (Sigman and Haug 2006). The sinking velocity of the organic carbon is about 1 m per day, so that it may take more than 10 years to reach the ocean floor assuming an average depth of about 3600 m in the oceans. Formation of calcium carbonate shells of plankton such as coccolithophorids and foraminifera, animals such as worms and molluscs sequesters more CO_2 from the water column which also sediments to the deep seafloor; this process is called carbonate pump (Hain et al. 2014). The carbon which has reached the deep sea bottom will remain there for millions of years. With 36,000 million tons, this carbon reservoir is the largest on our planet.

1.1 The Aquatic Food Web

The main primary producers in aquatic habitats are phytoplankton which live in the photic zone. The lower limit is defined by the depth where 99% of the surface irradiance is absorbed. The top layer of the water column, that is in contact with the atmosphere, is called upper mixed layer (UML) in which most of the phytoplankton dwell. The lower boundary of the UML is the thermocline (pycnocline) below which the temperature abruptly drops and density increases. This stratification reduces the upwelling of nutrients from deeper layers into the UML (Fischer et al. 2017). Increasing temperatures fueled by global climate change augment the strength of the stratification and tend to diminish the depth of the UML. Organisms living in the UML are thus exposed to high solar visible and UV radiation. The global distribution of phytoplankton is not even, and the highest concentrations are found near the poles and coastal areas that hold the highest concentration of higher organisms in the food web such as fish and mammals. On the other hand, the central gyres of the oceans at mid-latitudes are deserts as can be visualized in pseudocolor images of the surface chlorophyll distribution (Fig. 1.1).

The growth of macroalgae is almost exclusively restricted to rocky shores of the oceans (with the notable exception of *Sargassum*) (Richmond and Stevens 2014). Algal beds and sea grass meadows are important habitats for larval invertebrates and vertebrates. Even though their productivity contributes only a fraction of that of phytoplankton, macroalgae provide a significant contribution of ecological and economic importance. Red, green and brown macroalgae can be either harvested in the wild or grown in aquaculture. They are used as food, fodder, fertilizer and biofuels (Seghetta et al. 2016; Mac Monagail et al. 2017). They are also a source of

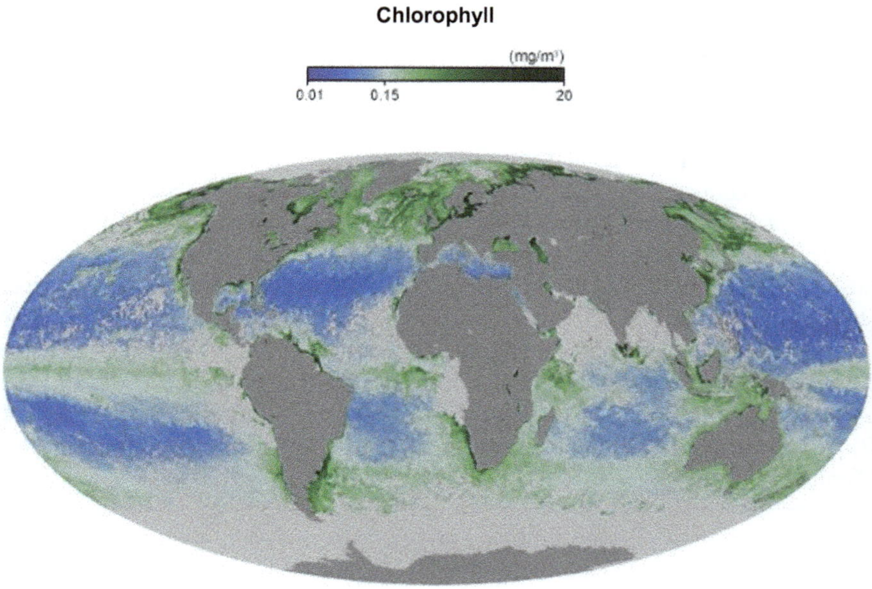

Fig. 1.1 Chlorophyll concentrations in the oceans as measured by satellite imaging in August 2020. Imagery produced by the Earth Observatory Group in coordination with Gene Feldman and Norman Kuring, NASA Goddard Ocean Color Group

polysaccharides, phycobiliproteins, mycosporine-like amino acids (Torres et al. 2019) and natural bioactive compounds for therapeutic uses (Ismail et al. 2020). Millions of tons are harvested every year for many technological applications (Konda et al. 2015; Garcia-Vaquero and Hayes 2016). Stress factors such as temperature and ocean acidification affect the algal communities and their interaction with grazers (Harley 2014; Sampaio et al. 2017). Terrestrial runoff carries dissolved organic matter (DOM) into coastal areas (Clark et al. 2016), changing the transparency and thus affecting the growth of macroalgae. In addition, increasing pollution by heavy metals, microplastics, fertilizers, fossil oil products as well as drugs and pharmaceuticals affect macroalgal species composition (Pinto 2015; Gubelit et al. 2016).

Prokaryotic and eukaryotic phytoplanktons as well as macroalgae and sea grasses form the basis of the branched marine food webs and feed the primary and secondary consumers (Hussey et al. 2014). These food webs are very dynamic and rapidly altered in response to changing environmental conditions and stress factors (Quillien et al. 2016; Ullah et al. 2018). Small changes in the assemblages of the primary producers can have considerable effects on the consumers. Increasing solar UV, higher temperatures and ocean acidification affect all elements in the marine food webs (Feister and Häder 2019; Gao et al. 2019).

1.2 Environmental Stress Factors

Since 1970, the atmospheric global temperatures have risen by 0.27 °C per decade, while the ocean surface temperatures have increased by 0.13 °C in the same time interval (Larson et al. 2020). This anthropogenic forcing has significant effects on the marine biota as evidenced, e.g., by massive coral bleaching (Sully et al. 2019) and increases in tropical cyclones and hurricanes (Halpern et al. 2008; Kossin et al. 2020). In addition, rising temperatures affect habitat selection and species composition of marine taxa (Thomas et al. 2012), such as diatoms and copepods (Jenkins and Black 2019), dinoflagellates (Brosnahan et al. 2020), coccolithophorids (Gafar et al. 2019) and cyanobacteria (Harada et al. 2019). Habitats defined by thermal niches will move toward the poles (Oziel et al. 2020); e.g., tropical radiolaria have been found in Arctic waters (http://earth.columbia.edu; Bjørklund et al. 2012), and in tropical phytoplankton species diversity declines (Thomas et al. 2012).

While the availability of CO_2 is essential for photosynthetic organisms, excessive concentrations are harmful and can be considered anthropogenic pollutants. Over the last millenium, the atmospheric CO_2 concentration has been relatively constant at about 270 ppm, but since the beginning of the Industrial Revolution around 1880, it steadily increased. Today, the carbon dioxide concentration in the atmosphere has risen to above 410 ppm (Boretti 2020). An increase to 650 ppm is considered as a red line, and passing it will increase the global temperature above the target of 2 °C (Knutti et al. 2016). Following a business-as-usual scenario in fossil fuel utilization will drive the atmospheric CO_2 concentration beyond 750 ppm by the end of the century (Santos et al. 2017).

The atmosphere and the water surface are in close equilibrium so that the oceans absorb CO_2 (Gill 2016) and take up about 25% of the anthropogenically emitted carbon dioxide (Gao et al. 2012b). The absorbed CO_2 forms carbonic acid with water which eventually dissociates into a carbonate ion and a proton which decreases the pH of the water causing ocean acidification (Doney et al. 2016).

$$CO_2 + H_2O \rightarrow H_2CO_3 \tag{1}$$

$$H_2CO_3 \rightarrow H^+ + HCO_3^- \tag{2}$$

$$HCO_3^- \rightarrow H^+ + CO_3^{2-} \tag{3}$$

If the release of CO_2 into the atmosphere continues at the current rate, a pH drop is predicted by 0.3–0.4 units (Feely et al. 2004). Because the CO_2 availability is a bottleneck for photosynthesis, initially, it was assumed that any increase could augment productivity in aquatic primary producers. But detailed analysis gave inconsistent results (Gao and Campbell 2014). Some species indeed showed an increase in productivity, while others did not, and some even had a reduced growth rate at elevated CO_2 concentrations (Chen and Gao 2004; Kim 2006). In some diatoms, increased CO_2 levels augment growth and productivity (Low-Décarie et al. 2011; King et al. 2011). A study that compiled 69 experiments on ocean acidification (Bach and Taucher 2019) concluded that diatoms responded to high CO_2 in ca 60% of the studies, and from these almost 56% responded in a positive way (benefitted), while in 32% it had a negative impact. Ocean acidification can affect orientation and survival in fish larvae and reduce the fecundity of copepods (Siebeck et al. 2015; Thor and Dupont 2015). Thus, it may affect the community structure of phytoplankton populations with negative consequences for the food web.

Some green and red macroalgae have incrustations of $CaCO_3$ in their thallus, and some phytoplankton such as coccolithophorids and radiolaria have an exoskeleton of calcium carbonate (Marszalek 1975; Gao et al. 1993). Also, many zoological taxa, such as worms, mollusks, corals and cephalopods, have a calcified exo- or endoskeleton (Marin et al. 2012; Ries et al. 2009). Ocean acidification hampers the incorporation of calcium carbonate (Doney et al. 2016; Jokiel et al. 2016), and the organisms are exposed to higher solar UV radiation which is absorbed by the calcium carbonate skeleton (Gao et al. 2012a, b; Häder et al. 2015b).

Even though solar radiation is an essential requirement for all photosynthetic producers, excessive irradiances are a stress factor for the biota. In addition to extreme visible irradiances, high-energetic solar UV damages biomolecules, organelles, cells and organisms. UV-C radiation is quantitatively absorbed by oxygen and ozone in the atmosphere, and UV-B is strongly reduced by stratospheric and tropospheric ozone (Häder and Tevini 1987). In contrast, a large fraction of UV-A reaches the surface of the Earth (Steinbrecht et al. 2009). Anthropogenic release of chlorofluorocarbons (CFCs) and other trace gases catalytically damaged the stratospheric ozone layer resulting an increase in solar UV-B radiation starting in the late 1970s (Hoffmann et al. 2014). However, the ratification of the Montreal Protocol and its later amendments reversed this trend (Abbasi and Abbasi 2017), but because of the long lifetimes of the CFCs in the stratosphere, it is assumed that it will take half a century to reach pre-1980s values (Hoffmann et al. 2014). Climate change will affect future ozone depletion and solar UV levels (UNEP EEAP 2016).

UV-B radiation is absorbed by many biological targets such as proteins, lipids and membranes. It damages the DNA in organisms living in the photic zone and affects photosynthesis in primary producers which are exposed to solar radiation. Furthermore, it induces the production of reactive oxygen species when absorbed by cellular targets and organic matter in the water (Pallela 2014; Amado et al. 2015). Because of the shoaling of the UML, organisms experience higher UV radiation, even though the surface irradiance does not increase (Häder and Gao 2017; Gao and Häder 2017). Solar UV has been found to modify pollutants and often increase their toxicity such as crude oil products.

1.3 Aquatic Pollution

Given the fast-growing population, our species is not seriously protective of the limited resources on our planet (Campbell et al. 2017). Anthropogenic pollution of aquatic ecosystems threatens both the ecosystems and the dwindling food production in the oceans (Kirkman et al. 2020). Rather than recycling used materials, most end products are dumped either on land or at sea. Depositing excavated soil or material from torn-down buildings may not be seriously disastrous, but runoff from agricultural areas changes the ecostructure of coastal habitats and the food web (Myles 2017). Increasing concentrations of fertilizers and pesticides favor the expansion of marine dead zones with anoxic waters (Gattuso et al. 2018). Pharmaceuticals, drugs and household cleansers affect the coastal flora and fauna (Pariatamby and Bhatti 2020). Deliberate dumping of hazardous chemicals into the ocean results in biochemical and anatomical malformation and affects the size, growth and reproduction of invertebrates and vertebrates including sensitive fish species for human consumption (Clark et al. 2019). In addition, toxic materials such as heavy metals are waste products of industry which are accidentally or deliberately discarded into rivers and eventually reach a final destination in the oceans (Al Naggar et al. 2018).

Deposited arms and ammunition especially after the two world wars pose serious threats to the biota and human populations (Voie and Mariussen 2017). Nuclear wastes from power plants and warfare arsenal are difficult to dispose of, and many countries are still searching for safe and permanent repositories for radioactive materials. One periculous solution was found again by dumping the material in the oceans (Stolberg 2016). This strategy is also used to get rid of outdated space machinery such as satellites, rockets and spacecraft.

Of growing concern is the disposal of plastic debris which either reaches the oceans by riverine transport or by winds and littering of coastal areas. This material is broken down by mechanical forces of wind and waves augmented by solar UV radiation (Díaz-Mendoza et al. 2020). Since the resulting microplastics have very long lifetimes on the order of decades to centuries, it accumulates both floating in the oceanic gyres or in the sediments. It affects the biota since it is mistaken for food by plankton and invertebrates and is bioaccumulated in the food web. Many fish samples in species destined for human consumption have been found to contain microplastic debris (Barboza et al. 2018).

This volume covers the major groups of pollutants in aquatic ecosystems and draws the conclusions on the damaging effects for the biota covering the individual taxa and the food web.

References

Abbasi S, Abbasi T (2017) Monitoring ozone loss and its consequences: past, present, and future. Ozone Hole. Springer, New York, NY, pp 121–131

Al Naggar Y, Khalil MS, Ghorab MA (2018) Environmental pollution by heavy metals in the aquatic ecosystems of Egypt. Open Acc J Toxicol 3:555603

Amado AM, Cotner JB, Cory RM, Edhlund BL, McNeill K (2015) Disentangling the interactions between photochemical and bacterial degradation of dissolved organic matter: amino acids play a central role. Microb Ecol 69(3):554–566

Bach T, Taucher J (2019) CO_2 effects on diatoms: a synthesis of more than a decade of ocean acidification experiments with natural communities. Ocean Sci J 15:1159–1175

Barboza LGA, Vethaak AD, Lavorante BR, Lundebye A-K, Guilhermino L (2018) Marine microplastic debris: an emerging issue for food security, food safety and human health. Mar Pollut Bull 133:336–348

Bjørklund KR, Kruglikova SB, Anderson OR (2012) Modern incursions of tropical radiolaria into the Arctic Ocean. J Micropalaeontology 31:139–158

Boretti A (2020) Covid 19 impact on atmospheric CO_2 concentration. Int J Glob Warming 21 (3):317–323

Brosnahan ML, Fischer AD, Lopez CB, Moore SK, Anderson DM (2020) Cyst-forming dinoflagellates in a warming climate. Harmful Algae 91:101728

Campbell BM, Beare DJ, Bennett EM, Hall-Spencer JM, Ingram JS, Jaramillo F, Ortiz R, Ramankutty N, Sayer JA, Shindell D (2017) Agriculture production as a major driver of the Earth system exceeding planetary boundaries. Ecol Soc 22(4)

Chen X, Gao K (2004) Characterization of diurnal photosynthetic rhythms in the marine diatom *Skeletonema costatum* grown in synchronous culture under ambient and elevated CO_2. Funct Plant Biol 31(4):399–404

Clark CD, De Bruyn WJ, Aiona PD (2016) Temporal variation in optical properties of chromophoric dissolved organic matter (CDOM) in Southern California coastal waters with nearshore kelp and seagrass. Limnol Oceanogr 61(1):32–46

Clark B, Longo SB, Clausen R, Auerbach D (2019) From sea slaves to slime lines: commodification and unequal ecological exchange in global marine fisheries. In: Ecologically Unequal Exchange. Springer, pp 195–219

Díaz-Mendoza C, Mouthon-Bello J, Pérez-Herrera NL, Escobar-Díaz SM (2020) Plastics and microplastics, effects on marine coastal areas: a review. Environ Sci Pollut Res:1–10

Doney SC, Fabry VJ, Feely RA, Kleypas JA (2016) Ocean acidification: the other CO_2 problem. Wash J Envtl L Pol'y 6:213

Falkowski P (2013) Primary productivity in the sea, vol 19. Springer Science & Business Media

Feely RA, Sabine CL, Lee K, Berelson W, Kleypas J, Fabry VJ, Millero FJ (2004) Impact of anthropogenic CO_2 on the $CaCO_3$ system in the oceans. Science 305(5682):362–366

Feister U, Häder D-P (2019) Solar UV radiation and penetration into water. In: Häder D-P, Gao K (eds) Aquatic Ecosystems in a Changing Climate. Taylor & Francis Group, Boca Raton, pp 12–31

Fischer CJ, Zhang J-Z, Baringer MO (2017) An estimate of diapycnal nutrient fluxes to the euphotic zone in the Florida Straits. Sci Reports 7:16098

Gafar N, Eyre B, Schulz K (2019) Particulate inorganic to organic carbon production as a predictor for coccolithophorid sensitivity to ongoing ocean acidification. Limnol Oceanogr Lett 4(3):62–70

Gao K, Campbell D (2014) Photophysiological responses of marine diatoms to elevated CO_2 and decreased pH: a review. Func Plant Biol 41:449–459. https://doi.org/10.1071/fp13247

Gao K, Häder D-P (2017) Effects of ocean acidification and UV radiation on marine photosynthetic carbon fixation. In: Kumar M, Ralph PJ (eds) Systems biology of marine ecosystems. Springer, Cham, Switzerland, pp 235–250

Gao K, Aruga Y, Asada K, Ishihara T, Akano T, Kiyohara M (1993) Calcification in the articulated coralline alga *Corallina pilulifera*, with special reference to the effect of elevated CO_2 concentration. Mar Biol 117:129–132

Gao K, Helbling EW, Häder D-P, Hutchins DA (2012a) Responses of marine primary producers to interactions between ocean acidification, solar radiation, and warming. Mar Ecol Prog Ser 470:167–189

Gao KS, Xu JT, Gao G, Li YH, Hutchins DA, Huang BQ, Wang L, Zheng Y, Jin P, Cai XN, Häder DP, Li W, Xu K, Liu NN, Riebesell U (2012b) Rising CO_2 and increased light exposure synergistically reduce marine primary productivity. Nat Clim Change 2(7):519–523. https://doi.org/10.1038/nclimate1507

Gao K, Beardall J, Häder DP, Hall-Spencer JM, Gao G, Hutchins DA (2019) Effects of ocean acidification on marine photosynthetic organisms under the concurrent influences of warming, UV radiation and deoxygenation. Front Mar Sci 6:322

Garcia-Vaquero M, Hayes M (2016) Red and green macroalgae for fish and animal feed and human functional food development. Food Rev Int 32(1):15–45

Gattuso J-P, Magnan AK, Bopp L, Cheung WW, Duarte CM, Hinkel J, Mcleod E, Micheli F, Oschlies A, Williamson P (2018) Ocean solutions to address climate change and its effects on marine ecosystems. Front Mar Sci 5:337

Gill AE (2016) Atmosphere—ocean dynamics. Elsevier

Gubelit Y, Polyak Y, Dembska G, Pazikowska-Sapota G, Zegarowski L, Kochura D, Krivorotov D, Podgornaya E, Burova O, Maazouzi C (2016) Nutrient and metal pollution of the eastern Gulf of Finland coastline: Sediments, macroalgae, microbiota. Sci Total Environ 550:806–819

Häder D-P, Gao K (2017) The impacts of climate change on marine phytoplankton. In: Phillips BF, Pérez-Ramírez M (eds) Climate change impacts on fisheries and aquaculture. A global analysis, vol 2. Wiley, Hoboken, NJ, pp 901–928

Häder D-P, Horneck G (2018) Bioassays for solar UV radiation. In: Häder D-P, Erzinger GS (eds) Bioassays. Advanced methods and applications. Elsevier, Atlanta, GA, pp 331–346

Häder D-P, Tevini M (1987) General photobiology. Pergamon, Oxford, UK

Häder D-P, Williamson CE, Wängberg S-Å, Rautio M, Rose KC, Gao K, Helbling EW, Sinha RP, Worrest R (2015b) Effects of UV radiation on aquatic ecosystems and interactions with other environmental factors. Photochem Photobiol Sci 14(1):108–126

Hain MP, Sigmal D, Haug GH (2014) The biological pump in the past. Reference module in earth systems and environmental sciences, treatise on geochemistry (2nd edn). The oceans and marine geochemistry, vol 8. Elsevier, The Netherlands, pp 485–517

Halpern BS, Walbridge S, Selkoe KA, Kappel CV, Micheli F, D'Agrosa C, Bruno JF, Casey KS, Ebert C, Fox HE (2008) A global map of human impact on marine ecosystems. Science 319 (5865):948–952

Harada M, Nagano A, Yagi S, Furukawa R, Yokobori S-i, Yamagishi A (2019) Planktonic adaptive evolution to the sea surface temperature in the Neoproterozoic inferred from ancestral NDK of marine cyanobacteria. Earth Planet Sci Lett 522:98–106

Harley C (2014) Seaweed responses to climate change: predictions, observations, and knowledge gaps. In: Salish sea ecosystem conference

Hoffmann L, Hoppe C, Müller R, Dutton G, Gille J, Griessbach S, Jones A, Meyer C, Spang R, Volk C (2014) Stratospheric lifetime ratio of CFC-11 and CFC-12 from satellite and model climatologies. Atmos Chem Phys 14(22):12479–12497

http://earth.columbia.edu, http://earth.columbia.edu/articles/view/2993

Hussey NE, MacNeil MA, McMeans BC, Olin JA, Dudley SF, Cliff G, Wintner SP, Fennessy ST, Fisk AT (2014) Rescaling the trophic structure of marine food webs. Ecol Lett 17(2):239–250

Ismail MM, Alotaibi BS, El-Sheekh MM (2020) Therapeutic uses of red macroalgae. Molecules 25(19):4411

Jenkins GP, Black KP (2019) Contrasting impact of diatoms on dominant calanoid and cyclopoid copepods in a temperate bay. Estuar Coast Shelf Sci 217:211–217

Jokiel PL, Jury CP, Kuffner IB (2016) Coral calcification and ocean acidification. In: Hubbard D, Rogers C, Lipps J, Stanley JG (eds) Coral reefs at the crossroads. Coral reefs of the world, vol 8. Springer, Dordrecht, pp 7–45

Khyade VB, Swaminathan MS (2016) Water: the pacemaker for life of Earth. World Sci News 44:93–125

Kim J-H (2006) The effect of seawater CO_2 concentration on growth of a natural phytoplankton assemblage in a controlled mesocosm experiment. Limnol Oceanogr 51:1629–1636

King AL, Sanudo-Wilhelmy SA, Leblanc K, Hutchins DA, Fu F (2011) CO_2 and vitamin B_{12} interactions determine bioactive trace metal requirements of a subarctic Pacific diatom. The ISME J 5(8):1388–1396

Kirkman SP, Baliwe NG, Nhleko J, Pfaff MC (2020) Ecosystem health and human wealth–a comparison of sub-Saharan African large marine ecosystems. Environ Dev:100551

Knutti R, Rogelj J, Sedláček J, Fischer EM (2016) A scientific critique of the two-degree climate change target. Nat Geosci 9(1):13–18

Konda NM, Singh S, Simmons BA, Klein-Marcuschamer D (2015) An investigation on the economic feasibility of macroalgae as a potential feedstock for biorefineries. BioEnergy Res 8 (3):1046–1056

Kossin JP, Knapp KR, Olander TL, Velden CS (2020) Global increase in major tropical cyclone exceedance probability over the past four decades. Proc Nat Acad Sci USA 117(22):11975–11980

Larson E, Portmann R, Solomon S, Murphy D (2020) Decadal attribution of historic temperature and ocean heat content change to anthropogenic emissions. Geophys Res Lett 47(3): e2019GL085905

Laurenceau-Cornec EC, Trull T, Davies DM, Bray SG, Doran J, Planchon F, Carlotti F, Jouander M, Cavagna A-J, Waite A (2015) The relative importance of phytoplankton aggregates and zooplankton fecal pellets to carbon export: insights from free-drifting sediment trap deployments in naturally iron-fertilised waters near the Kerguelen Plateau. Biogeosciences 12:1007–1027

Low-Décarie E, Fussmann GF, Bell G (2011) The effect of elevated CO_2 on growth and competition in experimental phytoplankton communities. Glob Change Biol 17(8):2525–2535

Mac Monagail M, Cornish L, Morrison L, Araújo R, Critchley AT (2017) Sustainable harvesting of wild seaweed resources. Eur J Phycol 52(4):371–390

Marin F, Le Roy N, Marie B (2012) The formation and mineralization of mollusk shell. Front Biosci 4(1099):125

Marszalek D (1975) Calcisphere ultrastructure and skeletal aragonite from the alga *Acetabularia antillana*. J Sediment Res 45(1)

Myles PB (2017) Maritime clusters and the ocean economy: an integrated approach to managing coastal and marine space. Routledge

Oziel L, Baudena A, Ardyna M, Massicotte P, Randelhoff A, Sallée J-B, Ingvaldsen RB, Devred E, Babin M (2020) Faster Atlantic currents drive poleward expansion of temperate phytoplankton in the Arctic Ocean. Nat Commun 11(1):1–8

Pallela R (2014) Antioxidants from marine organisms and skin care. In: Laher I (ed) Systems biology of free radicals and antioxidants. Springer, Berlin, Heidelberg, pp 3771–3783

Pariatamby A, Bhatti MS (2020) Waste management challenges in Malaysia. In: Sustainable waste management challenges in developing countries. IGI Global, pp 186–222

Pinto MIA (2015) Pesticides in water, sediments and biota of semi-closed coastal lagoons: sources, pathways and impact on aquatic organisms. Universidade Nova

Quillien N, Nordström MC, Schaal G, Bonsdorff E, Grall J (2016) Opportunistic basal resource simplifies food web structure and functioning of a highly dynamic marine environment. J Exp Mar Biol Ecol 477:92–102

Richmond S, Stevens T (2014) Classifying benthic biotopes on sub-tropical continental shelf reefs: how useful are abiotic surrogates? Estuar Coast Shelf Sci 138:79–89

Ries JB, Cohen AL, McCorkle DC (2009) Marine calcifiers exhibit mixed responses to CO_2-induced ocean acidification. Geology 37(12):1131–1134

Sampaio E, Rodil IF, Vaz-Pinto F, Fernández A, Arenas F (2017) Interaction strength between different grazers and macroalgae mediated by ocean acidification over warming gradients. Mar Environ Res 125:25–33

Santos J, Cesarin A, Sales C, Triano M, Martins P, Braga A, Neto N, Barroso A, Alves P, Huaman C (2017) Increase of atmosphere CO_2 concentration and its effects on culture/weed interaction. World Academy of Science EaT. Int J Biol Biomol Agri Food Biotechnol Eng 11:419–426

Seghetta M, Hou X, Bastianoni S, Bjerre A-B, Thomsen M (2016) Life cycle assessment of macroalgal biorefinery for the production of ethanol, proteins and fertilizers–a step towards a regenerative bioeconomy. J Clean Prod 137:1158–1169

Siebeck UE, O'Connor J, Braun C, Leis JM (2015) Do human activities influence survival and orientation abilities of larval fishes in the ocean?. Integr Zool 10(1):65–82

Sigman DM, Haug GH (2006) The biological pump in the past. Treatise on geochemistry, vol 6. Pergamon Press, New York, pp 491–528

Steinbrecht W, Claude H, Schönenborn F, McDermid IS, Leblanc T, Godin-Beekman S, Keckhut P, Hauchecorne A, van Gijsel JAE, Swart DPJ, Bodeker GE, Parrish A, Boyd IS, Kämpfer N, Hocke K, Stolarski R, Frith SM, Thomason LW, Remsberg EE, von Savigny C, Rozanov A, Burrows JP (2009) Ozone and temperature trends in the upper stratosphere at five stations of the network for the detection of atmospheric composition change. Int J Remote Sens 30(15–16):1989–2009

Stolberg E-M (2016) 'From icy backwater to nuclear waste ground': the Russian Arctic ocean in the twentieth century. In: Sea narratives: cultural responses to the sea, 1600–present. Springer, pp 111–137

Sully S, Burkepile D, Donovan M, Hodgson G, Van Woesik R (2019) A global analysis of coral bleaching over the past two decades. Nat Commun 10(1):1–5

Thomas MK, Kremer CT, Klausmeier CA, Litchman E (2012) A global pattern of thermal adaptation in marine phytoplankton. Science 338(6110):1085–1088

Thor P, Dupont S (2015) Transgenerational effects alleviate severe fecundity loss during ocean acidification in a ubiquitous planktonic copepod. Glob Change Biol 21(6):2261–2271

Torres MD, Flórez-Fernández N, Domínguez H (2019) Integral utilization of red seaweed for bioactive production. Marine Drugs 17(6):314

Ullah H, Nagelkerken I, Goldenberg SU, Fordham DA (2018) Climate change could drive marine food web collapse through altered trophic flows and cyanobacterial proliferation. PLoS Biol 16 (1):e2003446

UNEP EEAP (2016) Environmental effects of ozone depletion and its interactions with climate change: progress report, 2015. Photochem Photobiol Sci 15(2):141–174. https://doi.org/10.1039/c6pp90004f

Voie ØA, Mariussen E (2017) Risk assessment of sea dumped conventional munitions. Propellants, Explos, Pyrotech 42(1):98–105

Chapter 2
Pollution Affecting Cyanobacteria in Aquatic Habitats

Abha Pandey, Sonal Mishra, Neha Kumari, Vidya Singh, and Rajeshwar P. Sinha

Abstract Aquatic ecosystems comprise the largest portion ($\sim$72%) of the biosphere and play a crucial role in stabilizing the global climate as well as providing a large array of services for a fast-growing human population. Nowadays, there is a growing amount of data that prove that many anthropogenic pollutants from agricultural, urban and industrial wastes are dumped in the aquatic ecosystems and accumulate in many environments, including the habitats of marine and freshwater cyanobacteria. Cyanobacteria, as a natural part of phytoplankton assemblages, are known for their role as dominant primary producers and base of aquatic food webs. These microorganisms are continuously exposed to various concentrations of the pollutants that are present in their habitats and affect cyanobacterial communities at different levels such as abundance, growth strategies, succession patterns and dominance. Even if no direct changes in cyanobacterial communities are visible, the pollutants may accumulate in cyanobacteria and be passed onto higher trophic levels in a cascading manner, resulting in biomagnifications of certain pollutants. The most common pollutants in aquatic ecosystems are heavy metals, herbicides, pesticides, nutrients, pharmaceuticals, polycyclic aromatic hydrocarbons and microplastics. It is important to determine their concentrations in cyanobacterial cells and in their environment to know the possibility of contaminants that might be transferred to higher trophic levels. However, some strains of cyanobacteria are capable of metabolizing these pollutants that makes them less toxic or sometimes they even remove pollutants from the environment. This chapter highlights the toxic effects of different types of pollution (both point sources and non-point sources) on cyanobacterial communities in aquatic habitats. The anthropocentric concept of "pollution" and the links between pollution, eutrophication and harmful algal blooms (HABs) are also analyzed. An understanding of the synergistic interactions between these aspects and climate change effects will be useful to devise suitable remediation strategies for future use.

A. Pandey · S. Mishra · N. Kumari · V. Singh · R. P. Sinha (✉)
Laboratory of Photobiology and Molecular Microbiology, Centre of Advanced Study in Botany, Institute of Science, Banaras Hindu University, Varanasi 221005, India

© The Author(s), under exclusive license to Springer Nature Switzerland AG 2021

D.-P. Häder et al. (eds.), *Anthropogenic Pollution of Aquatic Ecosystems*,
https://doi.org/10.1007/978-3-030-75602-4_2

Keywords Aquatic ecosystems · Cyanobacteria · Eutrophication · Heavy metal · Pollutants

2.1 Introduction

Cyanobacteria are ubiquitous across various environments ranging from terrestrial to aquatic ecosystems. An aquatic environment can be classified into two major divisions—freshwater environment and marine environment based on salinity. Cyanobacteria are prevalently found in both these types of aquatic environments and contributing to ecosystem-level processes by playing a major role in the biogeochemical cycle. Marine environments consist of approximately 71% of the Earth's surface and are made up of oceans, estuaries, coral reefs and coastal ecosystems, whereas freshwater ecosystems cover less than 1% of the Earth's surface and are comprised of lentic, lotic, river water, lakes, aquifers and wetland ecosystems. Together these environments are responsible for 50% of the world's net primary production (Alexander and Fairbridge 1999). Even though the standing crop of aquatic primary producers is only about 1% when compared to that of primary producers from terrestrial ecosystems, the former rival their terrestrial counterparts in producing 50% of the biomass, incorporating 50% of the anthropogenically released CO_2 and being responsible for 50% of the oxygen emission on this planet (Longhurst et al. 1995; Rousseaux and Gregg 2014). About 90% of anthropogenically released CO_2 is taken up by the oceans (Reid and Hill 2016) and by funneling excess CO_2 to the deep-sea sediments via the biological pump they help to mitigate global change (Hain et al. 2014).

Aquatic ecosystems are exploited to provide food, transportation and recreation. Marine ecosystems also provide ingredients for fertilizers, additives and cosmetics, whereas freshwater ecosystems are utilized to provide water for drinking, sanitation and agricultural and industrial uses. Anthropogenic activities have also a severe impact on aquatic ecosystems, which are also increasingly affected by global change, urban and tourism developments and the unsustainable exploitation of aquatic resources. Pollution of water bodies from agricultural, industrial and urban runoffs and waste disposal threatens the supply of clean drinking water and has negative impacts on marine and freshwater organisms (Beiras 2018; Verhougstraete et al. 2015). Pollution in aquatic habitats may inhibit cyanobacterial growth (Fig. 2.1). Coastal areas make up approximately 7% of marine environments, but due to their high primary productivity, they provide over 50% of food for oceanic ecosystems (Alexander and Fairbridge 1999). However, they are also highly impacted by pollutants. The aquatic environment is contaminated by a variety of synthetic compounds due to anthropogenic discharges like chemical and industrial processes. For example, organic loads can enter riverine or estuarine water, thereby altering the organic pool, and affect degradation processes undertaken by in situ aquatic microbial communities. Microbial communities are affected by various chemicals that are used in industries as they are discharged in the aquatic ecosystem. Many chemical compounds used for industrial enterprises are usually toxic and persistent, but microbial communities are capable of degrading these toxic chemical compounds. The extent of degradation of these chemical compounds by microbes is influenced by the physical and chemical factors persisting in the aquatic systems. For example, aquatic contaminants of hydrophobic nature can result in physical portioning by interacting with dissolved organic matter and particulate organic matter present in the water column (Akkanen et al. 2004; Lin et al. 2003) resulting in a closer association of susceptible substrate and the microbial communities participating in the degradation of chemical compounds. However, partitioning can also result in the concentration of the contaminant to toxic levels, thus leading to suppression of microbes-mediated biodegradation processes, and this can ultimately affect overall biological communities in an aquatic ecosystem.

Cyanobacteria are one of the largest, extremely diverse, phylogenetically oldest lineages and have continuously existed in the biosphere of the Earth for billions of years. These organisms increased the oxygen concentration in the Earth's atmosphere and changed the atmosphere toward the oxygen-based life. Cyanobacteria are Gram-negative prokaryotes, and together with bacteria, they were among the first organisms capable of converting solar light energy into chemical bond energy by managing the flow of electrons in specifically adapted photosystems. They have a unique property of photosynthesis in the whole plant kingdom according to the endosymbiosis theory. Cyanobacteria are found in marine and freshwater environments and form a disparate polyphyletic group of unicellular or simple multicellular microorganisms. These organisms can also be found in several terrestrial habitats, encompassing caves, rocks and soil and therefore are recurrently exposed to varying natural and anthropogenic stressors. Cyanobacteria have evolved various

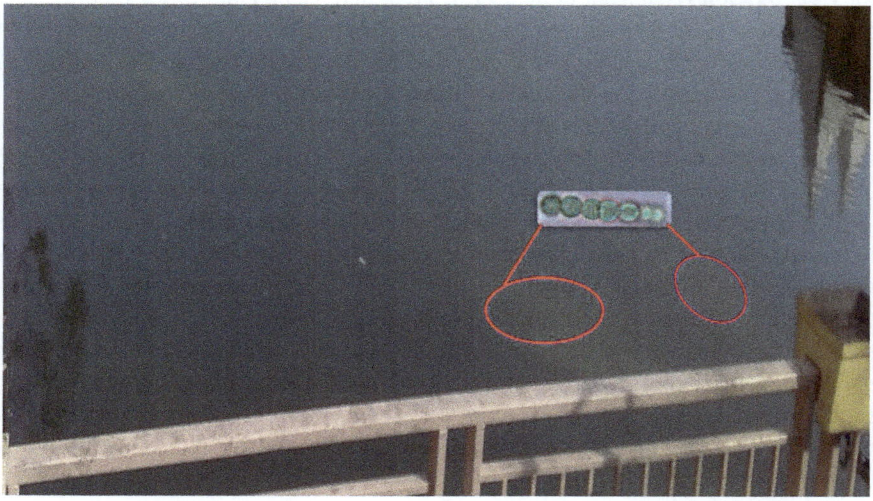

Fig. 2.1 Pollution-driven inhibition of cyanobacterial growth in a large Kund (water body) in Varanasi, India (Photograph by S. Mishra)

sensory mechanisms and adaptive systems to overcome changes occurring in the environment in which they are present.

This chapter focuses on pollutants that routinely enter freshwater and marine environments and attempts to decipher the effects of different types of pollution on cyanobacteria. Furthermore, many mechanisms that let the cyanobacterial species to survive in harsh conditions are emphasized in the context of the presence of chemical pollutants (concerning their determination) in cyanobacterial habitats and inside cells, due to the disquieting transfer of these substances to higher trophic levels.

2.2 Effects of Atmospheric Trace Gases and Pollutants on Cyanobacteria

2.2.1 Atmospheric Carbon Dioxide

Cyanobacteria contribute significantly to the equilibrium of the Earth's atmosphere by absorbing carbon dioxide and producing oxygen. The CO_2 absorbs strongly in the infrared region of the electromagnetic spectrum and thus involved in the greenhouse effect. The increasing concentration of atmospheric CO_2 and high temperature affect the growth of several photosynthetic organisms. At present, the global efforts are conducted toward reducing the production, increasing the utilization of CO_2 and sequestering of carbon in living biomass and soils (Wang and Hsieh 2002). The photosynthetic rate and growth of cyanobacteria are altered by the concentration of CO_2. It has been reported that cyanobacteria have a great potential for CO_2 fixation (Hanagata et al. 1992). Fogg and Than-Tun (1960) reported that 5% CO_2 was toxic for *Anabaena cylindrica*. The supply of an increased level of CO_2 enhances the biomass production of algae to a certain extent. The survival of cyanobacteria depends on their ability to acclimate to environmental changes.

Global warming increases the surface temperature. Environmental parameters such as temperature and pH affect cyanobacterial growth. Mishra et al. (2019a) observed the effect of temperature and pH on cyanobacteria. At high temperature (>32 °C) and low pH (<6.5), the solubility of O_2 decreases and with heavy metals in aquatic habitats increases which is responsible for increasing toxicity and ultimately death of organisms (Fig. 2.2). Staal et al. (2003) have reported that a high temperature had negatively impacted the N_2-fixing heterocystous cyanobacteria in the tropical oceans. Idso et al. (1989) observed a similar effect of CO_2 enrichment in *Azolla*, a fern harboring a cyanobacterium, *Anabaena azollae*. Strother (2008) suggested that there was a continuous procurement of carbon concentration mechanisms that is required for inorganic carbon uptake by the anabolic physiology of cyanobacteria and algae.

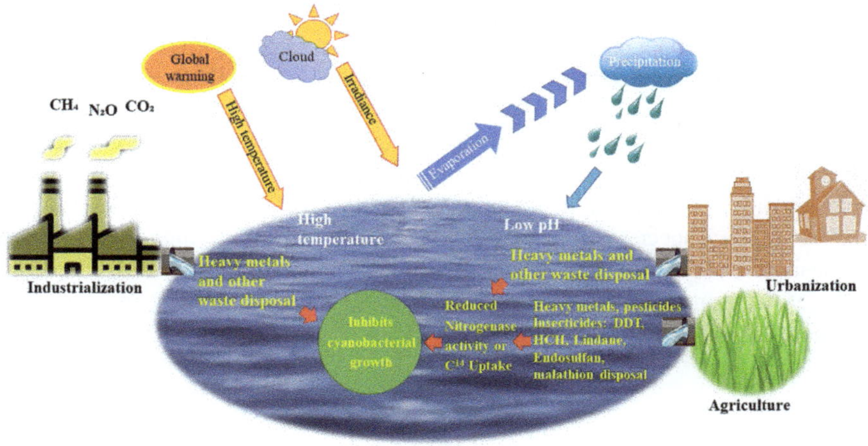

Fig. 2.2 Schematic illustration depicting the waste disposal of heavy metals, pesticides, insecticides (DDT, HCH, lindane, endosulfan, malathion) and other pollutants through industrialization, urbanization and agricultural process. These disposals inhibit cyanobacterial growth by reducing the nitrogenase activity or carbon uptake. Ozone depleting gases (e.g., CO_2, N_2O, CH_4) emanating from industries are responsible for global climate change or global warming and are also responsible for the decrease in pH through acidification by CO_2. Under these conditions, heavy metals become more soluble in aquatic habitat and inhibit cyanobacterial growth

2.2.2 *Atmospheric Oxygen*

Walker (1984) suggested that initially, Earth's atmosphere was anaerobic until the evolution of a form of photosynthesis that was able to utilize water as a reductant. The climatic devastation of glaciations and consequent changes in atmospheric composition lead to a major biological revolution in 3.0–2.7 Ga, characterized by the emergence of new low-temperature adapted taxa cyanobacteria (Lowe and Tice 2007). Cyanobacteria increased the atmospheric oxygen (O_2) from zero to the present 0.2 atmospheres. Water blooms are formed in several fish ponds where cyanobacteria are the major part of phytoplanktonic biomass that results in massive mortality among the fish due to the depletion of O_2 after the bloom collapses (Sevrin-Reyssac and Pletikosic 1990). Hirata et al. (2003) reported that the production of a violet pigment was increased by the cultivation of *Nostoc spongiaeforme* TISTR 8169 with 1 or 2 mM of H_2O_2. Oxidative stress, higher temperature and more intense light are responsible for the synthesis of Nostocine A that generate toxic reactive oxygen species in the cells of target organisms. The reactive oxygen species were involved in the oxidative damage of the cyanobacterium *Anabaena* sp. and *Nostoc* sp. caused by UV-B radiation (He and Häder 2002; Pandey et al. 2020).

Holzinger and Lütz (2006) and Björn (2007) reviewed the UV effects on cyanobacteria including the destruction in chloroplasts and mitochondria and the occurrence of structures that were likely to be related to the UV stress. Wang et al. (2008) showed that the photosynthetic activity is decreased and lipid peroxidation is enhanced in a desert cyanobacterium (*Nostoc* sp.) by UV-B radiation. The exogenous supply of antioxidants such as ascorbate acid and N-acetylcysteine had protective effects on cells of *Nostoc* sp. directly against reactive oxygen species initiated by UV-B radiation, as did sodium nitroprusside, which served as a signal molecule in the formation of algal cell protection of Photosystem II and as a reactive oxygen species scavenger. Sinha and Häder (2008) suggested that the synthesis of UV-absorbing/screening compounds such as mycosporine-like amino acids (MAAs) and scytonemin is involved in the prevention of UV-induced photodamage by absorbing UV-A and/or UV-B.

2.3 Effects of Organic Pollutants on Cyanobacteria

Many cyanobacterial species are sensitive to organochlorine chemicals, and the effects depend on the type and nature of the chemicals, the organisms and the environmental conditions. Most common organic pollutants are based on aliphatic, alicyclic or aromatic structures and most of the organic carbon accessible to microorganisms. In the environment, the majority of organic carbons available to microorganisms are photosynthetically fixed carbon compounds. Hence, several man-made organic chemicals resembling naturally occurring organic carbon can be easily degraded. However, the man-made chemicals may change the carrying capacity of the environment (i.e., the maximum level of microbial activity that can be expected under a particular environmental condition). Organic pesticides are a series of diverse chemical compounds used to control major pests, but their inappropriate application may affect non-target organisms like cyanobacteria.

2.3.1 Organochlorine Insecticides

Organochlorines are carbon-based substances that have one or more chlorine atoms. In these compounds, chlorine is highly reactive, and many organochlorines persist for a longer period in the environment after their initial use, and they are readily stored in fatty tissues and can bioaccumulate through the aquatic food chain.

2.3.1.1 DDT

Among the organic contaminants of great concern, the persistent organochlorine insecticide, DDT (1,1,1-trichloro-2,2-bis(4-chlorophenyl)ethane) and its principal metabolites, DDE (1,1-dichloro-2,2-bis(*p*-chlorophenyl)ethylene) and DDD (1,1-dichloro-2,2-bis(*p*-chlorophenyl)ethane) are most widely distributed in the environment worldwide, despite its poor water solubility. The commercial use of DDT was banned four decades ago, although they exist for a long time in several parts of the world. Lal et al. (1987a) reported that DDT hampered the growth of *Anabaena* sp. but stimulated the growth of *Aulosira* sp. When these organisms were treated for 35 days, then they recovered from the toxic effect. Photosynthesis in *Anabaena* sp. and *Aulosira* sp. was inhibited at all concentrations of DDT. In another report, Lal et al. (1987b) also observed that *Anabaena* sp. and *Aulosira fertilissima* showed a remarkable potential to accumulate DDT. The bioconcentration of DDT in *Anabaena* sp. and *A. fertilissima* varied from 3 to 1568 $\mu g\ g^{-1}$ and from 6 to 1429 $\mu g\ g^{-1}$, respectively. Additionally, *Anabaena* sp. and *A. fertilissima* metabolized DDT to DDD as a major metabolite and DDE as a minor metabolite.

2.3.1.2 HCH

The isomeric mixture of HCH (1,2,3,4,5,6-hexachlorocyclohexane) is a first-generation organochlorine insecticide used widely in agriculture and public health. γ-HCH (lindane) is the most common insecticide among the isomers. HCH wastes that are generated during the production of lindane contain more than 85% of non-insecticidal isomers. Although HCH is known to be less persistent than DDT, HCH isomers persist in the soil and water at heavily contaminated sites after receiving wastes for a long time near the production sites. The use of the commercial formulation (isomeric mixture) of HCH has been banned or restricted in many countries due to the biomagnification of HCH isomers in the food chain. The concentrations of α-and γ-hexachlorocyclohexane were found to be high in the northern waters in the Arctic Ocean due to lower water temperature that reduced its transport to the atmosphere. However, the levels of heavy metals, alkanes, polyaromatic hydrocarbons (PAH) and organochlorines in northern waters were comparable to those in uncontaminated ocean waters in the mid-latitudes (Muir et al. 1992). The bioaccumulation of organochlorines and heavy metals in Arctic marine food chains begins with exotic ice algae or phytoplankton in surface waters. In the trophic compartments of Mar Menor, an oligotrophic coastal lagoon located in Spain, residues of organochlorine insecticides and/or any of their analyzed metabolites were detected in 70.5% of the total of 78 samples of water, sediment and benthic organisms, albeit at small concentrations. Among the detected compounds, HCH stood out second (47.4%) to endosulfan (51.3%) in the analyzed samples (Pérez-Ruzafa et al. 2000). Lindane acts as an insecticide, acaricide and rodenticide. Due to biomagnification of β-isomer, the isomeric mixture of HCH formulations has been banned in most countries, but lindane is used in some countries. Because cyanobacteria are a primary base of the aquatic food web and act as an important biofertilizer for rice cultivation, their protection from the grazers or tolerance to residual effects of toxicants is necessary for the enrichment of soil fertility. In 1983, Grant and coworkers examined the potential of lindane to control ostracods grazing on blue-green algae that play a major role in nitrogen fixation in flooded rice soils. An inhibition in the biomass production and the photosynthetic rate was seen in *Anabaena* sp. exposed to different concentrations (0.5, 1.0, 1.5, and 2.0 μg mL^{-1}) of lindane (Babu et al. 2001). The cyanobacterial species capable of controlling pollution were found in freshwater ecosystems like the Egyptian Lakes Qaroun and Mariut, as some species showed toxicity or stimulation due to lindane (5 and 10 μg mL^{-1}) resulting from agricultural runoff (El-Bestawy et al. 2007).

2.3.1.3 Endosulfan

The toxic forms of endosulfan (6,7,8,10,10-hexachloro-1,5,51,6,9,9-hexahydro-6,9-methano-2,4,3-benzadioxathiepin 3-oxide) are α or β endosulfan and endosulfan sulfate. Endosulfan applied to non-flooded soils at levels close to the recommended rates resulted in an increment of the total population of cyanobacteria (Muralikrishna and Venkateswarlu 1984). Endosulfan has an adverse effect on the growth of *Anabaena* sp. and *A. fertilissima* even at 1 μg mL^{-1} with inhibition being significantly above 50% at 20 μg mL^{-1} (Tandon et al. 1988). The application of 20 μg mL^{-1} endosulfan caused filaments clumping in *A. fertilissima* and also lowered ^{14}C uptake and nitrogenase activities. However, the amounts of ethylene produced from acetylene used for an indirect assay of nitrogen fixation were equal to or above control levels and did not affect the nitrogen fixation (Tandon et al. 1988). Endosulfan or endosulfan sulfate chemicals eliminated *Gloeocapsa* sp. and *Nostoc* sp. in non-flooded as well as flooded conditions. In an earlier study, *Anabaena* sp. has been shown to degrade endosulfan, but endosulfan sulfate was not detected in this case (Shivaramaiah 2000). The stress responses in terms of oxidative stress, the role of proline and antioxidant enzymes involved in free radical detoxification in three cyanobacterial strains (viz., *A. fertilissima*, *Anabaena variabilis* and *Nostoc muscorum*) exposed to endosulfan were studied (Kumar et al. 2008). The growth, photosynthetic pigments and carbohydrate contents were reduced, while the total protein, proline, malondialdehyde, superoxide dismutase, ascorbate peroxidase and catalase were increased at higher doses of endosulfan while photosynthetic pigments were enhanced at a lower concentration of endosulfan. This increased level of malondialdehyde indicates that free radicals, catalase, ascorbate peroxidase, superoxide dismutase and proline were involved in the free radical-scavenging mechanism.

2.3.2 Organophosphorus Insecticides

2.3.2.1 Monocrotophos and Quinalphos

Monocrotophos was toxic to *Nostoc linckia* at 100 μg mL^{-1} and to *Phormidium tenue* at 50 and 100 μg mL^{-1}, while the growth of these two cyanobacteria was enhanced at a lower concentration (Megharaj et al. 1987). Rajendran et al. (2007)

reported that photosynthetic oxygen evolution as well as nitrogenase and glutamine synthetase activities was reduced in *Tolypothrix scytonemoides* due to monocrotophos at a lower concentration of 0.2–0.5 μg mL^{-1}. However, quinalphos was significantly toxic to cyanobacteria after two successive additions at 0.5–2 kg ha^{-1}. Two successive additions of monocrotophos and a single application of quinalphos were toxic to cyanobacteria even at close to field doses (0.5–2 kg ha^{-1}) to flooded soil (Megharaj et al. 1988).

2.3.2.2 Fenitrothion and Malathion

Lal et al. (1987a) found that a higher concentration (10 μg mL^{-1}) of fenitrothion (*O*,*O*-dimethyl *O*-4-nitro-*m*-tolyl phosphorothioate) was extremely toxic to *Anabaena* sp. and *A. fertilissima*, as it reduced photosynthesis by more than 75%. Malathion (diethyl (dimethoxythiophosphorylthio) succinate) is an aliphatic organothiophosphorus insecticide, and it adversely affected the growth of *Anabaena* sp. and *A. fertilissima* (Tandon et al. 1988). Malathion decreased $^{14}CO_2$ uptake and nitrogenase activities in *A. fertilissima*, while nitrogen fixation was not affected in *Anabaena* sp. as the amounts of ethylene produced were equal to or above control levels. *Anabaena* sp. survived up to 500 μg of malathion mL^{-1} indicating that *A. fertilissima* was more sensitive to malathion.

2.3.2.3 Phosphamidon

Phosphamidon ((EZ)-2-chloro-2-diethylcarbamoyl-1-methylvinyl dimethyl phosphate) has an adverse impact on nitrogen-fixing cyanobacterial species and affects dinitrogen fixation. The dinitrogen fixation in *Anabaena* PCC 7119 was reduced after 24 h of treatment with phosphamidon at 10 μg mL^{-1} (Perona et al. 1991) resulting in a delayed effect on cellular composition, while the photosynthetic pigments, protein, nucleic acids and carbohydrates were adversely affected at dosages of 40–60 μg mL^{-1}. Photosynthetic O_2 evolution could be significantly reduced after 72 h treatment of phosphamidon of concentration of 50 μg mL^{-1} due to adverse effects on the photosynthetic pigments. Phosphamidon also leads to vegetative cell swelling and heterocyst deterioration without altering heterocyst frequency. In an earlier report (Marco et al. 1990), the addition of trichlorfon does not affect the photoevolution of O_2 per unit chlorophyll, but the activity per unit dry weight was reduced due to a decrease in photosynthetic pigments of *Anabaena* PCC 7119.

2.3.3 Substituted Phenols

Phenol and its derivatives are one of the largest groups of environmental pollutants due to their presence in many industrial effluents (such as those from oil and petroleum refineries: coking plants; coal processing; resin, paint, dye and petrochemical production; textile and paper mills) and act as antibacterial and antifungal agents (Krastanov et al. 2013; Zhang et al. 2013). Shashirekha et al. (1997) reported that the chlorophyll content of *Phormidium valderianum* was reduced by the addition of 25 and 50 mg mL^{-1} phenol as compared to that of control treatments and these polyphenols probably influence growth restriction by negatively impacting both the activity of enzymes and photosynthetic processes.

2.3.4 Aromatic Hydrocarbons

The effects of aromatic industrial pollutants, such as benzene, toluene, xylene and 4-nitrophenol, on various cyanobacterial strains have also been reported. Sundaram et al. (2011) showed that changes in the structures of *N. muscorum* and *Synechococcus* PCC7942 cells are altered at concentrations up to 300 μM of these stressors, and among the organic stressors examined, xylene highly reduced the growth (in terms of specific growth rate) of both microorganisms. Toluene showed a similar effect as xylene but only in the case of *Synechococcus* cells. Both benzene and paranitrophenol moderately affected the growth rate of the cyanobacteria. The impact of benzene and its derivatives on *Arthrospira platensis* and *A. cylindrica* has also been studied, and all examined organic solvents were highly toxic to the halophilic cyanobacterium. 300 μM benzene, toluene and xylene significantly inhibited the growth rate in *A. cylindrica*, while paranitrophenol had less effect even at a higher dose (400 μM) (Sundaram and Soumya 2011). Other groups of aromatic environmental pollutants are the phthalate esters (PEs) broadly used as plasticizers. Acey et al. (1987) showed that the growth rate of *Synechococcus lividus* was reduced by di-n-butyl phthalate (DBP) which also promotes cell aggregation. The effect of DBP, diethyl phthalate (DEP) and dimethyl phthalate (DMP) on the growth (based on the absorbance at 663 nm) of three cyanobacteria (*Anabaena flos-aquae, Microcystis aeruginosa* 2396 and *M. aeruginosa* 4141) was studied by Babu and Wu (2010). The influence of DMP on the growth rate (based on the absorbance at 730 nm) was also studied in three other cyanobacteria (*Synechocystis* sp., *Synechococcus* sp. and *Cyanothece* sp.); cyanobacterial growth was induced at the lower concentration of DMP (20 mg mL^{-1}) while reduced at a higher concentration of DMP (50–500 mg mL^{-1}) (Zhang et al. 2016). Di-, tri- and polycyclic aromatic hydrocarbons are commonly found to be toxic, but the growth (based on protein and chlorophyll *a* concentrations) of the saline cyanobacterium *P. tenue* was increased by naphthalene and anthracene at a concentration of 0.05% (w:v). Petroleum hydrocarbons had an adverse effect on all life forms and were harmful for both aquatic and terrestrial ecosystems. Oil spills affect marine life and people whose career depends on the utilization of marine resources. Marine petroleum fuel spills are hazardous that occur by transportation accidents and several other industrial and mining activities (Bossert et al. 1984) and known as the most recurrent organic pollutants of aquatic ecosystems (Margesin and Schinner 1997).

2.3.5 Other Organic Pollutants

Due to anthropogenic activities, organic pollutants belonging to various other chemical classes can occur in the terrestrial environments, and their effects on cyanobacteria have been examined by many researchers. Recently, Rajendran et al. (2007) reported that the growth of *T. scytonemoides* was increased at lower concentrations of the fungicide bavistin (50 μg mL^{-1}) while chlorophyll content and photosynthetic oxygen evolution rate and respiratory oxygen consumption were reduced at higher concentration of bavistin (100–130 μg mL^{-1}). The chlorophyll *a*, protein content, photosynthetic oxygen evolution rate and activities of nitrogenase and glutamine synthetase of *T. scytonemoides* were reduced when it was treated with the biopesticide nimbicidin with a concentration ranging between 0.5 and 3 μg mL^{-1}. Wang and Xie (2007) reported that superoxide dismutase (SOD), glutathione-S-transferase (GST) activities and glutathione (GSH) levels were enhanced in *M. aeruginosa* strains when exposed to high concentrations of nonylphenols (1 and 2 mg mL^{-1}) indicating that *M. aeruginosa* were resistant to nonylphenol toxicity while the growth and toxin production were enhanced at low concentrations of nonylphenol (0.02–0.5 mg L^{-1}).

2.4 Effects of Metals and Metalloids on Cyanobacteria

The term "heavy metal" refers to those metallic elements which are toxic even at low concentration having a relatively high density of 5 g cm^{-3} and atomic mass ranging between 54.63 and 200.59 (Dias et al. 2002). Metals and metalloids are ubiquitously present in the environment due to natural processes and anthropogenic activities such as contributions from industrial, agricultural and domestic wastes to the toxic levels, in the aquatic ecosystem (Table 2.1). The eight most common pollutant heavy metals listed by the Environment Protection Agency (EPA) are arsenic (As), cadmium (Cd), chromium (Cr), copper (Cu), mercury (Hg), nickel (Ni), lead (Pb) and zinc (Zn) (Athar and Vohora 1995). The influence of heavy metals on the structure of the community and activities of microorganisms in aquatic systems varies depending on their concentration and available chemical forms (Muller et al. 2005; Pinto et al. 2003) (Fig. 2.2). At optimum concentration, heavy metals like Cu, Zn and Mn support major life processes such as maintenance of osmotic balance, stabilization of cellular organelles, and they function as electron acceptors in many processes and cofactors for metalloenzymes (Gadd 2010; Jensen 2013; Vašková et al. 2012). The direct effect of heavy metal pollution on the microbial ecosystem includes the alterations in the physiology, diversity and abundance of microorganisms, which indirectly affect the biogeochemical cycles and ocean productivity (Bong et al. 2010; Chakravarty and Banerjee 2008; Hoostal et al. 2008). The toxicity effects of heavy metals such as Cu, Cd, Ni and Zn have been studied earlier in cyanobacteria (Polyak et al. 2013; Zeng and Wang 2011).

Table 2.1 Selected metal (loid)s and their background levels (μM) in freshwater aquatic systems (adapted and modified from Goldman and Horne 1983 and Lindsay 1979)

Metal(loid)	Freshwater
Chromium	Trace
Mercury	Trace
Copper	0.010
Cadmium	0.00053
Zinc	0.30
Arsenic	Trace

Estuaries that form the transition zone between fresh and marine water system receive heavy metals and other pollutants of anthropogenic origin and help in the regulation of their supply to the coastal environment.

Cyanobacteria are the major picoplankton distributed in the estuarine environment where they occupy a key position at the base of the food web and play a central role in carbon, nutrient and oxygen cycling (Day et al. 2012). Approximately 20–40% of global carbon fixation is performed by *Prochlorococcus* and *Synechococcus* alone although they account for only one percent of photosynthetic biomass (Blindauer 2008). Debelius et al. (2011) showed that the general growth, pigment composition, photosynthesis and enzyme expression of cyanobacteria are influenced by trace metals above their optimum levels. Many cyanobacterial strains have the potential to accumulate, detoxify or metabolize heavy metals up to a certain concentration range (Baptista et al. 2014). Cyanobacteria act as biological metal-sorbents which represent an important sink for metals in aquatic environments and are capable of controlling the speciation of many biologically active metals (Baptista et al. 2014; Kumar et al. 2015; Yee et al. 2004). Although metal accumulation is a promising strategy for the survival of cyanobacteria in metal-polluted environments, their impact on the food web will be disturbing. *Oscillatoria* sp. (filamentous) and *Synechocystis* sp. (unicellular) could tolerate up to 200 fg K_2CrO_4 mL^{-1} in BG11 medium under laboratory conditions (Hameed and Hasnain 2005).

Several metals are required for cyanobacteria like copper for thylakoidal plastocyanin, zinc for carboxysomal carbonic anhydrase, cobalt in cobalamine, magnesium for chlorophyll, molybdenum for heterocystous nitrogenase and manganese for the thylakoidal water-splitting oxygen-evolving complex (Gadd 2010; Jensen 2013; Vašková et al. 2012). Thus, essential and non-essential metals require homeostatic systems for creating the cellular environments in these species for their uptake. Correct metals are acquired by metalloproteins, while the incorrect ones are somehow avoided (Cavet et al. 2003). In a recent report, El-Sheekh et al. (2005) showed that cyanobacterial cultures of *N. muscorum* and *Anabaena subcylindrica* were capable of the removal of heavy metals, and copper, cobalt, lead and manganese were removed by 12.5–81.8, 11.8–33.7, 26.4–100 and 32.7–100%, respectively, from wastewater. The metal sorption efficiency of the cyanobacteria depends on the type of biosorbent, the physiological state of the cells, availability of heavy metals, heavy metal concentration and chemical composition of the wastewater. Olafson et al. (1979) provided the first evidence for the existence of metallothionein that is a cadmium inducible metal-binding protein and was isolated from cadmium-exposed *Synechococcus* sp.

Polyak et al. (2013) showed that the growth of two strains of *M. aeruginosa* isolated from a lake in Russia was reduced by Cu, Ni and Zn. As observed earlier, initial exposure to metals like lead, copper and zinc in *Spirulina platensis*-S5 reduced the growth and also causes yellowing and fragmentation of filaments and reduced number of spirals (Choudhary et al. 2007). Similarly, growth and chlorophyll content of *Oscillatoria* sp. were reduced in the presence of heavy metals (Jayashree et al. 2012) in a dose-dependent manner, and growth rate of nitrogen-fixing cyanobacteria *Anabaena doliolum* (Sultan et al. 2007) and *A. variabilis* (Aftab and Ahmad 2013) was also decreased in the presence of heavy metals. A lower concentration of nickel promoted the growth in *A. doliolum*, while a higher concentration suppressed the growth (Shukla et al. 2009). Apart from the reduction of growth, heavy metals at higher doses also produce reactive oxygen species (ROS) like superoxide, hydrogen peroxide and hydroxyl radicals, and heavy metal toxicity increases 30 times more ROS production (Mittler 2002). The quantity of lipid peroxidation (LPO) estimated in terms of malondialdehyde (MDA) was enhanced in *S. platensis* treated with Pb, Cu and Zn (Choudhary et al. 2007). Enhanced amount of MDA was indicative of free radicals formation in the test microorganism under heavy metals treatments (Choudhary et al. 2007). Cu acts as a redox-active transition metal that catalyzes the generation of OH^-, most toxic and reactive to lipids, from O_{2-} by the Fenton reaction and thus results in an increase in MDA content. Heavy metal stress increases the MDA content in the nitrogen-fixing cyanobacteria *A. doliolum* (Sultan et al. 2007) and *Anabaena* sp. PCC7120 (Singh et al. 2012).

Several antioxidant defense systems have evolved in the organelles of cyanobacteria to counteract the toxic effect of ROS, and it mainly involves two types of antioxidant defense mechanisms, i.e., enzymatic and non-enzymatic reactions. Superoxide dismutase (SOD), catalase (CAT), ascorbate peroxidase (APX) and glutathione reductase (GR) are enzymatic antioxidants, while non-enzymatic antioxidants consist of nutrient-antioxidants like carotenoids, α-tocopherol, ascorbic acid, glutathione, flavonoids, uric acid and plasma proteins such as transferrin, albumin and metalothionein (Luximon-Ramma et al. 2002). Choudhary et al. (2007) studied the effect of heavy metals (Pb, Cu and Zn) in *S. platensis* and showed that increased concentrations of metals enhance the SOD activity indicating scavenging phenomena of ROS (O_{2-}) and capability of resistance increases in *S. platensis*. Similarly, SOD activity was also enhanced in the nitrogen-fixing cyanobacterium *A. variabilis* in the presence of metals (Aftab and Ahmad 2013). Sultan et al. (2007) found that heavy metals increase the CAT activity in *A. doliolum*, but it was decreased at a higher concentration of heavy metals. Similarly, APX and GR activity was increased in the nitrogen-fixing cyanobacterium *A. doliolum*, but GR activity was reduced at a higher concentration of metals, and GR and APX were highly correlated with each other. Table 2.2 summarizes the response of antioxidant enzymes induced in cyanobacteria upon heavy metal exposure.

Table 2.2 Changes in antioxidant enzymes activity induced in cyanobacteria upon heavy metal exposure (adapted and modified from Shilpi et al. 2015)

S. No.	Heavy metals	Cyanobacteria	Antioxidant enzymes	Response	References
1	Pb, Cu and Zn	*Spirulina platensis*	SOD	Increased	Choudhary et al. (2007)
2	Cd	*Anabaena* sp.	SOD, CAT, APX and GR	Increased	Singh et al. (2012)
3	Cd and Cu	*Anabaena doliolum*	SOD, CAT, APX and GR	Increased	Sultan et al. (2007)
4	Mg, Mn, Cd, Hg, Zn, Co and Pb	*Anabaena variabilis*	SOD, CAT	Increased Decreased	Aftab and Ahmad (2013)

Among 17 cyanobacterial strains tested by Inthorn et al. (1996), the filamentous cyanobacterium *Tolypothrix tenuis* exhibited a high level of cadmium tolerance and had the highest potential for cadmium removal through adsorption onto the cell surface. Cr is a non-essential, highly toxic metal to microorganisms and plants and is considered as a serious environmental pollutant due to its widespread industrial use especially in the leather industry. Cu may form complexes with organic compounds that greatly modify their properties. More heterocysts are produced in cyanobacteria when fixed nitrogen is not available. Cupric chloride inhibited nitrogenase activity resulting in the nitrogen-starvation and induced more vegetative cells to differentiate into heterocysts. The basis for Cu toxicity in algae is largely unknown. Copper binds strongly with the sulfhydral groups of proteins (Porter and Sheridan 1981). *N. linckia* acts as a pollution indicator due to its sensitivity toward Cu. The experiments reported here suggest that increased levels of copper present in sewage sludge effluents of various factories can disrupt the natural balance of nitrogen fixation and cell division rates in the whole water community of phytoplankton in culture (Brand et al. 1986). Elevated free Cu^{2+} can reduce photosynthetic rates (Barón et al. 1995) and interrupt the uptake of other essential trace metals such as manganese (Sunda 1989). Several isolates of the cyanobacterium *Synechococcus* could not survive at a free Cu^{2+} of 11 pM (Brand et al. 1986).

The effect of $HgCl_2$ was studied by Murthy et al. (1990) in two cyanobacteria, *S. platensis* and *Anacystis nidulans*, by monitoring the changes in fluorescence yield of chlorophyll. Three different types of changes in the fluorescence yield of chlorophyll *a* were noticed, based on the concentration of $HgCl_2$, 1. at a low concentration (1.5 μM), $HgCl_2$ behaved similarly to diuron, and in that, weak modulated light increases the fluorescence intensity F_0 possibly by blocking the flow of electrons on the reducing side of photosystem II; 2. a moderate concentration (3 μM) results in quenching of the variable fluorescence of chlorophyll showing a decrease in electron flow on the donor side of PS II; 3. at a high concentration (18 μM), a marked quenching of the chlorophyll fluorescence was seen due to both the blocking of PS II on the donor side and structural changes in the antenna pigments.

The response of *Phormidium* sp. to inorganic arsenic dissolved in media was examined by Shigeki et al. (1984). This cyanobacterium accumulated arsenic and showed inherent capabilities of survival to a high concentration of arsenic stress. Additionally, the strain rapidly reduced the absorbed arsenate to arsenite [As(III)] within their tissues. The arsenic was rapidly accumulated in plankton at water concentrations of about 5 μg L^{-1} when added as arsenate (AsO_4) to a freshwater model ecosystem (Reuther 1992). Table 2.3 gives the possible toxic influences of selected metal(loid)s and mechanisms of resistance and detoxification by cyanobacteria and microalgae.

Table 2.3 Possible toxic influences of selected metal(loid)s and mechanisms of resistance and detoxification by cyanobacteria and microalgae (adapted and modified from Ramakrishnan et al. 2010)

S. No.	Metal (loid)	Mechanisms of resistance/ detoxification	Toxic influences	Reference
1	Chromium	Adsorption; precipitation; reduction of Cr(VI) to Cr(III)	Inhibition of enzymatic activity, cell membrane disruption	Cervantes et al. (2001)
2	Mercury	Volatilization (Hg(II) → CH_3Hg^+/$(CH_3)_2Hg$), reduction (Hg(II) → Hgo)	Cell membrane disruption, inhibition of enzymatic activity; protein denaturation, inhibition of cell division, inhibition of transcription, inhibition of translation	Deng et al. (2008)
3	Copper	Reduction (Cd^{2+} → $CdPO_4$), copper homeostasis	Cell membrane disruption, inhibition of enzymatic activity	Bossuyt and Janssen (2005)
4	Cadmium	Outer membrane/cell wall binding Sequestration by exopolysaccharides, efflux pumps, precipitation as metal salts (Cd^{2+} → CdS; Cd^{2+} → $CdPO_4$)	Cell membrane disruption, inhibition of the enzymatic activity	Olafson et al. (1979); Pawlik and Skowroński (1994)
5	Zinc	Adsorption, sequestration by metallothionein; phytochelations (heavy metal-sequestering peptides	Cell membrane disruption	Cavet et al. (2003)
6	Arsenic	Reduction, oxidative-methylation and adenosylation	Inhibition of enzymatic activity; DNA damage	Dembitsky and Levitsky (2004)

2.5 Effects of Nutrients on Cyanobacteria

Nutrient enrichment of water bodies by urban, agricultural and industrial development has promoted the occurrence of harmful algal blooms (Paerl and Huisman 2008). The risk of intoxication from cyanobacteria and their toxins is an important consideration when dealing with water use from harmful algal blooms (Carmichael 2001; Mishra et al. 2019b). Chemical pollutants, such as endocrine disrupting compounds (EDCs), are of great concern due to their industrial and domestic applications and potential adverse effects on metabolism, development, growth and reproduction in exposed freshwater and marine wildlife (Tan et al. 2007). EDCs reduced fertility, feminization, reproductive organ anomalies and alter sexual behavior. The occurrence of EDCs was seen in wastewaters (Yu et al. 2013), natural waters (Yang et al. 2014) and drinking waters (Benotti et al. 2009) all around the world including synthetic steroid hormones, pharmaceutical drugs, polycyclic aromatic hydrocarbons, polychlorinated biphenyls, alkylphenols, pesticides but also natural products such as phytoestrogens. Anthropogenic pollutants released from manufacturing processes can enter the aquatic environment as they are not fully removed by sewage treatment systems (Kidd et al. 2007; Pal et al. 2010). In addition to chemical pollution, aquatic ecosystems are exposed to multiple stressors (Ormerod et al. 2010). Increased anthropogenic nutrient input into receiving water bodies resulted in the eutrophication of freshwaters and coastal marine ecosystems (Smith 2003). Harmful algal blooms are enriched with nutrients (particularly nitrogen and phosphorus). Severe cyanobacterial blooms in eutrophic water bodies have been noticed in Lake Taihu in China (Paerl et al. 2011), Lake Biwa in Japan (Nalewajko and Murphy 2001), Lake Erie and Michigan in North America (Rinta-Kanto et al. 2005), Lake Winnipeg in Canada (Schindler et al. 2012), Lake Victoria in Africa (Verschuren et al. 2002), the Baltic Sea in Northern Europe (Suikkanen et al. 2007) and other lakes, rivers and estuaries having ecological and economic importance (Huisman et al. 2005). The turbidity of eutrophic waters is enhanced by dense cyanobacterial blooms, also the growth of aquatic plants is suppressed, and thus, it adversely affects the aquatic environment for many invertebrates and fish species (Scheffer 2004). In the last few decades, the main concern is the increase in toxin-producing strains of the cyanobacterium *Microcystis* sp. that produce the hepatotoxin microcystin (Michalak et al. 2013). More than 100 structural variants of microcystins have been identified and classified as "probably carcinogenic to humans" (Grosse et al. 2006).

2.6 Effects of Crude Oil Spills on Cyanobacteria

Crude oil spills affect phytoplankton particularly in shallow water marine environments like the Arctic Ocean. Pyrene is one of the most toxic components of heavy oil which gets accumulated in sediments. Moreover, this toxicity will get increased in the presence of solar UV-B and high temperature (Grenvald et al. 2013). Oil exploration will enhance in several coastal habitats like Western Greenland (Hylander et al. 2014). The effect of oil pollution on growth rates and physiological activities has been studied in cyanobacterial conglomerates in the river Nile (Egypt) by comparing a polluted with a non-polluted site (El-Sheekh et al. 2000). At lower concentrations, both crude oil and its refinery products increase the physiological activities like growth, protein content and nucleic acids, while at higher doses, the physiological activities are decreased. Polychlorinated biphenyls (PCBs) are lipophilic, so they can easily enter the cell membranes of phytoplankton leading to considerable bioconcentration affecting both marine and freshwater ecosystems. PCBs are also important for air–water exchange and vertical fluxes. It was observed that PCB concentrations in phytoplankton were affected by the air–water exchange in inaccessible environments which are not directly affected by terrific pollution (Dachs et al. 2000).

Responses of microbes to marine oil spills are determined by several factors like the composition of oil and extent of weathering as well as the environmental conditions, mainly temperature and nutrient concentrations. Crude oil alters the microbial community consisting of multiple co-existing species when it enters seawater. Several cyanobacterial mats were formed and colonized mostly of the oil-polluted coasts during the 1991 Gulf War environmental disaster, but most of the oil spill affected cyanobacterial mats were capable of firstly re-colonizing the destroyed habitats (Sorkhoh et al. 1992). Excessive cyanobacterial growth was observed on the sites where they were not present before the oil spills, showing that cyanobacterial mats grow without bioturbation (i.e., destabilizing the sediment surface caused by crabs and polychaetes) of sediments. Cyanobacterial growth is inhibited by the grazing pressure of benthic animals. The massive growth of cyanobacteria at the coast of oil pollution can be carried out by damaging several crab colonies in the mudflats that prevent the bioturbation process, and also, grazing of arthropods is prevented. This colonization is described through three different processes (Barth 2003). The first one is the removal of the uppermost part of the oiled sediment by desiccation, cracking and peeling of the cyanobacterial mats, the second includes the relocation of burrowing macrofauna like crabs and benthic animals like gastropods which outcompete the cyanobacteria again, and the third involves further massive growth of cyanobacteria forming thick-layered mats. The surface is completely blocked by these layers resulting in the formation of an anaerobic condition which inhibits oil degradation. Cyanobacterial blooms restrict microbial oil degradation and also prevent any resettlement by macrofauna.

Statistics estimate that 3.2 million tonnes of oil per year are released from general shipping and industrial activities among all sources into the environment (ITOPF 1990). Oil dispersants, chemical ingredients approved for use by the US EPA, are commonly used after oil spills in marine environments, help break up oil slicks on the water surface and enhance the rate of biodegradation of oil, and hence, they are used when other methods like oil containment and removal are not sufficient. However, the aftermath of the toxicity of oil spill dispersants alone or in the presence of oil must be calculated. Usually, undispersed oil affects mostly shorelines and surface-dwelling organisms, while most dispersed oil affects particularly pelagic and benthic organisms as they remain in the water column (NRC 2005).

2.7 Effects of Plastics on Cyanobacteria

Plastic pollution is threatening marine ecosystems at the global level (cf Chap. 17). Several substances are leached by plastic litter into marine environments, but the mode of action is not known for photosynthetic bacteria at the base of the marine food web. The cyanobacterial genus *Prochlorococcus* constitutes the most common photosynthetic organisms in the ocean (Partensky et al. 1999) with a calculated mean global population of $\sim 10^{27}$ cells and primary productivity of 4 Gt C y^{-1} in some oligotrophic ocean regions (Flombaum et al. 2013). Two strains of *Prochlorococcus* having different ecotypes were exposed to leachate from common plastic items: high-density polyethylene bags and polyvinyl chloride matting. In vitro growth and photosynthetic activity gets impaired in leachate-exposed *Prochlorococcus* resulting in altered genome-wide transcription. The strains responded differently to each leachate in terms of degree and time. The composition of the community of *Prochlorococcus* and the productivity of ocean phytoplankton communities is affected by the exposure of plastic leachates. The growth and photophysiological responses of *Prochlorococcus* exposed to PVC leachates are more affected than HDPE leachates. Oceanic phytoplankton are adversely affected when polar and nonpolar organic pollutants are present as mixtures. This negative effect is more prominent in picocyanobacteria such as *Synechococcus* and *Prochlorococcus* (Echeveste et al. 2010, 2016). Echeveste et al. (2016) have reported that persistent organic pollutants (POPs) are more toxic in complex mixtures than single or simple POP mixtures. It has also been observed that low levels of organic pollutants affect photosynthetic activity in *Prochlorococcus* and decrease transcription of genes encoding the large subunit of RuBisCO and PSII D1 protein in marine populations (Fernández-Pinos et al. 2017). Zinc enriched with both plastic and PVC leachates and more often enriched with PVC leachates. Zinc is a known constituent of many plastic additives, involving slip agents, colorants, fillers and heat stabilizers, with the latter used mainly in PVC products (Hahladakis et al. 2018). When *Synechococcus* was exposed to zinc having a concentration of 713 μg L^{-1}, the growth was inhibited (Debelius et al. 2011).

2.8 Effects of Antibiotic Pollution on Cyanobacteria

Antibiotic pollution also affects the aquatic systems either directly or indirectly. Antibiotics are used to control pathogenic bacteria, but indigenous microorganisms are at a hazard by the application of these antibiotics (Flaherty and Dodson 2005). The non-target microorganisms play important roles in ecosystems like nutrient cycling, mineralization of organic matter and deterioration of pollutants (Näslund et al. 2008). Thus, both primary producers and decomposers are affected by antibiotics by disruption of the ecosystem processes. Antibiotics are introduced in the environment through many routes like wastes of pharmaceutical plants and hospitals, used as fertilizers and applied in the fields (Heberer 2002). Growth of aquatic animals as well as cyanobacterial mats is also affected that are important for nutrient cycling and the evolution of oxygen (Breitholtz et al. 2006). Pan et al. (2008) had studied the phytotoxicity effect of amoxicillin in *Synechocystis* sp. at a higher dose (50 mg mL^{-1}). The effect of β-lactam antibiotics and erythromycin, ciprofloxacin and sulfamethoxazole on cyanobacteria has also been studied (Białk-Bielińska et al. 2011; Liu et al. 2011, 2012) and showed that growth rate decreased and photosystem II was inhibited. The phytotoxic effects of penicillin, ciprofloxacin and tylosin were studied in cyanobacterial mats singly or in mixtures of antibiotics by El-Nahhal and Alshanti (2015). Single antibiotics had a more toxic effect on cyanobacterial mats. The EC_{50} value of individual tests for penicillin, ciprofloxacin and tylosin was 0.13, 0.71 and 5.28 mg mL^{-1}, respectively. It was found that penicillin, ciprofloxacin and tylosin were more toxic to cyanobacterial mats than Diuron (standard toxic material). EC_{50} values for a mixture of penicillin and tylosin, ciprofloxacin and tylosin and ciprofloxacin and penicillin were found to be 0.077, 0.103, and 0.292 TU, respectively. Cyanobacteria were more affected by penicillin followed by ciprofloxacin, and tylosin showed the least effect. So, antibiotics impose adverse effects on cyanobacterial mats that have an important role in the ecosystem.

2.9 Conclusions

Human activities on the natural environment are some of the important forces that drive and therefore control that environment. Table 2.4 gives a brief comparison of the challenges assessed by the different pollutants discussed in this chapter. With respect to major groups of pollutants in aquatic environment, cyanobacterial communities have shown a potential to degrade and eventually bioremediate the environment from toxic pollutants. However, our interpretation of the cumulative effects of these pollutants on cyanobacterial community in aquatic ecosystems is not very clear till date and specifically from the context of toxicology. For instance, the synergistic response of pollutants along with heavy metal pollutants can have deleterious effects on prokaryotic systems, and ultimately pollutants can persist in the environment for a longer time. Therefore, in future, laboratory-based manipulation experiments including cyanobacterial communities isolated from cold and warm aquatic environments should be investigated in the presence of a combination of persistent pollutants. Amalgamation of current methodologies such as transcriptomics and metabolomics can significantly increase our comprehension of the effects of pollutants on cyanobacterial cell systems and eventually long-term impacts on ecosystem processes.

Table 2.4 Brief comparison of the challenges assessed by the different pollutants in aquatic environments discussed in this chapter (adapted and modified from Nogales et al. 2011)

Human activity	Pollutants	Associated impact and risks
Agriculture and livestock farming	Fertilizers, pesticides, antibiotics, manure	Nutrient enrichment, eutrophication, hypoxia/anoxia, development of harmful algal blooms, toxicity, bioaccumulation, pathogens, the spread of antibiotic resistances
Urban development	Domestic waste, sewage sludge	Organic enrichment, eutrophication, pathogens (bacterial and viral), hypoxia/anoxia, development of harmful algal blooms, toxicity by chemicals and heavy metals, loss of amenity and recreational value, litter (solid wastes)
Industry	Industrial waste, organic pollutants (xenobiotics), heavy metals, radionuclides	Toxicity, bioaccumulation, increased atmospheric deposition
Maritime transport and shipping	Hydrocarbons, xenobiotics (antifouling agents and heavy metals), ballast water, litter	Toxicity, bioaccumulation, the introduction of exotic species, pathogens
Oil extraction and refining	Hydrocarbons	Toxicity, bioaccumulation
Fossil fuel combustion	Hydrocarbons, heavy metals, heat, CO_2	Toxicity, increased atmospheric deposition, warming
Tourism (including recreational nautical activities)	Litter (plastic, cigarette buds), fecal waste, nutrients, hydrocarbons, xenobiotics	Death of animal wildlife, accumulation in sediments, hypoxia/anoxia, loss of amenity and recreational value, nutrient enrichment, development of algal blooms, pathogens
Sediment load and dredging	Particles, nutrients, organic pollutants, heavy metals	Turbidity, decrease in light penetration, nutrient enrichment, toxicity, bioaccumulation
Aquaculture	Organic load, fecal waste, antibiotics	High organic load, hypoxia/anoxia in sediments, pathogens, introduction of exotic species, spread of antibiotic resistances
Fishing	Discharged fishes, pathogens, hydrocarbons, xenobiotics	Habitat destruction, decrease of fish stocks pathogens, toxicity, bioaccumulation
Land claiming and deforestation		Erosion, increased terrestrial runoff

Acknowledgements A. Pandey (09/013(0619)/2016-EMR-I), N. Kumari (09/013(0819)/2018-EMR-I) and V. Singh (09/013(0568)/2014-EMR-I) are thankful to the Council of Scientific and Industrial Research (CSIR), New Delhi, India, for financial support in the form of fellowships. S. Mishra (Joint CSIR-UGCJRF-2019/NTA Ref. No.:191620046790) is thankful to the University Grants Commission (UGC), New Delhi, India, for providing the fellowship in the form of JRF.

References

Acey R, Healy P, Unger TF, Ford CE, Hudson RA (1987) Growth and aggregation behavior of representative phytoplankton as affected by the environmental contaminant di-n-butyl phthalate. Bull Environ Contam Toxicol 39(1):1–6

Aftab KU, Ahmad IZ (2013) Alterations in antioxidative defense system of *Anabaena variabilis* in the presence of heavy metals. APCBEE Procedia 5:491–496

Akkanen J, Vogt RD, Kukkonen JV (2004) Essential characteristics of natural dissolved organic matter affecting the sorption of hydrophobic organic contaminants. Aquat Sci 66(2):171–177

Alexander DE, Fairbridge RW (1999) Encyclopedia of environmental science. Springer Science & Business Media

Athar M, Vohora SB (1995) Heavy metals and environment. New Age International

Babu B, Wu J (2010) Biodegradation of phthalate esters by cyanobacteria. J Phycol 46(6):1106–1113

Babu GS, Hans RK, Singh J, Viswanathan PN, Joshi PC (2001) Effect of lindane on the growth and metabolic activities of cyanobacteria. Ecotoxicol Environ Saf 48(2):219–221

Baptista MS, Vasconcelos VM, Vasconcelos MT (2014) Trace metal concentration in a temperate freshwater reservoir seasonally subjected to blooms of toxin-producing cyanobacteria. Microb Ecol 68(4):671–678

Barón M, Arellano JB, Gorgé JL (1995) Copper and photosystem II: a controversial relationship. Physiol Plant 94(1):174–180

Barth HJ (2003) The influence of cyanobacteria on oil polluted intertidal soils at the Saudi Arabian Gulf shores. Mar Pollut Bull 46(10):1245–1252

Beiras R (2018) Marine pollution: sources, fate and effects of pollutants in coastal ecosystems. Elsevier

Benotti MJ, Trenholm RA, Vanderford BJ, Holady JC, Stanford BD, Snyder SA (2009) Pharmaceuticals and endocrine disrupting compounds in US drinking water. Environ Sci Technol 43(3):597–603

Białk-Bielińska A, Stolte S, Arning J, Uebers U, Böschen A, Stepnowski P, Matzke M (2011) Ecotoxicity evaluation of selected sulfonamides. Chemosphere 85(6):928–933

Björn LO (2007) Stratospheric ozone, ultraviolet radiation, and cryptogams. Biol Conserv 135 (3):326–333

Blindauer CA (2008) Zinc-handling in cyanobacteria: an update. Chem Biodivers 5(10):1990–2013

Bong CW, Malfatti F, Azam F, Obayashi Y, Suzuki S (2010) The effect of zinc exposure on the bacteria abundance and proteolytic activity in seawater. In: Hamamura N, Suzuki S, Mendo S, Barroso CM, Iwata H, Tanabe S (eds) Interdisciplinary studies on environmental chemistry-biological responses to contaminants. Tokyo, TERRAPUB, pp 57–63

Bossert I, Kachel WM, Bartha R (1984) Fate of hydrocarbons during oily sludge disposal in soil. Appl Environ Microbiol 47(4):763–767

Bossuyt BT, Janssen CR (2005) Copper regulation and homeostasis of *Daphnia magna* and *Pseudokirchneriella subcapitata*: influence of acclimation. Environ Pollut 136(1):135–144

Brand LE, Sunda WG, Guillard RR (1986) Reduction of marine phytoplankton reproduction rates by copper and cadmium. J Exp Mar Biol Ecol 96(3):225–250

Breitholtz M, Rudén C, Hansson SO, Bengtsson BE (2006) Ten challenges for improved ecotoxicological testing in environmental risk assessment. Ecotoxicol Environ Saf 63(2):324–335

Carmichael WW (2001) Health effects of toxin-producing cyanobacteria: the CyanoHABs. Hum Ecol Risk Assess: An Int J 7(5):1393–1407

Cavet JS, Borrelly GP, Robinson NJ (2003) Zn, Cu and Co in cyanobacteria: selective control of metal availability. FEMS Microbiol Rev 27(2–3):165–181

Cervantes C, Campos-García J, Devars S, Gutiérrez-Corona F, Loza-Tavera H, Torres-Guzmán JC, Moreno-Sánchez R (2001) Interactions of chromium with microorganisms and plants. FEMS Microbiol Rev 25(3):335–347

Chakravarty R, Banerjee PC (2008) Morphological changes in an acidophilic bacterium induced by heavy metals. Extremophiles 12(2):279–284

Choudhary M, Jetley UK, Khan MA, Zutshi S, Fatma T (2007) Effect of heavy metal stress on proline, malondialdehyde, and superoxide dismutase activity in the cyanobacterium *Spirulina platensis*-S5. Ecotoxicol Environ Saf 66(2):204–209

Dachs J, Eisenreich SJ, Hoff RM (2000) Influence of eutrophication on air-water exchange, vertical fluxes, and phytoplankton concentrations of persistent organic pollutants. Environ Sci Technol 34(6):1095–1102

Day JW Jr, Kemp WM, Yáñez-Arancibia A, Crump BC (2012) Estuarine ecology. In: John Day W, Byron Crump C, Michael Kemp W, Alejandro YA (eds) John Wiley & Sons, pp 554

Debelius B, Forja JM, Lubián LM (2011) Toxicity of copper, nickel and zinc to *Synechococcus* populations from the Strait of Gibraltar. J Mar Syst 88(1):113–119

Dembitsky VM, Levitsky DO (2004) Arsenolipids. Prog Lipid Res 43(5):403–448

Deng L, Wu F, Deng N, Zuo Y (2008) Photoreduction of mercury (II) in the presence of algae, *Anabaena cylindrical*. J Photochem Photobiol B 91(2–3):117–124

Dias MA, Lacerda IC, Pimentel PF, De Castro HF, Rosa CA (2002) Removal of heavy metals by an *Aspergillusterreus* strain immobilized in a polyurethane matrix. Lett Appl Microbiol 34 (1):46–50

Echeveste P, Dachs J, Berrojalbiz N, Agustí S (2010) Decrease in the abundance and viability of oceanic phytoplankton due to trace levels of complex mixtures of organic pollutants. Chemosphere 81(2):161–168

Echeveste P, Galbán-Malagón C, Dachs J, Berrojalbiz N, Agustí S (2016) Toxicity of natural mixtures of organic pollutants in temperate and polar marine phytoplankton. Sci Total Environ 571:34–41

El-Bestawy EA, Abd El-Salam AZ, Mansy AE (2007) Potential use of environmental cyanobacterial species in bioremediation of lindane-contaminated effluents. Int Biodeterior Biodegradation 59(3):180–192

El-Nahhal Y, Alshanti A (2015) Toxicity of single and mixtures of antibiotics to cyanobacteria. J Environ Anal Toxicol 5(3):1000274

El-Sheekh MM, El-Naggar AH, Osman ME, Haieder A (2000) Comparative studies on the green algae *Chlorella homosphaera* and *Chlorella vulgaris* with respect to oil pollution in the river Nile. Water Air Soil Pollut 124(1–2):187–204

El-Sheekh MM, El-Shouny WA, Osman ME, El-Gammal EW (2005) Growth and heavy metals removal efficiency of *Nostoc muscorum* and *Anabaena subcylindrica* in sewage and industrial wastewater effluents. Environ Toxicol Pharmacol 19(2):357–365

Fernández-Pinos MC, Vila-Costa M, Arrieta JM, Morales L, González-Gaya B, Piña B, Dachs J (2017) Dysregulation of photosynthetic genes in oceanic *Prochlorococcus* populations exposed to organic pollutants. Sci Rep 7(1):1–9

Flaherty CM, Dodson SI (2005) Effects of pharmaceuticals on *Daphnia* survival, growth, and reproduction. Chemosphere 61(2):200–207

Flombaum P, Gallegos JL, Gordillo RA, Rincón J, Zabala LL, Jiao N, Karl DM, Li WK, Lomas MW, Veneziano D, Vera CS (2013) Present and future global distributions of the marine Cyanobacteria *Prochlorococcus* and *Synechococcus*. Proc Natl Acad Sci USA 24:9824–9829

Fogg GE, Than-Tun (1960) Interrelations of photosynthesis and assimilation of elementary nitrogen in a blue-green alga. Philos Trans R Soc Lond B Biol Sci 153(950):111–127

Gadd GM (2010) Metals, minerals and microbes: geomicrobiology and bioremediation. Microbiology 156(3):609–643

Goldman CR, Horne AJ (1983) Limnology. McGraw-Hill Publishing Company, New York

Grant IF, Egan EA, Alexander M (1983) Pesticides to control ostracods grazing on blue-green algae. Soil Biol Biochem 15(2):193–197

Grenvald JC, Nielsen TG, Hjorth M (2013) Effects of pyrene exposure and temperature on early development of two co-existing Arctic copepods. Ecotoxicology 22(1):184–198

Grosse Y, Baan R, Straif K, Secretan B, Ghissassi FE, Cogliano V (2006) Carcinogenicity of nitrate, nitrite, and cyanobacterial peptide toxins. Lancet Oncol 7(8):628–629

Hahladakis JN, Velis CA, Weber R, Iacovidou E, Purnell P (2018) An overview of chemical additives present in plastics: migration, release, fate and environmental impact during their use, disposal and recycling. J Hazard Mater 344:179–199

Hain MP, Sigman DM, Haug GH (2014) The biological pump in the past. In: Sigman DM, Haug GH (eds) Reference module in earth systems and environmental sciences, treatise on geochemistry, the oceans and marine geochemistry. Elsevier, The Netherlands, pp 485–517

Hameed A, Hasnain S (2005) Cultural characteristics of chromium resistant unicellular cyanobacteria isolated from local environment in Pakistan. Chinese J Oceanol Limnol 23 (4):433–441

Hanagata N, Takeuchi T, Fukuju Y, Barnes DJ, Karube I (1992) Tolerance of microalgae to high CO_2 and high temperature. Phytochem 31(10):3345–3348

He YY, Häder D-P (2002) UV-B-induced formation of reactive oxygen species and oxidative damage of the cyanobacterium *Anabaena* sp.: protective effects of ascorbic acid and N-acetyl-L-cysteine. J Photochem Photobiol B 66(2):115–124

Heberer T (2002) Occurrence, fate, and removal of pharmaceutical residues in the aquatic environment: a review of recent research data. Toxicol Lett 131(1–2):5–17

Hirata K, Yoshitomi S, Dwi S, Iwabe O, Mahakhant A, Polchai J, Miyamoto K (2003) Bioactivities of nostocine A produced by a freshwater cyanobacterium *Nostoc spongiaeforme* TISTR 8169. J Biosci Bioeng 95(5):512–517

Holzinger A, Lütz C (2006) Algae and UV irradiation: effects on ultrastructure and related metabolic functions. Micron 37(3):190–207

Hoostal MJ, Bidart-Bouzat MG, Bouzat JL (2008) Local adaptation of microbial communities to heavy metal stress in polluted sediments of Lake Erie. FEMS Microbiol Ecol 65(1):156–168

Huisman JM, Matthijs HC, Visser PM (2005) Harmful cyanobacteria. In: Aquatic ecology series 3. Springer, Dordrecht

Hylander S, Grenvald JC, Kiørboe T (2014) Fitness costs and benefits of ultraviolet radiation exposure in marine pelagic copepods. Funct Ecol 28(1):149–158

Idso SB, Allen SG, Anderson MG, Kimball BA (1989) Atmospheric CO_2 enrichment enhances survival of *Azolla* at high temperatures. Environ Exp Bot 29(3):337–341

International Tanker Owners Pollution Federation (ITOPF) (1990) Response to marine oil spills. Witherby, London

Inthorn D, Nagase H, Isaji Y, Hirata K, Miyamoto K (1996) Removal of cadmium from aqueous solution by the filamentous cyanobacterium *Tolypothrix tenuis*. J Biosci Bioeng 82(6):580–584

Jayashree S, Thangaraju N, Gnanadoss JJ (2012) Toxic effects of chromium on the aquatic cyanobacterium *Oscillatoria* sp. and removal of chromium by biosorption. Indian J Exp Biol 3 (5):28–34

Jensen F (2013) Polarization consistent basis sets VIII. The transition metals Sc–Zn. J Chem Phys 138(1):14107

Kidd KA, Blanchfield PJ, Mills KH, Palace VP, Evans RE, Lazorchak JM, Flick RW (2007) Collapse of a fish population after exposure to a synthetic estrogen. Proc Natl Acad Sci USA 104(21):8897–8901

Krastanov A, Alexieva Z, Yemendzhiev H (2013) Microbial degradation of phenol and phenolic derivatives. Eng Life Sci 13(1):76–87

Kumar S, Habib K, Fatma T (2008) Endosulfan induced biochemical changes in nitrogen-fixing cyanobacteria. Sci Total Environ 403(1–3):130–138

Kumar KS, Dahms HU, Won EJ, Lee JS, Shin KH (2015) Microalgae-a promising tool for heavy metal remediation. Ecotoxicol Environ Saf 113:329–352

Lal S, Saxena DM, Lal R (1987a) Effects of DDT, fenitrothion and chlorpyrifos on growth, photosynthesis and nitrogen fixation in Anabaena (Arm 310) and Aulosira fertilissima. Agric Ecosyst Environ 19(3):197–209

Lal S, Lal R, Saxena DM (1987b) Bioconcentration and metabolism of DDT, fenitrothion and chlorpyrifos by the blue-green algae *Anabaena* sp. and *Aulosira fertilissima*. Environ Pollut 46 (3):187–196

Lin CH, Pedersen JA, Suffet IH (2003) Influence of aeration on hydrophobic organic contaminant distribution and diffusive flux in estuarine sediments. Environ Sci Technol 37(16):3547–3554

Lindsay WL (1979) Chemical equilibria in soils. John Wiley & Sons, Chichester, UK

Liu B, Liu W, Nie X, Guan C, Yang Y, Wang Z, Liao W (2011) Growth response and toxic effects of three antibiotics on *Selenastrum capricornutum* evaluated by photosynthetic rate and chlorophyll biosynthesis. J Environ Sci 23(9):1558–1563

Liu Y, Gao B, Yue Q, Guan Y, Wang Y, Huang L (2012) Influences of two antibiotic contaminants on the production, release and toxicity of microcystins. Ecotoxicol Environ Saf 77:79–87

Longhurst A, Sathyendranath S, Platt T, Caverhill C (1995) An estimate of global primary production in the ocean from satellite radiometer data. J Plankton Res 17(6):1245–1271

Lowe DR, Tice MM (2007) Tectonic controls on atmospheric, climatic, and biological evolution 3.5–2.4 Ga. Precambrian Res 158(3–4):177–197

Luximon-Ramma A, Bahorun T, Soobrattee MA, Aruoma OI (2002) Antioxidant activities of phenolic, proanthocyanidin, and flavonoid components in extracts of *Cassia fistula*. J Agric Food Chem 50(18):5042–5047

Marco E, Martinez F, Orus MI (1990) Physiological alterations induced by the organophosphorus insecticide trichlorfon in *Anabaena* PCC 7119 grown with nitrates. Environ Exp Bot 30 (2):119–126

Margesin R, Schinner F (1997) Efficiency of indigenous and inoculated cold-adapted soil microorganisms for biodegradation of diesel oil in alpine soils. Appl Environ Microbiol 7:2660–2664

Megharaj M, Venkateswarlu K, Rao AS (1987) Metabolism of monocrotophos and quinalphos by algae isolated from soil. Bull Environ Contam Toxicol 39(2):251–256

Megharaj M, Venkateswarlu K, Rao AS (1988) Microbial degradation and algal toxicity of monocrotophos and quinalphos in flooded soil. Chemosphere 17(5):1033–1039

Michalak AM, Anderson EJ, Beletsky D, Boland S, Bosch NS, Bridgeman TB, Chaffin JD, Cho K, Confesor R, Daloğlu I, DePinto JV (2013) Record-setting algal bloom in Lake Erie caused by agricultural and meteorological trends consistent with expected future conditions. Proc Natl Acad Sci USA 110(16):6448–6452

Mishra S, Pandey A, Sinha RP (2019a) Characterization of photoprotective and pharmaceutically important compounds from cyanobacteria and algal assemblages from historical Kunds of Varanasi, India. J Chem Pharm Res 11(9):24–36

Mishra S, Pandey A, Ahmed H, Kumar D, Singh V, Sinha RP (2019b) Mycosporine-like amino acids (MAAs) profile of cyanobacteria from different historical Kunds of Varanasi, India. Int J Appl Sci Biotechnol 7(3):317–326

Mittler R (2002) Oxidative stress, antioxidants and stress tolerance. Trends Plant Sci 7(9):405–410

Muir DC, Wagemann R, Hargrave BT, Thomas DJ, Peakall DB, Norstrom RJ (1992) Arctic marine ecosystem contamination. Sci Total Environ 122(1–2):75–134

Muller FL, Larsen A, Stedmon CA, Søndergaard M (2005) Interactions between algal-bacterial populations and trace metals in fjord surface waters during a nutrient-stimulated summer bloom. Limnol Oceanogr 50(6):1855–1871

Muralikrishna PV, Venkateswarlu K (1984) Effect of insecticides on soil algal population. Bull Environ Contam Toxicol 33(1):241–245

Murthy SDS, Bukhov NG, Mohanty P (1990) Mercury-induced alterations of chlorophyll a fluorescence kinetics in cyanobacteria: multiple effects of mercury on electron transport. J Photochem Photobiol B Biol 6(4):373–380

Nalewajko C, Murphy TP (2001) Effects of temperature, and availability of nitrogen and phosphorus on the abundance of *Anabaena* and *Microcystis* in Lake Biwa, Japan: an experimental approach. Limnol 2(1):45–48

National Research Council (NRC) (2005) Oil spill dispersants: efficacy and effects. National Academy Press, Washington, DC

Näslund J, Hedman JE, Agestrand C (2008) Effects of the antibiotic ciprofloxacin on the bacterial community structure and degradation of pyrene in marine sediment. Aquat Toxicol 90(3):223–227

Nogales B, Lanfranconi MP, Piña-Villalonga JM, Bosch R (2011) Anthropogenic perturbations in marine microbial communities. FEMS Microbiol Rev 35(2):275–298

Olafson RW, Abel K, Sim RG (1979) Prokaryotic metallothionein: preliminary characterization of a blue-green alga heavy metal-binding protein. Biochem Bioph Res Co 89(1):36–43

Ormerod SJ, Dobson M, Hildrew AG, Townsend C (2010) Multiple stressors in freshwater ecosystems. Freshw Biol 55:1–4

Paerl HW, Huisman J (2008) Blooms like it hot. Science 320(5872):57–58

Paerl HW, Xu H, McCarthy MJ, Zhu G, Qin B, Li Y, Gardner WS (2011) Controlling harmful cyanobacterial blooms in a hyper-eutrophic lake (Lake Taihu, China): the need for a dual nutrient (N & P) management strategy. Water Res 45(5):1973–1983

Pal A, Gin KY, Lin AY, Reinhard M (2010) Impacts of emerging organic contaminants on freshwater resources: review of recent occurrences, sources, fate and effects. Sci Total Environ 408(24):6062–6069

Pan X, Deng C, Zhang D, Wang J, Mu G, Chen Y (2008) Toxic effects of amoxicillin on the photosystem II of *Synechocystis* sp. characterized by a variety of in vivo chlorophyll fluorescence tests. Aquat Toxicol 89(4):207–213

Pandey A, Pathak J, Singh DK, Ahmed H, Singh V, Kumar D, Sinha, RP (2020) Photoprotective role of UV-screening pigment scytonemin against UV-B-induced damages in the heterocyst-forming cyanobacterium *Nostoc* sp. strain HKAR-2. Braz J Bot 43(1):67–80

Partensky F, Hess WR, Vaulot D (1999) *Prochlorococcus*, a marine photosynthetic prokaryote of global significance. Microbiol Mol Biol Rev 63(1):106–127

Pawlik B, Skowroński T (1994) Transport and toxicity of cadmium: its regulation in the cyanobacterium *Synechocystis aquatilis*. Environ Exp Bot 34(2):225–233

Pérez-Ruzafa A, Navarro S, Barba A, Marcos C, Cámara MA, Salas F, Gutiérrez JM (2000) Presence of pesticides throughout trophic compartments of the food web in the Mar Menor lagoon (SE Spain). Mar Pollut Bull 40(2):140–151

Perona E, Marco E, Orús MI (1991) Alteration of dinitrogen fixation and metabolism in cyanobacterium *Anabaena* PCC 7119 by phosphamidon. Environ Exp Bot 31(4):479–488

Pinto G, Pollio A, Previtera L, Stanzione M, Temussi F (2003) Removal of low molecular weight phenols from olive oil mill wastewater using microalgae. Biotechnol Lett 25(19):1657–1659

Polyak Y, Zaytseva T, Medvedeva N (2013) Response of toxic cyanobacterium *Microcystis aeruginosa* to environmental pollution. Water Air Soil Pollut 224(4):1494

Porter JR, Sheridan RP (1981) Inhibition of nitrogen fixation in *alfalfa* by arsenate, heavy metals, fluoride, and simulated acid rain. Plant Physiol 68(1):143–148

Rajendran UM, Kathirvel E, Narayanaswamy A (2007) Effects of a fungicide, an insecticide, and a biopesticide on *Tolypothrix scytonemoides*. Pestic Biochem Phys 87(2):164–171

Ramakrishnan B, Megharaj M, Venkateswarlu K, Naidu R, Sethunathan N (2010) The impacts of environmental pollutants on microalgae and cyanobacteria. Crit Rev Env Sci Tec 40(8):699–821

Reid PC, Hill C (2016) Ocean warming: setting the scene. In: Laffoley D, Baxter JM (eds) Explaining ocean warming: causes, scale, effects and consequences. IUCN Gland, Switzerland, pp 17–45

Reuther R (1992) Arsenic introduced into a littoral freshwater model ecosystem. Sci Total Environ 115(3):219–237

Rinta-Kanto JM, Ouellette AJ, Boyer GL, Twiss MR, Bridgeman TB, Wilhelm SW (2005) Quantification of toxic *Microcystis* sp. during the 2003 and 2004 blooms in western Lake Erie using quantitative real-time PCR. Environ Sci Technol 39(11):4198–4205

Rousseaux CS, Gregg WW (2014) Interannual variation in phytoplankton primary production at a global scale. Remote Sens 6(1):1–19

Scheffer M (2004) The story of some shallow lakes. In: Scheffer M (ed) Ecology of shallow lakes. Springer, Dordrecht, pp 1–19

Schindler DW, Hecky RE, McCullough GK (2012) The rapid eutrophication of Lake Winnipeg: greening under global change. J Great Lakes Res 38:6–13

Sevrin-Reyssac J, Pletikosic M (1990) Cyanobacteria in fish ponds. Aquaculture 88:1–20

Shashirekha S, Uma L, Subramanian G (1997) Phenol degradation by the marine cyanobacterium *Phormidium valderianum* BDU 30501. J Ind Microbiol Biotechnol 19(2):130–133

Shigeki M, Hiroyasu K, Hiroshi O, Akira T (1984) Accumulation of arsenic in blue-green alga, *Phormidium* sp. Comp Biochem Physiol Part C: Toxicol Pharmacol 78(2):377–382

Shilpi G, Shilpi S, Sunita S (2015) Tolerance against heavy metal toxicity in cyanobacteria: role of antioxidant defense system. Int J Pharm Pharm Sci 7(2):9–16

Shivaramaiah HM (2000) Organochlorine pesticides in agroecosystems: monitoring residues with suitable strategies for their management and remediation. Ph.D. thesis, University of Sydney, Sydney, Australia

Shukla MK, Tripathi RD, Sharma N, Dwivedi S, Mishra S, Singh R, Shukla OP, Rai UN (2009) Responses of cyanobacterium *Anabaena doliolum* during nickel stress. J Environ Biol 30 (5):871

Singh PK, Rai S, Pandey S, Agrawal, C, Shrivastava, AK, Kumar S, Rai LC (2012) Cadmium and UV-B induced changes in proteome and some biochemical attributes of *Anabaena* sp. PCC7120. Phykos 42(1):39–50

Sinha RP, Häder DP (2008) UV-protectants in cyanobacteria. Plant Sci 174(3):278–289

Smith VH (2003) Eutrophication of freshwater and coastal marine ecosystems a global problem. Environ Sci Pollut Res 10(2):126–139

Sorkhoh N, Al-Hasan R, Radwan S, Höpner T (1992) Self-cleaning of the Gulf. Nature 359 (6391):109

Staal M, Meysman FJ, Stal LJ (2003) Temperature excludes N_2-fixing heterocystous cyanobacteria in the tropical oceans. Nature 425:504–507

Strother PK (2008) A speculative review of factors controlling the evolution of phytoplankton during Paleozoic time. Rev Micropaléontol 51(1):9–21

Suikkanen S, Laamanen M, Huttunen M (2007) Long-term changes in summer phytoplankton communities of the open northern Baltic Sea. Estuar Coast Shelf Sci 71(3–4):580–592

Sultan P, Shah SM, Williams P, Jan A, Ahmad N (2007) Biochemical basis of heavy metal induced stress tolerance in the N_2-fixing cyanobacterium *Anabaena doliolum*. Afr J Clin Exp Microbiol 8(1):8–22

Sunda WG (1989) Trace metal interactions with marine phytoplankton. Biol Oceanogr 6(5–6):411–442

Sundaram S, Soumya KK (2011) Study of physiological and biochemical alterations in cyanobacterium under organic stress. Am J Plant Sci 6(1):1–16

Sundaram S, Soumya KK, Ramgopal PJK, Rahman A (2011) Impact of organic stress on growth, photosynthetic and physiological response of some cyanobacterial isolates. J Environ Sci Technol 4:264–283

Tan BL, Hawker DW, Müller JF, Leusch FD, Tremblay LA, Chapman HF (2007) Comprehensive study of endocrine disrupting compounds using grab and passive sampling at selected wastewater treatment plants in South East Queensland, Australia. Environ Int 33(5):654–669

Tandon RS, Lal R, Rao VN (1988) Interaction of endosulfan and malathion with blue-green algae *Anabaena* and *Aulosira fertilissima*. Environ Pollut 52(1):1–9

Vašková J, Vaško L, Kron I (2012) Oxidative processes and antioxidative metaloenzymes. Antioxidant Enzyme 2:19–58

Verhougstraete MP, Martin SL, Kendall AD, Hyndman DW, Rose JB (2015) Linking fecal bacteria in rivers to landscape, geochemical, and hydrologic factors and sources at the basin scale. Proc Natl Acad Sci USA 112(33):10419–10424

Verschuren D, Johnson TC, Kling HJ, Edgington DN, Leavitt PR, Brown ET, Talbot MR, Hecky RE (2002) History and timing of human impact on Lake Victoria, East Africa. Proc R Soc B 269(1488):289–294

Walker JC (1984) How life affects the atmosphere. BioScience:486–491

Wang G, Chen K, Chen L, Hu C, Zhang D, Liu Y (2008) The involvement of the antioxidant system in protection of desert cyanobacterium *Nostoc* sp. against UV-B radiation and the effects of exogenous antioxidants. Ecotoxicol Environ Saf 69(1):150–157

Wang J, Xie P (2007) Antioxidant enzyme activities of *Microcystis aeruginosa* in response to nonylphenols and degradation of nonylphenols by *M. aeruginosa*. Environ Geochem Health 29 (5):375–383

Wang Y, Hsieh YP (2002) Uncertainties and novel prospects in the study of the soil carbon dynamics. Chemosphere 49(8):791–804

Yang J, Li H, Ran Y, Chan K (2014) Distribution and bioconcentration of endocrine disrupting chemicals in surface water and fish bile of the Pearl River Delta, South China. Chemosphere 107:439–446

Yee N, Benning LG, Phoenix VR, Ferris FG (2004) Characterization of metal-Cyanobacteria sorption reactions: a combined macroscopic and infrared spectroscopic investigation. Environ Sci Technol 38(3):775–782

Yu Y, Wu L, Chang AC (2013) Seasonal variation of endocrine disrupting compounds, pharmaceuticals and personal care products in wastewater treatment plants. Sci Total Environ 442:310–316

Zeng J, Wang WX (2011) Temperature and irradiance influences on cadmium and zinc uptake and toxicity in a freshwater cyanobacterium Microcystis aeruginosa. J Hazard Mater 190(1–3):922–929

Zhang X, Liu L, Zhang S, Pan Y, Li J, Pan H, Xu S, Luo F (2016) Biodegradation of dimethyl phthalate by freshwater unicellular cyanobacteria. In: Xia C (ed) BioMed Res Int:1–8

Zhang Y, Lu D, Ju T, Wang L, Lin S, Zhao Y, Wang C, He H, Du Y (2013) Biodegradation of phenol using Bacillus cereus WJ1 and evaluation of degradation efficiency based on a graphene-modified electrode. Int J Electochem Sci 8(1):504–519

Chapter 3
Effects of Pollution on Fish

Donat-P. Häder

Abstract Fish are an increasing major food supply for most human populations derived from freshwater and marine fishing and aquaculture. Fish and other marine organisms are threatened by a plethora of environmental stress factors such as acidification, global climate change and solar UV radiation. In addition to overfishing, many types of pollution decimate the amount of available fish production. Heavy metal pollution is a major threat in freshwater habitats, such as lakes and rivers, as well as in coastal ecosystems. Mercury is a notorious toxin which has been found to be responsible for fish loss and food poisoning. Persistent organic pollutants enter the sea by terrestrial run-off and are carried by rivers or transported through the air. These substances include flame retardants, fluorinated polymers, and perfluoroalkyl compounds, which are found in the water column and the sediment. Pesticides such as herbicides and insecticides also reach aqueous habitats by terrestrial run-off. They tend to bioaccumulate in the food web and are found in fish and mammals used for human consumption. Crude oil contamination from ship wrecks or oil platform disasters add to accidental spills and seepage from natural sources. Hydrocarbon components are toxic to the environment and pose persistent threats for the health of humans and animals. Also, the dispersants used to facilitate the dissolution of oil in the water are toxic to the biota. Pharmaceuticals and drugs are usually not removed in wastewater treatment plants and reach downstream rivers and are discharged into the ocean. Plastic pollution has developed over the last 70 years. The debris accumulates on land and is transported by wind and rivers to marine ecosystems. Broken down by mechanical forces and UV on the beach, the resulting microplastic is a persistent pollutant in the water and in the sediment affecting microorganisms. The debris is bioaccumulated in the food web and finally found in fish, birds and mammals.

D.-P. Häder (✉)
Department Biology, Friedrich-Alexander Universität, 91054 Erlangen, Germany
e-mail: donat@dphaeder.de

© The Author(s), under exclusive license to Springer Nature Switzerland AG 2021
D.-P. Häder et al. (eds.), *Anthropogenic Pollution of Aquatic Ecosystems*,
https://doi.org/10.1007/978-3-030-75602-4_3

Keywords Fish · Aquaculture · Persistent organic pollutants · Pesticides · Crude oil pollution · Bioaccumulation · Microplastics

3.1 Introduction

The global fish production is a major contribution to the food supply for the fast-growing human population on our planet. The Food and Agriculture Organization (FAO) states that the world commercial fishing in the wild amounted to about 171 million tonnes in 2016 of which 151 million tonnes (88%) were for direct human consumption (http://www.fao.org/fisheries/en/). This amount exceeds that of meat from all terrestrial animals combined. An additional 48 million tonnes were produced by aquaculture. The volume of globally captured fish has increased by 14% between 1990 and 2018, but that of global aquaculture production has skyrocketed by 527% during the same time (Fig. 3.1). Integrated over the total fish consumption, this amounts to an increase by 122%. In comparison, inland aquaculture produced 51 million tonnes of aquatic animals in rivers, lakes and fish farms (http://www.fao.org/state-of-fisheries-aquaculture 2020). In 2018, the value of global aquaculture production reached a sale value (at the farm) of US$ 263. 6 billion for aquatic animals, algae and ornamental seashells and pearls. Asia dominates the global aquaculture production (89%) with China, India, Indonesia,

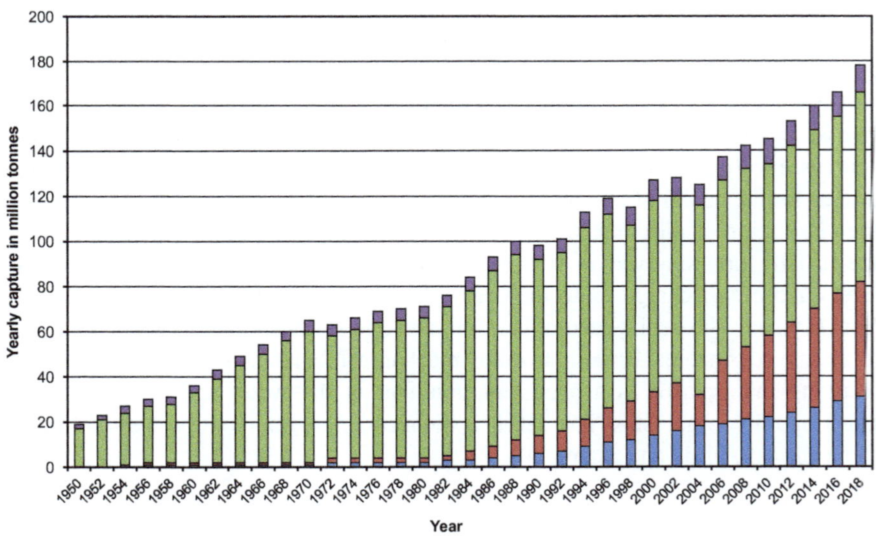

Fig. 3.1 Global fisheries and aquaculture production (includes fish, crustaceans, mollusks and other aquatic animals, but excludes mammals, reptiles, etc). Blue—marine aquaculture, red—inland aquaculture, green—marine capture, violet—inland capture, in million tonnes. Based on numbers published by FAO.org/state of fisheries-aquaculture (2020)

Vietnam and Bangladesh being the major producers. Other key global players are Egypt, Norway and Chile. The global number of employed people in fishery and fish farming has increased from 37.5 million in 1995 to 59.5 million in 2018.

Numerous studies indicate that overfishing is a major problem for marine capture even though the global capture has stabilized since 2000 (Fig. 3.1) (Garcia et al. 2018). In contrast, marine and even more so freshwater aquacultures have exploded since the 1950s (Zhou 2017; Ahmed et al. 2019).

Anthropogenic emissions of carbon dioxide, methane and other greenhouse gases accumulate in the atmosphere, resulting in global climate change with rising temperatures. The atmospheric CO_2 concentration has increased from about 280 ppm before the industrial revolution to more than 410 ppm today (Gao and Häder 2020). Simultaneously, the mean global water temperature has risen by about 1 °C over the last century (Fischetti 2013). An even stronger increase in temperature over the Arctic has resulted in a strong decrease in Arctic sea ice on the order of 9.1% per decade between 1979 and 2006 (Stroeve et al. 2007) with far-reaching effects for the biota including fish.

The oceans are a sink for more than 30% of the CO_2 released by human activities. The estimated uptake of 25 million tons per day (Sabine et al. 2004) results in an acidification of the surface layer of the water column. The proton concentration has increased by 30% resulting in a decrease of the water pH by 0.1 units (Caldeira and Wickett 2003); an increase of $[H^+]$ by 100–150% is predicted for the end of the century (Houghton et al. 2001; Zeebe and Wolf-Gladrow 2001; Gattuso et al. 2015). The resulting drop in the water pH by 0.3–0.4 units (Feely et al. 2004) threatens the productivity of freshwater and marine organisms (Hasler et al. 2018; Russell et al. 2013). Both freshwater and marine fish are affected by water acidification. For example, decreasing pH values in the water negatively affects larval development and growth rates (Watson et al. 2009; Parker et al. 2011; Lannig et al. 2010). Water acidification impairs olfactory discrimination and homing ability (Munday et al. 2009) and erodes crucial auditory behavior in marine fish (Simpson et al. 2011).

Other stress factors are increasing solar UV radiation. Anthropogenic emissions of trace gases such as organobromides and chlorocarbons have decreased the stratospheric ozone concentration since the 1980s resulting in increased UV-B (280–315 nm) irradiances hitting the surface of the Earth. The Montreal Protocol and its amendments have limited further increases but a recovery to pre-1980 levels is assumed to be reached only by or after mid-century because of the long lifetimes of chlorofluorocarbons (CFCs) in the stratosphere which can be decades (Hoffmann et al. 2014). In addition, global climate change alters total column ozone, and therefore, UV irradiances (Schnell et al. 2016; Meul et al. 2016; Williamson et al. 2014). Solar UV-B radiation can affect different fish species by damaging the goblet cells (mucus secreting cells) in the dorsal epidermis (Kaweewat and Hofer 1997) and the eyes (Sharma et al. 2005). The larval development has been found to be impaired by exposure to UV-B radiation in surface dwelling fish species and increase their mortality (Kumar et al. 2019; Hunter et al. 1981, 1982).

Pollution is another major environmental stress factor for both freshwater and marine fish species. Heavy metal pollution is affecting freshwater and coastal habitats. Many rivers in rapidly developing industrial regions carry high concentrations of Zn, Al, Cd, Pb, Cu, Ni, Fe, Mn, Cr, Co (cf. Chap. 13, this volume) which is of major concern for fish populations as well as for human consumers (Bhuyan et al. 2017; Xia et al. 2018). These heavy metals are discharged into the ocean where they pollute coastal areas, but the plumes from the river mouths can extend hundreds of kilometers into the oceans (Zhen et al. 2016; Lavrova et al. 2016). In addition, deliberate dumping of toxic wastes such as heavy metals from chemical processes reaches wider distances in the oceans (cf. Chap. 16, this volume, Dumping of Toxic Waste into the Oceans) (Rathoure 2020).

Inland water resources such as lakes, reservoirs and rivers are also endangered by pollution from organic materials such as detergents and surfactants, pesticides and persistent organic pollutants (PAHs) (Fikarová et al. 2018; Dahshan et al. 2016). Other organic pollutants which are found in increasing concentrations in natural freshwater, aquaculture installments and coastal habitats are pharmaceuticals which are of emerging concern (cf. Chap. 11, this volume, Pharmaceutical pollutants in aquatic ecosystems) (Biel-Maeso et al. 2018; Brodin et al. 2014), since they can alter the behavior of aquatic organisms (Sundin et al. 2019). In addition, illicit drugs are found in increasing concentrations in aquatic ecosystems (Rosi-Marshall et al. 2015).

Since the 1940s, plastic foils and bags have changed our life standard. Mainly based on crude oil components, these materials are inexpensive, easy to produce and durable. The reverse side of this is that these materials litter the countryside around villages and cities, as well as beaches, and are transported into rivers which carry the debris to the oceans (Jambeck et al. 2015). Millions of tons have already accumulated in the environment (Thompson et al. 2004). On the beaches, the material becomes brittle when exposed to solar UV radiation and is fragmented by mechanical forces of wind and waves (Song et al. 2017). These microplastic particles (<5 mm diameter) (Thompson 2015) float in the water and accumulate in the sediments (do Sul and Costa 2014). Since most plastic materials do not deteriorate easily and are not biodegradable, they will persist in the environment for very long times (Worm et al. 2017). The mid-oceanic gyres in the Atlantic, Pacific and Indian Oceans concentrate this debris (Cole et al. 2011; Eriksen et al. 2016). In addition to choking, larger plastic materials block the stomachs and intestines of aquatic invertebrates, fish, amphibians, mammals and birds (Sigler 2014). However, the transport across trophic levels and the impact on human health still needs to be evaluated (Andrady 2011).

3.2 Heavy Metal Pollution

In 1956, a Japanese medical doctor reported an epidemic disease of the central nervous system in people living in the district of Minamata, Kumamoto Prefecture. The cause for this disease was unknown at that time. Later on, it was found that a nearby chemical factory (Chisso Corporation) producing acetaldehyde had discharged mercury through its chemical waste pipe into the nearby bay. Bacteria in the water metabolized the Hg into methyl mercury which accumulated through the food web and ended up in the muscles of fish. The first effects were seen in cats which ate the fish after which they walked unsteady and died. Later on, the local human population who also used the fish in their staple diet was affected. More than 2000 people died and thousands more were crippled since fetuses and children were particularly affected (Hachiya 2006). In order to quantify the problem in 1973, sediment samples and fish were collected from the Minamata Bay and analyzed for Hg pollution (Fujiki and Tajima 1992). There was a clear correlation between the Hg concentration in the sediment and in a croaker (*Micropogonias*) which suggests a pathway from suspended particles via zooplankton into the fish (Nishimura and Kumagai 1983). A similar mercury pollution has been found in indigenous Amazonian people, who have a high fish consumption with an average of 243 g per capita per day (Boischio and Henshel 2000). Mercury pollution induces mutagenic effects as found in three Amazonian fish species (*Prochilodus nigricans* (detritivore), *Mylossoma duriventris* (omnivore) and *Hoplias alabaricus* (piscivore)) as shown by the micronucleus test (MNT) (Porto et al. 2005). Mercury is often used for amalgamation of gold in small gold mines, especially in the Amazon region (Brazil and Bolivia) (Maurice-Bourgoin et al. 1999). Also, in Lake Titicaca, increased concentrations of Hg and other heavy metals have been found in the water, sediment and the muscles of several fish attributed to contamination from artisanal gold mining (Gammons et al. 2006). The concentration of Hg can be measured in the water, in fish and in human hair. Similar problems of mercury pollution in fish were reported from Sweden (Ackefors et al. 1970) and Haifa Bay (Hornung et al. 1984). In Israel, it was determined that the Hg contamination of the shallow water zone was due to anthropogenic pollution from a mercury-cell chlor-alkali plant. In the vicinity of the plant, the pollution was about 1 μg Hg/g dry weight sediment which was about 157 times higher than in an unpolluted area. In carnivorous gastropod mollusks (*Arcularia circumcinta* and *A. gibbosula*), concentrations of 38.7 and 18.2 μg Hg/g dry weight were analyzed, respectively. In the South China Sea, large inter-species differences in total mercury and methyl mercury were identified which were found to be due to feeding habitats and fish size. Higher levels were found in carnivorous fish than in omnivorous and herbivorous fish (Liu et al. 2014). The sources of Hg in the water were assumed to be from atmospheric deposition and anthropogenic pollution. The limits for mercury in fish have been established in various countries as 0.2–1 mg/kg for total mercury and 0.3–1 mg/kg for methyl mercury (Cheng and Hu 2012).

Anthropogenic heavy metal pollution is often the result of mining, forestry, waste disposal and fuel combustion (Olsson et al. 1998). Several heavy metal pollutants, including Cd, Cr, Cu, Fe, Pb and Zn, have been found to decrease the length and weight of six fish species from the northeast Mediterranean Sea (*Sparus auratus*, *Atherina hepsetus*, *Mugil cephalus*, *Trigla cuculus*, *Sardina pilchardus* and *Scombereseox saurus*) (Canli and Atli 2003; Kalay et al. 1999). In most samples metal concentrations were highest in the liver with 4.5 (Cd), 17.1 (Cr) and

41.2 (Pb) µg/g dry weight as analyzed in *T. cuculus*, *S. pilchardus* and *A. hepsetus*, respectively. In the liver of *M. cephalus,* a very high concentration of Cu was found (202.8 µg/g dry weight), and the gills of *S. saurus* had an iron concentration of 885.5 µg/g dry weight. A similar correlation between heavy metal concentrations and fish size has been found in several carp and catfish species in the Yangtze River, China (Yi and Zhang 2012).

Similar heavy metal concentrations were found in muscle, gill and liver tissues of *Leuciscus cephalus* and *Lepomis gibbosus* caught at two stations off Saricay, SW Anatolia (Yılmaz et al. 2007). However, in these samples, the heavy metal concentrations were below the limits set by EAO, WHO and the EU. Likewise, fish from the Cochin area (India) showed concentrations of Cd, Pb, Hg, Cr, As, Zn, Cu, Co, Mn, Ni and Se which were below the maximum residual levels prescribed by the EU and USFDA and thus considered safe for human consumption (Sivaperumal et al. 2007). In contrast, some of 200 fish samples from the Tuzla Lagoon, Turkey showed Zn, Cd and Pb concentrations in the muscle tissue during spring which exceeded the levels set by law (Dural et al. 2007). Investigations of heavy metal concentrations in fish (whitefish, perch, pike trout, burbot and vendace) collected from three lakes in the region between Russia and Norway indicated that most metals did not show bioaccumulation with the exception of Hg (Amundsen et al. 1997). Other studies indicated considerable bioaccumulation of heavy metals in the liver and gills as well as in muscle tissue as shown for the common carp (*Cyprinus carpio*) (Vinodhini and Narayanan 2008). The degree of accumulation followed a sequence of Cd > Pb > Ni > Cr in the gills.

Heavy metals can induce oxidative stress in fish as shown in the Nigerian cat fish (*Clarias gariepinus*) caught in the Ogun River at a site which is close to six major industries (Farombi et al. 2007). The concentrations of Zn, Cu, Pb, As and Cd in the heart, gills, kidney and liver had been determined with an atomic absorption spectrophotometer. The oxidative stress was deduced from the activities of superoxide dismutase, catalase, glutathione S-transferase and glutathione which are enzymatic antioxidants. The effects of global climate change and pollutants have been discussed independently but recent research indicates that these two stressors are linked and impact aquatic ecosystems and the biota since future climate change may alter risks from pollution due to altered contaminant pathways, emission, bioavailability and remobilization (Schiedek et al. 2007).

3.3 Organic Pollutants in Fish

Many organic chemicals are persistent organic pollutants (POPs) which are not easily degraded and stay in the environment for long periods (Harrad 2009). Typical examples are fluorotelomer alcohols (FTOHs, used in textile, leather and paper products), brominated flame retardants, fluorinated polymers (e.g., Teflon), perfluoroalkyl compounds such as polyfluorinated sulfonamides (FSAs, used in photographic papers, medical applications and pesticides), perfluoroalkylsulfonic acids (FPSAs, used to make paper, textiles and carpets waterproof, in older fire extinguishers and hydraulic fluids), perfluorocarboxylic fluorotelomer carboxylic acids (PFCAs, used as surfactants, emulsifiers and for the production of peptides) and fluorotelomer sulfonic acids (used as flame retardants and in electroplating). These pollutants are found in air, rain, soil, water and sediments (Shoeib et al. 2005; Ashraf 2017). The Stockholm Convention of May 2001 summarizes the sources, characteristics, fate and effects and became binding in 2004 after 50 countries had ratified it and by late 2008 180 parties had accepted the convention. It lists DDT, aldrin, chlordane, dieldrin, hexachlorobenzene, haptachlor, mirex, polychlorinated dibenzofurans, polychlorinated biphenyls, polychlorinated dibenzo-*p*-dioxins and toxaphene but is open for further substances (Xu et al. 2013). A study by Ogata et al. (2009) compared samples from 30 beaches in 17 countries and analyzed them for organochlorine compounds. The highest concentrations were found in samples from North American coasts and lower concentrations in Western Europe and Japan, while Australia, tropical Asia and Southern Africa had even lower concentrations. One exception was DDT which was detected in high concentrations in samples from Vietnam where it is used as pesticide to control malaria. Table 3.1 summarizes the concentrations of several POPs found in samples from the open Atlantic Ocean (Jaward et al. 2004).

Persistent organochlorines including α-HCH, DDE, heptachlor, endosulfan and methoxychlor were found in porewater and surface water samples from Minjiang River Estuary, Fujian, China (Zhang et al. 2003). The added concentrations of 18 organochlorines were analyzed as 214–1819 ng/L in surface water, 4541–13,699 ng/L in porewater and 29–52 ng/g in sediments.

POPs are taken up by phytoplankton with small picoplankton being the most affected as shown for plankton communities from the Mediterranean, Atlantic and Southern Ocean (Echeveste et al. 2016). These toxins accumulate in the food web and

Table 3.1 Concentration of POPs measured in air over the open Atlantic Ocean (pg m^{-3})

Compound	Minimum	Mean	Maximum
PCB 28	<2.5	9.3	42.0
PCB 52	1.7	6.0	24.0
PCB 90/101	0.8	3.7	16.0
PCB 118	<0.6	1.1	6.0
PCB 138	<0.7	2.2	9.4
PCB 153/132	1.5	5.0	21.0
PCB 180	<0.3	1.0	3.6
Σ_{79} PCB	12.0	79.0	360.0
HCB	4.8	23.0	100.0
α-HCH	<0.1	3.0	11.0
γ-HCH	<3.6	22.0	100.0
p,p'-DDE	<1.5	20.0	47.0
p,p'-DDT	<2.2	2.2	5.4

Data from Jaward et al. (2004)

are found in oysters (*Crassostea gigas*) (Luna-Acosta et al. 2015). Heavy polycyclic aromatic hydrocarbons, polychlorobisphenyls, dichrorodiphenytrichloroethane and lindane were found to compromise the antioxidant and immune-defense capabilities and led to high oyster mortalities. Ten edible marine species (fish and invertebrates) from the Gulf of Naples, Italy, showed POPs including hexachlorobenzene (HCB), DDT and polychlorinated bisphenyls (Naso et al. 2005). The highest concentrations were detected in animals from shallow coastal waters which led to the conclusion that these pesticides stemmed from agricultural, industrial and municipal activities in the area. While the concentrations for HCB and DDT were below the allowed maximum limits, the concentrations for PCBs were much higher that the limit of 200 ng/g fresh weight set by the European Union for meat.

Organochlorine compounds (hexachlorobenzene, *p,p'*-DDE and mirex) were found in three fish species caught off Elephant Island, Antarctica, while PCB were present only in low concentrations (Weber and Goerke 2003). These compounds were taken up through the benthic food chain. The levels of the POPs were similar to those in the Northern Hemisphere indicating an equilibrated global distribution. While there is a substantial research on POPs in the marine environment, only few results are published for aquaculture systems. One study reports the concentrations of polychlorinated bisphenyls (145–460 ng/g), organochlorine pesticides (5–250 ng/g) and polybrominated diphenyl ethers (1–85 ng/g) in both wild and farmed Atlantic salmon (*Salmo salar*) (Jacobs et al. 2002). Also, freshwater lakes are contaminated by POPs. DDTs (0.04–4.25 μg/g lipid) and PCBs (0.25–40.8 μg/g lipid) were the major organochlorine pollutants in samples from Lake Como, Italy (Villa et al. 2011). Ecological factors, such as hydrophobicity, site, season, feeding strategies, species and weight, affect the chemical and biological mechanisms which govern the biomagnifications within the aquatic food webs. This was shown for three fish species (Bluntnose minnows, *Pimephales notatus*, spottail shiners, *Notropis hudsonius* and emerald shiners, *Notropis atherinoides)*, caught at three locations in Detroit River, Ontario, Canada (McLeod et al. 2014).

POPs in edible fish pose an environmental and health problem for humans. PCBs are assumed to be carcinogens and pose hazards to the brain, nervous, immune and reproductive systems. Data based on several commercial fish species (bluefin tuna *Thunnus thynnus*, swordfish *Xiphias gladius*, Atlantic mackerel *Scomber scombrus* from the Mediterranean and of the Antarctic toothfish *Dissostichus mawsoni* from the Ross Sea, Antarctica) indicate that the amount of toxic equivalents exceeded the tolerable weekly intake, if 400 g of fish per week were consumed for the Mediterranean species but not for the Antarctic toothfish (Corsolini et al. 2005).

3.4 Pesticides

As early as 1986, Murty summarized the knowledge on toxicity of pesticides to fish in a two-volume book (Murty 1986) and an update can be found in a review 27 years later describing the effects of these extremely toxic agricultural pesticides on fish metabolism sometimes inducing mortality (Murthy et al. 2013). Most pesticides reach the marine ecosystems by terrestrial run-off (Häder 2019). The common herbicides atrazine, diuron, Irgarol®1051 and isoproturon are found in coastal waters because they are transported by terrestrial run-off. Most of them do not pose a high risk for the biota, but occasionally they may be concentrated to toxic levels (Sjollema et al. 2014). Samples from the Caribbbean and Pacific showed organochlorine and other cyclodiene pesticides on the surface of sediments at concentrations from <2 ng/L to 18.45 μg/L, which are many times higher than in the open water (Menzies et al. 2013). Since nematodes, cestodes and acanthocephalans and other parasites bioaccumulate these pollutants, which they take up with their food, they can be used to monitor the effects on the environment (Nachev and Sures 2016). Eggs of herring and guillemot are used in Europe to detect dioxin-like polychlorinated biphenyls (dl-PCBs) and polychlorinated dibenzo-*p*-dioxins and dibenzofurans (PCDD/Fs) (Miller et al. 2014). In the tropical Great Barrier Reef lagoons agricultural herbicides are degraded by light and sediment (Mercurio et al. 2016). Bivalves such as *Cerastoderma edule* and *Scrobicularia plana* are employed in ecotoxicological bioassays to quantify the toxic effects of the frequently used herbicide Primextra Gold TZ (Gonçalves et al. 2016). The Water Framework Directive of the European Union puts a number of new pesticides on their priority list such as the herbicides aclonifen and bifenox, the biocides cybutryn (Irgarol) and terbutryn and the insecticides heptachlor/heptachlor epoxide and cypermethrin (Vorkamp et al. 2014). Danish water and fish samples showed concentrations of cypermethrin and heptachlor which were higher than environmental quality standard, but did not exceed the maximal allowed concentrations. Concentrations of other pesticides were below the detection limits.

In the coastal waters off Mumbai, India, organochlorine pesticides have been analyzed which accumulate in the food chain. However, it is interesting that in Swedish samples of herring and several seal species enantio-selective accumulations of α-hexachlorocyclohexane, chlordanes and their metabolites were found (Wiberg et al. 1998). Chiral preference of bioaccumulated chlordane has been used to determine the source, pathway and fate of this organochlorine pesticide (Singh et al. 2016).

The Indus River supports about 270 million people in Tibet, India, Pakistan and China. However, most of the water is being exploited for irrigation and other parts are lost by evaporation and natural losses so that only 14.6% reach the Arabic Sea (Albinia 2020). In the Thar Desert in Rajasthan, water, which is brought over a distance of 300 km from the Sukkur reservoir, is used to grow the famous Dunicut Chili. This water carries fertilizers and pesticides from the fields in addition to arsenic leached from the soil and poisons the drinking water. The result is that many farmers have gall bladder and kidney problems. Often a kidney has to be removed by an operation. The Indus River also feeds the Manchar Lake in Sindh, the largest freshwater lake in Pakistan. However, because a large amount of water is diverted upstream, only a small amount of freshwater enters the lake, and the agricultural residues become so concentrated that almost the complete fish population has been destroyed and the water needs to be processed to be drinkable.

3.5 Crude Oil

The Deepwater Horizon blowout has been a spectacular disaster which released millions of barrels of crude oil into aquatic ecosystems resulting in tremendous ecological impacts (Forth et al. 2017b; Camilli et al. 2010; Smith 1993). The grounding of the Exxon Valdez spilled "only" 37,000 tonnes (about 20% of its total cargo) on the Alaskan shoreline. Large and ultra-large carriers can transport cargos of up to 306,000 tonnes (Fingas 2016). In contrast to these singular accidents, accidental spills resulting from quakes, storms, human errors or mechanical failure accumulate too much larger spills. Operations on oil platforms and during loading and unloading contribute to substantial pollution. In addition, seepage from natural sources contributes about 47% to the total oil spilling (equal to about 600,000 metric tons per year), but the error bar on these estimates is very large (Kvenvolden and Cooper 2003).

Crude oil components are toxic to aquatic wild life (cf. Chap. 15 on Petroleum Hydrocarbons in Coastal Patagonia), and these chemicals are very persistent in the environment threatening the biota for many years (Peterson et al. 2003). Two years after an oil spill of 600 metric tons in Buzzards Bay, Massachusetts, most of the hydrocarbons were still found in the march and sediments of the bay (Blumer and Sass 1972). Analyses after the Deepwater Horizon blowout showed that 21% of the samples were toxic to bacteria (shown with the Microtox assay), 34% were toxic to phytoplankton (QwikLite assay) and 43% resulted in DNA damage (L-Microscreen Prophage induction assay) due to the concentration of hydrocarbons (Paul et al. 2013). The genotoxic potential of the water and the sediment porewater samples persisted for at least one and a half years after capping the well as shown by the Microscreen Prophage test.

Crude oil components are highly toxic to fish. Polycyclic aromatic hydrocarbons (PAHs) have been found to affect the embryo-larval development in fish and result in developmental abnormalities (González-Doncel et al. 2008) and alkyl-PAHs to cause chronic toxicity in early stages of fish (Hodson et al. 2007). They retard the heart development in large predatory pelagic fish (Incardona et al. 2014). Later in life, it affects the excitation–contraction coupling in the fish heart (Brette et al. 2014). The molecular mechanism of this toxic effect is based on some polycyclic aromatic compounds which operate as effective agonists of the aryl hydrocarbon receptor (Incardona 2017) resulting in edema and craniofacial and body axis defects. A similar mechanism has been found for dioxins. In addition, tricyclic

PAHs induce a disruption of cardiovascular functions and morphogenesis independent of the aryl hydrocarbon receptor pathway (Incardona et al. 2005). Weathering and chemical dispersion of spilled crude oil reduce the concentration of volatile fractions which decreases the mortality of shrimp and some fish (e.g., sand lance, *Ammodytes hexapterus*) (Anderson et al. 1987).

Crude oil spills are mostly treated with dispersants to facilitate the dissolution of oil in water. Subsequently, the smaller vesicles (5–10 μm) are broken down by microbial degradation (Forth et al. 2017a; Kleindienst et al. 2015a). However, not only the crude oil itself but also the dispersant (e.g., Corexit 9500A used after the Deepwater Horizon blowup) were found to be highly toxic to ciliates and dinoflagellates as well as coral reef larvae, after almost 7 million liters were released in the Gulf of Mexico (Negri et al. 2016; Almeda et al. 2014). The dispersant used in the Gulf of Mexico was also genotoxic and cytotoxic to sperm whale skin cells (Wise et al. 2014). Oil-spill removers such as detergents and emulsifiers have also been found to be toxic to several species of shellfish and fish (Portmann and Connor 1968). The dispersant Corexit 9527 has also been found to show lethal toxicity in Arctic marine fish and invertebrates (Foy 1982), and the related Corexit EC9500A was found to increase the toxicity of PAHs in early and juvenile life stages of marine fish by inducing hypoxia (Dasgupta et al. 2015). The toxic effect changes the zooplankton species composition and is relayed through the food web to higher trophic levels. The toxicity even increases by exposure to solar UV radiation.

The final step in the degradation of spilled crude oil is a breakdown by hydrocarbonoclastic bacteria such as *Marinobacter* (Ron and Rosenberg 2014). In order to speed up the process, the intoxicated waters are seeded with these bacteria augmented by simultaneous application of fertilizers such as nitrogen and phosphorous. However, the dispersants negatively affect the activity of oil-degrading bacteria (Kleindienst et al. 2015b). But bacteria such as *Colwellia* have the ability to break down the dispersant.

3.6 Plastic Pollution

Plastic production has been increasing over the last 70 years and with it the plastic pollution in the environment. The uncontrolled usage of foils and bags contribute to the garbage accumulation in populated areas, especially in developing countries in Africa, Asia and South America. Plastic pollutants accumulate both in freshwater (Klein et al. 2015) and marine ecosystems (Lusher 2016). Large amounts of plastic debris (an estimated 1.15–2.41 million tonnes every year) are carried by rivers toward the oceans (Lebreton et al. 2017). The ten most polluting rivers are located in Asia, such as the Yangtze and the Pearl River and are responsible for about two-thirds of the plastic that reaches the oceans. On the beaches, the plastic becomes brittle and breaks down to microplastic particles (<5 mm in diameter) when exposed to solar UV radiation and mechanical forces from wind and waves

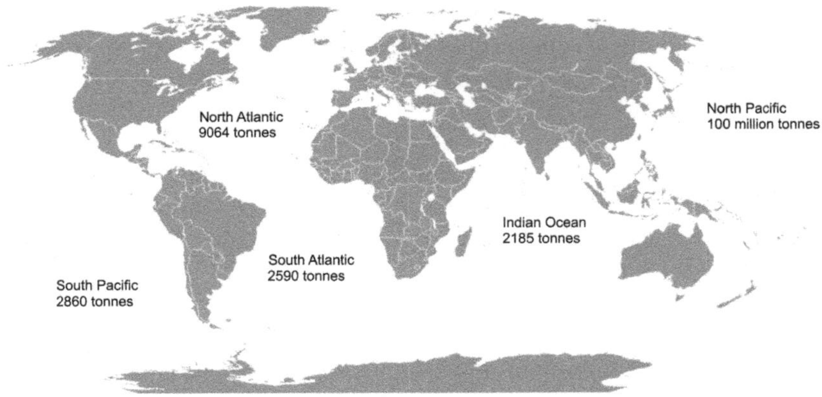

Fig. 3.2 Plastic debris concentrated in subtropical gyres. Data after Eriksen et al. (2016)

(Müller et al. 2018; Hermsen et al. 2018; Thompson 2015). Increasing temperatures due to climate change increase the UV-induced fragmentation (Duis and Coors 2016). Microplastics accumulate in the great gyres in the North and South Atlantic, Pacific and the Indian Ocean (Fig. 3.2).

Since most of the material is not biodegradable, this debris accumulates in the water or sediment as persistent organic pollutant. The most common plastics found in the sea are polyethylene, polypropylene and polystyrene in two abundant size ranges of 1–500 μm and 500 μm–5 mm. Other sources of microplastics are rub-off from tires, artificial turf as well as paint from boat hulls (Magnusson et al. 2016) which reach the marine environment by stormwater, wastewater or by wind. Nanoplastics (1–1000 nm) originate from the cosmetics industry (Barboza et al. 2018a). Personal care products, synthetic fibers from household dust and laundry add to the microplastics pollution which is not only found in coastal areas but also in the mid-oceanic gyres (Cole et al. 2011). The numbers on plastic pollution in aquatic environments show a wide uncertainty due to different sampling strategies, sample preparations and analysis methods (Stock et al. 2020). Methods to collect, identify and quantify microplastics include selective, volume-reduced and bulk sampling (Hidalgo-Ruz et al. 2012). Plastics are separated by floatation, filtration, sieving or visual methods.

Since many invertebrates and vertebrates feed by size and not by taste, the microplastic particles are ingested and bioaccumulated in the food web (do Sul and Costa 2014; Botterell et al. 2019). Microplastics are easily bioavailable and taken up by aquatic organisms such as echinoderms, corals, mollusks, larval stages of vertebrates, green sea turtles, cetaceans and seabirds causing adverse effects on their health (Caron et al. 2018). After ingestion by phytoplankton and zooplankton, microplastics are known to inhibit biological processes. Filter feeders such as mussels take up microplastics (Browne et al. 2015; Van Cauwenberghe et al. 2015; Wesch et al. 2016). Setälä et al. (2018) found that one-third of the gooseneck barnacles (*Lepas* spp.) contained 1–30 microplastic particles. The particles are relayed and concentrated in the food web and finally reach fish, birds and mammals. In a study off the Portuguese coast, 26 species of commercial fish were analyzed and microplastics were found in 17 species (Neves et al. 2015). In the British Channel, 2.9% of fish larvae of 347 individuals sampled had ingested microplastics, mostly blue fibers (Steer et al. 2017). The incidence increased toward the coast. Another study of 10 fish species from the British Channel found microplastics in the gastrointestinal tract in 36.5% of 504 individuals (Lusher et al. 2013). Out of 417 individuals of striped red mullet (*Mullus surmuletus*), almost one-third was found to have ingested microplastics (mostly cellophane and PET components) (Alomar et al. 2017). However, analysis of the fish's liver did not show indications of oxidative stress or cellular damage even though a small increase in activity of glutathione S-transferase was found. In contrast, a more recent study on larval and juvenile Medaka fish indicated that microplastics ingestion induces sublethal effects on growth and behavior (Pannetier et al. 2020). Microplastics were not only detected in the intestine but also in the muscle of fish (Akhbarizadeh et al. 2018). The concentration was higher in benthic fish than in pelagic species in the Northeast Persian Gulf. High consumption of contaminated fish may pose a health threat. A recent review documents microplastics in global waters, and ingestion by fish occurs widely and could result in various health problems (Wang et al. 2020). Thirty-three commercial fish species (out of 34) in the South Pacific gyre had ingested microplastic particles which were then transferred from prey to predators (Markic et al. 2018). In addition, microplastics are potential carries of adhering contaminants such as toxic metals (including mercury), styrene, phthalates, polychlorinated biphenyls, bisphenol A and polycyclic aromatic hydrocarbons (Barboza et al. 2018b, c).

Ingestion of microplastics by marine organisms can result in mortality (Luis et al. 2015; Gray and Weinstein 2017) or reduced feeding, body mass and metabolic rate (Welden and Cowie 2016). It can interfere with predatory performance (de Sá et al. 2015), behavioral responses and slower swimming (Barboza et al. 2018c). Decreased fertilization and larval abnormalities have been found (Martínez-Gómez et al. 2017). It may also induce neurotoxicity by acetylcholinesterase inhibition and oxidative damage (Oliveira et al. 2013; Avio et al. 2015; Ribeiro et al. 2017) and other negative effects (Barboza et al. 2018b).

Even though substantial information on the effects of plastics on marine organisms has been gathered, our knowledge of possible effects on human health is limited and controversial (Barboza et al. 2018a). Particles >150 μm are probably not taken up by the human body, but smaller ones (≤ 20 μm) could be transferred into inner organs and even smaller (0.1–10 μm) might cross cell membranes, the blood–brain barrier and the placenta and thereby induce immunosuppression and inflammations (Lusher 2016).

3.7 Conclusions

Fish has always been an abundant and important staple diet for most human societies. Even though the supply is threatened by overfishing, the consumption has been rising over the recent past. In addition to fishery in freshwater and marine waters, a growing share is contributed by aquaculture in lakes and oceans. Global climate change, ocean acidification and solar UV are considerable stress factors for fish, and pollution is another important stressor; e.g., heavy metals such as mercury threaten both freshwater and marine habitats. Persistent organic pollutants are derived from terrestrial run-off and are found in coastal areas, but also in the open ocean. In order to protect marine ecosystems and the biota, wastewater treatments need to be installed to remove herbicides, insecticides and other pesticides which enter aquatic ecosystems via terrestrial run-off. Crude oil pollution stems from seepage from natural sources and from anthropogenic activities, shipwrecks and accidents on oil platforms. Better security measures should be installed to minimize anthropogenic oil pollution. Fossil oil-derived hydrocarbon components are toxic and persist for many years in the environment. Since many pollutants bioaccumulate in the food web and are found in fish and mammals used for human consumption, rigorous testing for potential health risks should be mandatory. Plastic pollution has increased tremendously over the last decades. Single-use plastic litters the landscape and is transported to the sea. It is broken down by the action of wind and waves supported by solar UV exposure. The resulting microplastics are persistent in the water column and sediment, taken up by microorganisms and accumulate in the food web so that they are ultimately found in fish, birds and mammals. There are several strategies to reduce the burden of single-use plastics: Many materials can be replaced by paper, organic material and biodegradable plastic. Plastic debris should be prevented from entering the terrestrial habitats and the sea. Some countries already prohibit the use of plastic bags, etc. Finally, ships are being built and tested to collect plastic debris in the sea and use the material as fuel to power the vessels.

References

Ackefors H, Lofroth G, Rosen C-G (1970) A survey of the mercury pollution problem in Sweden with special reference to fish

Ahmed N, Thompson S, Glaser M (2019) Global aquaculture productivity, environmental sustainability, and climate change adaptability. Environ Manage 63(2):159–172

Akhbarizadeh R, Moore F, Keshavarzi B (2018) Investigating a probable relationship between microplastics and potentially toxic elements in fish muscles from northeast of Persian Gulf. Environ Pollut 232:154–163

Albinia A (2020) Lebensadern in Gefahr. National Geographic (Deutschland) July 2020:66–91

Almeda R, Hyatt C, Buskey EJ (2014) Toxicity of dispersant Corexit 9500A and crude oil to marine microzooplankton. Ecotoxicol Environ Saf 106:76–85

Alomar C, Sureda A, Capó X, Guijarro B, Tejada S, Deudero S (2017) Microplastic ingestion by *Mullus surmuletus* Linnaeus, 1758 fish and its potential for causing oxidative stress. Environ Res 159:135–142

Amundsen P-A, Staldvik FJ, Lukin AA, Kashulin NA, Popova OA, Reshetnikov YS (1997) Heavy metal contamination in freshwater fish from the border region between Norway and Russia. Sci Total Environ 201(3):211–224

Anderson JW, Riley R, Kiesser S, Gurtisen J (1987) Toxicity of dispersed and undispersed Prudhoe Bay crude oil fractions to shrimp and fish. In: International oil spill conference, vol 1. American Petroleum Institute, pp 235–240

Andrady AL (2011) Microplastics in the marine environment. Mar Pollut Bull 62(8):1596–1605

Ashraf MA (2017) Persistent organic pollutants (POPs): a global issue, a global challenge. Environ Sci Pollut Res 24:4223–4227

Avio CG, Gorbi S, Milan M, Benedetti M, Fattorini D, d'Errico G, Pauletto M, Bargelloni L, Regoli F (2015) Pollutants bioavailability and toxicological risk from microplastics to marine mussels. Environ Pollut 198:211–222

Barboza LGA, Vethaak AD, Lavorante BR, Lundebye A-K, Guilhermino L (2018a) Marine microplastic debris: an emerging issue for food security, food safety and human health. Mar Pollut Bull 133:336–348

Barboza LGA, Vieira LR, Branco V, Figueiredo N, Carvalho F, Carvalho C, Guilhermino L (2018b) Microplastics cause neurotoxicity, oxidative damage and energy-related changes and interact with the bioaccumulation of mercury in the European seabass, *Dicentrarchus labrax* (Linnaeus, 1758). Aquat Toxicol 195:49–57

Barboza LGA, Vieira LR, Guilhermino L (2018c) Single and combined effects of microplastics and mercury on juveniles of the European seabass (*Dicentrarchus labrax*): changes in behavioural responses and reduction of swimming velocity and resistance time. Environ Pollut 236:1014–1019

Bhuyan MS, Bakar MA, Akhtar A, Hossain MB, Ali MM, Islam MS (2017) Heavy metal contamination in surface water and sediment of the Meghna river, Bangladesh. Environ Nanotechnol Monit Manage 8:273–279

Biel-Maeso M, Baena-Nogueras RM, Corada-Fernández C, Lara-Martín PA (2018) Occurrence, distribution and environmental risk of pharmaceutically active compounds (PhACs) in coastal and ocean waters from the Gulf of Cadiz (SW Spain). Sci Total Environ 612:649–659

Blumer M, Sass J (1972) Oil pollution: persistence and degradation of spilled fuel oil. Science 1120–1122

Boischio AAP, Henshel D (2000) Fish consumption, fish lore, and mercury pollution—risk communication for the Madeira river people. Environ Res 84(2):108–126

Botterell ZL, Beaumont N, Dorrington T, Steinke M, Thompson RC, Lindeque PK (2019) Bioavailability and effects of microplastics on marine zooplankton: a review. Environ Pollut 245:98–110

Brette F, Machado B, Cros C, Incardona JP, Scholz NL, Block BA (2014) Crude oil impairs cardiac excitation-contraction coupling in fish. Science 343(6172):772–776

Brodin T, Piovano S, Fick J, Klaminder J, Heynen M, Jonsson M (2014) Ecological effects of pharmaceuticals in aquatic systems—impacts through behavioural alterations. Philos Transact Royal Soc London B 369(1656):20130580

Browne MA, Underwood AJ, Chapman MG, Williams R, Thompson RC, van Franeker JA (2015) Linking effects of anthropogenic debris to ecological impacts. Proc Roy Soc B Biol Sci 282 (1807):20142929. https://doi.org/10.1098/rspb.2014.2929

Caldeira K, Wickett ME (2003) Oceanography: anthropogenic carbon and ocean pH. Nature 425 (6956):365

Camilli R, Reddy CM, Yoerger DR, Van Mooy BA, Jakuba MV, Kinsey JC, McIntyre CP, Sylva SP, Maloney JV (2010) Tracking hydrocarbon plume transport and biodegradation at Deepwater Horizon. Science 330(6001):201–204

Canli M, Atli G (2003) The relationships between heavy metal (Cd, Cr, Cu, Fe, Pb, Zn) levels and the size of six Mediterranean fish species. Environ Pollut 121(1):129–136

Caron AG, Thomas CR, Berry KL, Motti CA, Ariel E, Brodie JE (2018) Ingestion of microplastic debris by green sea turtles (*Chelonia mydas*) in the Great Barrier Reef: validation of a sequential extraction protocol. Mar Pollut Bull 127:743–751

Cheng H, Hu Y (2012) Understanding the paradox of mercury pollution in China: high concentrations in environmental matrix yet low levels in fish on the market. Environ Sci Technol 46:4695–4696

Cole M, Lindeque P, Halsband C, Galloway TS (2011) Microplastics as contaminants in the marine environment: a review. Mar Pollut Bull 62(12):2588–2597

Corsolini S, Ademollo N, Romeo T, Greco S, Focardi S (2005) Persistent organic pollutants in edible fish: a human and environmental health problem. Microchem J 79(1–2):115–123

Dahshan H, Megahed AM, Abd-Elall AMM, Abd-El-Kader MA-G, Nabawy E, Elbana MH (2016) Monitoring of pesticides water pollution-the Egyptian river Nile. J Environ Health Sci Eng 14 (1):15

Dasgupta S, Huang IJ, McElroy AE (2015) Hypoxia enhances the toxicity of Corexit EC9500A and chemically dispersed Southern Louisiana Sweet crude oil (MC-242) to sheepshead minnow (*Cyprinodon variegatus*) larvae. PLoS ONE 10(6):e0128939

de Sá LC, Luís LG, Guilhermino L (2015) Effects of microplastics on juveniles of the common goby (*Pomatoschistus microps*): confusion with prey, reduction of the predatory performance and efficiency, and possible influence of developmental conditions. Environ Pollut 196:359–362

do Sul JAI, Costa MF (2014) The present and future of microplastic pollution in the marine environment. Environ Pollut 185:352–364

Duis K, Coors A (2016) Microplastics in the aquatic and terrestrial environment: sources (with a specific focus on personal care products), fate and effects. Environ Sci Europe 28(1):2. https://doi.org/10.1186/s12302-015-0069-y

Dural M, Göksu MZL, Özak AA (2007) Investigation of heavy metal levels in economically important fish species captured from the Tuzla lagoon. Food Chem 102(1):415–421

Echeveste P, Galbán-Malagón C, Dachs J, Berrojalbiz N, Agustí S (2016) Toxicity of natural mixtures of organic pollutants in temperate and polar marine phytoplankton. Sci Total Environ 571:34–41

Eriksen M, Thiel M, Lebreton L (2016) Nature of plastic marine pollution in the subtropical gyres. In: Hazardous chemicals associated with plastics in the marine environment. Springer, pp 135–162

Farombi EO, Adelowo O, Ajimoko Y (2007) Biomarkers of oxidative stress and heavy metal levels as indicators of environmental pollution in African cat fish (*Clarias gariepinus*) from Nigeria Ogun River. Int J Environ Res Public Health 4(2):158–165

Feely RA, Sabine CL, Lee K, Berelson W, Kleypas J, Fabry VJ, Millero FJ (2004) Impact of anthropogenic CO_2 on the $CaCO_3$ system in the oceans. Science 305(5682):362–366

Fikarová J, Kříženecká S, Elznicová J, Faměra M, Lelková T, Matkovič J, Grygar TM (2018) Spatial distribution of organic pollutants (PAHs and polar pesticides) in the floodplain of the Ohře (Eger) River, Czech Republic. J Soils Sediments 18(1):259–275

Fingas M (2016) Oil spill science and technology. Gulf Professional Publishing

Fischetti M (2013) Deep heat threatens marine life. Sci Am 308(April):92

Forth HP, Mitchelmore CL, Morris JM, Lay CR, Lipton J (2017a) Characterization of dissolved and particulate phases of water accommodated fractions used to conduct aquatic toxicity testing in support of the Deepwater Horizon natural resource damage assessment. Environ Toxicol Chem 36(6):1460–1472

Forth HP, Mitchelmore CL, Morris JM, Lipton J (2017b) Characterization of oil and water accommodated fractions used to conduct aquatic toxicity testing in support of the Deepwater Horizon oil spill natural resource damage assessment. Environ Toxicol Chem 36(6):1450–1459

Foy MG (1982) Acute lethal toxicity of Prudhoe Bay crude oil and Corexit 9527 to Arctic marine fish and invertebrates. EPS 4-EC-82-3 July 1982, Department of the Environment, Ottowa (Ontario) 62

Fujiki M, Tajima S (1992) The pollution of Minamata Bay by mercury. Water Sci Technol 25 (11):133–140

Gammons CH, Slotton DG, Gerbrandt B, Weight W, Young CA, McNearny RL, Cámac E, Calderón R, Tapia H (2006) Mercury concentrations of fish, river water, and sediment in the Río Ramis-Lake Titicaca watershed, Peru. Sci Total Environ 368(2–3):637–648

Gao K, Häder D-P (2020) Photosynthetic performances of marine microalgae under influences of rising CO_2 and solar UV radiation. In: Wang Q (ed) Microbial photosynthesis. Springer Nature, Singapore, pp 139–150

Garcia SM, Ye Y, Rice J, Charles A (2018) Rebuilding of marine fisheries Part 1: global review. FAO Fisheries and Aquaculture Technical Paper (630/1):I-274

Gattuso J-P, Magnan A, Billé R, Cheung W, Howes E, Joos F, Allemand D, Bopp L, Cooley S, Eakin C (2015) Contrasting futures for ocean and society from different anthropogenic CO_2 emissions scenarios. Science 349 (6243):aac4722

Gonçalves A, Mesquita A, Verdelhos T, Coutinho J, Marques J, Gonçalves F (2016) Fatty acids' profiles as indicators of stress induced by of a common herbicide on two marine bivalves species: *Cerastoderma edule* (Linnaeus, 1758) and *Scrobicularia plana* (da Costa, 1778). Ecol Ind 63:209–218

González-Doncel M, González L, Fernández-Torija C, Navas JM, Tarazona JV (2008) Toxic effects of an oil spill on fish early life stages may not be exclusively associated to PAHs: studies with Prestige oil and medaka (*Oryzias latipes*). Aquat Toxicol 87(4):280–288

Gray AD, Weinstein JE (2017) Size- and shape-dependent effects of microplastic particles on adult daggerblade grass shrimp (*Palaemonetes pugio*). Environ Toxicol Chem 36:3074–3080

Hachiya N (2006) The history and the present of Minamata disease. Japan Med Assoc J 49:112–118

Häder D-P (2019) Pollution dynamics of organic contaminants in marine ecosystems. Mar Pollut Curr Status Impacts Remedies 1:91–113

Harrad S (2009) Persistent organic pollutants. Wiley

Hasler CT, Jeffrey JD, Schneider EV, Hannan KD, Tix JA, Suski CD (2018) Biological consequences of weak acidification caused by elevated carbon dioxide in freshwater ecosystems. Hydrobiologia 806(1):1–12

Hermsen E, Mintenig SM, Besseling E, Koelmans AA (2018) Quality criteria for the analysis of microplastic in biota samples: a critical review. Environ Sci Technol 52(18):10230–10240

Hidalgo-Ruz V, Gutow L, Thompson RC, Thiel M (2012) Microplastics in the marine environment: a review of the methods used for identification and quantification. Environ Sci Technol 46(6):3060–3075

Hodson PV, Khan CW, Saravanabhavan G, Clarke L, Brown RS, Short J, Hollebone B, Wang Z, Lee K, King T (2007) Alkyl PAH in crude oil cause chronic toxicity to early life stages of fish. In: Arctic and marine oilspill program technical seminar, vol 1. Environment Canada; 1999, p 291

Hoffmann L, Hoppe C, Müller R, Dutton G, Gille J, Griessbach S, Jones A, Meyer C, Spang R, Volk C (2014) Stratospheric lifetime ratio of CFC-11 and CFC-12 from satellite and model climatologies. Atmos Chem Phys 14(22):12479–12497

Hornung H, Krumgalz BS, Cohen Y (1984) Mercury pollution in sediments, benthic organisms and inshore fishes of Haifa Bay, Israel. Mar Environ Res 12(3):191–208

Houghton JT, Ding Y, Griggs DJ, Noguer M, van der Linden PJ, Dai X, Maskell K, Johnson C-A (2001) Climate change 2001: the scientific basis, vol 881. Cambridge University Press, Cambridge, UK

Hunter JR, Kaupp SE, Taylor JH (1981) Effects of solar and artificial ultraviolet-B radiation on larval northern anchovy, *Engraulis mordax*. Photochem Photobiol 34(4):477–486

Hunter JR, Kaupp SE, Taylor JH (1982) Assessment of effects of UV radiation on marine fish larvae. In: Calkins J (ed) The role of solar ultraviolet radiation in marine ecosystems. Plenum Press, New York, pp 459–497

Incardona JP (2017) Molecular mechanisms of crude oil developmental toxicity in fish. Arch Env Contamin Toxicol 73(1):19–32

Incardona JP, Carls MG, Teraoka H, Sloan CA, Collier TK, Scholz NL (2005) Aryl hydrocarbon receptor–independent toxicity of weathered crude oil during fish development. Environ Health Perspect 113(12):1755–1762

Incardona JP, Gardner LD, Linbo TL, Brown TL, Esbaugh AJ, Mager EM, Stieglitz JD, French BL, Labenia JS, Laetz CA (2014) Deepwater Horizon crude oil impacts the developing hearts of large predatory pelagic fish. Proc Nat Acad Sci USA 111(15):E1510–E1518

Jacobs MN, Covaci A, Schepens P (2002) Investigation of selected persistent organic pollutants in farmed Atlantic salmon (*Salmo salar*), salmon aquaculture feed, and fish oil components of the feed. Environ Sci Technol 36(13):2797–2805

Jambeck JR, Geyer R, Wilcox C, Siegler TR, Perryman M, Andrady A, Narayan R, Law KL (2015) Plastic waste inputs from land into the ocean. Science 347(6223):768–771

Jaward FM, Barber JL, Booij K, Dachs J, Lohmann R, Jones KC (2004) Evidence for dynamic air–water coupling and cycling of persistent organic pollutants over the open Atlantic Ocean. Environ Sci Technol 38(9):2617–2625

Kalay M, Ay Ö, Canli M (1999) Heavy metal concentrations in fish tissues from the Northeast Mediterranean Sea. Bull Environ Contam Toxicol 63(5):673–681

Kaweewat K, Hofer R (1997) Effect of UV-B radiation on goblet cells in the skin of different fish species. J Photochem Photobiol B Biol 41(3):222–226

Klein S, Worch E, Knepper TP (2015) Occurrence and spatial distribution of microplastics in river shore sediments of the Rhine-Main area in Germany. Environ Sci Technol 49(10):6070–6076. https://doi.org/10.1021/acs.est.5b00492

Kleindienst S, Paul JH, Joye SB (2015a) Using dispersants after oil spills: impacts on the composition and activity of microbial communities. Nat Rev Microbiol 13(6):388–396

Kleindienst S, Seidel M, Ziervogel K, Grim S, Loftis K, Harrison S, Malkin SY, Perkins MJ, Field J, Sogin ML (2015b) Chemical dispersants can suppress the activity of natural oil-degrading microorganisms. Proc Nat Acad Sci USA 112(48):14900–14905

Kumar S, Malik D, Rathi P (2019) Effect of UV-B radiations on mortality and enzymes level in fish larvae of *Tor tor* and *Schizothorax richardsonii* on laboratory scale. J Appl Nat Sci 11 (2):462–467

Kvenvolden K, Cooper C (2003) Natural seepage of crude oil into the marine environment. Geo-Mar Lett 23(3–4):140–146

Lannig G, Eilers S, Pörtner HO, Sokolova IM, Bock C (2010) Impact of ocean acidification on energy metabolism of oyster, *Crassostrea gigas*—changes in metabolic pathways and thermal response. Mar Drugs 8(8):2318–2339

Lavrova OY, Soloviev DM, Strochkov MA, Bocharova TY, Kashnitsky AV (2016) River plumes investigation using Sentinel-2A MSI and Landsat-8 OLI data. In: Remote sensing of the ocean, sea ice, coastal waters, and large water regions 2016. International Society for Optics and Photonics, p 99990G

Lebreton LC, Van der Zwet J, Damsteeg J-W, Slat B, Andrady A, Reisser J (2017) River plastic emissions to the world's oceans. Nat Commun 8:15611

Liu J, Xu X, Yu S, Cheng H, Hong Y, Feng X (2014) Mercury pollution in fish from South China Sea: levels, species-specific accumulation, and possible sources. Environ Res 131:160–164

Luis LG, Ferreira P, Fonte E, Oliveira M, Guilhermino L (2015) Does the presence of microplastics influence the acute toxicity of chromium(VI) to early juveniles of the common goby (*Pomatoschistus microps*)? A study with juveniles from two wild estuarine populations. Aquat Toxicol 164:163–174

Luna-Acosta A, Bustamante P, Budzinski H, Huet V, Thomas-Guyon H (2015) Persistent organic pollutants in a marine bivalve on the Marennes-Oléron Bay and the Gironde Estuary (French Atlantic Coast)—part 2: potential biological effects. Sci Total Environ 514:511–522

Lusher A (2016) Microplastics in the marine environment: distribution, interactions and effects. In: Bergmann M, Gutow L, Klages M (eds) Marine anthropogenic litter. Springer International Publishing, pp 245–307. https://doi.org/10.1007/978-3-319-16510-3_10

Lusher AL, Mchugh M, Thompson RC (2013) Occurrence of microplastics in the gastrointestinal tract of pelagic and demersal fish from the English Channel. Mar Pollut Bull 67(1–2):94–99

Magnusson K, Eliasson K, Fråne A, Haikonen K, Hultén J, Olshammar M, Stadmark J, Voisin A (2016) Microplastics to the marine environment. IVL Swedish Environmental Research Institute, Stockholm, Sweden

Markic A, Niemand C, Bridson JH, Mazouni-Gaertner N, Gaertner J-C, Eriksen M, Bowen M (2018) Double trouble in the South Pacific subtropical gyre: increased plastic ingestion by fish in the oceanic accumulation zone. Mar Pollut Bull 136:547–564

Martínez-Gómez C, León VM, Marina SC, Gomáriz-Olcina M, Vethaak AD (2017) The adverse effects of virgin microplastics on the fertilization and larval development of sea urchins. Mar Environ Res 130:69–76

Maurice-Bourgoin L, Quiroga I, Guyot I, Malm O (1999) Mercury pollution in the upper Beni River. Amazonian Basin Bolivia Ambio 28:302–306

McLeod AM, Paterson G, Drouillard KG, Haffner GD (2014) Ecological factors contributing to variability of persistent organic pollutant bioaccumulation within forage fish communities of the Detroit River, Ontario, Canada. Environ Toxicol Chem 33(8):1825–1831

Menzies R, Quinete NS, Gardinali P, Seba D (2013) Baseline occurrence of organochlorine pesticides and other xenobiotics in the marine environment: Caribbean and Pacific collections. Mar Pollut Bull 70(1):289–295

Mercurio P, Mueller JF, Eaglesham G, O'Brien J, Flores F, Negri AP (2016) Degradation of herbicides in the tropical marine environment: influence of light and sediment. PLoS ONE 11 (11):e0165890

Meul S, Dameris M, Langematz U, Abalichin J, Kerschbaumer A, Kubin A, Oberländer-Hayn S (2016) Impact of rising greenhouse gas concentrations on future tropical ozone and UV exposure. Geophys Res Lett 43(6):2919–2927. https://doi.org/10.1002/2016GL067997

Miller A, Nyberg E, Danielsson S, Faxneld S, Haglund P, Bignert A (2014) Comparing temporal trends of organochlorines in guillemot eggs and Baltic herring: advantages and disadvantage for selecting sentinel species for environmental monitoring. Mar Environ Res 100:38–47

Müller A, Becker R, Dorgerloh U, Simon F-G, Braun U (2018) The effect of polymer aging on the uptake of fuel aromatics and ethers by microplastics. Environ Pollut 240:639–646

Munday PL, Dixson DL, Donelson JM, Jones GP, Pratchett MS, Devitsina GV, Døving KB (2009) Ocean acidification impairs olfactory discrimination and homing ability of a marine fish. Proc Nat Acad Sci USA 106(6):1848–1852

Murthy KS, Kiran B, Venkateshwarlu M (2013) A review on toxicity of pesticides in Fish. Int J Open Sci Res 1(1):15–36

Murty AS (1986) Toxicity of pesticides to fish, vol I, II. CRC press

Nachev M, Sures B (2016) Environmental parasitology: parasites as accumulation bioindicators in the marine environment. J Sea Res 113:45–50

Naso B, Perrone D, Ferrante MC, Bilancione M, Lucisano A (2005) Persistent organic pollutants in edible marine species from the Gulf of Naples, Southern Italy. Sci Total Environ 343(1): 83–95

Negri AP, Brinkman DL, Flores F, Botté ES, Jones RJ, Webster NS (2016) Acute ecotoxicology of natural oil and gas condensate to coral reef larvae. Sci Rep 6:21153

Neves D, Sobral P, Ferreira JL, Pereira T (2015) Ingestion of microplastics by commercial fish off the Portuguese coast. Mar Pollut Bull 101(1):119–126

Nishimura H, Kumagai M (1983) Mercury pollution of fishes in Minamata Bay and surrounding water: analysis of pathway of mercury. Water Ari Soil Pollut 20(4):401–411

Ogata Y, Takada H, Mizukawa K, Hirai H, Iwasa S, Endo S, Mato Y, Saha M, Okuda K, Nakashima A (2009) International pellet watch: global monitoring of persistent organic pollutants (POPs) in coastal waters. 1. Initial phase data on PCBs, DDTs, and HCHs. Mar Pollut Bull 58(10):1437–1446

Oliveira M, Ribeiro A, Hylland K, Guilhermino L (2013) Single and combined effects of microplastics and pyrene on juveniles (0 + group) of the common goby *Pomatoschistus microps* (Teleostei, Gobiidae). Ecol Ind 34:641–647

Olsson P-E, Kling P, Hogstrand C (1998) Mechanisms of heavy metal accumulation and toxicity in fish. In: Metal metabolism in aquatic environments. Springer, pp 321–350

Pannetier P, Morin B, Le Bihanic F, Dubreil L, Clérandeau C, Chouvellon F, Van Arkel K, Danion M, Cachot J (2020) Environmental samples of microplastics induce significant toxic effects in fish larvae. Environ Int 134:105047

Parker LM, Ross PM, O'Connor WA (2011) Populations of the Sydney rock oyster, *Saccostrea glomerata*, vary in response to ocean acidification. Mar Biol 158(3):689–697

Paul JH, Hollander D, Coble P, Daly KL, Murasko S, English D, Basso J, Delaney J, McDaniel L, Kovach CW (2013) Toxicity and mutagenicity of Gulf of Mexico waters during and after the Deepwater Horizon oil spill. Environ Sci Technol 47(17):9651–9659

Peterson CH, Rice SD, Short JW, Esler D, Bodkin JL, Ballachey BE, Irons DB (2003) Long-term ecosystem response to the Exxon Valdez oil spill. Science 302(5653):2082–2086

Portmann J, Connor P (1968) The toxicity of several oil-spill removers to some species of fish and shellfish. Mar Biol 1(4):322–329

Porto JI, Araujo CS, Feldberg E (2005) Mutagenic effects of mercury pollution as revealed by micronucleus test on three Amazonian fish species. Environ Res 97(3):287–292

Rathoure AK (2020) Heavy metal pollution and its management: bioremediation of heavy metal. In: Waste management: concepts, methodologies, tools, and applications. IGI Global, pp 1013–1036

Ribeiro F, Garcia AR, Pereira BP, Fonseca M, Mestre NC, Fonseca TG, Ilharco LM, Bebianno MJ (2017) Microplastics effects in *Scrobicularia plana*. Mar Pollut Bull 122:379–391

Ron EZ, Rosenberg E (2014) Enhanced bioremediation of oil spills in the sea. Curr Opin Biotechnol 27:191–194

Rosi-Marshall E, Snow D, Bartelt-Hunt SL, Paspalof A, Tank J (2015) A review of ecological effects and environmental fate of illicit drugs in aquatic ecosystems. J Hazard Mater 282:18–25

Russell BD, Connell SD, Findlay HS, Tait K, Widdicombe S, Mieszkowska N (2013) Ocean acidification and rising temperatures may increase biofilm primary productivity but decrease grazer consumption. Philos Transact Royal Soc London B 368(1627):20120438

Sabine CL, Feely RA, Gruber N, Key RM, Lee K, Bullister JL, Wanninkhof R, Won CS, Wallace DWR, Tilbrook B, Millero FJ, Peng T-H, Kozyr A, Ono T, Rios AF (2004) The oceanic sink for anthropogenic CO_2. Science 305:367–371

Schiedek D, Sundelin B, Readman JW, Macdonald RW (2007) Interactions between climate change and contaminants. Mar Pollut Bull 54(12):1845–1856

Schnell JL, Prather MJ, Josse B, Naik V, Horowitz LW, Zeng G, Shindell DT, Faluvegi G (2016) Effect of climate change on surface ozone over North America, Europe, and East Asia. Geophys Res Lett 43(7):3509–3518. https://doi.org/10.1002/2016GL068060

Setälä O, Lehtiniemi M, Coppock R, Cole M (2018) Microplastics in marine food webs. In: Microplastic contamination in aquatic environments. Elsevier, pp 339–363

Sharma J, Masuda R, Tanaka M (2005) Ultrastructural study of skin and eye of UV-B irradiated ayu *Plecoglossus altivelis*. J Fish Biol 67(6):1646–1652

Shoeib M, Harner T, Wilford BH, Jones KC, Zhu J (2005) Perfluorinated sulfonamides in indoor and outdoor air and indoor dust: occurrence, partitioning, and human exposure. Environ Sci Technol 39(17):6599–6606

Sigler M (2014) The effects of plastic pollution on aquatic wildlife: current situations and future solutions. Water Air Soil Pollut 225(11):2184

Simpson SD, Munday PL, Wittenrich ML, Manassa R, Dixson DL, Gagliano M, Yan HY (2011) Ocean acidification erodes crucial auditory behaviour in a marine fish. Biol Lett 7(6):917–920

Singh K, Hegeman WJ, Laane RW, Chan HM (2016) Review and evaluation of a chiral enrichment model for chlordane enantiomers in the environment. Environ Rev 24(4):363–376

Sivaperumal P, Sankar T, Nair PV (2007) Heavy metal concentrations in fish, shellfish and fish products from internal markets of India vis-a-vis international standards. Food Chem 102 (3):612–620

Sjollema SB, Martínez-García G, van der Geest HG, Kraak MH, Booij P, Vethaak AD, Admiraal W (2014) Hazard and risk of herbicides for marine microalgae. Environ Pollut 187:106–111

Smith C (1993) News sources and power elites in news coverage of the Exxon Valdez oil spill. Journalism Q 70(2):393–403

Song YK, Hong SH, Jang M, Han GM, Jung SW, Shim WJ (2017) Combined effects of UV exposure duration and mechanical abrasion on microplastic fragmentation by polymer type. Environ Sci Technol 51(8):4368–4376

Steer M, Cole M, Thompson RC, Lindeque PK (2017) Microplastic ingestion in fish larvae in the western English Channel. Environ Pollut 226:250–259

Stock F, Narayana VKB, Scherer R, Löder MGJ, Brennholt N, Laforsch C, Reifferscheid G (2020) Pitfalls and limitations in microplastic analyses. https://doi.org/10.1007/698_2020_654

Stroeve J, Holland MM, Meier W, Scambos T, Serreze M (2007) Arctic sea ice decline: faster than forecast. Geophys Res Lett 34(9). https://doi.org/10.1029/2007gl029703

Sundin J, Jutfelt F, Thorlacius M, Fick J, Brodin T (2019) Behavioural alterations induced by the anxiolytic pollutant oxazepam are reversible after depuration in a freshwater fish. Sci Total Environ 665:390–399

Thompson RC (2015) Microplastics in the marine environment: sources, consequences and solutions. In: Bergmann M, Gutow L, Klages M (eds) Marine anthropogenic litter. Springer International Publishing, pp 185–200. https://doi.org/10.1007/978-3-319-16510-3_7

Thompson RC, Olsen Y, Mitchell RP, Davis A, Rowland SJ, John AW, McGonigle D, Russell AE (2004) Lost at sea: where is all the plastic? Science (Washington) 304(5672):838

Van Cauwenberghe L, Devriese L, Galgani F, Robbens J, Janssen CR (2015) Microplastics in sediments: a review of techniques, occurrence and effects. Mar Environ Res 111:5–17. https://doi.org/10.1016/j.marenvres.2015.06.007

Villa S, Bizzotto EC, Vighi M (2011) Persistent organic pollutant in a fish community of a sub-alpine lake. Environ Pollut 159(4):932–939

Vinodhini R, Narayanan M (2008) Bioaccumulation of heavy metals in organs of fresh water fish *Cyprinus carpio* (Common carp). Int J Environ Sci Technol 5(2):179–182

Vorkamp K, Bossi R, Bester K, Bollmann UE, Boutrup S (2014) New priority substances of the European Water Framework Directive: biocides, pesticides and brominated flame retardants in the aquatic environment of Denmark. Sci Total Environ 470:459–468

Wang W, Ge J, Yu X (2020) Bioavailability and toxicity of microplastics to fish species: a review. Ecotoxicol Environ Saf 189:109913

Watson S-A, Southgate PC, Tyler PA, Peck LS (2009) Early larval development of the Sydney rock oyster *Saccostrea glomerata* under near-future predictions of CO_2-driven ocean acidification. J Shellfish Res 28(3):431–437

Weber K, Goerke H (2003) Persistent organic pollutants (POPs) in Antarctic fish: levels, patterns, changes. Chemosphere 53(6):667–678

Welden NAC, Cowie PR (2016) Long-term microplastic retention causes reduced body condition in the langoustine, *Nephrops norvegicus*. Environ Pollut 218:895–900

Wesch C, Bredimus K, Paulus M, Klein R (2016) Towards the suitable monitoring of ingestion of microplastics by marine biota: a review. Environ Pollut 218:1200–1208. https://doi.org/10.1016/j.envpol.2016.08.076

Wiberg K, Oehme M, Haglund P, Karlsson H, Olsson M, Rappe C (1998) Enantio selective analysis of organochlorine pesticides in herring and seal from the Swedish marine environment. Mar Pollut Bull 36(5):345–353

Williamson CE, Zepp RG, Lucas RM, Madronich S, Austin AT, Ballaré CL, Norval M, Sulzberger B, Bais A, McKenzie RL, Robinson SA, Häder D-P, Paul ND, Bornman JF (2014) Solar ultraviolet radiation in a changing climate. Nat Clim Change 4:434–441

Wise CF, Wise JT, Wise SS, Thompson WD, Wise JP (2014) Chemical dispersants used in the Gulf of Mexico oil crisis are cytotoxic and genotoxic to sperm whale skin cells. Aquat Toxicol 152:335–340

Worm B, Lotze HK, Jubinville I, Wilcox C, Jambeck J (2017) Plastic as a persistent marine pollutant. Annu Rev Environ Res 42:26

Xia F, Qu L, Wang T, Luo L, Chen H, Dahlgren RA, Zhang M, Mei K, Huang H (2018) Distribution and source analysis of heavy metal pollutants in sediments of a rapid developing urban river system. Chemosphere 207:218–228

Xu W, Wang X, Cai Z (2013) Analytical chemistry of the persistent organic pollutants identified in the Stockholm Convention: a review. Anal Chim Acta 790:1–13

Yi Y, Zhang S (2012) The relationships between fish heavy metal concentrations and fish size in the upper and middle reach of Yangtze River. Procedia Environ Sci 13:1699–1707

Yılmaz F, Özdemir N, Demirak A, Tuna AL (2007) Heavy metal levels in two fish species *Leuciscus cephalus* and *Lepomis gibbosus*. Food Chem 100(2):830–835

Zeebe RE, Wolf-Gladrow DA (2001) CO_2 in seawater: equilibrium, kinetics, isotopes, vol 65. Gulf Professional Publishing

Zhang Z, Hong H, Zhou J, Huang J, Yu G (2003) Fate and assessment of persistent organic pollutants in water and sediment from Minjiang River Estuary, Southeast China. Chemosphere 52(9):1423–1430

Zhen G, Li Y, Tong Y, Yang L, Zhu Y, Zhang W (2016) Temporal variation and regional transfer of heavy metals in the Pearl (Zhujiang) River, China. Environ Sci Pollut Res 23(9):8410–8420

Zhou X (2017) An overview of recently published global aquaculture statistics. FAO Aquac Newsl 56:6

Chapter 4
Effects of Pollution in Aquatic Food Chains

Raúl González, Cristian Durante, Marina Arcagni, Romina Juncos, Juan Seco Pon, Enrique Crespo, and Maite Narvarte

Abstract Aquatic ecosystems usually receive pollutants from near industrialized and highly populated zones, through riverine inflow, discharges of sewage or other wastes or the atmosphere. The trophic webs of such ecosystems can be affected by those pollutants, with the inhabitant species manifesting different vulnerabilities for certain compounds. Pollutants can bioaccumulate when ingested at a higher rate than excreted, and some can also be biomagnified if accumulated at such higher rates throughout the food chain. Biomagnifying substances, such as organochlorines, are usually lipophilic or, like methylmercury, have a high affinity for proteins. Other pollutants, as for example heavy metals, are mostly associated with water and usually do not biomagnify at all, though they may bioaccumulate to very high concentrations in certain organisms. Bivalve mollusks and other filter-feeders usually bioaccumulate substances at higher rates because they receive pollutants from suspended particles. Top predators (fish, seabirds and mammals) in aquatic trophic chains can be affected by biomagnified compounds. In this chapter, we summarize the conceptual basis of and terminology used in ecotoxicological studies and review regional investigations on the effects and modes of propagation of contaminants through the food webs in the aquatic ecosystems in Argentina.

R. González (✉) · M. Narvarte
Escuela Superior de Ciencias Marinas, Universidad Nacional del Comahue and CONICET (CIM Almirante Storni), San Martín 247, R8520CXV San Antonio Oeste, Río Negro, Argentina

C. Durante · E. Crespo
Centro para el Estudio de Sistemas Marinos (CESIMAR)—Centro Nacional Patagónico—CONICET, Bv. Brown 2915, U9120ACD Puerto Madryn, Chubut, Argentina

M. Arcagni · R. Juncos
Centro Atómico Bariloche, Comisión Nacional de Energía Atómica, Av. Bustillo 9500, R8402AGD Bariloche, Argentina

J. Seco Pon
Instituto de Investigaciones Marinas y Costeras, IIMyC—Universidad Nacional de Mar del Plata—CONICET, Funes 3250, B7602AYJ Mar del Plata, Argentina

© The Author(s), under exclusive license to Springer Nature Switzerland AG 2021
D.-P. Häder et al. (eds.), *Anthropogenic Pollution of Aquatic Ecosystems*,
https://doi.org/10.1007/978-3-030-75602-4_4

Keywords Pollution · Food chains · Marine · Southwest Atlantic Ocean · Freshwater · Argentina

4.1 Introduction

Marine coastal areas, lakes and rivers usually receive pollutants through riverine inflow, sewage discharges or the atmosphere (Häder et al. 2020). More than 90% of the nutrients transported by rivers and the atmosphere have been estimated to remain trapped in the margins of oceans, while 90% of the inputs of trace elements entering the sea accumulate in coastal environments (Beiras 2018). Contaminants may enter food webs through benthic or planktonic organisms. Owing to the lipophilic nature of organic contaminants (e.g., persistent organic pollutants—POPs, methylmercury), they have the potential to increase their concentrations in higher trophic levels.

Heavy metals are toxic pollutants commonly distributed in the aquatic environment that can cause severe biologic hazards in organisms when the concentrations exceed tolerable levels (Liu et al. 2019). Such metals affect animal health, in certain instances acting as carcinogens through oxidative mechanisms, damaging DNA and inactivating enzymes (Bal and Kasprzak 2002) and also affecting the nuclear chromatin (Le Croizier et al. 2018).

Examples of heavy metals present in marine environments at different biotoxic levels are As, Cd, Cr, Cu, Hg, Ni, Pb and Zn. Whereas Cr, Cu, Ni and Zn are considered essential elements for the construction of biomacromolecules, As, Cd, Hg and Pb are nonessential and possess a strong toxicity at low concentrations (Ishaque et al. 2006). Hg is considered to be a neurotoxin because of the ability to potentially damage the cell membranes of gonad and brain cells (Sinkus et al. 2017), whereas Cd can induce irreversible damage to the skeletal structures, gonads and kidneys (Bahnasawy et al. 2011). Even though both Cr and Cu are biologically essential through their active participation in carbohydrate metabolism and protein synthesis, those metals may become toxic when accumulated at high concentrations in organisms (Liu et al. 2019).

POPs are compounds that, to a varying degree, resist photolytic, biologic and chemical degradation. These substances are usually amphipathic and as such characterized by their hydrophilic and hydrophobic properties, which favor the potential of those molecules for bioaccumulation through permeation of the surface of an organism (e.g., gills, digestive tracts, skins), when the POP penetrates a lipid-rich environment of such a biologic membrane from the surrounding highly polar water (Wilhelmsson et al. 2013). POPs are also semivolatile, enabling them to move long distances in the atmosphere before deposition occurs (Ritter et al. 1995).

POPs are considered hormone disrupters, causing a malfunctioning of the endocrine and reproductive systems in animals and humans, as well as learning disabilities, birth defects and behavioral, neurologic, reproductive and immunologic disorders (Ritter et al. 1995).

Twelve organic compounds were initially cataloged as POPs by the Stockholm Convention on Persistent Organic Pollutants (UNEP 2001). The list includes a wide number of commercial substances (e.g., aldrin, chlordane, dieldrin, among others) used as pesticides to control insect-borne diseases, insects living in agricultural soils and rodents as well as to prevent fungal growth on food crops. DDT is one of the most known POPs, having been widely used as insecticide against malaria and typhus and as agricultural insecticide. Other POPs (e.g., Mirex) are used as flame retardants in plastics, rubber, paints, paper and household appliances; as heat-exchange fluids in electrical transformers and capacitors; and as additives in paint, carbon-free copy paper and plastics (e.g., PCB; Ritter et al. 1995).

Other POPs (e.g., dioxins and furans) are unintentionally generated as by-products of high-temperature processes, such as incomplete combustion and pesticide production. Dioxins are produced from the burning of hospital, municipal and hazardous wastes or peat, coal and wood as well as from automobile emissions. Furans are the by-products of high-temperature processes, such as incomplete combustion after waste incineration or in automobile-fuel ignition in addition to pesticide and PCB production (Ritter et al. 1995).

4.2 Trophic Transfer of Pollutants in Food Chains

Pollutants, whether heavy metals or POPs, enter organisms primarily via three routes: breathing, osmosis and food digestion, which are physiologically mediated by the gills, skin and digestive tract, respectively (Connell 1988). Those chemicals —in the example of the essential elements—are subsequently transported to the internal organs for utilization, storage or release, with the components of the blood and/or hemolymph being the carriers for that transport (Deb and Fukushima 1999). The route that results in the fastest rate of accumulation is the direct absorption of chemicals across respiratory surfaces or the cuticle.

Bioaccumulation is generally referred to as a process in which the chemical concentration in an organism achieves a level that exceeds that occurring in the respiratory medium (i.e., water for a fish or air for bird or a mammal), the diet or both (Connell 1988; Arnot and Gobas 2006; Ali and Khan 2019). The extent to which chemicals bioaccumulate is expressed by several quantities, comprising the bioconcentration factor (BCF), the bioaccumulation factor (BAF), the biomagnification factor (BMF) and the trophic or food-web magnification factor (TMF and FWMF; Gobas et al. 2009).

Bioconcentration is the process by which a chemical substance is absorbed by an organism from the abiotic environment only through its respiratory and dermal surfaces, i.e., chemical exposure from the diet is not included (Arnot and Gobas 2006). The degree to which bioconcentration occurs is expressed as the BCF and is calculated as the ratio of the steady-state chemical concentration in an aquatic water-respiring organism (C_{BC}, in g/kg ww [wet weight]) over the concentration in the water (C_W, in g/L), i.e., C_{BC}/C_W (Arnot and Gobas 2006; Gobas et al. 2009).

Since bioaccumulation refers to the accumulation of a contaminant in an organism as a result of its uptake from both the abiotic environment and the organism's food and/or diet, that process describes the total ingestion and enrichment of a contaminant in an organism, relative to the concentration in the environment (Borgå 2013). Therefore, bioaccumulation is the net result of all uptake and loss (Fig. 4.1), namely, respiratory and dietary uptake and loss by egestion, passive diffusion, metabolism, transfer to offspring and growth dilution (Arnot and Gobas 2006; Borgå 2013). The degree to which bioaccumulation occurs can be expressed as the BAF = C_{BA}/C_W, which is the ratio of the steady-state chemical concentration in an aquatic water-respiring organism (C_{BA}, in g/kg ww*)* and the water (C_W, in g/L) determined from field data consisting of the concentrations of the chemical in the organism sampled, its diet and the water (Gobas et al. 2009). Bioaccumulation differs from bioconcentration because chemical exposure in the diet—and therefore potential biomagnification—is included.

Biomagnification is the result of the uptake of a contaminant from the diet leading to higher concentrations in the feeder than in the diet (Connell 1988). Subsequently, biomagnification leads to increased concentrations of the chemical at higher trophic positions in the food web (Borgå 2013). Biomagnification can be determined both under field conditions and in laboratory-feeding experiments, and expressed by BMF = C_{BM}/C_D, defined as the ratio of the steady-state chemical concentrations in a water- or air-respiring organism (C_{BM}, in g/kg ww) and in the diet of the organism (C_D, in g/kg dry weight [dw]) determined in a controlled laboratory experiment in which the test organisms are exposed to a chemical in the diet (but not in the water or air). Under field conditions, both units of C_{BM} and C_D are expressed as g/kg ww.

The most relevant measurement of biomagnification in food webs is the TMF or the FWMF (Fisk et al. 2001; Hoekstra et al. 2003; Gobas et al. 2009). A TMF is determined empirically through field measurements of both the contaminant concentration and the relative trophic level (estimated from stable N-isotope ratios [$\delta^{15}N$] obtained by means of tissue measurements of $^{15}N/^{14}N$) in food webs and can be viewed as an "average food-web biomagnification factor" (Borgå et al. 2011). The TMF value is calculated by correlating normalized (expressed as the $\log_{10}$) chemical concentrations in the biota with the trophic positions of the biota sampled.

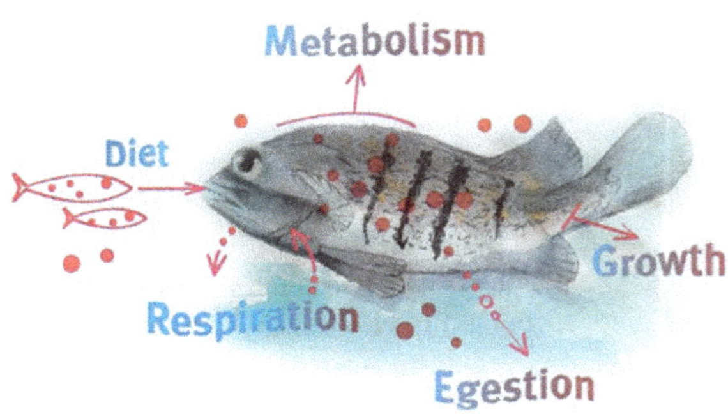

Fig. 4.1 Paths of input and output of contaminants from the environment to and from a fish (see the text for interpretation). Adapted from Borgå (2013)

A linear regression is then used to calculate the slope, *m*, which value is used to derive the TMF as the TMF = 10^{m} (Gobas et al. 2009).

In the example of trace metals, mostly of the nonessential heavy metals do not biomagnify in food chains (Ali and Khan 2019). Metals such as Cd, Cu, Ni, Pb and Zn generally do not biomagnify in food chains when starting with primary producers, macroinvertebrate consumers and fish; but a biomagnification of those metals does occur in certain specific food chains (Cardwell et al. 2013). A biomagnification of Cd and Hg, though not of Pb or As, was reported in benthic food chains by Zhao et al. (2013). Using stable-isotope techniques to assess the biomagnification of heavy metals in a benthic trophic chain, Ikemoto et al. (2008) found a biomagnification of Hg but neither a biomagnification nor a biodilution of Cd, Pb, As, Cr, Cu and Zn. The element As, however, has been reported to biodilute in freshwater and terrestrial food chains, but to biomagnify in marine ecosystems (Huang 2016).

The biomagnification of POPs in trophic webs has been widely assessed as well. Gobas et al. (1993) reported a more than fivefold biomagnification of PCBs, hexachlorobenzene and Mirex in fish under experimental conditions, pointing to food digestion and absorption in the digestive tract as the essential driving force of biomagnification. Hoekstra et al. (2003) reported FWMF values from 0.65 to 9.30 in Arctic food webs for different organochlorine contaminants, including chlorinated-benzene isomers, e.g., HCB, hexachlorocyclohexanes (HCHs), Chlordane, DDTs and PCBs. Fisk et al. (2001) reported FWMFs ranging among 2.7 and 10.8 for several POPs in Arctic marine food webs involving birds and mammals.

4.3 Pollution in Aquatic Ecosystems of Argentina: A Summary Review

The following section constitutes a summary review of the most relevant studies of pollution in both marine and freshwater aquatic ecosystems in Argentina, with an emphasis on publications that have focussed on the effects and propagation of contaminants through the aquatic food webs.

4.3.1 Marine Ecosystems

4.3.1.1 Invertebrates

The assessment of bioaccumulation in invertebrates is essential because of the need to understand their health and role in ecosystem functioning and since they can also be vectors of the sediment-associated contaminants of coastal fish. Several studies on metal concentrations in the Southwestern Atlantic Ocean (SWAO) coastal systems were mostly carried out on the mollusks (Gil et al. 2019 and references therein), crabs (Giarratano et al. 2016) and polychaetes (Dolagaratz Carricavur et al. 2018). Other studies described the accumulation patterns of metals in bivalves (Buzzi and Marcovecchio 2018) and their biomagnification through food webs in the Bahía Blanca estuary (BBE) (La Colla et al. 2019). Bioaccumulation for Fe, Cr and Ni was reported by Vilches et al. (2019) in the muscle of the longfin squid in the area of Necochea-Quequén (Buenos Aires province).

Metal levels in the burrowing crab *Neohelice granulata* have been studied in the Mar Chiquita coastal lagoon (MCcl), Samborombón bay (Smb) and the BBE for over the last 30 years. Simonetti et al. (2018) detected the presence of heavy metals in eggs of that crab, which is of great relevance since the eggs are preyed on by several species of fish and seabirds. In view of the presence of metals in eggshells of the American oystercatcher *Haematopus palliatus* at surprisingly high concentrations of Cd, a possible transfer of the metals available in the environment may be occurring through the food chain (Simonetti et al. 2015).

With respect to sites located in Patagonia, the concentration of Hg, Cd, Cu, Pb and Zn in marine sediments was studied at 19 sampling points along 4,400 km, from San Antonio bay to the mouth of the Gallegos river (Fig. 4.2). The ports Comodoro Rivadavia (CRp) and San Antonio Oeste (SAOp) manifested the highest levels, reflecting anthropic pollution (Gil et al. 2006, 2019). In the example of SAOp, mine tailings were identified as the principal sources of heavy-metal pollution, especially the Pb and Zn, from past mining activities (Häder et al. 2020). Bioaccumulation was consequently found in *N. granulata* (Giarratano et al. 2016), as evidence of the leaching of metals into the environment. The metals Pb, Hg and Cd in the soft tissues of bivalves suggested a low-to-moderate pollution in some coastal points (Vázquez et al. 2007). Studies on the tissues of organisms at high trophic levels reported low heavy-metal concentrations, thus providing no evidence for biomagnification (Frías et al. 2012; Rosas et al. 2012). The content of several metals in the zooplankton from Argentine and Uruguayan coastal waters was studied in by Scarlatto et al. (1997). Except for Cd, all the metals studied evidenced maximal values on the Uruguayan shores. Fernández Severini et al. (2013) reported that in the BBE the zooplankton accumulated considerable concentrations of Cd, Cu and Pb, whereas the mesozooplankton from that area must have incorporated a sizeable proportion of Zn from the suspended particulate matter (SPM) and the microplankton (Marcovecchio et al. 2016).

Arias et al. (2009) found an association between the phytoplankton microbiologic seasonal cycling and the dissolved and/or suspended polycyclic aromatic hydrocarbons (PAHs) in the BBE. Sedimentary coastal survey in Patagonia performed over the last 30 years, evidenced the presence of biogenic and anthropic contributions and both chronic and more recent pollution resulting from small oil spills (Commendatore et al. 2012). Even with concentration level of the POPs in mussels has been low, their presence in sediments denotes that the contributions there would not be recent (Commendatore et al. 2015; Gil et al. 2019).

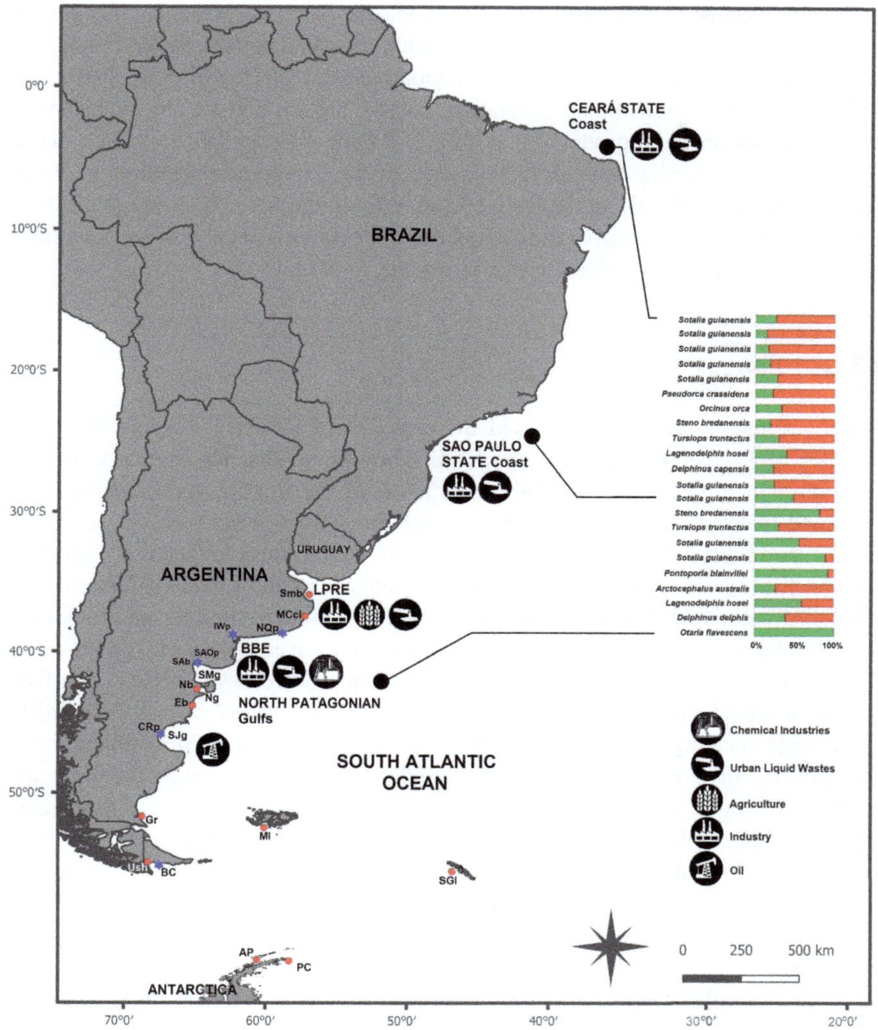

Fig. 4.2 Location of coastal sites (red circles), ports (blue stars) and geographical areas mentioned in the text. The ratio between the organochlorine pesticides (green bar color) and the total polychlorinated biphenyls (ΣPCB, the red bar color) detected in marine mammal species in the SWAO (black circles) is indicated. Abbreviations: BBE (Bahía Blanca estuary), LPRE (La Plata river estuary), Smb (Samborombon Bay), MCcl (Mar Chiquita coastal lagoon), NQp (Necochea-Quequén port), IWp (Ingeniero White port), SAOp (San Antonio Oeste port), SAb (San Antonio bay), SMg (San Matías gulf), Ng (Nuevo gulf), Nb (Nueva bay), Eb (Engaño bay), CRp (Comodoro Rivadavia port), SJg (San Jorge gulf), Gr (Gallegos river), MI (Malvinas Islands), SGI (South Georgia Islands), Ush (Ushuaia), BC (Beagle Channel), AP (Antarctic Peninsula), PC (Potter cove)

Finally, despite the worldwide prohibition since 2008, the presence of organotin compounds (OTCs, used as antifouling biocide in vessel paints) on the Patagonian coast is another issue of concern (Commendatore et al. 2015; Gil et al. 2019). It is well known that their bioaccumulation causes the emergence of male sexual characteristics on the reproductive system of female gastropods (imposex), occurring in snails inhabiting in main fishing ports of the Patagonia (Bigatti et al. 2009). OTCs were found mainly in the form of dibutyltin in the SAb (Narvarte et al. 2017) and as monobutyltin in native mussels in the BBE (Quintas et al. 2019). No studies were reported for OTCs transference in marine food chains.

4.3.1.2 Fish

Marcovecchio et al. (1988) assessed the presence of total Hg, Cd and Zn in the muscles of 17 fish species from the BBE. They found that Hg increased in concentration in direct proportion to the increase in the trophic level, which result indicated that Hg became biomagnified throughout the trophic web.

The pollution by total Hg, Cd and Zn in the edible muscle tissue of two commercial fish species was studied in the Smb (located at the LPRE) (Marcovecchio 2004). A marked relationship between the metal contents of the species studied and their trophic and ecologic habits was reported. Marcovecchio et al. (2019) found that the total Hg contents in the abiotic matrix and in three fish species from BBE had been significantly reduced after fifteen years, which was probably caused by technological changes, by a more adequate management of industrial effluents or by the removal of a substantial amount of sediment volume through dredging and refilling. However, concentrations of Cr and Mn in the muscle tissues of fish exceed the allowable levels for human consumption (La Colla et al. 2017).

Reports of pollution by heavy metals in the Patagonian coastal ecosystem date back to the 1980s. Although mollusks, crustaceans, seabirds and marine mammals were included (Gil et al. 1996, 1999), the pollution in fish species was never assessed.

In the example of POPs, the PCBs constitute in general the major organochlorine residues, followed by the DDTs and the HCHs. Fish from the LPRE exhibited the highest POP levels, principally PCBs and polychlorinated dibenzo-p-dioxins and furans, when compared to samples collected upstream on the Paraná and Iguazú rivers (Barra et al. 2006). In that area, Cappelletti et al. (2015) found a 30- to 40-fold increase in dry weight of dioxin-like polychlorinated biphenyls and polybrominated biphenyl ethers (PBDEs) concentrations between the settling particles and fish, with biota-sediment–accumulation factors ranging between 5 and 20. Based on a long-term-dated sediment samples from the outer LPRE, Colombo et al. (2018) found strong evidence for evaluating the effectiveness of control measures for PCB emissions in this area.

Lanfranchi et al. (2006) found that endosulfan sulfate, chlordanes, HCH isomers and DDT compounds predominated in the tissues and the ingested food of the fish *C. guatucupa* in the BBE with some pollutants being highly bioaccumulated and biomagnified. Oliva et al. (2017) found that the low-molecular-weight PAHs generated from petrogenic pollution prevailed in muscles of four fish species inhabiting the BBE.

Lana et al. (2014) found that some Antarctic notothenioid-fish species in the Potter cove (South Shetland Islands, Antarctica) were more susceptible to the accumulation of the PCBs, DDT and its metabolites, PBDEs and HCH. Working at the same site, Ríos et al. (2019) noted the strong influence of several biologic conditions on POP accumulation and stressed that each particular one should be considered when choosing some of those fish as sentinels of POP pollution in Antarctic marine environments.

4.3.1.3 Seabirds

The extent to which heavy metals bioaccumulate in coastal seabirds has been mainly approached with Hg as the element studied and sub-Antarctic and Antarctic penguins. The majority of the studies refer to TMFs based on Hg levels and penguin species-specific foraging habitats and obtained by $\delta^{15}N$ values (Cipro et al. 2016; Polito et al. 2016; among others). Moreover, in the Antarctic region, Calle et al. (2015) directly related concentrations of Hg detected in a predator organism to those measured in its prey. Other studies in that marine region reported elemental biotransference rates—mainly of Hg, but also of Fe and Zn—between trophic links including coastal seabirds like penguins and seagulls (dos Santos et al. 2006).

In the example of the pelagic seabirds, studies included a range of species that could be lumped together within the order Procellariiformes (albatrosses and petrels). Cephalopods, fish and—to a lesser extent—crustaceans are basic foods for albatrosses, while carrion and gelatinous plankton are of a lower preference (Cherel and Klages 1997).

As with the coastal seabirds, the extent to which heavy metals bioaccumulate in the pelagic seabirds has been mainly analyzed by means of TMF with Hg as the element studied and the large-to-medium-sized petrels as the birds under investigation (Anderson et al. 2009, 2010; Cipro et al. 2016). The few studies (Table 4.1) that applied TMFs analyzed the Hg isotopic values and petrel species-specific foraging habitats on the basis of $\delta^{15}N$ values (Anderson et al. 2009, 2010; Cipro et al. 2016).

The presence of a range of POPs has been cited in different matrices of coastal seabirds inhabiting the Patagonian Shelf. In particular, pollutants such as OCPs, which class encompassed contaminants like DDTs, HCHs and cyclopentadienes, as well as PCBs and PBDEs are the main compounds that have been studied whereas other contaminants have been scarcely analyzed. Regarding OCPs, PCBs and PBDEs, the diving species like penguins are the most frequently studied along the coast (Schiavone et al. 2009; Baldassin et al. 2016; Ellis et al. 2018, among others). Research on OCPs, PCBs and PBDEs has involved other coastal diving birds, including cormorants (Taniguchi et al. 2009; van den Steen et al. 2011), while additional seabirds such as skuas (Kumar et al. 2002; Corsolini et al. 2011, among

Table 4.1 Summary of the most relevant studies referring to pollution by taxa and pollutant type for Argentine marine and freshwater ecosystems. Dots indicates the approach used in each study

Taxonomic group	Pollutants	Ecosystem	Scope of the study			References
			BAC	BMF	TMF/ FWMF	
Zooplankton	Zn, Cu, Cd, Fe, Pb and Mn	Marine/ Argentine and Uruguay coasts	•	–	–	Scarlatto et al. (1997)
Zooplankton	Cd, Cu and Pb	Marine/ BBE	•	–		Fernández Severini et al. (2013)
Zooplankton	Zn	Marine/ BBE	•	•	–	Marcovecchio et al. (2016)
Mollusks, crustaceans, seabirds and marine mammals	Hg, Cd, Pb, Cu and Zn	Marine/ Patagonian coast of Argentina	•	–	–	Gil et al. (2006)
Mollusks, crustaceans and polychaetes	Pb, Hg and Cd	Marine/ Patagonian coast of Argentina	•	–	–	Giarratano et al. (2016), Buzzi and Marcovecchio (2018), Dolagaratz Carricavur et al. (2018), Gil et al. (2019), La Colla et al. (2019)
Mollusks, fishes and cetaceans	Cd, Cr, Cu, Fe, Ni, Pb and Zn	Marine/ BBE and NQp	•	–	–	Vilches et al. (2019)
Mollusks	Cd, Cu, Pb, Zn, Ni and Cr	Marine/ Coast of Buenos Aires province	•	–	–	Buzzi and Marcovecchio (2018)
Mollusks	Cd, Zn and Cu	Marine/BC, Ush	•	–	–	Comoglio et al. (2011)
Crustaceans	Hg, Zn, Ni and Pb	Marine/ MCcl, Smb and BBE	•	–	–	De Marco et al. (2006)
Mollusk, crustaceans and birds	Cd, Cu, Pb, Ni, Zn, Mn, Cr and Fe	Marine/ BBE	•	•	–	Simonetti et al. (2015), (2018)
Crustaceans	Pb, Hg an Cd	Marine/ Patagonian coast of Argentina	•	–	–	Vázquez et al. (2007), Giarratano et al. (2016), Gil et al. (2019)

(continued)

Table 4.1 (continued)

Taxonomic group	Pollutants	Ecosystem	Scope of the study			References
			BAC	BMF	TMF/ FWMF	
Mollusks	Organotin compounds	Marine/ Patagonian coast of Argentina	•	–	–	Bigatti et al. (2009), Commendatore et al. (2015), Narvarte et al. (2017)
Mollusks	Organotin compounds	Marine/ BBE	•	–	–	Quintas et al. (2019)
Fishes	Hg, Cd and Zn	Marine/ BBE, LPRE	•	•	–	Marcovecchio et al. (1988), Marcovecchio (2004)
Crustaceans, fishes and halophytes	Hg, Cd and Zn	Marine/ LPRE, MCcl	•	•	–	Marcovecchio et al. (2001)
Fishes	Cr, Pb, Fe and Mn	Marine/ BBE	•	•	–	La Colla et al. (2017)
Fishes	PCBs, PBDEs	Marine/ LPRE	•	•	–	Colombo et al. (2018)
Fishes	HCH, DDT	Marine/ BBE	•	•	–	Lanfranchi et al. (2006)
Mollusks and fishes	PAHs	Marine/ BBE	•	–	–	Arias et al. (2009), Oliva et al. (2017)
Fishes	PAHs, PCBs, DDT PBDEs HCH	Marine/PC	•	–	–	Lana et al. (2014), Ríos et al. (2019)
Birds	Hg, Cd	Marine/AP	•	•	–	Cipro et al. (2016), Polito et al. (2016)
Birds	Hg	Marine/AP	•	•	•	Calle et al. (2015)
Birds	Hg	Marine/SGI	•	•	–	Anderson et al. (2009)
Birds	Hg, Zn	Marine/AP	•	–	–	dos Santos et al. (2006)
Birds	Al, As, Ba, Be, Cd, Co, Cs, Cu, Fe, Li, Mn, Mo, Ni, Pb, Rb, Sb, Sc, Se, U, V, W and Zn	Marine/SGI	•	•	–	Anderson et al. (2010)
Birds	PCBs, DDTs, HCB, HCHs,	Marine/AP	•	•	–	Goerke et al. (2004), Cipro et al. (2010)

(continued)

Table 4.1 (continued)

Taxonomic group	Pollutants	Ecosystem	Scope of the study			References
			BAC	BMF	TMF/ FWMF	
	Chlordanes, Drins and Mirex					
Birds	PCBs, HCBs, HCHs, Chlordanes, Cyclodines, DDTs and Mirex	Marine/AP	•	•	–	Kim et al. (2015)
Birds	HCB, Chlordanes, DDTs, Drins, PCBs and Mirex	Marine/ Southern Brazil	•	•		Colabuono et al. (2014)
Birds and cetaceans	PCBs, HCB, OCS, HCHs, DDTs, Chlordane and Drins	Marine/MI	•	•	–	de Boer and Wester (1991)
Marine mammals	HgT, Cd and Ag	Marine/ SWAO	•	–	–	Baraj et al. (2009), Cáceres-Saez et al. (2013b), Durante et al. (2020), Gerpe et al. (2009), Lailson-Brito et al. (2002), Marcovecchio et al. (1990)
Marine mammals	HgT and Cd	Marine/ Buenos Aires province	•	•	–	Peña et al. (1988)
Marine mammals	PCBs, DDTs, HCB	Marine/ Southern Brazil	•	–	–	Lailson-Brito et al. (2012)
Marine mammals	PCBs, DDTs, Mirex, HCHs	Marine/ SWAO	•	–	–	Durante et al. (2016), Santos-Neto et al. (2014)
Marine mammals	PBDEs	Marine/ Brazilian coast	•	•	–	Llorca et al. (2017)
Food web	Ag, Hg	Freshwater/ LNH	•	•	–	Juncos et al. (2017), Arcagni et al. (2018)
Food web	Ag, Hg	Freshwater/ LMO	•	•	–	Arcagni et al. (2013)

(continued)

Table 4.1 (continued)

Taxonomic group	Pollutants	Ecosystem	Scope of the study			References
			BAC	BMF	TMF/ FWMF	
Plankton, shrimp, fish	Ag, Al, As, Cd, Cr, Cu, Fe, Hg, Mn, Ni, Pb, Se, U, Zn	Freshwater/ LML	•	•	•	Griboff et al. (2018)
Plankton, shrimp, fish	Ag, Al, As, Cd, Cr, Cu, Fe, Hg, Mn, Ni, P, Pb, Sem U, Zn	Freshwater/ RTR	•	•	•	Griboff et al. (2020)
Plankton, shrimp, fish	Ag, Al, As, Cd, Ce, Cr, cu, Fe, Hg, Mn, Mo, Nd, Ni, Pb, Pt, Sr, Zn	Freshwater/ SRR	•	•	•	Monferrán et al. (2016)
Macrophyte, shrimp, fishes	PCBs	Freshwater/ LPL	•	–	•	González Sagrario et al. (2002)
Bivalves, fishes	PCBs, OCPs, Ag, As, Cd, Cu, Cr, Hg, Ni, Pb, Se, U, V, Zn	Freshwater/ LPr	•		•	Colombo et al. (1990), Lombardi et al. (2010), Avigliano et al. (2019)
Fishes	PCBs, OCPs, PBDEs	Freshwater/ MCL	•	–	–	Ballesteros et al. (2014)
Fishes, birds	OCPs, PCBs, PBDEs, As, Cd, Co, Cr, Cu, Fe, Hg, Ni, Pb, Zn	Freshwater/ PR	•	–	–	Colombo et al. (2011), Avigliano et al. (2016)
Amphibians, birds	OCPs, Cd, Pb	Freshwater/ LFR	•	–	–	Cid et al. (2007), Jofré et al. (2008)
Fishes	PCBs, OCPs, PBDEs, Ag, As, Ba, Co, Cr, Cs, Hg, Ni, Sb, Se, U, Zn	Freshwater/ NR	•	–	–	Arribére et al. (2003), Ondarza et al. (2010), (2014)
Fishes	Pharmaceuticals	Freshwater/ Pr and AR	•	–	–	Ondarza et al. (2019)

References: BAC (bioaccumulation), BMF (biomagnification factor), TMF/WFMF (trophic or food-web magnification factor), BBE (Bahía Blanca estuary), LPRE (La Plata river estuary), MCcl (Mar Chiquita coastal lagoon), NQp (Necochea-Quequén port), MI (Malvinas Islands), PC (Potter cove, Antarctica), SGI (South Georgia Islands), LNH (Lake Nahuel Huapi), LMO (Lake Moreno), RTR (Rio Tercero Reservoir), LML (Los Molinos lake), SSR (San Roque reservoir), LPL (Los Padres lake), LPr (La Plata river), QGR (Quequen Grande river), MCL (Mar Chiquita lake), Pr (Paraná river), LFR (La Florida reservoir), NR (Negro river), AR (Acaragua river), QR (Quemquemtreu river) others) and seagulls (Łukowski et al. 1987; Quadri-Adrogué et al. 2020, among others) have also been analyzed.

TMFs based on certain groups of POPs in coastal seabirds such as sub-Antarctic and Antarctic penguins are available, which parameter enables a more comprehensive understanding of the persistent serial bioaccumulation of these contaminants in the coastal avifauna. Several studies approached that topic by directly relating concentrations of contaminants such as PCBs, DDTs, HCB, HCHs, chlordane and aldrin (with certain studies also including Mirex) analyzed in a predatory organism and in its dietary prey (de Boer and Wester 1991; Goerke et al. 2004; Cipro et al. 2010; Kim et al. 2015). Very few studies have focussed on the relationship between particular groups of POPs and penguin-specific foraging habitats according to the foraging ecology of those birds on the basis of the $\delta^{15}N$ (Mello et al. 2016).

The POPs most frequently reported in pelagic seabirds are the OCPs, PCBs and PBDEs. As to these groups of contaminants, the most commonly studied pelagic species are the medium-sized and giant petrels (van den Brink 1997; Colabuono et al. 2012; Quadri-Adrogué et al. 2019, among others). Most of these studies were performed on Brazilian coastal grounds, followed by investigations conducted along the AP and Argentine coasts. A single study (Colabuono et al. 2014) revealed the extent to which POPs are bioaccumulated in pelagic seabirds. These authors calculated the TMFs for OCPs throughout the specific foraging habitats of albatrosses and petrel species using the $\delta^{15}N$ values. Studies on the trophic transfer of OCPs and other chlorinated hydrocarbons in additional pelagic seabirds inhabiting other areas within the Patagonian Shelf are virtually zero.

4.3.1.4 Marine Mammals

Marine pollution became a growing problem affecting the species and populations of mammals (Borrell et al. 2010), which occupy the top of food chain, constitute long-living species and accumulate contaminants through their diet (Das et al. 2003). Those features point to a consideration of the marine mammals as good indicators of pollutant bioavailability and biomagnification throughout the food chain. In general, the contaminant levels in marine mammals of the SWAO are lower than those reported for the Northern Hemisphere and are below the values associated with either adverse sublethal effects or lethality in mammals (Crespo et al. 2010).

Organobromine compounds are included in the list of emerging contaminants that are widely distributed around the world. Despite this recognition, Llorca et al. (2017) identified a big gap in the information on these contaminants in Latin America. Only a few studies have been conducted on marine mammals from Brazil, which reported the highest levels of methoxylated PBDEs in Latin America (Dorneles et al. 2007). Organochlorine concentrations have been reported from different species of marine mammals throughout the SWAO (Fig. 4.2) between the Ceará (Brazil) and Tierra del Fuego (Argentina) (Borrell et al. 2010; Lailson-Brito et al. 2012; Santos-Neto et al. 2014; Durante et al. 2016, and references therein). The species from Brazil harbor concentrations one order of magnitude higher than those from Argentina with the exception of the killer whales from Rio-de-Janeiro coast, which show the highest levels of organochlorine compounds in the SWAO (Lailson-Brito et al. 2012). Killer whales prey mainly on marine mammals, so that those high values of this pollutant would thus be expected as a result of biomagnification. In Argentina, the concentrations of organochlorines in marine mammals in the fatty tissues either were not detectable or negligible (Gil et al. 1996), while the most recent studies still have indicated low levels of these contaminants (Borrell et al. 2010; Torres et al. 2015; Durante et al. 2016).

The industrial pollution seems to be the main source in the north of Brazil (between the Sao-Paulo-State coast and Ceará-State coast) where the total PCBs were predominant, whereas from the São Paulo State to the mouth of the Negro river, the contributions appear to be agricultural as well as industrial (Fig. 4.2; Borrell et al. 2010; Durante et al. 2016; Lailson-Brito et al. 2012; Santos-Neto et al. 2014, and references therein).

In the example of trace elements, studies focussed on the concentration of Hg and Cd in South American sea lion (Peña et al. 1988), southern right whale and Commerson's and dusky dolphins (Gil et al. 1996). Studies involving metals were carried out in several species including South American sea lions and fur seals (Baraj et al. 2009; Gerpe et al. 2009; Rodrigues et al. 2019) along with certain cetaceans (Lailson-Brito et al. 2002; Dorneles et al. 2007; Polizzi et al. 2013; Cáceres-Saez et al. 2013a, b, 2019; Durante et al. 2020, among others). These studies provide information about the availability of the contaminants in different levels of the trophic chain according to each diet.

As to the possible routes for the transmission of pollutants, the prey species that are eaten by marine mammals include the shortfin squid *Illex argentinus*, the anchovy *Engraulis anchoita* and the Argentine hake *Merluccius hubbsi* (Table 4.2). Marine mammals that feed on squids usually accumulate high levels of Cd and Ag, since cephalopods are one of the main sources of both elements (Cáceres-Saez et al. 2019), whereas in the mammals that feed on fish the tendency is to harbor a higher concentration of total Hg. Table 4.2 illustrates that the shortfin squid falls at a value of usually between 15 and 65% within the Index of Relative Importance (%IRI, a composite measure of trophic preferences based on prey numbers, prey volume, and the frequency of occurrence; Pinkas et al. 1971) for most top predators in the SWAO, implying that a large proportion of squid biomass is ingested by terminal members of the food chain.

Dorneles et al. (2007) reported high levels of Cd in shortfin squids from southern Brazil, indicating the high availability of this metal in one of the most widely consumed preys of the top predators in the SWAO and explaining the accumulation in those animals. Piscivorous feeders tend to present a higher concentration of total Hg, since that metal is usually accumulated by fish. The hake is preyed upon mostly by male sea lions, fur seals and Commerson's dolphins. The %IRI for the hake varies between 10 and 85% of the biomass ingested by those predators. Anchovies are mainly taken by pelagic feeders like fur seals and dusky and common dolphins

Table 4.2 Index of relative importance (%IRI) of marine mammals' prey from Southwestern Atlantic Ocean

Preys	FsP	MsP	SfP	SfB	DSJ	DSM	CSM	Cmd	Fkw	Frd	Bsi	Bsm	Ts	Smm	Sfm	Sdi
Illex argentinus	17.6	6.7	**45.9**		24.9	2.8	0.2	12.3	31.0		11.9	16.9	32.6	**58.7**	**65.4**	**42.4**
Merluccius hubbsi	11.8	**68.5**	21.6		10.7	3.2	0.6	**85.9**			24.9	**62.5**	**37.4**	13.7	15.1	17.2
Enteroctopus megalocyathus	**55.1**	3.4	0.01									1.5	21.0		1.4	
Patagonotothen ramsayi								0.2			**43.8**	10.4	1.9	13.8	4.3	0.8
Engraulis anchoita	4.3	1.4	21.2		**48.3**	**85.1**	**82.6**	0.04		2.5	2.7	0.0	0.2		6.5	
Ctenophora														7.9	0.9	27.7
Doryteuthis gahi	1.08	9.10	2.08		14.96		0.31	0.92					1.3		0.4	
Raneya brasiliensis	6.8	5.0	1.5				0.3			0.05	0.02	5.4	0.8		4.8	0.1
Serolis schythei											12.9			0.2		
Munida gregaria			4.8													
Mixine sp.			1.7													
Seriolella porosa						3.3	2.75									
Loligo sanpaulensis				41.8		3.5	12.1			41.1						
Cynoscion guatucupa							0.6			**46.7**						
Trichiurus lepturus				**55.0**												
Martialia hyadesii									**50.4**							
Macruronus magellanicus									14.0							
Artemesia longinaris										4.1						
Other species	3.0	5.7	1.0	3.0	1.0	1.9	0.4	0.5	4.4	5.3	3.5	3.0	4.5	5.5	0.8	11.6

The highest %IRI is highlighted in bold font. Abbreviations: FsP (female sea lion Patagonia), MsP (male sea lion Patagonia), SfP (South American fur seal Patagonia), SfB (South American fur seal Southern Brazil), DSJ (dusky dolphin San Jorge gulf), DSM (dusky dolphin San Matias gulf), CSM (common dolphin San Matias gulf), Cmd (Commerson's dolphin), Fkw (false killer whale), Frd (Franciscana dolphin), beaked skate immature (Bsi); beaked skate mature (Bsm); tope shark (Ts); spiny dogfish male mature (Smm); spiny dogfish female mature (Sfm), spiny dogfish immature (Sim) and principally to the south of Península Valdés. The %IRI for anchovies varies between 6 and 85% of the biomass ingested by those species' predators.

Sea lion males are mostly demersal and benthic feeders, with the common hake as a main prey, while the females are demersal pelagic predators, feeding along the coast and much more on a great variety of prey, not just one dominant prey type. Fur seals are pelagic feeders consuming mainly anchovies either as solitary individuals or in small groups and with those seals no difference is found between the sexes. The dusky, common and Commerson's dolphins are likewise pelagic feeders, herding shoals of pelagic fish such as anchovies, small hakes and small squids. False killer whales are deep divers that feed on deep-sea squids like *Martialia hyadesii*. The Franciscana dolphin is pelagic and coastal, feeding on very small juvenile stages of fish (mostly sciaenids), squid and crustaceans. In view of the heterogeneity, variety and complexity of the mammalian marine prey items, no overall pattern exists that might reveal a relationship between a particular fish prey and the accumulation of total Hg in that fish's mammalian predators (Crespo et al. 2007).

4.3.2 Freshwater Ecosystems

Freshwater ecosystems in Argentina comprise three major basins, with some important rivers and several closed endorheic watersheds hosting more than 400 deep lakes and more than 10,000 shallow lakes in the pampas region alone plus thousands of rivers and streams (Fig. 4.3). Most of the population in Argentina is concentrated in the provinces of Buenos Aires, Córdoba and Santa Fe, and therefore, the greatest productive development and sources of contamination are found there.

Studies that involve the trophic transfer of heavy metals through food chains along with the assessment of biomagnification are scarce (e.g., Monferrán et al. 2016; Juncos et al. 2017; Arcagni et al. 2018). The Reconquista river (La Plata river basin) has been considered one of the most polluted watercourses in Latin America, ever since several anthropic activities (e.g., industry, urbanization, agriculture) introduced heavy metals into the environment. Bioaccumulation at levels above the maximum concentrations permitted of Cd, Cu, Hg, Pb, Zn and Hg in muscle, liver and gills from several fish species have been reported (Lombardi et al. 2010; Avigliano et al. 2016).

Pollution caused by sewage effluents of Córdoba city as well as the runoff from agriculture and industrial emissions has been assessed in the Suquía river, where Pb, Cu, Cr and Zn concentration in sediments exceed the levels considered hazardous for aquatic life (Monferrán et al. 2011). Griboff et al. (2018, 2020) reported biodilution of Al, Cr, Mn, Fe, Ni, Zn, As, Se and U throughout a food web in Los Molinos and Río Tercero reservoirs (Córdoba province), whereas biomagnification was observed for Hg and Cu. Biomagnification of Hg and Zn and biodilution of Ni,

Cd, Cr, Al and other metals, throughout the food web, were reported by Monferrán et al. (2016) for San Roque lake (Córdoba).

Patagonian watersheds in general receive low anthropic impact, and the presence of heavy metals in water, sediments and food webs is usually attributed to volcanic activity (Daga et al. 2014). Anthropic pollution sources in this area are mainly associated with industry, untreated urban sewages or human-mediated atmospheric transport. In the upper Negro river, the concentrations of heavy metals (especially Hg) released by a chloralkali-electrolysis factory that operated until 1995 were detected in macrophytes and fish, suggesting that the contaminants were still entering the river (Arribére et al. 2003). The extended use of pesticides in the Negro river valley has been suggested as responsible for the observed increased riverine input of Cu to the Atlantic Ocean. A similar but less intense activity that seems to have a minor effect is observed in the Chubut river (Gaiero et al. 2002).

Other studies in Nahuel Huapi lake showed elevated concentrations of As, Cr and Hg—related to untreated sewage—in the mussel *Diplodon chilensis* and elevated Ag concentrations in mussels and fish (Ribeiro-Guevara et al. 2004, 2005). A decrease in Ag concentrations in the sediments and biota between 1998 and 2011 was related to the replacement of the silver halide film photography by digital imaging technology (Ribeiro-Guevara et al. 2005; Juncos et al. 2017). Variable Ag concentrations were recorded in lower trophic level organisms (plankton and benthic macroinvertebrates) and undetectable levels in fish, indicating food-web biodilution (Juncos et al. 2017).

Despite being far from the point sources of anthropic pollution of heavy metals, concentrations of total Hg exceeding the threshold for human consumption were detected in phytoplankton and native fishes from lakes Nahuel Huapi and Moreno (Arribére et al. 2008; Arcagni et al. 2013, 2017). Moreover, a food-web analysis of the trophodynamics of methyl Hg (MeHg) using carbon ($\delta^{13}C$) and nitrogen ($\delta^{15}N$) stable isotopes demonstrated that this neurotoxic Hg species biomagnified in the food web of Nahuel Huapi lake where the percentage of MeHg to THg reached 100% in sportfish (Arcagni et al. 2018).

Pollution by POPs is a concern issue for freshwater ecosystems in Argentina, regarding the extended development of the agriculture and the related use of pesticides. Studies have focused on assessing the contamination status and/or the environmental risk through the analysis of POPs levels in abiotic matrices and/or biotic compartments (e.g., Colombo et al. 2011; Ballesteros et al. 2014).

The trophic transfer of PCBs through the calculation of the BMF in a short food chain was reported in Los Padres lake (González Sagrario et al. 2002). Nevertheless, despite the general tendency toward biomagnification, we could not find studies that had evaluated the trophodynamics of POPs through food webs.

Along the Paraguay-Paraná rivers and tributaries, a region devoted to both extensive and intensive agriculture, the pesticides endosulfans and chlorpyrifos were found being ubiquitous compounds in both abiotic and biotic matrices (Etchegoyen et al. 2017). In the Reconquista river, Rovedatti et al. (2001) found that pesticide levels (i.e., DDTs, HCHs, heptachlor, chlordane) were 40 to 400 times higher than the legal limits established for the protection of aquatic life.

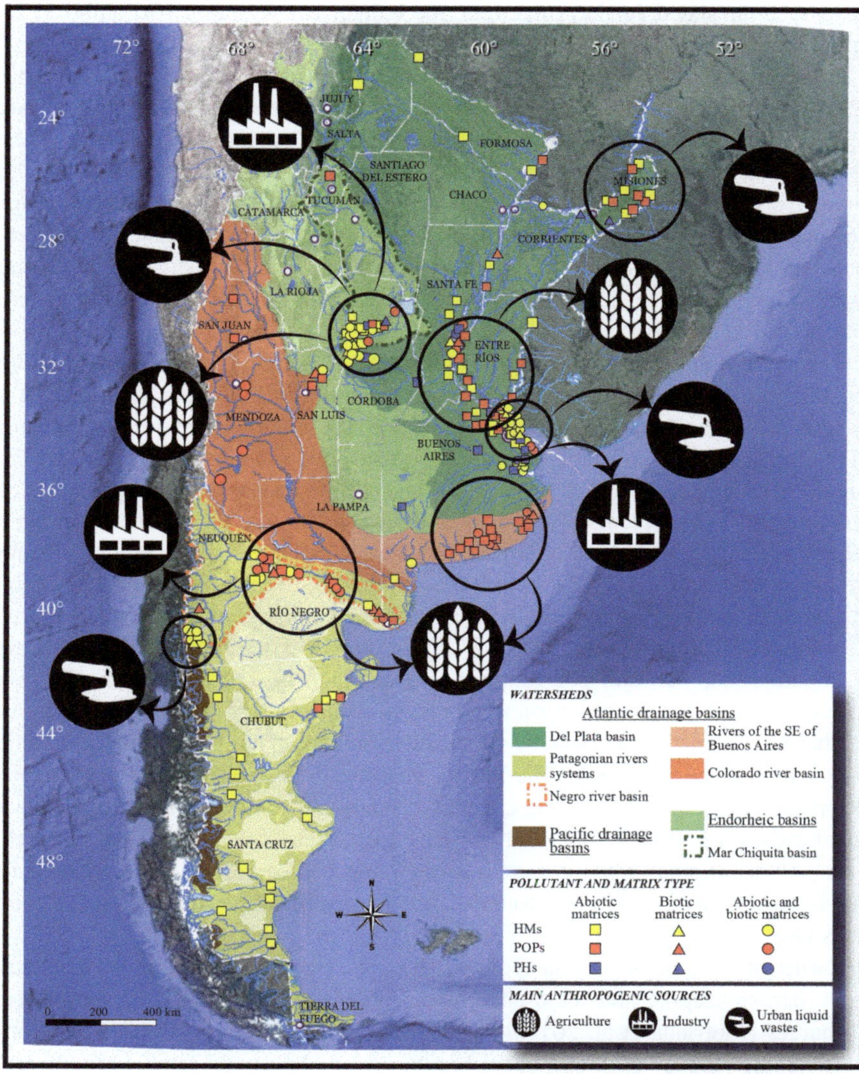

Fig. 4.3 Distribution of studies dealing with pollution in freshwater ecosystems in Argentina. The pollutant type, the environmental matrix under consideration and the principal anthropic pollution sources are indicated according to the symbols in the lower-right corner. Also, main drainage basins and most studied sub-basins are indicated. Abbreviations: HMs (heavy metals), POPs (persistent organic pollutants), PHs (pharmaceuticals)

Muscle tissue of fish *P. lineatus* from the polluted Metropolitan coast of Buenos Aires contained extremely high concentrations of organic pollutants (e.g., PCBs, dioxins, DDTs, chlordanes, chlorobenzenes, PBDEs, Mirex, among others). The concentrations of PCB are among the highest reported for fish, being sixfold higher than the limit stated in the guidelines for human consumption (Colombo et al. 2011).

The endosulfans, DDTs, PCBs and PBDEs were detected in water, sediment, SPM and the tissues of the silverside *Odontesthes bonariensis* in Mar Chiquita lake, in the Córdoba province. Endosulfan, HCH and PCB levels were detected in the lake, although the use of OCPs and PCBs has been prohibited since 1998 and 2005, respectively. Moreover, the PCB levels in fish surpassed the levels acceptable for human consumption, thus constituting a risk to human health. The lake's tributaries, such as Suquía river, which pass major urban centers, have been suggested as being responsible for the levels of PBDEs registered within this agricultural system (Ballesteros et al. 2014).

Throughout the Negro river basin (in northern Patagonia), where long-standing and historically intensive fruit cultures have been established, thus representing a hotspot for legacy-pesticide deposition within the environment, high levels of DDTs in agricultural soils and macrophyte's roots and of endosulfan in SPM were found (Miglioranza et al. 2013). PCBs and PBDE related to the presence of hydroelectric power plants, dams and dumping sites were also detected in agricultural soils (Miglioranza et al. 2013). Ondarza et al. (2014) reported a significant predominance of OCPs, PCBs and PBDEs in muscle, liver, gills and gonads of the silverside *O. hatcheri* from this river, suggesting also agriculture as a main source of pollutants. The concentrations of PCBs in the silverside muscles exceeded the maximum limits for safe consumption, indicating a possible human health risk related to that fish's consumption (Ondarza et al. 2014).

Pharmaceuticals and personal-care products, including a diverse array of organic chemicals widely used for public health, veterinary, cosmetic and sanitary applications, have raised an increasing concern as emerging contaminants in the last two decades (Sauvé and Desrosiers 2014). Being such a relatively new area of research, a lot of uncertainties remain regarding their mechanisms of accumulation, their effects on organisms and their potential for biomagnification throughout the food webs (Xie et al. 2017).

Earlier studies in Argentina reported the presence of caffeine, ibuprofen, atenolol, carbamazepine and diclofenac in the order of several μg/L in wastewater discharges (Elorriaga et al. 2013a, b; Valdés et al. 2014, 2015). Studies that contemplate the impact of those pharmaceuticals on the entire food web are still completely lacking.

Valdés et al. (2016) evaluated the presence of 20 pharmaceuticals in gills, intestine, liver, brain and muscle of two fish species from polluted areas in the Suquía river basin. They reported that the concentration of the analgesic codeine reached values even higher than those reported for pharmaceuticals in the homogenates of fish captured at wastewater-treatment plants worldwide.

Ondarza et al. (2019) reported the presence of caffeine and the antibiotics erythromycin, sulfonamides and trimethoprim in muscle, liver and gills of several fish species from the Paraná river. The authors identified the presence of benzoylecgonine for the first time in Argentina, a primary metabolite of the illicit drug cocaine, and emphasized the need to analyze contamination in protected urbanized areas or in areas undergoing urbanization.

4.4 Conclusions

The review of the existing literature on the regional marine and freshwater ecosystems reveals an enormous gap in food-web–trophic-transfer analysis of pollutants, with very few studies having been undertaken using stable isotopes of N and C to assess the biomagnification or biodilution patterns. In summary, certain studies have analyzed the occurrence of a biomagnification of specific pollutants in trophic webs by means of descriptive information or by calculating some kind of biomagnification index or measurement. Although several studies have described trophic relationships through the analysis of stable isotopes for limited sectors of the Patagonian coastal ecosystem, studies using those tools for assessing biomagnification by estimating the TMF/WFMF ratio were scarce. Future research could take advantage of that approach in order to carry out an evaluation of the effect of biomagnification at the level of the aquatic trophic chains.

References

Ali H, Khan E (2019) Trophic transfer, bioaccumulation and biomagnification of non-essential hazardous heavy metals and metalloids in food chains webs—concepts and implications for wildlife and human health. Hum Ecol Risk Assess Int J. https://doi.org/10.1080/10807039.2018.1469398

Anderson O, Phillips R, McDonald R, Shore R, McGill R, Bearhop S (2009) Influence of trophic position and foraging range on mercury levels within a seabird community. Mar Ecol Progr Ser 375:277–288

Anderson O, Phillips R, McDonald R, Shore R, McGill R, Bearhop S (2010) Element patterns in albatrosses and petrels: influence of trophic position, foraging range and prey type. Environ Pollut 158:98–107

Arcagni M, Campbell L, Arribére M, Marvin-DiPasquale M, Rizzo A, Ribeiro Guevara S (2013) Differential mercury transfer in the aquatic food web of a double basined lake associated with selenium and habitat. Sci Total Environ 454:170–180

Arcagni M, Rizzo A, Juncos R, Pavlin M, Campbell L, Arribére M, Horvat M, Ribeiro Guevara S (2017) Mercury and selenium in the food web of Lake Nahuel Huapi, Patagonia, Argentina. Chemosphere 166:163–173

Arcagni M, Juncos R, Rizzo A, Pavlin M, Fajon V, Arribére M, Horvat M, Ribeiro Guevara S (2018) Species-and habitat-specific bioaccumulation of total mercury and methylmercury in the food web of a deep oligotrophic lake. Sci Total Environ 612:1311–1319

Arias A, Spetter C, Freije R, Marcovecchio J (2009) Polycyclic aromatic hydrocarbons in water, mussels (*Brachidontes* sp., *Tagelus* sp.) and fish (*Odonthestes* sp.) from Bahía Blanca Estuary, Argentina. Estuar Coast Shelf S 85(1):67–81

Arnot J, Gobas F (2006) A review of bioconcentration factor (BCF) and bioaccumulation factor (BAF) assessments for organic chemicals in aquatic organisms. Environ Rev 14:257–297

Arribére M, Ribeiro Guevara S, Sánchez R, Gil M, Ross G, Daurade L, Kestelman A (2003) Heavy metals in the vicinity of a chlor-alkali factory in the upper Negro River ecosystem, Northern Patagonia, Argentina. Sci Total Environ 301:187–203

Arribére M, Ribeiro Guevara S, Bubach D, Arcagni M, Vigliano P (2008) Selenium and mercury in native and introduced fish species of patagonian lakes, Argentina. Biol Trace Elem Res 122:42–63

Avigliano E, Lozano C, Plá R, Volpedo A (2016) Toxic element determination in fish from Paraná River Delta (Argentina) by neutron activation analysis: tissue distribution and accumulation and health risk assessment by direct consumption. J Food Compos Anal 54:27–36

Avigliano E, Monferrán M, Sánchez S, Wunderlin D, Gastaminza J, Volpedo A (2019) Distribution and bioaccumulation of 12 trace elements in water, sediment and tissues of the main fishery from different environments of the La Plata basin (South America): risk assessment for human consumption. Chemosphere 236:124394

Bahnasawy M, Khidr A, Dheina N (2011) Assessment of heavy metal concentrations in water, plankton, and fish of Lake Manzala, Egypt. Turk J Zool 35:271–280

Bal W, Kasprzak K (2002) Induction of oxidative DNA damage by carcinogenic metals. Toxicol Lett 127(1–3):55–62

Baldassin P, Taniguchi S, Gallo H, Maranho A, Kolesnikovas C, Amorim D, Mansilla M, Navarro R, Tabeira L, Bicego M, Montone R (2016) Persistent organic pollutants in juvenile Magellanic penguins (*Spheniscus magellanicus*) in South America. Chemosphere 149:391–399

Ballesteros M, Miglioranza K, Gonzalez M, Fillmann G, Wunderlin D, Bistoni M (2014) Multimatrix measurement of persistent organic pollutants in Mar Chiquita, a continental saline shallow lake. Sci Total Environ 490:73–80

Baraj B, Niencheski L, Windom H, Hermanns L (2009) Trace metal concentration in liver, kidney and heart in South American fur seal (*Arctocephalus australis*) from Southern Brazil. Mar Pollut Bull 58(12):1948–1952

Barra R, Colombo JC, Eguren G, Gamboa N, Jardim WF, Mendoza G (2006) Persistent organic pollutants (POPs) in Eastern and Western South American countries. In: Ware GW, Nigg HN, Doerge DR (eds) Reviews of environmental contamination and toxicology. Reviews of Environmental Contamination and Toxicology, vol 185. Springer, New York

Beiras R (2018) Marine pollution. Sources, fate and effects of pollutants in coastal ecosystems. Elsevier, p 408

Bigatti G, Primost M, Cledón M, Averbuj A, Theobald N, Gerwinski W (2009) Biomonitoring of TBT contamination and imposex incidence along 4700 km of Argentinean shoreline (SW Atlantic: from 38 °S to 54 °S). Mar Pollut Bull 58:695–701

Borgå K (2013) Ecotoxicology: bioaccumulation. Encycl Ecol 346–348

Borgå K, Kidd K, Muir D, Berglund O, Conder J, Gobas F, Kucklick J, Malm O, Powell D (2011) Trophic magnification factors: considerations of ecology, ecosystems and study design. Int Environ Assess Manag 8(1):64–84

Borrell A, García-Solà A, Aguilar A, García N, Crespo E (2010) Organochlorine residues in South American Sea Lions, *Otaria flavescens* (Shaw, 1800): bioaccumulation and Time Trends. B Environ Contam Tox 84:731–737

Buzzi N, Marcovecchio J (2018) Heavy metal concentrations in sediments and in mussels from Argentinean coastal environments, South America. Environ Earth S 77:321

Cáceres-Saez I, Dellabianca N, Goodall R, Cappozzo H, Guevara S (2013a) Mercury and selenium in subantarctic Commerson's Dolphins (*Cephalorhynchus commersonii*). Biol Trace Elem Res 151(2):195–208

Cáceres-Saez I, Guevara S, Dellabianca N, Goodall R, Cappozzo H (2013b) Heavy metals and essential elements in Commerson's dolphins (*Cephalorhynchus commersonii*) from the southwestern South Atlantic Ocean. Environ Monit Assess 185(7):5375–5386

Cáceres-Saez I, Haro D, Blank O, Aguayo-Lobo A, Dougnac C, Arredondo C, Guevara S (2019) Stranded false killer whales, *Pseudorca crassidens*, in Southern South America reveal potentially dangerous silver concentrations. Mar Pollut Bull 145:325–333

Calle P, Alvarado O, Monserrate L, Cevallos JM, Calle N, Alava J (2015) Mercury accumulation in sediments and seabird feathers from the Antarctic Peninsula. Mar Pollut Bull 91:410–417

Cappelletti N, Speranza E, Tatone L (2015) Bioaccumulation of dioxin-like PCBs and PBDEs by detritus-feeding fish in the Rio de la Plata estuary, Argentina. Environ Sci Poll R 22:7093–7100

Cardwell R, Deforest D, Brix K, Adams W (2013) Do Cd, Cu, Ni, Pb and Zn biomagnify in aquatic ecosystems? Rev Environ Contam T 226:101–122

Cherel Y, Klages N (1997) A review of the food of albatrosses. In: Robertson G, Gales R (eds) Albatross biology and conservation. Chipping Norton, Australia, Surrey Beatty and Sons, pp 113–136

Cid F, Antón R, Caviedes-Vidal E (2007) Organochlorine pesticide contamination in three bird species of the Embalse La Florida water reservoir in the semiarid midwest of Argentina. Sci Total Environ 385:86–96

Cipro C, Taniguchi S, Montone R (2010) Occurrence of organochlorine compounds in *Euphasia superba* and unhatched eggs of *Pygoscelis* genus penguins from Admiralty Bay (King George Island, Antarctica) and estimation of biomagnification factors. Chemosphere 78:767–771

Cipro C, Montone R, Bustamante P (2016) Mercury in the ecosystem of Admiralty Bay, King George Island, Antarctica: occurrence and trophic distribution. Mar Pollut Bull 114:564–570

Colabuono F, Taniguchi S, Montone R (2012) Organochlorine contaminants in albatrosses and petrels during migration in South Atlantic Ocean. Chemosphere 86:701–708

Colabuono F, Barquete V, Taniguchi S, Ryan P, Montone R (2014) Stable isotopes of carbon and nitrogen in the study of organochlorine contaminants in albatrosses and petrels. Mar Pollut Bull 83:241–247

Colombo J, Khalil M, Arnac M, Horth A, Catoggio J (1990) Distribution of chlorinated pesticides and individual polychlorinated biphenyls in biotic and abiotic compartments of the Rio de La Plata, Argentina. Environ Sci Technol 24:498–505

Colombo J, Cappelletti N, Williamson M, Migoya M, Speranza E, Sericano J, Muir D (2011) Risk ranking of multiple-POPs in detritivorous fish from the Río de la Plata. Chemosphere 83:882–889

Colombo J, Speranza E, Astoviza M, Migoya M, Skorupka C, Morrone M, Heguilor S, Tatone L, Bilos C (2018) Long-term (1970–2017) temporal trends of Polychlorinated Biphenyls in fish, settling material and sediments from populated and remote sites in Río de la Plata Estuary, Argentina. Environ Sci Technol 52:12412–12418

Commendatore M, Nievas M, Amin O, Esteves J (2012) Sources and distribution of aliphatic and polyaromatic hydrocarbons in coastal sediments from the Ushuaia Bay (Tierra del Fuego, Patagonia, Argentina). Mar Environ Res 74:20–31

Commendatore M, Franco M, Gomes Costa P, Castro I, Fillmann G, Bigatti G (2015) Butyltins, polyaromatic hydrocarbons organochlorine pesticides and polychlorinated biphenyls in sediments and bivalve mollusks in a mid-latitude environment from the Patagonian coastal zone. Environ Toxicol Chem 34(12):2750–2763

Comoglio L, Amin O, Botté S, Marcovecchio J (2011) Use of biomarkers in resident organisms as a tool for environmental monitoring in a cold coastal system, Tierra del Fuego Island. Ecotox Environ Safe 74:382–393

Connell D (1988) Bioaccumulation behavior of persistent organic chemicals with aquatic organisms. In: Ware GW (eds) Reviews of environmental contamination and toxicology. Reviews of Environmental Contamination and Toxicology, vol 102. Springer

Corsolini S, Borghesi N, Ademollo N, Focardi S (2011) Chlorinated biphenyls and pesticides in migrating and resident seabirds from East and West Antarctica. Environ Inter 37:1329–1335

Crespo EA, Lewis MN, Campagna C (2007) Mamíferos marinos: pinnipedios y cetáceos. En: El Mar Argentino y sus recursos pesqueros. El Ecosistema Marino. INIDEP. Tomo 5:127–150

Crespo E, Borobia M, Alarcon D, Alonso M, Bazzalo M, Cremer M, Filla G, Magalhães F, Marigo F, Queiróz H, Reynolds J, Schaeffer Y, Wetzel D, Dorneles P, Lailson-Brito J (2010) Report of the working group on major threats and conservation. Lat Am J Aquat Mamm (Spec Issue) 8(1–2):47–56

Daga R, Ribeiro-Guevara S, Poire D, Arribére M (2014) Characterization of tephras dispersed by the recent eruptions of volcanoes Calbuco (1961), Chaitén (2008) and Cordón Caulle Complex (1960 and 2011), in Northern Patagonia. J S Am Earth Sci 49:1–14

Das K, Debacker V, Pillet S, Bouquegneau J (2003) Heavy metals in marine mammals. Toxicol Mar Mamm 3:135–167

de Boer J, Wester P (1991) Chlorobiphenyls and organochlorine pesticides in various Sub-Antarctic organisms. Mar Pollut Bull 22:441–447

Deb S, Fukushima T (1999) Metals in aquatic ecosystems: mechanisms of uptake, accumulation and release-ecotoxicological perspectives. Int J Environ Stud 56(3):385–417

De Marco S, Botte S, Marcovecchio J (2006) Mercury distribution in abiotic and biological compartments within several estuarine systems from Argentina: 1980–2005 period. Chemosphere 65:213–223

Dolagaratz Carricavur A, Chiodi-Boudet L, Romero M, Polizzi P, Marcovecchio J, Gerpe M (2018) Toxicological responses of *Laeonereis acuta* (Polychaeta, Nereididae) after acute, subchronic and chronic exposure to cadmium. Ecotox Environ Safe 149:217–224

Dorneles P, Lailson-Brito J, dos Santos R, da Costa P, Malm O, Azevedo A, Torres J (2007) Cephalopods and cetaceans as indicators of offshore bioavailability of cadmium off Central South Brazil Bight. Environ Pollut 148:352–359

dos Santos I, Silva-Filho E, Schaefer C, Sella S, Silva C, Gomes V, Passos M, Ngan P (2006) Baseline mercury and zinc concentrations in terrestrial and coastal organisms of Admiralty Bay, Antarctica. Environ Pollut 140:304–311

Durante C, Santos-Neto E, Azevedo A, Crespo E, Lailson-Brito J (2016) POPs in the South Latin America: bioaccumulation of DDT, PCB, HCB, HCH and Mirex in blubber of common dolphin (*Delphinus delphis*) and Fraser's dolphin (*Lagenodelphis hosei*) from Argentina. Sci Total Environ 572:352–360

Durante C, Reis B, Azevedo A, Crespo E, Lailson-Brito J (2020) Trace elements in trophic webs from South Atlantic: the use of cetaceans as sentinels. Mar Pol Bull 150:110674

Ellis D, Cipro C, Ogletree C, Smith K, Aronson R (2018) A 50-year retrospective of persistent organic pollutants in the fat and eggs of penguins of the Southern Ocean. Environ Pollut 241:155–163

Elorriaga Y, Marino D, Carriquiriborde P, Ronco A (2013a) Screening of pharmaceuticals in surface water bodies of the Pampas region of Argentina. Int J Environ Health 6:330–339

Elorriaga Y, Marino D, Carriquiriborde P, Ronco A (2013b) Human pharmaceuticals in wastewaters from urbanized areas of Argentina. B Environ Contam Tox 90:397–400

Etchegoyen M, Ronco A, Almada P, Abelando M, Marino D (2017) Occurrence and fate of pesticides in the Argentine stretch of the Paraguay-Paraná basin. Environ Monit Assess 189:63

Fernández Severini M, Hoffmeyer M, Marcovecchio J (2013) Heavy metals concentrations in zooplankton and suspended particulate matter in a Southwestern Atlantic temperate estuary (Argentina). Environ Monit Assess 185(2):1495–1513

Fisk A, Hobson K, Norstrom R (2001) Influence of chemical and biological factors on trophic transfer of persistent organic pollutants in the Northwater Polynya Marine Food Web. Environ Sci Technol 35:732–738

Frías J, Gil M, Esteves J, García Borboroglu P, Kane O, Smith J, Boersma P (2012) Mercury levels in feathers of Magellanic penguins. Mar Pollut Bull 64:1265–1269

Gaiero D, Probst J, Depetris P, Lelyter L, Kempe S (2002) Riverine transfer of heavy metals from Patagonia to the southwestern Atlantic Ocean. Reg Environ Change 3:51–64

Gerpe M, de León A, Bastida R, Moreno V, Rodríguez D (2009) Sharp accumulation of heavy metals after weaning in the South American fur seal *Arctocephalus australis*. Mar Ecol Prog Ser 375:239–245

Giarratano E, Gil M, Marinho C, Malanga G (2016) Metals from mine waste as potential cause of oxidative stress in burrowing crab *Neohelice granulata* from San Antonio bay. Ecotox Environ Safe 132:68–76

Gil M, Harvey M, Beldoménico H, García S, Commendatore M, Gandini P, Frere E, Yorio P, Crespo E, Esteves J (1996) Contaminación por metales y plaguicidas organoclorados en organismos marinos de la zona costera Patagónica. Informes Técnicos del PMIZCP, Fundación Patagonia Natural N°, p 32

Gil M, Harvey M, Esteves J (1999) Heavy metals in intertidal sediments from Patagonian coast, Argentina. B Environ Contam Tox 63:52–58

Gil M, Torres A, Harvey M, Esteves J (2006) Heavy metals in marine organisms from the coastal zone of Continental Argentine Patagonia. Rev Biol Mar Oceanog 41(2):167–176

Gil M, Giarratano E, Barros V, Bortolus A, Codignotto J, Delfino Schenke R, Góngora M, Lovrich G, Monti A, Pascual M, Rivas A, Tagliorette A (2019) World seas: an environmental evaluation. Southern Argentina: the patagonian continental shelf (Sheppard C (ed)), 914 p. Academic Press. ISMBN 978-0-12-805068-2

Gobas F, Zhang X, Wells R (1993) Gastrointestinal magnification: the mechanism of biomagnification and food chain accumulation of organic chemicals. Environ Sci Tech 13:2855–2863

Gobas F, de Wolf W, Verbruggen E, Plotzke K, Burkhard L (2009) Revisiting bioaccumulation criteria for POPs and PBT assessments. Integr Environ Assess Manag 5(4):624–637

Goerke H, Weber K, Bornemann H, Ramdohr S, Plötz J (2004) Increasing levels and biomagnification of persistent organic pollutants (POPs) in Antarctic biota. Mar Pollut Bull 48:295–302

González Sagrario M, Miglioranza K, Aizpun de Moreno J, Moreno V, Escalante A (2002) Polychlorinated biphenyls in different trophic levels from a shallow lake in Argentina. Chemosphere 48:1113–1122

Griboff J, Horacek M, Wunderli D, Monferran M (2018) Bioaccumulation and trophic transfer of metals, As and Se through a freshwater food web affected by anthrophic pollution in Córdoba, Argentina. Ecotox Environ Safe 148:275–284

Griboff J, Wunderlin D, Horacek M, Monferrán M (2020) Seasonal variations on trace element bioaccumulation and trophic transfer along a freshwater food chain in Argentina. Environ Sci Poll R 27:40664–40678

Häder D, Banaszak A, Villafañe V, Narvarte M, González R, Helbling W (2020) Anthropogenic pollution of aquatic ecosystems: emerging problems with global implications. Sci Total Environ 713:136586

Hoekstra P, O´Hara T, Fisk A, Borgå K, Solomon K, Muir D (2003) Trophic transfer of persistent organochlorine contaminants (OCs) within an Arctic marine food web from the southern Beaufort-Chukchi Seas. Environ Pollut 124:509–522

Huang J (2016) Arsenic trophodynamics along the food chains/webs of different ecosystems: a review. Chem Ecol 32(9):803–828

Ikemoto T, Phuc Cam Tu N, Watanabe M, Okuda N, Omori K, Tanabe S, Cach Tuyen B, Takeuchi I (2008) Analysis of biomagnification of persistent organic pollutants in the aquatic food web of the Mekong Delta, South Vietnam using stable carbon and nitrogen isotopes. Chemosphere 72:104–114

Ishaque A, Johnson L, Gerald T, Boucaud D, Okoh J, Tchounwou P (2006) Assessment of individual and combined toxicities of four non-essential metals (As, Cd, Hg and Pb) in the microtox assay. Int J Environ Res Public Health 3:118–120

Jofré M, Antón R, Caviedes-Vidal E (2008) Organochlorine contamination in anuran amphibians of an artificial lake in the semiarid midwest of Argentina. Arch Environ Contam Toxicol 55:471–480

Juncos R, Campbell L, Arcagni M, Daga R, Rizzo A, Arribére M, Ribeiro-Guevara S (2017) Variations in anthropogenic silver in a large Patagonian lake correlate with global shifts in photographic processing technology. Environ Pollut 223:685–694

Kim J, Son M, Kang J, Kim J, Jung J, Chang Y (2015) Occurrence of legacy and new persistent organic pollutants in avian tissues from King George Island, Antarctica. Environ Sci Technol 49:13628–13638

Kumar K, Kannan K, Corsolini S, Evans T, Giesy J, Nakanishia J, Masunag S (2002) Polychlorinated dibenzo-p-dioxins, dibenzofurans and polychlorinated biphenyls in polar bear, penguin and south polar skua. Environ Pollut 119:151–161

La Colla N, Botté S, Oliva A, Marcovecchio J (2017) Tracing Cr, Pb, Fe and Mn occurrence in the Bahía Blanca estuary through commercial fish species. Chemosphere 175:286–293

La Colla N, Botté S, Fiori S (2019) First records of metal concentrations in the Pacific oyster (*Crassostrea gigas*) from a Southwest Atlantic estuary. Environ Geochem Health 41:1321–1338

Lailson-Brito J, Azeredo M, Malm O, Ramos R, Di Beneditto A, Saldanha M (2002) Trace metals in liver and kidney of the franciscana (*Pontoporia blainvillei*) from the northern coast of Rio de Janeiro State, Brazil. Lat Am J Aq Mamm 1(1):107–114

Lailson-Brito J, Dorneles P, Azevedo-Silva C, Bisi T, Vidal L, Legat L, Mal O (2012) Organochlorine compound accumulation in delphinids from Rio de Janeiro State, southeastern Brazilian coast. Sci Total Environ 433:123–131

Lana N, Berton P, Covaci A, Ciocco N, Barrera-Oro E, Atencio A, Altamirano J (2014) Fingerprint of persistent organic pollutants in tissues of Antarctic notothenioid fish. Sci Total Environ 499:89–98

Lanfranchi A, Menone M, Miglioranza K, Janiot L, Aizpún J, Moreno V (2006) Striped weakfish (*Cynoscion guatucupa*): a biomonitor of organochlorine pesticides in estuarine and near-coastal zones. Mar Pollut Bull 52:74–80

Le Croizier G, Lacroix C, Artigaud S, Le Floch S, Raffray J, Penicaud V, Coquillé V, Autier J, Rouget M, Le Bayon N (2018) Significance of metallothioneins in differential cadmium accumulation kinetics between two marine fish species. Environ Pollut 236:462–476

Liu J, Cao L, Doua S (2019) Trophic transfer, biomagnification and risk assessments of four common heavy metals in the food web of Laizhou Bay, the Bohai Sea. Sci Total Environ 670:508–522

Llorca M, Farré M, Eljarrat E, Díaz-Cruz S, Rodríguez-Mozaz S, Wunderlin D, Barcelo D (2017) Review of emerging contaminants in aquatic biota from Latin America: 2002–2016. Environ Toxicol Chem 36(7):1716–1727

Lombardi P, Peri S, Guerrero N (2010) Trace metal levels in *Prochilodus lineatus* collected from the La Plata River, Argentina. Environ Monit Assess 160:47

Łukowski A, Karlowski M, Górski T (1987) Polychlorinated biphenyls in the tissues of Antarctic marine migratory birds and penguins from the breeding colony on King George Island (South Shetland Islands). Pol Polar Res 8:179–187

Marcovecchio J (2004) The use of *Micropogonias furnieri* and *Mugil liza* as bioindicators of heavy metals pollution in La Plata river estuary, Argentina. Sci Total Environ 323(1–3):219–226

Marcovecchio J, Moreno V, Perez A (1988) Determination of heavy metal concentrations in biota of Bahia Blanca, Argentina. Sci Total Environ 75:181–190

Marcovecchio J, Moreno V, Bastida R, Gerpe M, Rodríguez D (1990) Tissue distribution of heavy metals in small cetaceans from the Southwestern Atlantic Ocean. Mar Pollut Bull 21(6):299–304

Marcovecchio J, Andrade S, Ferrer L, Asteasuain R, De Marco S, Gavio M, Scarlatto N, Freije R, Pucci A (2001) Mercury distribution in estuarine environments from Argentina: the detoxification and recovery of salt marshes after 15 years. Wetl Ecol Manag 9:317–322

Marcovecchio J, Botté S, Fernández Severini M (2016) Distribution and behavior of zinc in estuarine environments: an overview on Bahía Blanca estuary (Argentina). Environ Earth Sci 75:1168

Marcovecchio J, De Marco S, Gavio M, Narvarte M, Fiori S, Gerpe M, Rodríguez D, Abbate M, La Colla N, Oliva A (2019) The Northern Argentine Sea. World seas: an environmental evaluation. Elsevier, pp 759–781

Mello F, Roscales J, Guida Y, Menezes J, Vicente A, Costa E, Jiménez B, Torres J (2016) Relationship between legacy and emerging organic pollutants in Antarctic seabirds and their foraging ecology as shown by $\delta^{13}C$ and $\delta^{15}N$. Sci Total Environ 573:1380–1389

Miglioranza K, Gonzalez M, Ondarza P, Shimabukuro V, Isla F, Fillmann G, Aizpún J, Moreno V (2013) Assessment of Argentinean Patagonia pollution: PBDEs, OCPs and PCBs in different matrices from the Río Negro basin. Sci Total Environ 452:275–285

Monferrán M, Galanti L, Bonansea R, Amé M, Wunderlin D (2011) Integrated survey of water pollution in the Suquía River basin (Córdoba, Argentina). J Environ Monit 13:398–409

Monferrán M, Garnero P, Bistoni M, Anbar A, Gordon G, Wunderlin D (2016) From water to edible fish. Transfer of metals and metalloids in the San Roque Reservoir (Córdoba, Argentina). Implications associated with fish consumption. Ecol Indic 63:48–60

Narvarte M, Salas M, Avaca M, Roche M, Delucchi F, Tartara M, Marcovecchio J (2017) Beyond imposex: other sexual alterations in a nassariid snail from Patagonia. Mar Biol Res 13(4):429–438

Oliva A, La Colla N, Arias A, Blasina G, López-Cazorla A, Marcovecchio J (2017) Distribution and human health risk assessment of PAHs in four fish species from a SWAtlantic estuary. Environ Sci Pollut R 24:18979–18990

Ondarza P, Gonzalez M, Fillmann G, Miglioranza K (2014) PBDEs, PCBs and organochlorine pesticides distribution in edible fish from Negro River basin, Argentinean Patagonia. Chemosphere 94:135–142

Ondarza P, Haddad S, Avigliano E, Miglioranza K, Brooks B (2019) Pharmaceuticals, illicit drugs and their metabolites in fish from Argentina: implications for protected areas influenced by urbanization. Sci Total Environ 649:1029–1037

Ondarza P, Miglioranza K, González M, Shimabukuro V, Aizpún J, Moreno V (2010) Organochlorine compounds in common carp (Cyprinus carpio) from Patagonia Argentina. J Braz Soc Ecotoxicol 5:41–47

Peña N, Moreno V, Marcovecchio J, Pérez A (1988) Total mercury, cadmium and lead distribution in tissues of the southern sea lion (*Otaria flavescens*) in the ecosystem of Mar del Plata, Argentina. In: Metals in coastal environments of Latin America. Springer, Berlin, Heidelberg, pp 140–146

Pinkas L, Oliphant MS, Iverson ILK (1971) Food habits of albacore, bluefin tuna, and bonito in California waters. Fish Bull 152:1–105

Polito M, Brasso R, Trivelpiece W, Karnovsky N, Patterson W, Emslie S (2016) Differing foraging strategies influence mercury (Hg) exposure in an Antarctic penguin community. Environ Pollut 218:196–206

Polizzi P, Boudet L, Romero M, Denuncio P, Rodríguez D, Gerpe M (2013) Fine scale distribution constrains cadmium accumulation rates in two geographical groups of Franciscana dolphin from Argentina. Mar Pol Bull 72(1):41–46

Quadri-Adrogué A, Miglioranza S, Copello S, Favero M, Seco Pon J (2019) Pelagic seabirds as biomonitors of persistent organic pollutants in the southwestern Atlantic. Mar Pollut Bull 149:110516

Quadri-Adrogué A, Seco Pon J, García G, Castano M, Copello S, Favero M, Miglioranza K (2020) Chlorpyrifos and persistent organic pollutants in feathers of the near threatened Olrog's Gull in southeastern Buenos Aires Province, Argentina. Environ Pollut 115918. https://doi.org/10.1016/j.envpol.2020.115918

Quintas P, Fernández E, Spetter C, Arias A, Garrido M, Marcovecchio J (2019) Preliminary studies about the role of physicochemical parameters on the organotin compound dynamic in a South American estuary (Bahía Blanca, Argentina). Environ Monit Assess 191:127

Ribeiro-Guevara S, Bubach D, Vigliano P, Lippolt G, Arribére M (2004) Heavy metal and other trace elements in native mussel *Diplodon chilensis* from Northern Patagonia Lakes, Argentina. Biol Trace Elem Res 102:245–263

Ribeiro-Guevara S, Arribére M, Bubach D, Vigliano P, Rizzo A, Alonso M, Sánchez R (2005) Silver contamination on abiotic and biotic compartments of Nahuel Huapi National Park lakes, Patagonia, Argentina. Sci Total Environ 336:119–134

Ríos J, Ruggeri M, Poma G, Malarvannan G, Covaci A, Puliafito S, Ciocco N, Altamirano J (2019) Occurrence of organochlorine compounds in fish from freshwater environments of the central Andes, Argentina. Sci Total Environ 693:133389

Ritter L, Solomon K, Forget J, Stemeroff M, O'Leary C (1995) An assessment report on DDT, Aldrin, Dieldrin, Endrin, Chlordane, Heptachlor, Hexachlorobenzene, Mirex, Toxaphene,

Polychlorinated Biphenyls, Dioxins and Furans. UN Environment Programme, Prepared for the International Programme on Chemical Safety (IPCS)

Rodrigues M, Netto E, Cumplido R, de Melo U, Quintanilha D, Marsico E, Guimaraes C (2019) Records of Southern Sea Lion (*Otaria flavescens*) (Mammalia: Carnivora: Otariidae) and Total Mercury (THg) levels in muscle tissue of a specimen on the east coast of Rio de Janeiro State, Southeastern Brazil. Open Access Libr J 6(3):1–14

Rosas C, Gil M, Uhart M (2012) Trace metal concentrations in southern right whale (*Eubalaena australis*) at Península Valdés, Argentina. Mar Pollut Bull 64:1255–1260

Rovedatti M, Castañé P, Topalián M, Salibián A (2001) Monitoring of organochlorine and organophosphorus pesticides in the water of the Reconquista River (Buenos Aires, Argentina). Water Res 35:3457–3461

Santos-Neto E, Azevedo-Silva C, Bisi T, Santos J, Meirelles A, Carvalho V, Azevedo A, Guimarães J, Lailson-Brito J (2014) Organochlorine concentrations (PCBs, DDTs, HCHs, HCB and MIREX) in delphinids stranded at the northeastern Brazil. Sci Total Environ 472:194–203

Sauvé S, Desrosiers M (2014) A review of what is an emerging contaminant. Chem Cent J 8:15

Scarlatto N, Marcovecchio J, Pucci A (1997) Heavy metal distribution in zooplankton from Buenos Aires coastal waters (Argentina). Chem Spec Bioavailab 9(1):21–26

Schiavone A, Corsolini S, Borghesi N, Focardi S (2009) Contamination profiles of selected PCB congeners, chlorinated pesticides, PCDD/Fs in Antarctic fur seal pups and penguin eggs. Chemosphere 76:264–269

Simonetti P, Botté S, Marcovecchio J (2015) Exceptionally high Cd levels and other trace elements in eggshells of American oystercatcher (*Haematopus palliatus*) from the Bahía Blanca Estuary, Argentina. Mar Pollut Bull 100(1):495–500

Simonetti P, Botté S, Marcovecchio J (2018) Heavy metal bioconcentration factors in the burrowing crab *Neohelice granulata* of a temperate ecosystem in South America: Bahía Blanca estuary, Argentina. Environ Sci Pollut Res 25(34):34652–34660

Sinkus W, Shervette V, Ballenger J, Reed L, Plante C, White B (2017) Mercury bioaccumulation in offshore reef fishes from waters of the Southeastern USA. Environ Pollut 228:222–233

Taniguchi S, Montone R, Bícego M, Colabuono F, Weber R, Sericano J (2009) Chlorinated pesticides, polychlorinated biphenyls and polycyclic aromatic hydrocarbons in the fat tissue of seabirds from King George Island, Antarctica. Mar Pollut Bull 58:129–133

Torres P, Miglioranza K, Uhart M, González M, Commendatore M (2015) Organochlorine pesticides and PCBs in southern right whales (*Eubalaena australis*) breeding at Península Valdés, Argentina. Sci Total Environ 518:605–615

UNEP (2001) Final act of the conference of pleniopotentiaries on the stockholm convention on persistent organic pollutants. United Nations Environment Program, Geneva, p 44

Valdés M, Amé M, Bistoni M, Wunderlin D (2014) Occurrence and bioaccumulation of pharmaceuticals in a fish species inhabiting the Suquía River basin (Córdoba, Argentina). Sci Total Environ 472:389–396

Valdés M, Marino D, Wunderlin D, Somoza G, Ronco A, Carriquiriborde P (2015) Screening concentration of E1, E2 and EE2 in sewage effluents and surface waters of the "Pampas" region and the "Río de la Plata" Estuary (Argentina). B Environ Contam Tox 94:29–33

Valdés M, Huerta B, Wunderlin D, Bistoni M, Barceló D, Rodriguez-Mozaz S (2016) Bioaccumulation and bioconcentration of carbamazepine and other pharmaceuticals in fish under field and controlled laboratory experiments. Evidences of carbamazepine metabolization by fish. Sci Total Environ 557–558:58–67

van den Brink N (1997) Directed transport of volatile organochlorine pollutants to polar regions: the effect on the contamination pattern of Antarctic seabirds. Sci Total Environ 198:43–50

van den Steen E, PoiSmbleau M, Demongin L, Covaci A, Dirtu A, Pinxten E, van Noordwijk H, Quillfeldt P, Eens M (2011) Organohalogenated contaminants in eggs of rockhopper penguins (*Eudyptes chrysocome*) and imperial shags (*Phalacrocorax atriceps*) from the Falkland Islands. Sci Total Environ 409:2838–2844

Vázquez N, Gil M, Esteves J, Narvarte M (2007) Monitoring heavy metal pollution in San Antonio Bay, Río Negro, Argentina. B Environ Contam Tox 79:21–125

Vilches F, Bobinac M, Labudía A, Paso-Viola M, Marcovecchio J, Cappozzo H, Panebianco M (2019) Metals concentration and bioaccumulation in the marine-coastal trophic web from Buenos Aires province southern coast, Argentina. Chem Ecol 35:1–23

Wilhelmsson D, Thompson R, Holmström K, Lindén O, Eriksson-Hägg H (2013) Chapter 6. Marine pollution. In: Managing ocean environments in a changing climate sustainability and economic perspectives, pp 127–169

Xie Z, Lu G, Yan Z, Liu J, Wang P, Wang Y (2017) Bioaccumulation and trophic transfer of pharmaceuticals in food webs from a large freshwater lake. Environ Pollut 222:356–366

Zhao L, Yang F, Yan X (2013) Biomagnification of trace elements in a benthic food web: the case study of Deer Island (Northern Yellow Sea). Chem Ecol 29(3):197–207

Chapter 5
Pollution in the Arctic Ocean

Sten-Åke Wängberg and Göran Björk

Abstract The Arctic Ocean (AO) is, despite its isolated localization in the extreme north, where human activity is restricted, exposed to anthropogenic pollution that has, or threatens to have, impact on the ecology. In this review, we discuss the characteristics of the AO that favors transport of pollutants from lower latitudes to the AO. We point out how important the physical qualities of the pollutants are, which routes they are taking into the AO, if by water currents, air or by rivers, and how they are circulating within the AO and entering the biological food web. It is based on knowledge we have on geography, air and water currents, water masses and the biological communities. Ice and ice cover are given special attention, both as a physical barrier for water/air exchange and how ice cover production affects the concentration of pollutants. By using examples from well-known pollutants (e.g., legacy organochlorine pesticides, currently used pesticides and mercury), we show, from measured field data, how these factors determine the distribution of pollutants and will summarize what can be learned from the analysis of contents in animals (birds, mammals and fish). We present assessments of the biological, chemical and physical effects the pollutants have. The paper finishes with a description of the new threats that are to be expected by new pollution sources related to climate change that opens passages for expanding traffic and exploration in the AO.

Keywords Arctic Ocean (AO) · BioConcentration Factor (BCF) · Fugacity · Mercury · Persistand Organic Pollutants (POP) · Pesticides · Polybrominated diphenyl ethers (PBDE) · Sea ice · Transport routes · Ocean circulation

S.-Å. Wängberg (✉) · G. Björk
Dept. Marine Sciences, University of Gothenburg, Gothenburg, Sweden
e-mail: sten-ake.wangberg@marine.gu.se

© The Author(s), under exclusive license to Springer Nature Switzerland AG 2021
D.-P. Häder et al. (eds.), *Anthropogenic Pollution of Aquatic Ecosystems*,
https://doi.org/10.1007/978-3-030-75602-4_5

5.1 Introduction

The aim of this chapter is to give a broad picture of the prerequisites for pollutants to reach this remote place, their fate and the potential consequences for the marine environment.

The outline is that we first present the physical geography of the area together with the transport routes in water and air that bring pollutants to the area. After this we describe how pollutants interact with this special environment in a general sense, and how the chemical characteristics of the pollutant determine its fate. This is followed by the presentation of some specific pollutants in the AO, including their abundances, transportation and how they are modified by physical and biological factors. This includes observational data as well as model investigations, followed by a presentation of concentrations within the biota and the effect of pollutants on organisms.

An outlook on the changes that are to come in the Arctic then follows in the context of climate change and what it will mean for pollutants. A short conclusion ends the chapter.

5.2 What is the Central Arctic Ocean?

The Arctic Ocean (AO) is one of the world oceans, beside the Atlantic Ocean, Pacific Ocean, Indian Ocean and Antarctic Ocean. It is the smallest, having an area of 7.8 million km^2 corresponding to only about 2% of the global ocean. It has borders to five countries such as Norway, Russia, USA, Canada and Denmark (by Greenland).

The AO includes two major deep sea basins (Fig. 5.1) called the Eurasian Basin and the Amerasian Basin separated by the prominent Lomonosov Ridge. The Amerasian Basin has two sub-basins: the Canada Basin and Makarov Basin separated by the Alfa-Mendeleev Ridge. The Eurasian Basin is the deepest one with a maximum depth of more than 4000 m. The deep basins are surrounded by large shallow shelf seas (Kara, Laptev and East Siberian) at the Asian sector as well as a narrower shelf along the American sector. The Barents Sea is not included in the AO as defined here. Only in the Fram Strait, there is a break in the shelves and a deep ($\sim$2600 m) passage to waters at lower latitudes.

5.2.1 *Sea Ice*

The AO sea ice cover consists of moving pack ice which covers the entire basin during winter and melts away from the coast during summer, leaving areas of open water behind, until the sea ice minimum extent occurs in September. The ice cover has undergone a remarkable decline over the last decades with a decrease of the minimum summer extent with around 40% and a thickness reduction from about

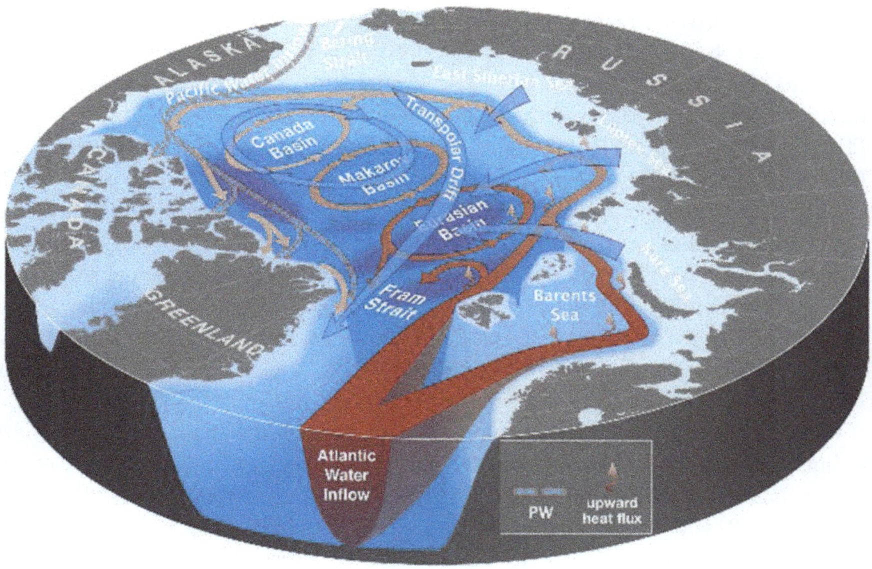

Fig. 5.1 Map of the Arctic Ocean with a schematic representation of major ocean currents and features of the internal circulation. Red to pink arrows indicate pathways of relatively warm and salty water of Atlantic origin entering at Fram Strait, then circulating the basin at intermediate depths and becoming successively colder during transit. Blue arrows show inflow (mainly runoff from land) and circulation of fresher surface waters. Gray dashed arrows show the major pathway of Pacific water entering through Bering Strait. Vertical arrows indicate where the largest heat flux from the ocean to the atmosphere occurs. *Source* Carmack et al. (2015)

3.6 m average winter thickness around 1980 to about 2.1 m in 2015–2018 (Kwok 2018). The type of ice has also shifted toward a dominance of the first year ice (FYI), compared to multiyear ice (MYI). MYI coverage has decreased by 50% from 2002 to 2017 such that MYI now only covers less than one-third of the AO (Kwok 2018). The thinning has also made the ice weaker, with less internal stress so it is more mobile in response to wind forcing, and the drift speed is therefore generally higher (Spreen et al. 2011).

5.3 Which are the Transport Ways for Pollutants Entering the AO?

5.3.1 Short Introduction to Ocean Circulation and Ways for Aquatic Entrance to the AO

Warm and salty water from the north Atlantic enters the AO via two major branches: through the Fram Strait and the Barents Sea (Fig. 5.1). The largest inflow to the AO is by the West Spitsbergen current at the eastern side of the Fram Strait with an estimated transport of about 7 Sv (1 Sv = 1×10^6 m^3 s^{-1}) (Beszczynska-Moller et al. 2012) of which a large part recirculates just north of the strait. The Atlantic water can be followed through the entire AO as a warm salty core at 300–600 m hugging the continental slopes and following deep sea ridges (Rudels et al. 2013). The Barents Sea branch with a transport of ~2 Sv (Ingvaldsen et al. 2002; Schauer et al. 2002) is cooled considerably during the passage through the relatively shallow Barents Sea, and it enters the AO as a much colder and somewhat fresher water mass when it meets the Fram Strait branch at the continental slope east of Severnaya Zemlya. The waters from the two branches then mix together during the further eastward route along the slope. After the Atlantic water has made its way through the AO via several internal circulation loops, it leaves the basin by the East Greenland current at the western side of the Fram Strait.

Water from the Pacific Ocean enters the AO through the shallow (50 m depth) Berings Strait. This inflow of about 0.8 Sv (e.g., Woodgate et al. 2012) is much fresher than the Atlantic inflow. It, therefore, makes up a relatively shallow circulation at 50–100 meters depth which is confined to the Alaskan and Canadian side of the basin and leaves partly through the Canadian Archipelago and partly through the Fram Strait.

The AO receives large amounts of freshwater, about 0.1 Sv, by the river runoff from Siberia and Alaska (Serreze et al. 2006). The river water spreads across most of the AO and makes up a 0–50 m low salinity surface layer. There is a prevailing circular flow of near surface water within the Amerasian Basin, in the so-called Beaufort Gyre, but eventually it is incorporated in the transpolar drift, which is a general drift pattern of surface waters toward the Fram Strait. Mechanical mixing due to ice motions and wind (in open areas) generates a well-mixed near surface layer (polar mixed layer, PML) which is generally shallow in summer (5–15 m) due to buoyancy supply from ice melt and deepens during winter to 20–50 m due to buoyancy loss by ice growth.

5.3.2 Air Bound Transport to the AO

The transport pattern of air masses from lower latitudes toward the AO is highly influenced by the generally cold temperature in the polar troposphere which is often referred to as the polar dome. The southward extent of the dome is defined by the Arctic front with a strong north–south temperature gradient. The horizontal extent of the polar dome has a strong seasonal variability, and it covers a much larger area during winter, down to 40 °N over Asia, while in summer it resides mostly over the AO north of 70 °N. The dome structure of isotherms is associated with three major transport routes of air masses from southern latitudes: (1) direct transport at low levels from sources inside the polar dome (which extent having a large seasonal variability), (2) rapid low-level transport starting south of the polar dome and uplift at the Arctic front and (3) ascent south of the Arctic front followed by high altitude transport or several cycles of uplift and descent, and eventually a slow final descent through the polar dome by radiative cooling (Stohl 2006).

Source regions of contaminants are dominated by high latitude Eurasia during winter and by low-level transport inside the polar dome while transport from North America and South East Asia is less important. During summer, AO is more shielded from continental sources, and there is little direct low-level transport but instead air is transported by routes 2 and 3 including descent of air from the free troposphere with no clear geographic distribution of the source areas (Hirdman et al. 2010).

The atmospheric circulation is stronger during winter resulting in a relatively short residence time of 7–9 days over the central AO of air in the lowest 100 m during winter, compared to 14–16 days during summer (Stohl 2006).

5.4 General Transport Routes of Pollutants Within the AO

Very small quantities of pollutants are spread into the environment in the AO, a pollutant to be found there has to be transported to the Arctic by one of the routes described above. It must then be persistent enough to not be broken down during this transport. A common name for many of these compounds is persistent organic pollutants (POP).

In Fig. 5.2, from Ma et al. (2016), the routes for transport of pollutants to and in the AO are displayed conceptually as well as more formally with the different storage compartments (boxes) and fluxes between them. Aquatic routes are presented above and in Fig. 5.1. For many substances, the major route is by the air from which it can reach the ocean either as wet or dry deposition.

If either water or air transport dominates, it depends, beside the mass transport in each medium, on the ability for substances to exchange between air and the aquatic system. The magnitude of this partitioning depends on air/water concentrations, physical qualities of the compound and temperature.

Dry deposition is either by sinking dry particles or by gas exchange between air and water. The gas exchange is directly dependent on the ratio of fugacities between water (f_w) and air (f_a), the fugacity ratio, FR. FR = f_w/f_a where $f_w = C_w * H'$, and $f_a = C_a * RT$, C_a = concentration in air, C_w = concentration in water and H' = Henry's law constant adjusted for temperature and salinity (Jantunen et al. 2015). Compounds will move from the compartment with the highest fugacity to the one with a lower fugacity. In Table 5.1, the H constant is given for some organochlorine pesticides (OCP) and currently used pesticides (CUP).

Wet deposition occurs when pollutants are absorbed into droplets and ice crystals and then precipitate as rain and snow. For substances transported this way, the partitioning between particles, air and snow/ice is controlled by the partitioning coefficients (K_p and K_s) (see Fig. 5.2).

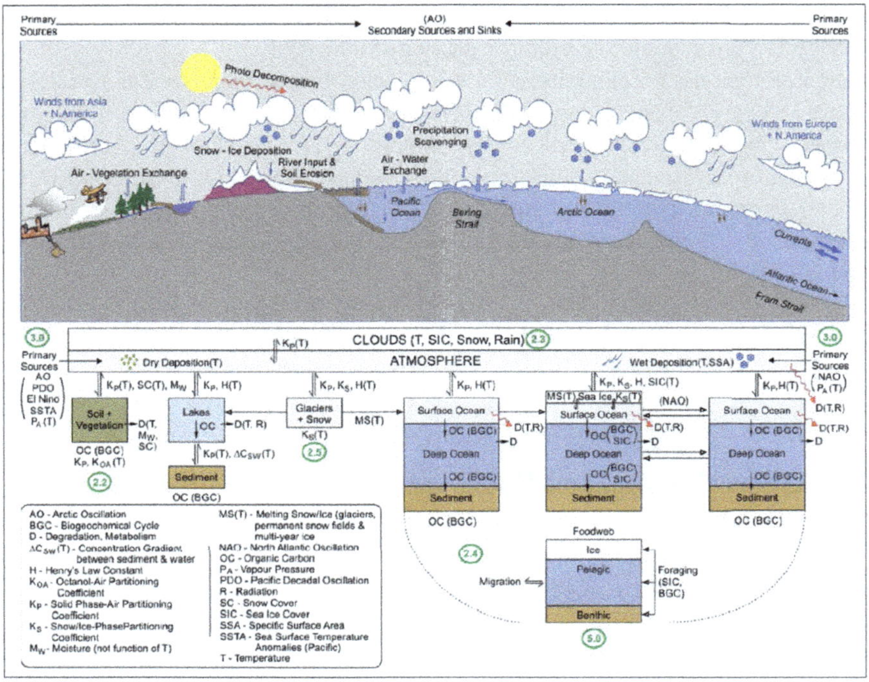

Fig. 5.2 Simplified schematic view of the Northern Hemisphere into which POPs are released (top panel). Release points for POPs lie predominantly in temperate regions from which the POPs spread out into the environment by winds and currents. A common approach to modeling represents the Earth system as a set of arrows to indicate POPs fluxes (releases, transport, losses) and boxes to indicate POPs reservoirs (air, water, soil, vegetation, ice) (bottom panel). The circled numbers in the bottom panel refer to the sections within Ma et al. (2016). *Source* Ma et al. (2016)

The fate of substances within the AO can be either degradation (chemical, physical or biological) or accumulation through biological/chemical/physical processes. They can be exported from the system either through evaporation and southward air transport or by ocean currents, where the outflow through the Fram Strait is dominating.

The uptake into the biological food web is determined by the biological concentration factor (BCF) which is the ratio between concentration in the biota and in water. For many substances, the BCF correlates with the hydrophobicity assessed as the octanol/water partitioning K_{OW}.

An example for how the partitioning between water and air transport directly determines the import into the high Arctic is shown by Li and Macdonald (2005) studying the transport of α-HCH and β-HCH through the Bering Strait. Data of concentration in surface water of these two species in the North Pacific and the western AO show that the α-HCH has similar or higher concentrations in the western AO, relative to the North Pacific, while β-HCH has significantly lower concentrations in the AO compared to south of the strait. Their explanation is that

Table 5.1 H's law constant (H) and K_{ow} constant for some organochlorine pesticides (OCP) and currently used pesticides (CUP)

		$H_{25\ °C}$ (Pa m^3 mol^{-1})	$H'_{0.5\ °C}$ (Pa m^3 mol^{-1})	K_{ow}
Legacy OCP	α-Hexachlorocyclohexane (α-HCH)	0.73	0.085	3.94
	β-Hexachlorocyclohexane (β-HCH)	0.022[a]	0.0054[a]	
	γ-Hexachlorocyclohexane (γ-HCH)	0.31	0.035	3.83
	Endosulfan I	0.71	0.203	4.94
	Trans-chlordane	6.9	1.379	6.27
	Cis-chlordane	5.8	1.238	6.2
	Trans-nonachlor	12	1.773	6.2
	Dieldrin	1.1	0.199	5.48
CUP	Dacthal	0.23	0.037	4.28
	Chlorpyrifos	0.57	0.054	4.96
	Chlorothalonil	0.025	0.004	4.81
	Trifluralin	0.005	0.002	5.34
	Pentachloronitrobenzene	7.8	1.176	4.64

Data from Pucko et al. (2017)
[a]Data from Sahsuvar et al. (2003) refering to 20 °C and 5 °C

the higher volatility air transport dominates for α-HCH, while β-HCH transport is dominated by water transport that is limited by the narrow and shallow Bering Strait.

5.5 Ice and Snow Effects

In the AO, there are special conditions for the air–sea exchange, transformation and accumulation of pollutants. The low temperature affects the stability, solubility and volatility of the compounds making them less reactive and more soluble in water.

A special feature of the AO is that it is covered by sea ice to a large degree that has a very strong influence on the partitioning of pollutants. The sea ice cover affects the transport and accumulation of pollutants by the following four mechanisms:

- When the ice is growing, pollutants in the water are rejected from the ice and enhance the concentration in the ocean surface layer.
- Formation of brine channels with enhanced concentrations during formation of first year ice (FYI).

- In melt ponds formed during summer on top of the ice floes there might be enhanced concentrations of pollutants either from melting snow or deposition from air.
- Ice is also functioning as a barrier against evasion of volatile pollutants.

When the first year ice (FYI) ice is produced, it is formed from seawater with its content of salt and pollutants. During freezing, sea salt and other substances in the water (like pollutants) are expelled out from the ice crystals and collect in brine channels, which can have very high concentrations of salt and other constituents. Garnett et al. (2019) studied how brine channels are fed by salt and POP compounds in a laboratory experiment. They found that there was a good correlation between salt (NaCl) and POPs and how they were supplied to the brine channels but that the timing depended on hydrophobicity of the POP. The more water soluble substances, as α-HCH, followed NaCl while more hydrophobic ones as BDEs were delayed. Since these channels are open to the water beneath the ice, they will increase the concentration of pollutants there. Interactions between pollutants and ice are presented schematically for chlorpyrifos in Box 5.1, but it is valid also for other dissolved pollutants. Without an ice cover, there is a development toward an equilibrium in concentration between air and water depending on physical and chemical qualities (see above). With an ice cover, this equilibrium is obstructed, and if there is a production of volatile compounds under the ice, the ocean top layer can be supersaturated up to very high relative concentrations (see for mercury below).

Box 5.1

Figure 5.3 describes a seasonal development that starts at the establishment of new ice in September-October where there is a freeze rejection of pollutants giving higher concentration in the water close to the ice. Later, when the ice cover is established, brine channels are formed which have higher pollutant concentrations in FYI than in MYI (old ice). A third process is the enrichment in melt ponds either from melting ice and snow or by gas exchange. While the two first processes are seen as uniform, independent of the physical properties of the compound, the exchange with air will be dependent on the FR between air and the pond water as described above. When the ice breaks up, or when the melt ponds drain, this enriched water will be found in a low buoyant surface layer at the oceanic top. As FYI has a higher concentration than MYI, ponds developed on FYI will have higher concentration and the export to PML from leakage, etc., will be higher than from MYI.

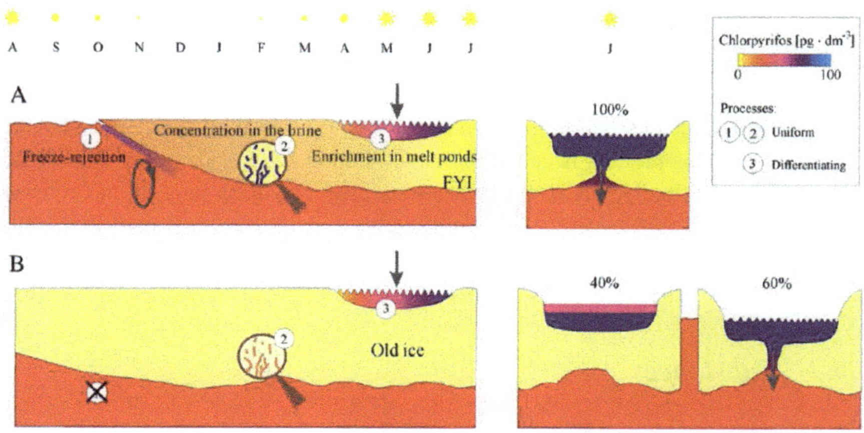

Fig. 5.3 Conceptual schematic illustrating processes leading to increased exposures to organic contaminants in FYI (A) and MYI (old ice) (B) using chlorpyrifos as an example; concentrations of chlorpyrifos in ice, brine, and interface seawater in October/November were estimated based on measured seawater concentrations and general relationships established for α-HCH; colors denote chlorpyrifos concentrations. *Source* Pucko et al. (2015) (including original references)

5.6 Pollutants in the AO and How and Where They are Observed and What is Their Fate

5.6.1 Measurements

The amount of pollutants in the environment can be determined or estimated in three ways:

- Direct measurements at various locations
- Modeling based on known physical and chemical factors combined with measurements at a specific location
- Estimated from the content in organisms or ice where compounds are accumulating.

Each of these has its advantages and disadvantages. When measuring in the field, you know exactly where and when the sample is taken. However, to be able to generalize to adequate space and timescales, samples need to be taken repeatedly at several positions and times of the year, to compensate for the strong seasonal variations described above. Modeling has the capacity to complement with higher space and time resolution than from direct sampling, as long as it gives realistic results when compared with available observations.

When dealing with persistent substances that accumulate in organisms, they can function as long-term integrated samplers and thus reduce the problem with sparse time resolution of instantaneous concentration measurement.

5.6.2 Biomagnification

Persistent substances, that are taken up into the biological food web and are bound to cell constituents and then transferred to the next trophic level, are concentrated in the food web. This process is called biomagnification. The tendency of a substance to magnify is, normally, estimated from its stability and accumulation in fat tissue. Accumulation into fat is estimated as the partitioning between octanol and water, the K_{ow} constant.

5.6.3 Substances

Here we present the anthropogenic pollutants observed in the Arctic in three groups depending on special characteristics.

- Mercury
- Legacy OCP and PCB–persistent compounds not in large use today but that are persistent and still circulating in the environment
- Currently used pesticides (CUP) and polybromated compounds

Mercury is treated as a separate group since this is one of the most hazardous compounds for the Arctic biota and which has been studied extensively; in addition, it undergoes complex environmental chemical transformations and transports. Other substances do not undergo the same transformations but are transported between compartments in a similar way such that those patterns we see for mercury are also relevant for other substances.

5.6.3.1 Mercury

Mercury has four dissolved species with different qualities: elemental mercury (Hg^0), inorganic Mercury (Hg^{2+}), monomethylated (MeHg) and dimethylated mercury (Me_2Hg) (Fig. 5.4). Of these, Hg^0 and Me_2Hg are the most volatile and are in exchange with air, while especially MeHg is taken up in biological food web and is biomagnified. Transformations between these forms are both biotic and abiotic, and some of these are phototransformations, depending on visible light and ultraviolet radiation. The environment is divided into four compartments: air, PML, deep ocean and benthos. Modifying factors are wind, ocean currents, precipitation (rain and snow) and uptake in biological communities. These factors, together with the physical qualities of the different pollutants including reactivity, determine the distribution and accumulation of mercury in the AO.

To quantify these transformations and concentrations, Soerensen et al. (2016) made a mass budget of mercury divided into PML and subsurface ocean. The net

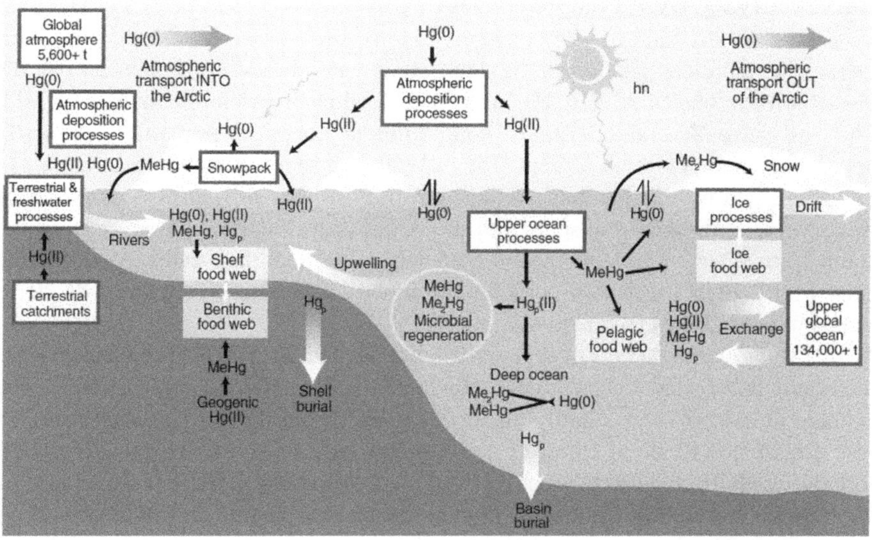

Fig. 5.4 Schematic diagram of major pathways and transformations in the Hg cycle for Arctic marine ecosystems. Large shaded arrows refer to atmospheric or oceanic transport pathways that exchange Hg with the global environment and move Hg between reservoirs within the Arctic Ocean. Small black arrows refer to biogeochemical processes that transform Hg into different species. *Source* Braune et al. (2015)

balance then shows that there is a net export of mercury in the aquatic phase, to the North Atlantic (both Hg and MeHg) of 15 ton y^{-1} (35 ton out and 20 ton in) The balance toward air is a net deposition of 24 ton y^{-1} (38 ton deposition and 14 ton evasion). The largest pool of mercury is Hg^{2+} (353 ton), but there is substantial transformation between it and the other species, especially in the PML that opens up for evasion of the volatile Hg^0. Transformation to the organic forms, which leads toward accumulation in the food web by MeHg and evasion by Me_2Hg, is larger at depth than in PML.

Included in the budget by Soerensen et al. (2016) is an input of mercury by rivers, in total 52.5 ton y^{-1}. Zolkos et al. (2020) have made a more detailed study of the contributions from rivers (2012–2017) and found a slightly lower amount (37 ton y^{-1}) where the river Lena dominated with an input of 6.6 ton y^{-1}.

As shown in this example, there are transformations to volatile species that have the potential to interact with the atmosphere. One aspect not included is the function of ice cover as a blockage against evasion (see above). An example of this is shown in (Anderson et al. 2008) from an expedition crossing the AO. The degree of oversaturation was generally on average 410%, due to methylation processes in the water, and could reach 1800% and did enhance during ice passages.

5.6.3.2 Legacy OCP Compounds and PCB

Legacy compounds included in this study are hexachlorobenzene (HCB), hexachlorocyclohexanes (HCH) and dichlorodiphenyltrichloroethanes (DDT). They all have several congeners with different chemical properties, primarily depending on the chlorine content. They are all hydrophobic, so they accumulate in organisms and have some, but varying, volatility so that air transport is a possible route for long-distance transport into the Arctic from the production areas in the south (see above). The atmospheric transport is the dominant pathway to the AO except for β-HCH that is transported primarily by ocean currents (Routti et al. 2019; Carlsson et al. 2018) (see above).

In a study in the AO north of Svalbard, uptake and fate of HCHs (α and γ) were measured, and a mass balance was established (Galbán-Maloagón et al. 2013). The average air content was slightly higher for α-HCH, but the net air/water transport was similar (-0.19 vs -0.18 ng m^{-2} d^{-1}, minus sign indicates flux towards water). Relative α-HCH concentration in the water was lower for γ-HCH (on average 0.4 vs 1 pg L^{-1}), and the content in phytoplankton was higher (on average 1.5 vs 0.5 ng g^{-1}). The lower concentration in the water for γ-HCH was explained by calculated larger removals from the water column through the biological pump and degradation.

An overview of DDT congeners (Σ_6DDT) concentration in Arctic water, sampled on three expeditions (AO-01, Beringa-05 and ISS-08), is presented in Carrizo et al. (2017). The variation in the polar mixed layer ranged from 0.10 to 66 pg L^{-1} with the highest values around Svalbard (Fig. 5.5a). The surface water was dominated by 4,4'-DDE. In the Makarov and Amundsen basins, the concentration increased with depth down to 2500 m reaching a concentration of 0.5 pg L^{-1}.

During the same expeditions, PCB congeners were measured (Fig. 5.5b) also showing low concentrations in the central AO but in contrast to DDTs, where the highest concentrations were found around Svalbard, higher concentrations for PCBs were found at the Siberian shelf. Carlsson et al. (2018) pointed out that trichlorinated PCB is an important fraction (50%) in the Chukchi and Beaufort seas, indicating that atmospheric deposition dominates while hexachlorinated are more important in the Greenland and Barents seas indicating waterborne transports.

Carlsson et al. (2018) describe the result from the ArcRisk project (www.arcrisk.eu) using PCB congeners as sentinels for pollution of the Arctic, and it has shown a continuous decrease over the past decades in the air as measured on the Zeppeline station on Svalbard. Deposition (wet and dry) from the atmosphere is the dominant process for transport of PCB into the marine Arctic but also rivers serve as an important source for transport of PCB to the Arctic.

5.6.3.3 Currently Used Pesticides and Brominated Compounds

There are several pollutants that are currently used and are persistent enough to reach the AO. One such group is what is called currently used pesticides Σ_8CUP

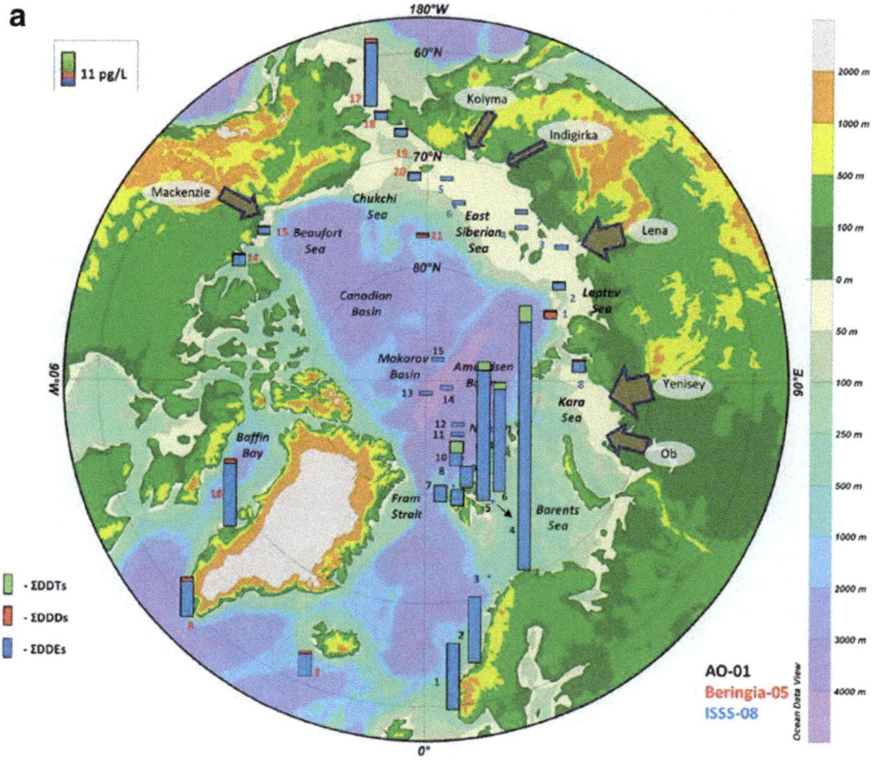

Fig. 5.5 a–c Concentrations of Σ_6DDT (**a**), Σ_{13}PCB (**b**) and Σ_{14}PBDE (**c**), measured during three expeditions. Major river inflows are indicated as arrows. *Sources* **a** Carrizo et al. (2017); **b** Carrizo and Gustafsson (2011); **c** Salvado et al. (2016)

(chloroneb, simazine, atrazine, alachlor, dacthal, chlorobenzilate, methoxychlor and permethrin). Another group is the polybrominated diphenyl ethers (PBDE) used as flame retardants. A thorough review of these and other new substances is given in AMAP (2017).

Gao et al. (2019) presented data on concentration of Σ_8CUP in water and air in the North Pacific, Chukchi Sea and Beaufort Sea in 2016 and 2017. The Σ_8CUP had a mean concentration in the water of 12.0 ± 7.61 ng/L with a dominance of chloroneb contributing >80% of the total. The atmospheric CUP composition differed from that in water. Chloroneb was still dominating in the air, but two CUPs that were abundant in seawater, alachlor and atrazin, were nearly absent in the air samples.

Concentrations of polybrominated diphenyl ethers (PBDE) in the AO from the same three expeditions listed above are given in Fig. 5.5c. Σ_{14}PBDE in the polar mixed layer ranged from 0.3 to 11.2 pg/L where the higher values were close to the outlet from the Lena and McKensey rivers and around Svalbard. There was a quite high concentration in the central AO, in the Beaufort gyre, a location for which no

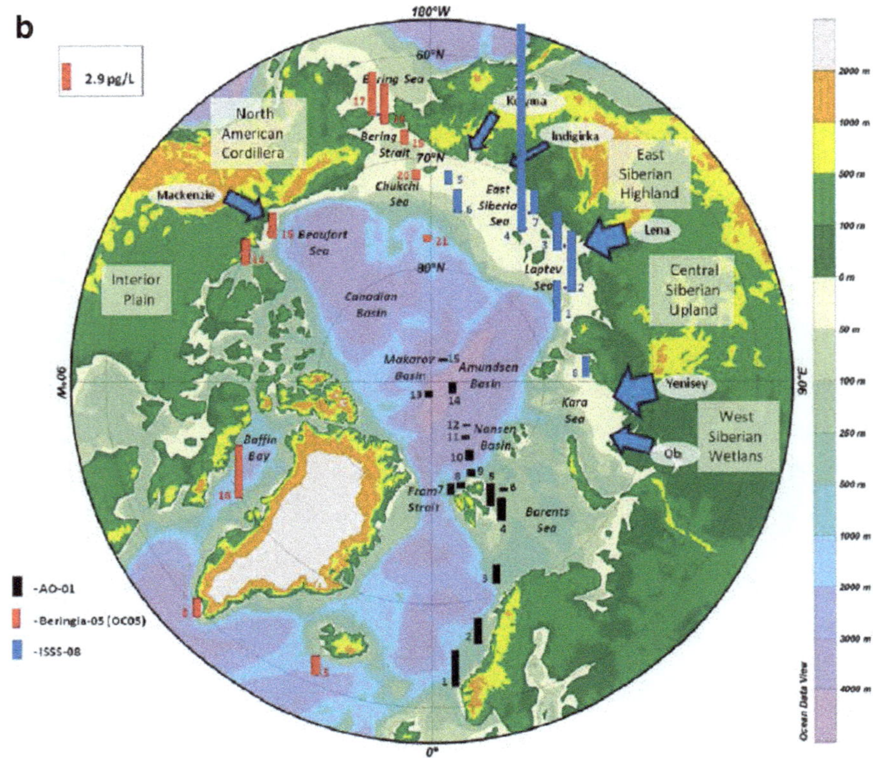

Fig. 5.5 (continued)

data are presented for PCBs and DDT in Fig. 5.5. BDE-209 was the dominating congener in most of the samples. In deep water, PBDE concentrations were up to ten times higher than in the PML. The depth-dependent variation in concentration shows that most congeners have higher concentrations in the deep water than in the Atlantic layer and even less in the PML. One explanation for this is that they are transported on sinking particles. The high brominated congener BDE-209 breaks this pattern which is explained by its hydrophobic and nonvolatile qualities. BDE-47 and BDE-99 are taken up easily by marine organisms than BDE-209 which explain their higher concentration in the deep water.

As a summary of the dispersal of pollutants in the AO, we see important factors being:

- The aquatic and aerial transport routes for pollutants to reach the AO.
- Physical qualities of the compounds that give the partitioning between air and water transport routes and their uptake and accumulation in organisms.
- Ice is an important factor since it accumulates pollutants, restricts the water/air exchange and also serves as an internal transport route in the AO, in addition to the water and air transport, as the ice cover is constantly in motion.

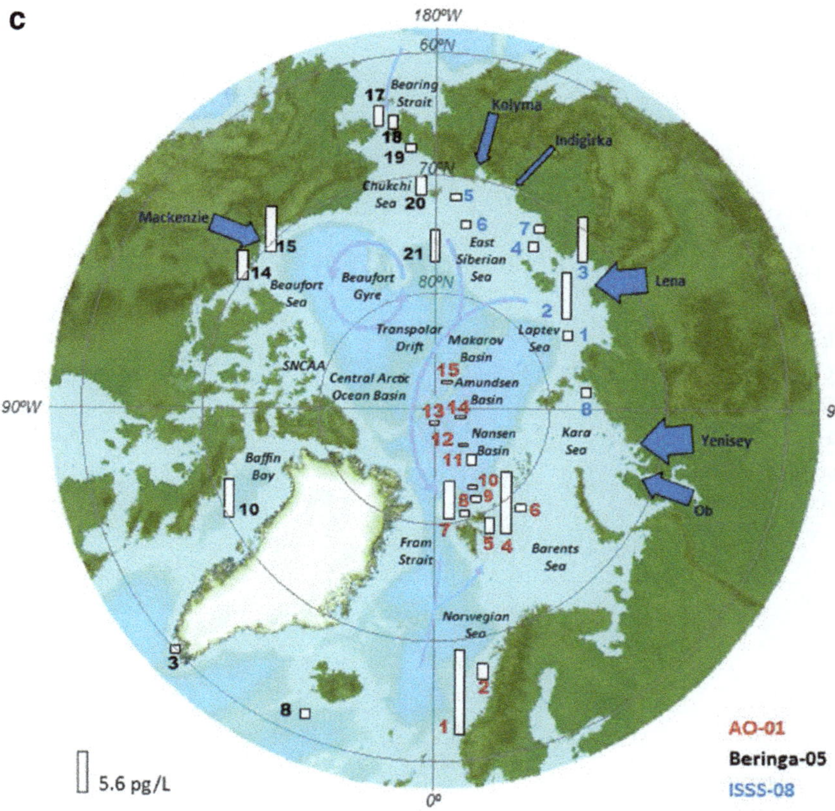

Fig. 5.5 (continued)

5.7 Pollutants in the Biota and Their Effects

All of the substances presented above have the potential to affect biological systems as described in Dietz et al. (2019) and Routti et al. (2019). Using concentration data to identify consequences by pollutants on the individual organism or ecosystem is difficult in most geographical areas, but in the high Arctic it is even more challenging due to the sparseness of observations, regarding both biological and chemical samples. For substances with a potential to biomagnify (see above), the concentration in the environment has not, in most cases, a direct impact but the critical factor is the concentration accumulated in the organism. Following this, Dietz et al. (2019) calculated a risk quotient (RQ) as the body residue/critical body residue (BR/CBR) ratio. CBR is based on the results from experimental systems where the needed dose is determined to have effects on several physiological processes. They did this calculation on a population of mammals (polar bears, killer whale, beluga, pilot whale) and marine birds (gull, puffin, murre/guillemot, fulmar,

Table 5.2 Pollution load in organisms at different trophic levels and calculated biomagnification factors

	Polar cod	Ringed seal	Polar bear	BMF	
	ng (g lipid)$^{-1}$	ng (g lipid)$^{-1}$	ng (g lipid)$^{-1}$	seal/ cod	Bear/ seal
Aldrin	n.a.	0.2	70		389
Chlordanes	20	160	1080	80	6.8
DDTs	50	206	1,006	4.1	0.5
Dieldrin	8.7	39	133	4.5	3.4
Endosulfan	2.9	0.14	8.1	0.05	58
Endrin	n.a.	0.4	8.0		20
HBCDs	3.1	7.6	4.8	2.5	0.6
HCB	11	7.5	92	0.68	12
HCHs	5.1	82	253	16	3.1
Heptachlor	0.02	0.23	2	11	8.7
Heptachlor-hepoxide	4.3	37	139	8.6	3.8
Mirex	52	3.9	7.4	0.08	1.9
PBDE	4.3	6.6	24	1.5	3.7
ΣPBDE	4				
PCB	29	197	4741	16	24
Σ_{10} PCB	22	618	2782		
Σ_7 PCDD	n.a.	0.008	0.044	6.6	
Σ_7 PCDD	n.a.	0.006	0.0035		
Σ_{10} PCDF	n.a.	0.020	0.012		0.6
Σ_{10} PCDF	n.a.	0.003	0.0019		
PCN	n.a.	0.16	4.3		27
PCP	n.a.	1	1		1
PFOA	0.17	1	25	5.9	25
PFOS	1.5	35	1182	23	34
Toxaophene		85	43		0.5

n.a. = data not available. Data from Villa et al. 2017

cormorant, sea ducks and shorebirds). They identified that polar bears analyzed from the Beaufort Sea were at high risk from mercury. When doing the same type of analysis for PCB-mediated effects, enhanced risk for mammals and seabirds was found in the north Atlantic but not in the AO (the number of observations is, however, small).

That high risk could not be identified by the RQ technique for many of the substances and populations should not be taken as a certification that these populations are not affected by the pollutants as both the BR and CBR values have uncertainties. The BR has been determined for only a small number of animals and individual variation can be substantial depending on location, sex and sampling time of the year. The CBR is estimated from many different physiological criteria, which might be based on measurements on other species that are weighted together which all are, and have to be, from experimental situations that might differ from the conditions in the field.

To investigate the tendency to biomagnify, Villa et al. (2017) compiled data from several sources on the concentrations in polar cod, ringed seal and polar bear data as a hypothetical food web (Cod > Seal > Bear) (Table 5.2). From this, they calculated the biomagnification factor (BMF) for a range of compounds. It confirms that for many of them, we see strong BMF, especially for the biocides PCB and PFOS.

Table 5.2

5.8 Pollution in the Arctic Ocean in the Future

The AO is in a rapidly changing mode as taking part, and even amplifying, the ongoing global climate change showing increasing temperature, reduction of the ice cover (see above) and changing ocean and atmospheric circulation, as presented in detail by the work of IPCC (www.ipcc.ch/). For pollutants, these changes will have importance as both the partitioning and stability of substances are controlled by various temperature-dependent parameters.

5.8.1 Time Series and Climate Change Effects on Uptake and Effects of Pollutants

Riget et al. (2019) summarized the trends of POP in the biota for Canada. Arctic Greenland, Norway, Iceland, Svalbard and Alaska show a strong negative trend for most substances, but not all, especially not the new brominated and fluorinated compounds (Fig. 5.6).

A support for decreased concentrations of POP substances was shown in Bolton et al. (2020), where ΣHCB, ΣHCH, ΣCHLs, dieldrin, ΣDDTs, ΣPCBs and ΣPBDEs were measured in blubber and muscle of bowhead whales landed at northern Alaska (Barrow) from 2006 until 2015 and compared with data from 1992 to 1993 presented by O'Hara et al. (1999). A significant ($p < 0.05$) reduction in concentrations of ΣHCH, ΣCHLs and ΣDDTs in blubber from male bowhead whales was seen, while for the others the decrease was not significant.

Other direct pollutant effects are that we will see new substances coming, and some other will decrease depending on the global production rates and changes in wind and ocean currents.

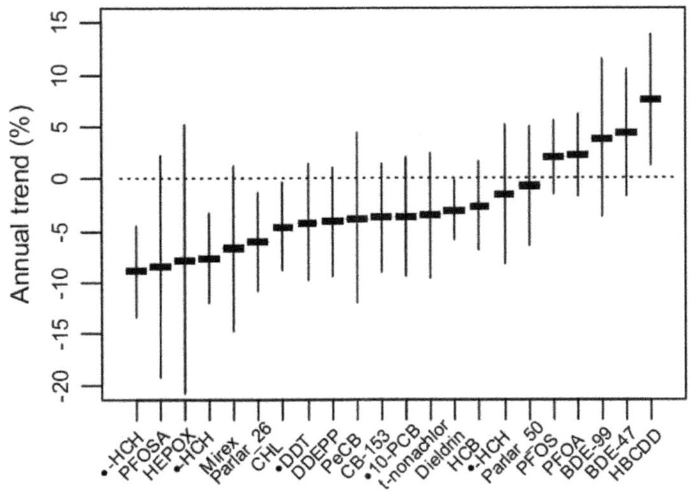

Fig. 5.6 Mean of annual change (±SD) of POPs from 1074 long-term Arctic time-series of compounds by increasing order. *Source* Riget et al. (2019)

5.8.2 Increased Exploration of the AO

The opening of the AO that follows from reduced ice cover will introduce new anthropogenic pollutants through increased shipping for transport, tourism and fishing. These include both the risks for effects from air pollution, oil spills and vessel discharges (Stevenson et al. 2019). Especially, oil spills can have large effects in the Arctic as the fate and transport of oil in ice-covered water are special as shown in Afenyo et al. (2016). In both of these papers, recommendations for how to reduce the risk and regulatory approaches are presented in Kirchner and Kleemola-Juntunen (2018).

5.9 Conclusions

The AO might seem to be far away without larger towns and industrial activities that usually are connected with high pollution and should therefore be spared from pollution problems. As we see in this overview, this is not entirely true. Many pollutants are persistent enough to be transported by wind or water to the AO. Cold conditions reduce the breakdown rate, and specific processes associated with the ice cover support the accumulation of pollutants. We have seen that many major pollutant groups are, despite a limited number of observations, identified in the environment and magnified in food webs and might reach that high concentration that the organisms are at risk.

More information about pollution and other factors, in the Arctic, are found in an easily assessed gateway to the presentations given by the Arctic Monitoring and Assessment Program (AMAP). On their Web site, www.amap.no, you find a large number of publications dealing with many aspects of the observations in the Arctic, including pollution. Very much new information will come within short time from two large programs: the MOSAIC (https://mosaic-expedition.org/) and the Synoptic Arctic Survey (https://synopticarcticsurvey.w.uib.no/).

References

Afenyo M, Veitch B, Khan F (2016) A state-of-the-art review of fate and transport of oil spills in open and ice-covered water. Ocean Eng 119:233–248

AMAP (2017) AMAP assessment (2016) chemicals of emerging Arctic concern. Arctic Monitoring Program (AMAP), Oslo, Norway. ISBN 978-82-7971-104-9

AMAP (2018) AMAP assessment (2018) biological effects of contaminants on Arctic wildlife and fish. Arctic Monitoring Program (AMAP), Tromsö, Norway ISBN 978-82-7971-106-3

Andersson ME, Sommar J, Gardfeldt Lindqvist O (2008) Enhanced concentrations of dissolved gaseous mercury in the surface waters of the Arctic Ocean. Mar Chem 110:190–194

Beszczynska-Moller A, Fahrbach E, Schauer U, Hansen E (2012) Variability in Atlantic water temperature and transport at the entrance to the Arctic Ocean, 1997–2010. ICES J Mar Sci 69:852–863

Bolton JL, Ylitalo GM, Chittaro P, George JC, Suydam R, Person BT, Gates JB, Baugh KA, Sformo T, Stimmelmayr R (2020) Multi-year assessment (2006–2015) of persistent organic pollutant concentrations in blubber and muscle from Western Arctic bowhead whales *(Balaena mysticetus)*, North Slope, Alaska. Mar Pollut Bull 151(2020):110857

Braune B, Chételat J, Marc A, Brown T, Clayden M, Evans M, Fisk A, Gaden A, Girard C, Hare A, Kirk J, Lehnherr I, Letcher R, Loseto L, Macdonald R, Mann E, McMeans B, Muir D, O'Driscoll N, Poulain A, Reimer K, Stern G (2015) Mercury in the marine environment of the Canadian Arctic: review of recent findings. Sci Total Environ 509–510:67–90

Carlsson P, Breivik K, Brorstrom-Lunden E, Cousins I, Christensen J, Grimalt JO, Halsall C, Kallenborn R, Abass K, Lammel G, Munthe J, MacLeod M, Odland JO, Pawlak J, Rautio A, Reiersen LO, Schlabach M, Stemmler I, Wilson S, Wohrnschimmel H (2018) Polychlorinated biphenyls (PCBs) as sentinels for the elucidation of Arctic environmental change processes: a comprehensive review combined with ArcRisk project results. Environ Sci Pollut Res 25:22499–22528

Carmack E, Polyakov I, Padman L, Fer I, Hunke E, Hutchings J, Jackson J, Kelley D, Kwok R, Layton C, Melling H, Perovich D, Persson O, Ruddick B, Timmermans ML, Toole J, Ross T, Vavrus S, Winsor P (2015) Toward quantifying the increasing role of oceanic heat in sea ice loss in the new Arctic. Bull Am Meteor Soc 96:2079–2105

Carrizo D, Gustafsson O (2011) Pan-Arctic river fluxes of polychlorinated biphenyls. Environ Sci Technol 45:8377–8384

Carrizo D, Sobek A, Salvado JA, Gustafsson O (2017) Spatial distributions of DDTs in the water masses of the Arctic Ocean. Environ Sci Technol 51:7913–7919

Dietz R, Letcher RJ, Desforges JP, Eulaers I, Sonne C, Wilson S, Andersen-Ranberg E, Basu N, Barst BD, Bustnes J O, Bytingsvik J, Ciesielski TM, Drevnick PE, Gabrielsen G, Haarr A, Hylland K, Jenssen BM, Levin M, McKinney MA, Norregaard RD, Pedersen KE, Provencher J, Styrishave B, Tartu S, Aars J, Ackerman J T, Rosing-Asvid A, Barrett R, Bignert A, Borns EW, Branigan M, Braune B, Bryan CE, Dam M, Eagles-Smith CA, Evans M, Evans TJ, Fisk AT, Gamberg M, Gustavson K, Hartman CA, Helander B, Herzog MP,

Hoekstra PF, Houde M, Hoydal K, Jackson AK, Kucklick J, Lie E, Loseto L, Mallory ML, Miljeteig C, Mosbech A, Muir DC G, Nielsen ST, Peacock E, Pedro S, Peterson SH, Polder A, Riget FF, Roach P, Saunes H, Sinding MHS, Skaare JU, Sondergaard J, Stenson G, Stern G, Treu G, Schuur SS Vikingsson G (2019) Current state of knowledge on biological effects from contaminants on arctic wildlife and fish. Sci Total Environ 696

Galban-Malagon CJ, Berrojalbiz N, Gioia R, Dachs J (2013) The "Degradative" and "Biological" pumps controls on the atmospheric deposition and sequestration of hexachlorocyclohexanes and hexachlorobenzene in the north Atlantic and Arctic Oceans. Environ Sci Technol 47:7195–7203

Gao Y, Zheng HY, Xia YY, Chen M, Meng XZ, Cai MH (2019) Spatial distributions and seasonal changes of current-use pesticides from the north Pacific to the Arctic Oceans. J Geophys Res-Atmos 124:9716–9729

Garnett J, Halsall C, Thomas M, France J, Kaiser J, Graf C, Leeson A, Wynn P (2019) Mechanistic insight into the uptake and fate of persistent organic pollutants in sea ice. Environ Sci Technol 53:6757–6764

Hirdman D, Sodemann H, Eckhardt S, Burkhart JF, Jefferson A, Mefford T, Quinn PK, Sharma S, Strom J, Stohl A (2010) Source identification of short-lived air pollutants in the Arctic using statistical analysis of measurement data and particle dispersion model output. Atmos Chem Phys 10:669–693

Ingvaldsen R, Loeng H, Asplin L (2002) Variability in the Atlantic inflow to the Barents Sea based on a one-year time series from moored current meters. Cont Shelf Res 22:505–519

Jakobsson M, Mayer L, Coakley B, Dowdeswell JA, Forbes S, Fridman B, Hodnesdal H, Noormets R, Pedersen R, Rebesco M, Schenke HW, Zarayskaya Y, Accettella D, Armstrong A, Anderson RM, Bienhoff P, Camerlenghi A, Church I, Edwards M, Gardner JV, Hall JK, Hell B, Hestvik O, Kristoffersen Y, Marcussen C, Mohammad R, Mosher D, Nghiem SV, Pedrosa MT, Travaglini PG, Weatherall P (2012) The international bathymetric chart of the Arctic Ocean (IBCAO) Version 3.0. Geophys Res Lett 39

Jantunen LM, Wong F, Gawor A, Kylin H, Helm PA, Stern GA, Strachan WMJ, Burniston DA, Bidleman TF (2015) 20 years of air-water gas exchange observations for pesticides in the Western Arctic Ocean. Environ Sci Technol 49:13844–13852

Kirchner S, Kleemola-Juntunen P (2018) Dumping and oil pollution: regulatory approaches for vessel operations in an ice-free Central Arctic Ocean. Rev Eur Comp Int Environ Law 27:28–34

Kwok R (2018) Arctic sea ice thickness, volume, and multiyear ice coverage: losses and coupled variability (1958–2018). Environ Res Lett 13

Li YF, Macdonald RW (2005) Sources and pathways of selected organochlorine pesticides to the Arctic and the effect of pathway divergence on HCH trends in biota: a review. Sci Total Environ 342:87–106

Ma JM, Hung H, Macdonald RW (2016) The influence of global climate change on the environmental fate of persistent organic pollutants: a review with emphasis on the Northern Hemisphere and the Arctic as a receptor. Global Planet Change 146:89–108

Mckinney MA, Pedro S, Dietz R, Sonne C, Fisk AT, Roy D, Jenssen BM, Letcher RJ (2015) A review of ecological impacts of global climate change on persistent organic pollutant and mercury pathways and exposures in Arctic marine ecosystems. Current Zool 61:617–628

O'hara TM, Krahn MM, Boyd D, Becker PR, Philo LM, Philo LM (1999) Organochlorine contaminant levels in Eskimo harvested bowhead whales of Arctic Alaska. J Wildl Dis 35:741–752

Pucko M, Stern GA, Burt AE, Jantunen LM, Bidleman TF, Macdonald RW, Barber DG, Geilfus NX, Rysgaard S (2017) Current use pesticide and legacy organochlorine pesticide dynamics at the ocean-sea ice-atmosphere interface in resolute passage, Canadian Arctic, during winter-summer transition. Sci Total Environ 580:1460–1469

Pucko M, Stern GA, Macdonald RW, Jantunen LM, Bidleman TF, Wong F, Barber DG, Rysgaard S (2015) The delivery of organic contaminants to the Arctic food web: Why sea ice matters. Sci Total Environ 506:444–452

Riget F, Bignert A, Braune B, Dam M, Dietz R, Evans M, Green N, Gunnlaugsdottir H, Hoydal KS, Kucklick J, Letcher R, Muir D, Schuur S, Sonne C, Stern G, Tomy G, VorkampV Wilson S (2019) Temporal trends of persistent organic pollutants in Arctic marine and freshwater biota. Sci Total Environ 649:99–110

Routti H, Atwood TC, Bechshoft T, Boltunov A, Ciesielski TM, Desforges JP, Dietz R, Gabrielsen G, Jenssen BM, Letcher RJ, McKinney MA, Morris AD, Riget FF, Sonne C, Styrishave B, Tartu S (2019) State of knowledge on current exposure, fate and potential health effects of contaminants in polar bears from the circumpolar Arctic. Sci Total Environ 664:1063–1083

Rudels B, Schauer U, Bjork G, Korhonen M, Pisarev S, Rabe B, Wisotzki A (2013) Observations of water masses and circulation with focus on the Eurasian Basin of the Arctic Ocean from the 1990s to the late 2000s. Ocean Sci 9:147–169

Sahsuvar L, Helm PA, Jantunen LM, Bidleman TF (2003) Henry's law constants for alpha-, beta-, and gamma-hexachlorocyclohexanes (HCHs) as a function of temperature and revised estimates of gas exchange in Arctic regions. Atmos Environ 37:983–992

Salvado JA, Sobek A, Carrizo D, Gustafsson O (2016) Observation-based assessment of PBDE loads in Arctic Ocean Waters. Environ Sci Technol 50:2236–2245

Schauer U, Loeng H, Rudels B, Ozhigin VK, Dieck W (2002) Atlantic water flow through the Barents and Kara Seas. Deep-Sea Res Part I-Oceanogr Res Pap 49:2281–2298

Serreze MC, Barrett AP, Slater AG, Woodgate RA, Aagaard K, Lammers RB, Steele M, Moritz R, Meredith M, Lee CM (2006) The large-scale freshwater cycle of the Arctic. J Geophys Res-Oceans 111

Soerensen AL, Jacob DJ, Schartup AT, Fisher JA, Lehnherr I, St Louis VL, Heimburger LE, Sonke JE, Krabbenhoft DP, Sunderland EM (2016) A mass budget for mercury and methylmercury in the Arctic Ocean. Global Biogeochem Cycles 30:560–575

Spreen G, Kwok R, Menemenlis D (2011) Trends in Arctic sea ice drift and role of wind forcing: 1992–2009. Geophys Res Lett 38

Stevenson TC, Davies J, Huntington HP, Sheard W (2019) An examination of trans-Arctic vessel routing in the Central Arctic Ocean. Marine Policy 100:83–89

Stohl A (2006) Characteristics of atmospheric transport into the Arctic troposphere. J Geophys Res-Atmos 111

Villa S, Migliorati S, Monti GS, Holoubek I, Vighi M (2017) Risk of pop mixtures on the Arctic food chain. Environ Toxicol Chem 36:1181–1192

Woodgate RA, Weingartner TJ, Lindsay R (2012) Observed increases in Bering Strait oceanic fluxes from the Pacific to the Arctic from 2001 to 2011 and their impacts on the Arctic Ocean water column. Geophy Res Lett 39

Zolkos S, Krabbenhoft DP, Suslova Tank SE, McClelland JW, Spencer RGM, Shiklomanov A, Zhulidov AV, Gurtovaya T, Zimov N, Zimov S, Mutter EA, Kutny L, Amos E, Holmes RM (2020) Mercury export from Arctic great rivers. Environ Sci Technol 54:4140–4148

Chapter 6
Contamination of Coral Reefs in the Mexican Caribbean

Anastazia T. Banaszak

Abstract This chapter describes the importance of coral reef ecosystems, particularly those of the Mexican Caribbean and the causes of coral reef degradation. Ironically, the explosion in tourism, while good for the economy, is devastating the natural resources in the region. The focus of the chapter is on the decrease in water quality as an overarching problem affecting Mexican Caribbean coral reefs, which are situated close to the coastline where tourism-associated growth has expanded unsustainably. Wastewater effluents are not adequately treated and eventually reach the sea through the underground aquifer system. The resulting contamination reduces water quality in coastal waters leading to the degradation of the once oligotrophic coral reef ecosystem over time. In recent years, contamination has been exacerbated by *Sargassum* blooms that accumulate on the coastline and decompose, leading to a further reduction in water quality. The biological effects of the different components of wastewater discharge, particularly, freshwater, nutrients, pathogens and sunscreens are detrimental to corals and can lead to mortality, diseases and bleaching. The synergistic effects of poor water quality, due to unsustainable growth associated with tourism and *Sargassum* blooms, with stressors related to climate change, will intensify coral reef ecosystem degradation by decreasing resilience to changes in the environment.

Keywords Ecosystem degradation · Karst hydrogeology · Mesoamerican Reef System · Reef insurance policy · Unsustainable tourism

---

A. T. Banaszak (✉)
Integrative Reef Conservation Research Laboratory, Unidad Académica de Sistemas Arrecifales, Instituto de Ciencias del Mar y Limnología, Universidad Nacional Autónoma de México, Puerto Morelos, Mexico
e-mail: banaszak@cmarl.unam.mx

© The Author(s), under exclusive license to Springer Nature Switzerland AG 2021

D.-P. Häder et al. (eds.), *Anthropogenic Pollution of Aquatic Ecosystems*,
https://doi.org/10.1007/978-3-030-75602-4_6

6.1 Importance of Coral Reef Ecosystems

The world's tropical coral reef ecosystems occupy approximately 255,000 km^2 (Spalding and Grenfell 1997) or about 0.09% of the total area occupied by the world's oceans (Spalding et al. 2001). A conservative estimate of the species richness associated with coral reefs globally is 830,000 species, meaning that these ecosystems provide habitats to approximately 25% of marine species (Fisher et al. 2015). Thus, these habitats are important in maintaining biodiversity. Coral reefs also provide protection to coastal communities and associated ecosystems by absorbing wave energy and reducing the impact of storms and hurricanes by up to 97%, thereby considerably reducing beach erosion and protecting coastal developments (Ferrario et al. 2014).

In recognition of the value of coral reefs, the state government of Quintana Roo insured Mexican Caribbean reefs against hurricane damage, the first insurance policy to protect a natural living structure in the world (https://www.nytimes.com/2020/12/05/climate/Mexico-reef-climate-change.html; accessed 5th December, 2020). The insurance policy paid off after Hurricane Delta made landfall in Puerto Morelos on October 7, 2020, with peak wind gusts of 175 km/h. The coral reef ecosystem sustained major damage because it absorbed wave energy generated by the storm, and that energy did not reach the coastline. Thus, the coral reef ecosystem protected the beaches from erosion and the coastal infrastructure, such as the luxury hotels built along the length of the Mexican Caribbean, from damage.

Coral reef ecosystems are also important for fisheries and tourism (Spalding et al. 2017). Tourism associated with coral reefs is one of the fastest-growing industries (Ong and Musa 2011). Taking into consideration coral reef-related tourism activities alone, a conservative estimate of the global value of coral reef services is $35.8 billion USD annually, thus providing an important source of income to many countries (Spalding et al. 2017) particularly low income and small island nations (Spalding et al. 2001).

6.2 Tourism in the Mexican Caribbean

Sun, sea and sand tourism is a key economic driver in the state of Quintana Roo, Mexico (Padilla 2015; Mata-Lara et al. 2018; Chávez et al. 2020), which along its entire length borders the Caribbean Sea and the northern section of the Mesoamerican Reef. Quintana Roo has received approximately 10 million tourists per year over the last four decades (Rioja-Nieto and Álvarez-Filip 2019) with 17 million tourists arriving in 2017 alone (Sedetur 2017). The development of the city of Cancun, in the northern part of Quintana Roo, initiated during the late 1960s and early 1970s, specifically to boost the Mexican economy through tourism. Since then, development has expanded along the entire coast northward reaching Isla Blanca and to the south through the Riviera Maya to Costa Maya (Murray 2007;

Pérez-Gómez et al. 2020). During the 1990s, a corridor of hotels was developed over a 130 km stretch of coast from Cancun to Playa del Carmen and Tulum (Aguilar-Maldonado et al. 2018). The projection for the number of hotel rooms by 2027 for this region is to reach 130,000 according to the Quintana Roo Tourism Minister (https://www.riviera-maya-news.com/cancun-riviera-maya-lead-way-as-state-surpasses-100000-hotel-rooms/2018.html; data retrieved on 28th November 2020). This is a 26-fold increase from the 4,918 hotel rooms available in 1998 (Murray 2007).

To provide services to the mass international and domestic tourism market, urban development and population growth have expanded rapidly along the coastal fringe of the Mexican Caribbean in just over a 50-year period. For example, the city of Cancun grew from 150 inhabitants in 1970 (Rioja-Nieto et al. 2019) to an estimated 943,922 inhabitants in 2020 (http://worldpopulationreview.com/world-cities/cancun-population/; data retrieved on 28th November 2020); a greater than 6000-fold increase in five decades.

One of the major drawcards of tourism to coral reefs is the transparent, turquoise-colored water and white sandy beaches characteristic of coral reef ecosystems (Chávez et al. 2020). The waters surrounding coral reefs are crystal clear because they are oligotrophic. This lack of nutrients inhibits algal and phytoplankton growth, which allows corals to settle and grow. The symbiotic relationship formed between corals and dinoflagellate microalgae involves recycling of nutrients and waste products between them, thus corals can be productive while living in a nutrient-poor environment (Muscatine and Porter 1977), which is the basis of the success of coral reef ecosystems. On the other hand, the white sand beaches, characteristic of the Mexican Caribbean, are the result of grazing by parrot fish and sea urchins (Ogden 1977).

While the influx of tourists to coral reef areas is considered positive for the economy of countries that border these ecosystems, with many being strongly dependent on direct and indirect income from tourism, this situation places a great deal of pressure on the environment. The fast rate of increase in tourism-associated infrastructure, along with the urban expansion required to build and staff hotels, particularly when not undertaken in a sustainable manner, brings with it numerous significant pressures on the coastal environment and its natural resources (Murray 2007; Padilla 2015). Ironically, although the tourism industry relies exclusively on the natural environment as the main attraction for visitors, it needs to alter that same environment to build the infrastructure to accommodate the massive influx of tourists and supply services to support the tourism industry.

6.3 Environmental Effects of Tourism in the Mexican Caribbean

As tourism-associated development has expanded along the Quintana Roo coast, negative impacts on the coastal ecosystems in the region are being increasingly documented (Camacho-Cruz et al. 2020; Rioja-Nieto et al. 2019; Pérez-Gómez et al. 2020). The major effects on coral reefs of development to support the tourism industry are the deforestation of mangroves and the increased output of wastewater effluents, causing eutrophication of a previously nutrient-poor system.

Mangrove cover declined by almost 30% over a 30-year period (1981–2010) along the heavily urbanized Riviera Maya corridor (Brenner et al. 2018). This is despite specific federal legislation to protect mangroves from activity that may cause their deterioration and prohibit contamination of these ecosystems by wastewater effluents (NOM-022-SEMARNAT-2003). One of the many functions of intact mangrove forests is to naturally purify water of nutrients and contaminants (Brenner et al. 2018) before it reaches the coral reef ecosystem.

Increased tourism brings with it higher outputs of anthropogenically derived pollutants such as wastewater effluents (Duarte et al. 2008; De'ath and Fabricius 2010; Burke et al. 2011; Wear and Thurber 2015). Due to the unique hydrogeology of the entire Yucatan Peninsula within which the state of Quintana Roo is located, all freshwater and any contaminants that are released into it are carried through an underground system that reaches the coast and the bordering coral reef ecosystem.

6.4 Hydrogeology of the Mexican Caribbean

A key characteristic of the Yucatan Peninsula, which along its eastern side borders the Mexican Caribbean, is its unique hydrogeology that consists of highly permeable limestone formations called karst topography (Rioja-Nieto et al. 2019). These limestone deposits have weathered and dissolved over time occasioning the formation of a network of caverns, dissolution cavities and channels that conduct water through an underground, rather than an aboveground river system, considered to be one of the most extensive in the world (Chávez et al. 2020). When this system reaches the surface, the resulting features are known locally as "cenotes" or sink holes (Reddell 1981) and as submarine springs, known locally as "ojos de agua" or blue holes (Null et al. 2014). The Yucatán Peninsula contains anywhere between 7,000 and 8,000 sink holes (Beddows et al. 2007).

The karst topography of the Yucatan Peninsula allows for rapid filtration of precipitation into the underground aquifer, which then transports the groundwater toward the coasts. For the entire region, including the state of Quintana Roo, this system is vital because it is the major source of freshwater (Polanco Rodríguez et al. 2017; Camacho-Cruz et al. 2020). Another complicating issue is that as groundwater is extracted, seawater can intrude and increase the salinity of the freshwater (Saint-Loup et al 2018).

The consequence of such a unique interconnected underground system of water transport is that it is vulnerable to inputs of contaminants, such as wastewater effluents, pesticides, heavy metals, fertilizers and microbes. Freshwater discharge through the underground aquifer system has been identified as an important source of nutrients to many coastal ecosystems including coral reefs in the area (Smith 2003; Null et al. 2014; Hernández-Terrones et al. 2015). Polanco-Rodríguez et al. (2014) found levels of banned pesticides that exceed limits established by Mexican law in twenty sink holes in the Yucatan Peninsula.

The most important impact of tourism-related development along the Quintana Roo coast is the reduction in the quality of water in the Yucatan Peninsula, principally due to a combination of a lack of planning for the inclusion of infrastructure associated with adequate drainage and sewerage networks combined with the suboptimal operation of wastewater treatment plants (Padilla 2015; Aguilar-Maldonado et al. 2018; Camacho-Cruz et al. 2020; Pérez-Gómez et al. 2020). An absence of regulations to protect the underground aquifer and coral reefs on the Caribbean coast of the Yucatan Peninsula has resulted in the contamination of the karst-based aquifer (Mutchler et al. 2010; Leal-Bautista et al. 2013; Sánchez et al. 2013; Hernández-Terrones et al. 2015).

Land-based sediment, nutrients, pollutants and contaminants usually enter coastal environments via surface waters such as rivers and streams (van Dam et al. 2011) as runoff resulting from agricultural, industrial and urban activity. However, given the geology of the Yucatan Peninsula, these components are absorbed into the underground aquifer and transported toward the coastal ecosystems. The inadequate processing by wastewater treatment plants results in most wastewater effluents being released in a partially treated or untreated form (UNEP 1994; Wear and Thurber 2015). Approximately 95% of wastewater is generated by the hotel industry with the remainder composed of household, municipal and industrial effluents (Padilla 2015). The wastewater treatment plants that do exist along the coast of Quintana Roo do not have the capacity to deal with the volume of wastewater produced such that it results in leakage or overflow and eventually is incorporated into the underground aquifer (Hernández-Terrones et al. 2011, 2015). Other factors that contribute to reduced water quality include leaking or overflow from aged household septic tanks into the karstic aquifer and recreational vehicles discharging wastewater into coastal waters (Padilla 2015).

A common practice is to inject wastewater effluents 90 m or more below the freshwater lens. However, given the karst topography, the effluents can travel through the underground aquifer system, and where the aquifer is connected hydraulically with the sea, these are released into coastal waters, thus depositing nutrients such as nitrogen (N) and phosphorus (P) into an oligotrophic system (Camacho-Cruz et al. 2020). Therefore, wastewater effluent is directly released through seeps and blue holes into the ocean close to the coral reefs that are part of the Mesoamerican Reef system as has been detected using radium isotopes (Hernández-Terrones et al. 2015).

The combination of pressures of increased tourism such as coastal developments for tourism, eutrophication and reduction of water quality is related to the degradation of coral reefs (Duarte et al. 2008; De'ath and Fabricius 2010; Burke et al. 2011; Wear and Thurber 2015). The inadequately treated, or worse, untreated wastewater effluents reach the Mexican Caribbean in due course. The degree to which these discharges impact coral reefs of the Mesoamerican Reef will depend on the adequacy of the level of wastewater treatment prior to discharge, the periodicity and the volume of the discharge as well as the predominant currents and circulation patterns governing the effluent sites (Pastorok and Bilyard 1985). When effluent drainage is continuous, high volume and in sites with poor circulation, the effects are greater (Pastorok and Bilyard 1985). In the case of the Mexican Caribbean where coral reefs are situated near the coastline, sometimes as close as 400 m distance, they are more easily impacted by effluent discharges (Häder et al. 2020). The repercussion of a contaminated underground transport system that can reach coastal environments, as is found in the Yucatan Peninsula, is the gradual degradation of the bordering coral reef ecosystem.

6.5 Degradation of the Mexican Caribbean Coral Reef Ecosystem

Caribbean coral reef ecosystems are undergoing unprecedented losses in coral cover, estimated at an 80% reduction over three decades (Gardner et al. 2003), while reef habitat degradation has been estimated at 1.5% annually from 1977 to 2001 (Jackson et al. 2014). Factors related to anthropogenic activities are driving this degradation. Rising sea surface temperatures due to climate change have provoked mass coral bleaching events (Hughes et al. 2017). Coral diseases such as White Band Disease (WBD, Aronson and Precht 2001) and more recently Scleractinian Coral Tissue Loss Disease (SCTLD, Álvarez-Filip et al. 2019) have caused major losses in coral cover. Indeed, SCTLD has spread along the Mexican Caribbean within a matter of months affecting 24 out of the 46 species recorded in this northern region of the Mesoamerican Reef system (Álvarez-Filip et al. 2019). Eleven of the affected species have been categorized as highly susceptible to the disease (Álvarez-Filip et al. 2019) and may face local extinction.

As a result of their high nutrient content, wastewater effluents have caused eutrophication of the oligotrophic (nutrient-poor) waters surrounding coral reefs (Null et al. 2014; Hernández-Terrones et al. 2015; Pérez-Gómez et al. 2020). By enhancing cyanobacterial and algal growth, these outcompete scleractinian corals through inhibition of larval recruitment, coral growth and survival over the short term (Fabricius 2005; Vega Thurber et al. 2014; Wagner et al. 2010). In the long term, eutrophication has provoked phase shifts from a scleractinian coral-dominated ecosystem to a macroalgal-dominated one (Mutchler et al. 2007; Bruno et al. 2009;

Martínez-Rendis et al. 2016; Suchley et al. 2016) and is related to the loss in coral and seagrass cover over the last two decades.

6.6 Composition of Wastewater Effluents in the Mexican Caribbean

Wastewater effluents are composed of freshwater, endocrine disruptors, sediments, nutrients and pathogens (Pastorok and Bilyard 1985; Wear and Thurber 2015). In a study undertaken by analyzing a series of aquifers that feed into coastal waters near Mexican Caribbean coral reef ecosystems, Metcalfe et al. (2011) found contaminants that included caffeine, pharmaceuticals, illicit drugs and personal care products, derived from domestic wastewater. Some of these components can be lethal to scleractinian corals (van Dam et al. 2011) or can have negative effects on processes such as reproduction, photosynthesis, calcification and growth (van Dam et al. 2011). For example, freshwater kills corals if exposed to reduced salinity for up to 24 h (Jokiel et al. 1993), whereas endocrine disruptors decrease coral growth rates and reproductive viability of corals (Pastorok and Bilyard 1985). Community and ecosystem level impacts in coral cover and accretion rates have also been found to decrease (Fabricius 2005) due to the presence of these contaminants.

6.6.1 Nutrients

Eutrophication, due to an excess of nutrients in the water column, favors the growth of cyanobacteria and fleshy macroalgae and inhibits settlement and growth of scleractinian corals. Wastewater effluents maintain a high nutrient load when the wastewater treatment process is inadequate (Mutchler et al. 2007, 2010) and are carried through submarine groundwater discharges via blue holes and seeps reaching the waters surrounding coral reefs (Rodríguez-Martínez et al. 2010).

Wastewater effluents are enriched with $\delta 15$ N and have been shown to become incorporated into the tissues of macroalgae, seagrasses and gorgonians, as well as skeletons of scleractinian corals (Sánchez et al. 2013; González-De Zayas et al. 2020). The nitrogen isotope ratio ($\delta 15$ N) in the seagrass *Thalassia testudinum*, gorgonians and fleshy macroalgae in coastal waters off Quintana Roo indicates that the source of N contamination is wastewater pollution from anthropogenic sources (Carruthers et al. 2005; Mutchler et al. 2007; Baker et al. 2010; González-De Zayas et al. 2020). Seasonal records show that $\delta 15$ N is present year-round (Camacho-Cruz et al. 2020). Such nutrient inputs are generally characterized by N:P ratios higher than 16, which are the optimal Redfield ratios for algal proliferation in coastal areas (Aguilar-Maldonado et al. 2018).

6.6.2 Pathogens

Pathogens including enteric bacteria have been detected in drinking water sources (Zarate Lomelí et al. 1999) and seawater close to shore in various sites along the Mexican Caribbean (Munro et al. 1999; Saint-Loup et al. 2018). The presence of pathogens has been related to the discharge of anthropogenically derived organic matter in wastewater effluents released into or close to coral reef ecosystems (Kaczmarsky et al. 2005; Castañeda-Chávez et al. 2018). The elkhorn coral *Acropora palmata*, one of the most important reef-building species in the Caribbean, which is on the red list of the International Union for Conservation of Nature (IUCN), is affected by White Pox Disease (Rodriguez-Martinez et al. 2001). White Pox Disease is caused by *Serratia marcescens* a human enterobacterium found in feces (Sutherland et al. 2010).

Thermal stress and eutrophication exacerbate the effect of pathogens on corals (Bruno et al. 2003; Maynard et al. 2015). Maynard et al. (2015) compared the potential rise in the frequency of disease outbreaks in species of reef-building corals using projections of temperature increases under moderate and high fossil fuel emissions. Their projections indicate that coral susceptibility to disease, pathogen abundance and pathogen virulence will increase and aggravate coral bleaching. Eutrophication has been shown to significantly increase the prevalence of Yellow Band Disease in the Caribbean reef-building mountainous star corals *Orbicella* (=*Montastraea*) *annularis* and *O. franksi* and of Aspergillosis in the Caribbean Sea fan *Gorgonia ventalina* (Bruno et al. 2003).

6.6.3 Other Components of Wastewater Effluents

Fertilizers, pesticides and herbicides are applied to golf courses and altered landscapes along the length of the Mexican Caribbean, especially in the hotel zones of Cancun and the Riviera Maya. These substances are also used in agriculture in the Yucatan Peninsula and can enter the underground aquifers and subsequently be released near coral reefs of the Mexican Caribbean.

Any of these substances have detrimental effects on corals including mortality, bleaching, tissue retraction and inhibition of photosynthesis in coral symbionts as well as reduced reproductive success through effects on fecundity, fertilization rates and metamorphosis of larvae from the free swimming to the benthic stage of the coral life cycle, yet are common in waters circulating near coral reefs (Lewis et al. 2009; van Dam et al. 2011).

6.7 Sunscreens

Sunscreens made from organic (chemical) UV filters are another component that can be found in effluents from wastewater treatment plants as they are not easily removed during processing (Schneider and Lim 2019).

The most probable sources of the components of commercially available sunscreen in the coral reef environment are due to wash off from human skin during aquatic activities (Sánchez-Quiles and Tovar-Sánchez 2015; Bell et al. 2017), as well as from effluents of wastewater treatment plants due to inadequate processing (Schneider and Lim 2019). An estimated range of between 6,000 and 14,000 tons of sunscreen is released annually into coral reef ecosystems (Downs et al. 2016). For the state of Quintana Roo, Casas-Beltran et al. (2020) estimated that an average of 243 tons of sunscreen is washed off annually into the waters that surround, or that reach, the coral reef ecosystem. This figure is based on the average time a tourist stays in Quintana Roo, the proportion of visitors that apply sunscreen and engage in aquatic activities as well as the wash off into the drainage system and the input by local residents.

Effects of sunscreens and their components on corals range from changed behavior (McCoshum et al. 2016), impaired swimming and deformed coral larvae (Downs et al. 2014, 2016), to genotoxic effects (Downs et al. 2016; He et al. 2019), as well as coral bleaching and mortality (Danovaro et al. 2008; Downs et al. 2009, 2016; DiNardo and Downs 2018), which can potentially affect the whole coral reef ecosystem (Downs et al. 2016). Uncoated zinc oxide nanoparticles, components of some inorganic sunscreens, have also been found to cause coral bleaching (Corinaldesi et al. 2018). Most of the studies are laboratory-based using high concentrations of the different components of organic sunscreens; therefore, the ecological relevance of the results has been questioned.

Despite this, Hawaii, Bonaire and Palau have banned the distribution and sale of any sunscreens that have oxybenzone (benzophenone-3, BP-3) and/or octinoxate (ethylhexyl methoxycinnamate, EHMC) as ingredients due to the potential negative impacts on coral reef ecosystems (Schneider and Lim 2019; Narla and Lim 2020). This is a particularly important issue for human health because the use of sunscreen prevents excessive exposure to ultraviolet radiation. UV-A (315–400 nm) and UV-B (280–315 nm) wavelengths can cause keratinocyte cancers and melanomas, induce eye damage, suppress immune systems and cause photoaging and sunburn (Pathak 1987; Narbutt et al. 2019; Narla and Lim 2020). This is especially important for tourists visiting coral reefs because UVR levels are high due to the combination of a naturally thin ozone layer, low solar zenith angles and crystal clear waters that transmit these wavelengths (Banaszak and Lesser 2009).

To protect humans from the potentially lethal effects of excess exposure to ultraviolet radiation while protecting and conserving coral reef ecosystems from damage requires the incorporation of alternative photoprotective ingredients such as UV-absorbing mycosporine-like amino acids (Chrapusta et al. 2017; Lawrence et al. 2018; Rosic 2019) found in many aquatic species (Banaszak 2003).

6.8 *Sargassum* Blooms

Localized sources of nutrients exacerbate outbreaks of *Sargassum* blooms, which have caused major problems throughout the Caribbean since 2011 (Oviatt et al. 2019; Wang et al. 2019). Pelagic *Sargassum*, which can form large floating rafts, functions as habitat and a refuge for a diverse array of biota (Fine 1970; Butler et al. 1983). In a survey of the motile macrofauna associated with *Sargassum* rafts, Monroy-Velazquez et al. (2019) found 32 species from eight phyla. Once these *Sargassum* rafts beach (Fig. 6.1), they start to decompose on the shoreline creating a brown tide (van Tussenbroek et al. 2017) for up to hundreds of meters offshore (Rodriguez-Martinez et al. 2019). Consequently, the brown discoloration of the water (Fig. 6.1) provokes reduced light conditions, changes in oxygen (anoxia and hypoxia) and pH resulting in mortality of fauna and flora associated with near-shore seagrass meadows including corals (van Tussenbroek et al. 2017; Chávez et al. 2020) and demersal neritic fish and crustaceans, due to hypoxic conditions and high ammonium and hydrogen sulfide concentrations (Rodriguez-Martinez et al. 2019). δ15 N values in the seagrass *Thalassia testudinum* are similar enough to the values in *Sargassum,*; hence, Pérez-Gómez et al. (2020) proposed that a significant proportion of the nitrogen taken up and assimilated by *T. testudinum* comes from decomposition of the beached *Sargassum*, indicating that *Sargassum* is an important source of nitrogen to the environment.

Heavy machinery has been used to remove beached *Sargassum* (Fig. 6.1) which can amount to 7,000 kg of per meter of shoreline (Salter et al. 2019; Chávez et al. 2020). Disposal of such large quantities of decaying *Sargassum* has been a major issue, and most has been disposed of on beaches, in mangroves and abandoned quarries (Chávez et al. 2020). No preventative measures, such as the use of geomembranes, have been put in place to stop degradation products, such as nutrients, salt, metals and other contaminants from leaching into the underground aquifer system (Chávez et al. 2020) including arsenic (Devault et al. 2021). This major source of contamination not only threatens the underground freshwater supply but also the coral reef ecosystem due to the connection with the underground network, thus deteriorating the coral reef ecosystem even further. The northern section of the Mesoamerican Reef system borders the state of Quintana Roo with the coral reef, in some places, separated from the coastline by a relatively narrow reef lagoon (Jordán-Dahlgren and Rodríguez-Martínez 2003). Due to the proximity of coral reefs to the coastline and the delivery of contaminants carried by the aquifer network freshwater through blue holes and seeps, these contaminants are destined to eventually reach the coral reef ecosystem (Null et al. 2014; Hernández-Terrones et al. 2015).

Fig. 6.1 Aerial view of heavy machinery removing *Sargassum* that has beached in Puerto Morelos, Mexico and caused brown discoloration and reduction in quality of the water column. Photo by Edgar Escalante Mancera and Miguel Angel Gomez Reali

6.9 Concluding Remarks

Chronic contamination of coral reefs in the Mexican Caribbean is due to poor water quality as a result of the unsustainable mass tourism model that has been applied in the state of Quintana Roo, exacerbated by the relatively recent *Sargassum* blooms that decay, thus intensifying the impact of other stressors on the ecosystem. In synergy with climate change stressors, such as higher sea surface temperatures, this contamination will aggravate the degradation of the coral reef ecosystem through decreased resilience of its component biota to thermal stress, diseases and ocean acidification.

To improve water quality and maintain white sandy beaches, the multiple point and non-point sources of contamination must be controlled because the mass tourism model is likely to continue given the projected future growth of tourism for the state of Quintana Roo. Strict management actions must be implemented to prevent further degradation, not only of the coral reef ecosystem through eutrophication, but also the potential negative effects on human health given that the groundwater system is used for human consumption. As tourism and the associated urban population increase, so does the need to dispose of wastewater into the underground aquifer, which could further contaminate the only freshwater supply used for human consumption. Greater contamination could result not only in the catastrophic breakdown of the Mexican Caribbean coral reef ecosystem, but also in risks to human health. Laws to ensure the tertiary level treatment of wastewater must be adopted, monitored and enforced to ensure a risk-free drinking water supply for humans and contaminant-free drainage through the aquifer to coastal environments.

The value of coral reef ecosystems in the Mexican Caribbean has been clearly recognized by authorities, but it is not enough to take out insurance policies to defray the costs of restoring coral reefs after hurricane damage. Water quality must be improved through investment in infrastructure to ensure adequate wastewater treatment. Only then reefs will be resilient enough to survive the impact of climate change.

Meanwhile, the tourism industry must understand the urgent need to protect the very environment that attracts visitors to the region. Greater emphasis should be placed on developing sustainable tourism projects, establishing a realistic carrying capacity for the industry in the region and implementing practices that protect the natural environment, both terrestrial and marine. This will better ensure the longevity of the tourist destinations in the Mexican Caribbean while reducing impacts on the natural resources, which, at the same time, are the region´s greatest attraction.

Acknowledgements Funding was provided by the Instituto de Ciencias del Mar y Limnología, Project 608. M. C. Laura Celis aided with the literature search. Helpful discussions with Sarah Connor improved the manuscript.

References

Aguilar-Maldonado JA, Santamaría-Del-Ángel E, González-Silvera A, Cervantes-Rosas OD, Sebastiá-Frasquet M-T (2018) Mapping satellite inherent optical properties index in coastal waters of the Yucatán Peninsula (Mexico). Sustainability 10:1894

Álvarez-Filip L, Estrada-Saldívar N, Pérez-Cervantes E, Molina-Hernández A, González-Barrios FJ (2019) A rapid spread of the stony coral tissue loss disease outbreak in the Mexican Caribbean. PeerJ 7:e8069

Aronson RB, Precht WF (2001) White-band disease and the changing face of Caribbean coral reefs. Hydrobiologia 460(1):25–38

Baker MD, Jordán-Dahlgren E, Maldonado MA, Harvell CD (2010) Sea fan corals provide a stable isotope baseline for assessing sewage pollution in the Mexican Caribbean. Limnol Oceanogr 55(5):2139–2149

Banaszak AT (2003) Photoprotective physiological and biochemical responses of aquatic organisms. In: Helbling EW, Zagarese H (eds) UV effects in aquatic organisms and ecosystems. Häder D-P, Jori G (Series eds) Comprehensive series in photochemical and photobiological sciences. The Royal Society of Chemistry, pp 329–356

Banaszak AT, Lesser MP (2009) Effects of solar ultraviolet radiation on coral reef organisms. Photochem Photobiol Sci 8(9):1276–1294

Beddows P, Blanchon P, Escobar E, Torres-Talamante O (2007) Los cenotes de la península de Yucatán. Arqueología Mexicana 83:32–35
Bell LJ, Ucharm G, Patris S, Diaz-Cruz MS, Roig MPS, Dawson MN (2017) Final report sunscreen pollution in Jellyfish Lake. Coral Reef Research Foundation Palau, p 26. https://coralreefpalau.org/wp-content/uploads/2017/10/CRRF-UNESCO-Sunscreen-in-Jellyfish-Lake-no.2732.pdf. Accessed 28 Nov 2020
Brenner L, Engelbauer M, Job H (2018) Mitigating tourism-driven impacts on mangroves in Cancún and the Riviera Maya, Mexico: an evaluation of conservation policy strategies and environmental planning instruments. J Coastal Conserv 22:755–767
Bruno FJ, Peters EL, Harvell DC, Hettinger A (2003) Nutrient enrichment can increase the severity of coral diseases. Ecol Lett 6(12):1056–1061
Bruno JF, Sweatman H, Precht WF, Selig ER, Schutte VGW (2009) Assessing evidence of phase shifts from coral to macroalgal dominance on coral reefs. Ecology 90(6):1478–1484
Burke L, Reytar K, Spalding M, Perry A (2011) Reefs at risk revisited. World Resources Institute, Washington DC, p 114
Butler JN, Morris F, Cadwallader J, Stoner AW (1983) Studies of *Sargassum* and the *Sargassum* community. Bermuda Biological Station, special publication no 22
Camacho-Cruz KA, Ortiz-Hernández Sánchez A, Carrillo L, Navarette ADJ (2020) Water quality in the eastern karst region of the Yucatan Peninsula: nutrients and stable nitrogen isotopes in turtle grass, *Thalassia testudinum*. Environ Sci Pollut Res 27(14):15967–15983
Carruthers TJB, van Tussenbroek BI, Dennison WC (2005) Influence of submarine springs and wastewater on nutrient dynamics of Caribbean seagrass meadows. Estuar Coast Shelf Sci 64:191–199
Casas-Beltran DA, Hernández-Pedraza M, Alvarado-Flores J (2020) Estimation of the discharge of sunscreens in aquatic environments of the Mexican Caribbean. Environments 7:15
Castañeda-Chávez MR, Lango-Reynoso F, García-Fuentes JL, Reyes-Aguilar AR (2018) Bacteria that affects (sic) coral health with an emphasis on the Gulf of Mexico and the Caribbean Sea. Latin Am J Aquatic Res 46(5):880–889
Chávez V, Uribe-Martínez A, Cuevas E, Rodríguez-Martínez RE, van Tussenbroek BI, Francisco V, Estévez M, Celis LB, Monroy-Velázquez LV, Leal-Bautista R, Álvarez-Filip L, García-Sánchez M, Masia L, Silva R (2020) Massive influx of pelagic *Sargassum* spp. on the coasts of the Mexican Caribbean 2014–2020: challenges and opportunities. Water 12 (10):2908
Chrapusta E, Kaminski A, Duchnik K, Bober B, Adamski M, Bialczyk J (2017) Mycosporine-like amino acids: potential health and beauty ingredients. Mar Drugs 15:326
Corinaldesi C, Marcellini F, Nepote E, Damiani E, Danovaro R (2018) Impact of inorganic UV filters contained in sunscreen products on tropical stony corals (*Acropora* spp.). Sci Total Environ 637–638:1279–1285
Danovaro R, Bongiorni L, Corinaldesi C, Giovannelli D, Damiani E, Astolfi P, Greci L, Pusceddu A (2008) Sunscreens cause coral bleaching by promoting viral infections. Environ Health Perspect 116(4):441–447
De'ath G, Fabricius K (2010) Water quality as a regional driver of coral biodiversity and macroalgae on the Great Barrier. Ecol Appl 20(3):840–850
Devault DA, Pierre R, Marfaing H, Dolique F, Lopez P-L (2021) *Sargassum* contamination and consequences for downstream uses: a review. J Appl Phycol 33:567–602
DiNardo JC, Downs CA (2018) Dermatological and environmental toxicological impact of the sunscreen ingredient oxybenzone/benzophenone-3. J Cosmet Dermatol 17(1):15–19
Downs CA, Kramarsky-Winter E, Martinez J, Kushmaro A, Woodley CM, Loya Y, Ostrander GK (2009) Symbiophagy as a mechanism for coral bleaching. Autophagy 5(2):211–216
Downs CA, Kramarsky-Winter E, Fauth JE, Segal R, Bronstein O, Jeger R, Lichtenfeld Y, Woodley CM, Pennington P, Kushmaro A, Loya Y (2014) Toxicological effects of the sunscreen UV filter, benzophenone-2, on planulae and in vitro cells of the coral, *Stylophora pistillata*. Ecotoxicology 23(2):175–191

Downs CA, Kramarsky-Winter E, Segal R, Fauth J, Knutson S, Bronstein O, Ciner FR, Jeger R, Lichtenfeld Y, Woodley CM, Pennington P, Cadenas K, Kushmaro A, Loya Y (2016) Toxicopathological effects of the sunscreen UV filter, Oxybenzone (Benzophenone-3), on coral planulae and cultured primary cells and its environmental contamination in Hawaii and the U. S. Virgin Islands. Arch Environ Contam Toxicol 70(2):265–288

Duarte CM, Dennison WC, Orth RJW, Carruthers TJB (2008) The charisma of coastal ecosystems: addressing the imbalance. Estuaries Coasts 31(2):233–238

Fabricius KE (2005) Effects of terrestrial runoff on the ecology of corals and coral reefs: review and synthesis. Mar Pollut Bull 50(2):125–146

Ferrario F, Beck M, Storlazzi C, Micheli F, Shepard CC, Airoldi L (2014) The effectiveness of coral reefs for coastal hazard risk reduction and adaptation. Nature Commun 5:3794

Fine ML (1970) Faunal variation on pelagic *Sargassum*. Mar Biol 7(2):112–122

Fisher R, O'Leary RA, Low-Choy S, Mengersen K, Knowlton N, Brainard RE, Caley MJ (2015) Species richness on coral reefs and the pursuit of convergent global estimates. Curr Biol 25 (4):500–505

Gardner TA, Cote IM, Gill JA, Grant A, Watkinson AR (2003) Long-term region-wide declines in Caribbean corals. Science 301(5635):958–960

González-De Zayas R, Rossi S, Hernández-Fernández L, Velázquez-Ochoa R, Soares M, Merino-Ibarra M, Castillo-Sandoval FS, Soto-Jiménez MF (2020) Stable isotopes used to assess pollution impacts on coastal and marine ecosystems of Cuba and México. Reg Stud Marine Sci 39:101413

Häder D-P, Banaszak AT, Villafañe VE, Narvarte MA, González RA, Helbling EW (2020) Anthropogenic pollution of aquatic ecosystems: emerging problems with global implications. Sci Total Environ 713:136586

He T, Tsui MMP, Tan CJ, Ng KY, Guo FW, Wang LH, Chen TH, Fan TY, Lam PKS, Burkhardt Murphy M (2019) Comparative toxicities of four benzophenone ultraviolet filters to two life stages of two coral species. Sci Total Environ 651(2):2391–2399

Hernández-Terrones L, Rebolledo-Vieyra M, Merino-Ibarra M, Soto M, Le-Cossec A, Monroy-Ríos E (2011) Groundwater pollution in a karstic region (NE Yucatan): baseline nutrient content and flux to coastal ecosystems. Water Air Soil Pollut 218:517–528

Hernández-Terrones LM, Null KA, Ortega-Camacho D, Paytan A (2015) Water quality assessment in the Mexican Caribbean: impacts on the coastal ecosystem. Cont Shelf Res 102:62–72

Hughes T, Kerry J, Álvarez-Noriega M, Álvarez-Romero JG, Anderson KD, Baird AH, Babcock RC, Beger M, Bellwood DR, Berkelmans R, Bridge TC, Butler IR, Byrne M, Cantin NE, Comeau S, Connolly SR, Cumming GS, Dalton SJ, Diaz-Pulido G, Eakin CM, Figueira WF, Gilmour JP, Harrison HB, Heron SF, Hoey AS, Hobbs J-PA, Hoogenboom MO, Kennedy EV, Kuo C, Lough JM, Lowe RJ, Liu G, McCulloch MT, Malcolm HA, McWilliam MJ, Pandolfi JM, Pears RJ, Pratchett MS, Schoepf V, Simpson T, Skirving WJ, Sommer B, Torda G, Wachenfeld DR, Willis BL, Wilson SK (2017) Global warming and recurrent mass bleaching of corals. Nature 543:373–377

Jackson JBC, Donovan MK, Cramer KL, Lam VV (2014) Status and trends of Caribbean coral reefs, 1970–2012. Global coral reef monitoring network, IUCN, Gland, Switzerland, p 304

Jokiel P, Hunter C, Taguchi S, Watarai L (1993) Ecological impact of a fresh-water "reef kill" in Kaneohe Bay, Oahu, Hawaii. Coral Reefs 12(3–4):177–184

Jordán-Dahlgren E, Rodríguez-Martínez RE (2003) The Atlantic coral reefs of Mexico. In: Cortés J (ed) Latin American coral reefs. Elsevier Science, Amsterdam, The Netherlands, pp 131–158

Kaczmarsky L, Draud M, Williams E (2005) Is there a relationship between proximity to sewage effluent and the prevalence of coral disease? Carib J Sci 41(1):124–137

Lawrence KP, Long PF, Young AR (2018) Mycosporine-like amino acids for skin photoprotection. Curr Med Chem 25(40):5512–5527

Leal-Bautista RM, Lenczewski M, Morgan C, Gahala A, McLain JE (2013) Assessing fecal contamination in groundwater from the Tulum region, Quintana Roo, Mexico. J Environ Protect 4(11):1272–1279

Lewis SE, Brodie JE, Bainbridge ZT, Rohde KW, Davis AM, Masters BL, Maughan M, Devlin MJ, Mueller JF, Schaffelke B (2009) Herbicides: a new threat to the Great Barrier Reef. Environ Pollut 157(8–9):2470–2484

Martínez-Rendis A, Acosta González A, Hernández-Stefanoni JL, Arias González JE (2016) Quantifying the reefscape transformation of a coastal Caribbean coral reef during a phase shift and the associated coastal landscape change. Mar Ecol 37(3):697–710

Mata-Lara M, Garza-Pérez JR, Aranda-Fragoso A, Afonso de Almeida PS (2018) Social alienation and environmental decline in a coral reef: challenges to coastal management in the Mexican Caribbean. Ocean Coast Manag 155:30–39

Maynard J, Van Hooidonk R, Eakin CM, Puotinen M, Garren M, Williams G, Heron SF, Lamb J, Weil E, Willis B, Harvell CD (2015) Projections of climate conditions that increase coral disease susceptibility and pathogen abundance and virulence. Nat Clim Change 5:688–694

McCoshum SM, Schlarb AM, Baum KA (2016) Direct and indirect effects of sunscreen exposure for reef biota. Hydrobiologia 776(1):139–146

Metcalfe CD, Beddows PA, Bouchot GG, Metcalfe TL, Li H, Van Lavieren H (2011) Contaminants in the coastal karst aquifer system along the Caribbean coast of the Yucatan Peninsula, Mexico. Environ Pollut 159(4):991–997

Monroy-Velazquez LV, Rodríguez-Martínez RE, van Tussenbroek BI, Aguiar T, Solis-Weiss V, Briones-Fourzan P (2019) Motile macrofauna associated with pelagic *Sargassum* in a Mexican reef lagoon. J Environ Manag 252:109650

Munro MH, Blunt JW, Dumdei EJ, Hickford SJ, Lill RE, Li S, Battershill CN, Duckworth AR (1999) The discovery and development of marine compounds with pharmaceutical potential. J Biotechnol 70(1–3):15–25

Murray G (2007) Constructing paradise: the impacts of big tourism in the Mexican coastal zone. Coastal Management 35(2–3):339–355

Muscatine L, Porter JW (1977) Reef corals: mutualistic symbioses adapted to nutrient-poor environments. Bioscience 27(7):454–460

Mutchler T, Dunton K, Townsend-Small A, Fredriksen S, Rasser M (2007) Isotopic and elemental indicators of nutrient sources and status of coastal habitats in the Caribbean Sea, Yucatan Peninsula, Mexico. Estua Coast Shelf Sci 74(3):449–457

Mutchler T, Mooney RF, Wallace S, Podsim L, Fredriksen S, Dunton KH (2010) Origins and fate of inorganic nitrogen from land to coastal ocean on the Yucatan Peninsula, Mexico. In: Kennish MJ, Paerl HW (eds) Coastal lagoons, critical habitats of environmental change. Taylor & Francis Group, Boca Raton, pp 283–305

Narbutt J, Philipsen PA, Harrison GI, Morgan KA, Lawrence KP, Baczynska KA, Grys K, Rogowski-Tylman M, Olejniczak-Staruch I, Tewari A, Bell M, O'Connor C, Wulf HC, Lesiak A, Young AR (2019) Sunscreen applied at ≥ 2 mg cm^{-2} during a sunny holiday prevents erythema, a biomarker of ultraviolet radiation-induced DNA damage and suppression of acquired immunity. Br J Dermatol 180(3):604–614

Narla S, Lim HW (2020) Sunscreen: FDA regulation, and environmental and health impact. Photochem Photobiol Sci 19(1):66–70

Null KA, Knee KL, Crook ED, de Sieyes NR, Rebolledo-Vieyra M, Hernández-Terrones L, Paytan A (2014) Composition and fluxes of submarine groundwater along the Caribbean coast of the Yucatan Peninsula. Cont Shelf Res 77:38–50

Ogden J (1977) Carbonate-sediment production by parrot fish and sea urchins on Caribbean reefs. Stud Geol 4:281–288

Ong TF, Musa G (2011) An examination of recreational divers' underwater behaviour by attitude —behaviour theories. Curr Issues Tourism 14(8):779–795

Oviatt CA, Huizenga K, Rogers CS, Miller WJ (2019) What nutrient sources support anomalous growth and the recent *Sargassum* mass stranding on Caribbean beaches? A review. Mar Pollut Bull 145:517–525

Padilla NS (2015) The environmental effects of tourism in Cancun, Mexico. Int J Environ Sci 6:282–294

Pastorok RA, Bilyard GR (1985) Effects of sewage pollution on coral-reef communities. Mar Ecol Prog Ser 21(1):175–189

Pathak M (1987) Sunscreens and their use in the preventative treatment of sunlight-induced skin damage. J Dermatol Surg Oncol 13(7):739–752

Pérez-Gómez JA, García-Mendoza E, Olivos-Ortiz A, Paytan A, Rebolledo-Vieyra M, Delgado-Pech B, Almazán-Becerril A (2020) Indicators of nutrient enrichment in coastal ecosystems of the northern Mexican Caribbean. Ecol Ind 118:106756

Polanco-Rodríguez AG, Alberto JA, Sánchez JS, Rejón GJ, Gómez JM, Del Valls Casillas TA (2014) Contamination by organochlorine pesticides in the aquifer of the Ring of Cenotes in Yucatán, México. Water Environ J 29(1):140–150

Polanco Rodríguez AG, Árcega Cabrera F, Araujo León JA, Oceguera Vargas I (2017) Perspectivas de la contaminación del acuífero en Yucatán, México: plaguicidas y metales pesados. In: Heredia Navarrete MR, Pérez Mutul J, Carrillo Espadas PI, Zavala Castro J (eds) Las ciencias biomédicas y las sociales: abriendo brecha en la Península de Yucatán 40 años después. Universidad Autónoma de Yucatán Mérida, Yucatán, México

Reddell JR (1981) A review of the cavernicole fauna of Mexico, Guatemala, and Belize. University of Texas at Austin Bulletin, Austin, Texas, Texas Memorial Museum Speleological Monographs 27:1–327

Rioja-Nieto R, Álvarez-Filip L (2019) Coral reef systems of the Mexican Caribbean: Status, recent trends and conservation. Mar Pollut Bull 140:616–625

Rioja-Nieto R, Garza-Pérez R, Álvarez-Filip L, Mariño-Tapia I, Enríquez C (2019) Chapter 27: the Mexican Caribbean: from Xcalak to Holbox. In: Sheppard C (ed) World seas: an environmental evaluation. Volume one: Europe, The Americas and West Africa, 2nd edn. Academic Press, pp 637–653

Rodríguez-Martínez RE, Banaszak AT, Jordán-Dahlgren E (2001) Necrotic patches affect *Acropora palmata* (Scleractinia: Acroporidae) in the Mexican Caribbean. Dis Aquatic Org 47 (3):229–234

Rodríguez-Martínez RE, Ruíz-Rentería F, van Tussenbroek B, Escalante-Mancera E, Jordán-Garza G, Jordán-Dahlgren E (2010) Environmental state and tendencies of the Puerto Morelos CARICOMP site, Mexico. Revista de Biología Tropical 58 (suppl.3):23–43

Rodriguez-Martinez R, Medina-Valmaseda AE, Blanchon P, Monroy-Velázquez LV, Almazán-Becerril A, Delgado-Pech B, Vásquez-Yeomans L, Francisco V, García-Rivas MC (2019) Faunal mortality associated with massive bleaching and decomposition of pelagic *Sargassum*. Mar Pollut Bull 146:201–205

Rosic NN (2019) Mycosporine-like amino acids: making the foundation for organic personalised sunscreens. Mar Drugs 17(11):638

Saint-Loup R, Felix T, Maqueda A, Schiller A, Renard P (2018) A survey of groundwater quality in Tulum region, Yucatan Peninsula, Mexico. Environ Earth Sci 77:644

Salter MA, Rodríguez-Martínez RE, Álvarez-Filip L, Jordán-Dahlgren E, Perry CT (2020) Pelagic Sargassum as an emerging vector of high rate carbonate sediment import to tropical Atlantic coastlines. Glob Planet Chang 195:103332

Sánchez A, Ortiz-Hernández MC, Talavera-Sáenz A, Aguíñiga-García S (2013) Stable nitrogen isotopes in the turtle grass *Thalassia testudinum* from the Mexican Caribbean: implications of anthropogenic development. Estuar Coast Shelf Sci 135:86–93

Sánchez-Quiles D, Tovar-Sánchez A (2015) Are sunscreens a new environmental risk associated with coastal tourism? Environ Int 83:158–170

Schneider SL, Lim HW (2019) Review of environmental effects of oxybenzone and other sunscreen active ingredients. J Am Acad Dermatol 80(1):266–271

SEDETUR (2017) Reporte Anual de Turismo Quintana Roo 2017, p 33. http://sedeturqroo.gob.mx/ARCHIVOS/REPORTE%20TURISMO%202017.pdf. Accessed on 28 Nov 2020

Smith VH (2003) Eutrophication of freshwater and coastal marine ecosystems. Environ Sci Pollut Res 10(2):126–139

Spalding M, Grenfell A (1997) New estimates of global and regional coral reef areas. Coral Reefs 16(4):225–230

Spalding MD, Ravilious C, Green EP (2001) World atlas of coral reefs, Prepared at the UNEP World Conservation Monitoring Centre. University of California Press, Berkeley, USA, p 424

Spalding M, Burke L, Wood SA, Ashpole J, Hutchison J, Zu Ermgassen P (2017) Mapping the global value and distribution of coral reef tourism. Mar Policy 82:104–113

Suchley A, McField MD, Alvarez-Filip L (2016) Rapidly increasing macroalgal cover not related to herbivorous fishes on Mesoamerican reefs. PeerJ 4:e2084

Sutherland KP, Porter JW, Turner JW, Thomas BJ, Looney EE, Luna TP, Meyers MK, Futch JC, Lipp EK (2010) Human sewage identified as likely source of white pox disease of the threatened Caribbean elkhorn coral. *Acropora palmata*. Environ Microbiol 12(5):1122–1131

UNEP (1994) Regional overview of land-based sources of pollution in the wider caribbean region. Caribbean Environment Programme Technical Report No. 33. United Nations Environment Programme, Kingston p 59

van Dam JW, Negri AP, Uthicke S, Mueller JF (2011) Chapter 9: chemical pollution on coral reefs: exposure and ecological effects. In: Sánchez-Bayo F, van den Brink PJ, Mann RM (eds) Ecological impacts of toxic chemicals. Bentham Science Publishers Ltd. pp 187–211

van Tussenbroek BI, Hernández Arana HA, Rodríguez-Martínez RE, Espinoza-Avalos J, Canizales-Flores HM, González-Godoy CE, Barba-Santos MG, Vega-Zepeda A, Collado-Vides L (2017) Severe impacts of brown tides caused by *Sargassum* spp. on near-shore Caribbean seagrass communities. Mar Pollut Bull 122(1–2):272–281

Vega Thurber RL, Burkepile DE, Fuchs C, Shantz AA, McMinds R, Zaneveld JR (2014) Chronic nutrient enrichment increases prevalence and severity of coral disease and bleaching. Glob Change Biol 20(2):544–554

Wagner DE, Kramer P, Van Woesik R (2010) Species composition, habitat, and water quality influence coral bleaching in southern Florida. Mar Ecol Prog Ser 408:65–78

Wang M, Hu C, Barnes BB, Mitchum G, Lapointe B, Montoya JP (2019) The great Atlantic *Sargassum* belt. Science 365(6448):83–87

Wear SL, Thurber RV (2015) Sewage pollution: mitigation is key for coral reef stewardship. Ann N Y Acad Sci 1355(1):15–30

Zarate Lomelí D, Saavedra Vázquez T, Rojas Galaviz JL, Yáñez-Arancibia A, Rivera E (1999) Terms of reference towards an integrated management policy in the coastal zone of the Gulf of Mexico and the Caribbean. Ocean Coast Manag 42(2):345–368

Chapter 7
Input of Terrestrial Material into Coastal Patagonian Waters and Its Effects on Phytoplankton Communities from the Chubut River Estuary (Argentina)

Juan I. Vizzo, Marco J. Cabrerizo, Virginia E. Villafañe, and E. Walter Helbling

Abstract Riverine and eolian-dust inputs to coastal waters are influenced by both rainfall and land-use (i.e., deforestation, agriculture, urbanization). Transport of sediments and nutrients from rivers to the oceans are well- documented worldwide, and it is of major management concern because coastal ecosystems provide a great variety of products and services to humankind. Episodic, but extreme rainfall or dust-storm events can generate abrupt pulsed riverine/eolian discharge events, thus increasing nutrients and sediments in surface waters. It is expected that they will be intensified in the upcoming decades by global change, but we scarcely know how they could impact on coastal planktonic communities. In this chapter we address the main terrestrial material transported via rivers and winds in Patagonian coastal waters of the South West Atlantic Ocean (SWAO) and their impact on the structure and functioning of phytoplankton communities. The rivers in Patagonia carry not only sewage from the cities that lie on their margins, but also heavy loads of nutrients, due to agricultural and cattle raising activities upstream from their mouths; extreme rainfall events increase these inputs manyfold. Eolic inputs are also important over Patagonia because this area has a strong prevalence of winds

J. I. Vizzo · V. E. Villafañe · E. W. Helbling (✉)
Estación de Fotobiología Playa Unión, Casilla de Correos 15, 9103 Rawson, Chubut, Argentina
e-mail: whelbling@efpu.org.ar

J. I. Vizzo · V. E. Villafañe · E. W. Helbling
Consejo Nacional de Investigaciones Científicas y Técnicas (CONICET), Buenos Aires, Argentina

M. J. Cabrerizo
Centro de Investigación Mariña da Universidade de Vigo (CIM-UVigo), Illa de Toralla s/n, 36331 Vigo, Spain

M. J. Cabrerizo
Departamento de Ecología y Biología Animal, Universidade de Vigo, Facultad de Ciencias del Mar, Campus Lagoas Marcosende s/n, 36310 Vigo, Spain

© The Author(s), under exclusive license to Springer Nature Switzerland AG 2021
D.-P. Häder et al. (eds.), *Anthropogenic Pollution of Aquatic Ecosystems*,
https://doi.org/10.1007/978-3-030-75602-4_7

from the west, carrying a wide variety of particles that are blown far into the ocean, especially after volcanic eruptions. We used the Chubut river estuary (Patagonia) as a reference coastal ecosystem in the SWAO to assess how riverine and eolic inputs impact on phytoplankton which are the base of one of the most productive fisheries areas of the SWAO. Overall, nutrient inputs of terrestrial origin, improved phytoplankton growth, photosynthesis performance and changed the community structure towards a dominance of mostly nanoplanktonic diatom species . These responses, however, might vary due to the interaction of terrestrial material inputs with other global change drivers such as warming, acidification, vertical mixing or solar ultraviolet radiation.

Keywords Extreme events · Precipitation · Southwest Atlantic · Wind

7.1 Introduction

The Southwest Atlantic Ocean (SWAO), located between 23°S and 55°S, is an area that includes the Brazil-Malvinas Confluence region which is characterized by the cold subantartic water that flows toward the equator (Malvinas current) and the Brazil current with a southward flow, along the continental margin of South America. A front separates these water masses that have different temperature, salinity and nutrient characteristics (Piola et al. 2018). The Patagonian coastal waters, the northernmost limit of which is $\sim$38°S within the SWAO (Piola et al. 2018), together with the slope front over the continental drop-off are a "hot spot" for primary production (Martinetto et al. 2020). They are not only highly productive areas but also provide diverse ecosystem services such as production of seafood, climate regulation, recreation and scientific potential (Longhurst et al. 1995; González et al. 2007; Heileman 2009). Along coastal Patagonia, there is high heterogeneity in abiotic factors (i.e., temperature, wind speed, solar radiation and rainfall) and a plethora of diverse habitats such as salt marshes, rocky shores and intertidal formations (Gil et al. 2019). This heterogeneity is also observed in the aquatic system, for example, evidenced by the surface chlorophyll *a* (Chl-*a*) values over coastal areas of the SWAO (Fig. 7.1). Several cities in Patagonia are located near the coast (e.g., San Antonio Oeste, Puerto Madryn, Playa Unión, Comodoro Rivadavia, Caleta Olivia and Ushuaia) or on the margin of rivers (e.g., Viedma, Trelew, Rawson and Río Gallegos). Despite the low population density of the region (<2 inhabitants km^{-2}, Census 2010 INDEC, https://www.indec.gob.ar) the coastal aquatic systems are heavily influenced by human agricultural, industrial (including fish factories) and urban activities (Gil et al. 2019).

The coastal areas of Patagonia are further affected by continental inputs (i.e., terrigenous material) carried by two main sources i.e., riverine runoff (either dissolved or suspended matter), or eolian dust, as suspended particles, blown by the strong winds characteristics of this area. Inorganic and organic materials carried by rivers and wind exert a dual role on coastal ecosystems: They bring nutrients but at

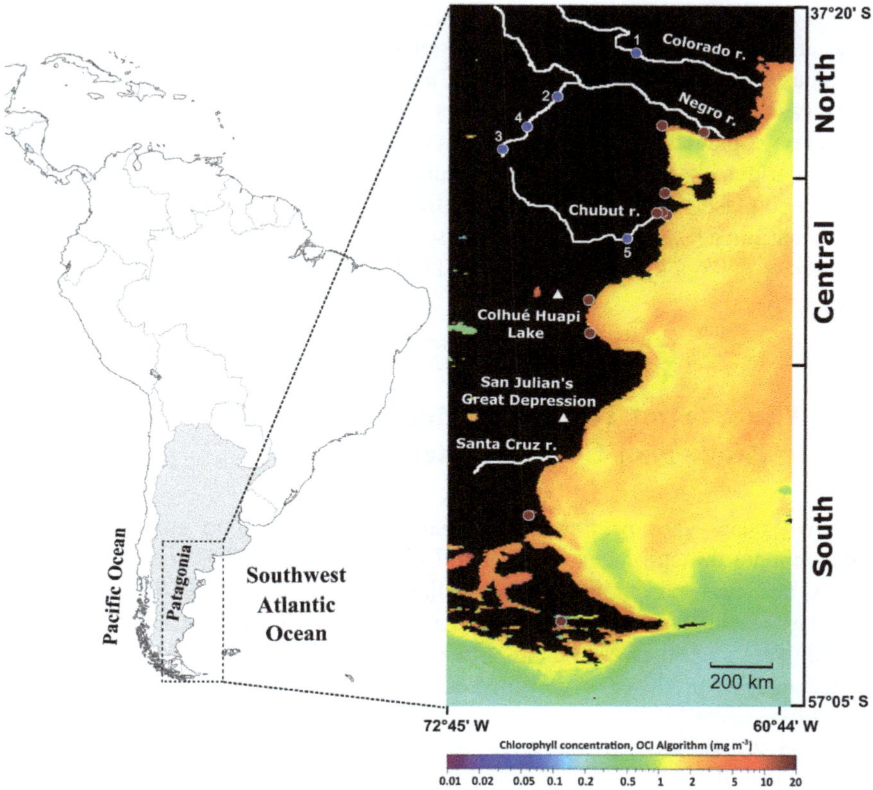

Fig. 7.1 Map of South America indicating the location of Patagonia. Enlarged is a satellite image showing mean surface Chl-*a* concentrations during the austral winter for the 2002–2019 period (data obtained from MODIS http://oceancolor.gsfc.nasa.gov; 4 km spatial resolution), location of the main rivers in the Patagonian region (North: Colorado and Negro rivers; Central: Chubut river and South: Santa Cruz river) and the main sources of dust (white triangles, Colhué Huapi lake and San Julian's Great Depression) ending in the SWAO. The blue circles with numbers indicate the main dam constructions: 1—Casa de Piedra; 2—El Chocón; 3—Alicurá; 4—Piedra del Águila; 5—Florentino Ameghino. Red circles indicate the main cities near the coast and rivers

the same time, they modify the underwater light field toward more turbid environments as they absorb solar radiation (Cloern et al. 2014), especially in the UV region, diminishing its penetration in the water column (Kirk 1994; Osburn and Morris 2003). These two effects obviously have an important impact on the metabolism and performance of organisms, especially in autotrophs such as phytoplankton, which depend on nutrients and solar radiation for photosynthesis and growth. Due to the ongoing global change—evidenced not only as a worldwide warming and acidification of water bodies (IPCC 2013) but also as local extreme rainfall events and changes in wind patterns (Stockwell et al. 2020) significant changes are occurring in coastal ecosystem services causing, for example, a decrease in fishery catches (Franco et al. 2020). In this chapter, we focus on the long-term trends and seasonal variations of riverine runoff and eolian dust inputs into coastal systems of Patagonia. Also, and using the Chubut River estuary as a study case, we address how these inputs impact the functioning and structure of coastal phytoplankton communities. We are also aware of the presence of pollutants in the terrigenous material reaching estuaries, e.g., persistent organic pollutants, metals and plastic debris, among others (Barletta et al. 2019) that may also cause a significant impact on phytoplankton; these topics, however, are addressed in other chapters of this book (e.g., Chaps. 4, 9, 13, 15); therefore, they are not considered here.

7.2 Transport of Terrestrial Material/Nutrients to Patagonian Coastal Waters

The two main ways of transport of terrigenous material to coastal waters of Patagonia are through: (1) riverine runoff (Depetris et al. 2005) and (2) wind, i.e., eolian dust transportation/deposition (Jickells and Moore 2015). Terrigenous materials reaching coastal areas by river runoff contain both particulate and dissolved matter, and they can be either inorganic (Granéli et al. 1996) and/or organic matter such as products of decomposition originated from plant tissue exudates (allochthonous; Willey et al. 2000) or released by phytoplankton (autochthonous; Thornton 2014). Most of the large particles are retained in lakes or reservoirs located along the river course, due to a decreased turbulence in these sites that allow for the settlement of heavy particles; thus, if this is the case, the input to coastal waters consists mostly of fine particles and dissolved matter (Gaiero et al. 2003). The main components of dissolved organic matter (DOM) are humic substances, which are organic compounds that form part of humus, the major organic fraction of peat, soil and coal (i.e., humic and fulvic acids; Osburn and Morris 2003).

Input of terrigenous material occurs also via the Patagonian dust originating from arid desert regions, i.e., those with rainfall less than 250 mm yr^{-1} (Prospero et al. 2002). Since Patagonia is an extended arid region, with a strong influence of westerly winds (Garreaud et al. 2013), the Patagonian dust is one of the main sources of terrigenous material to the SWAO (Johnson et al. 2011; Ginoux et al. 2012) as well to the Subantarctic Ocean (Cassar et al. 2007; Gassó and Stein 2007) and Antarctica (Li et al. 2010). The composition of Patagonian dust is mainly silicates and trace elements such as iron (Fe), aluminum (Al), sulfur (S) and magnesium (Mg) (Johnson et al. 2011; Browning et al. 2014; Crespi-Abril et al. 2018), but it also transports macronutrients, i.e., nitrate and silicic acid (Paparazzo et al. 2018). Although Patagonian dust comes from the entire Patagonian Plateau, the two main sources are the Colhué Huapi lake (45.5°S, 68 W; 90 km from the coast; Fig. 7.1) located in central Patagonia (Gassó and Torres 2019) and the San Julian's Great Depression (49°S, 69°W; Fig. 7.1). In the case of the Colhué Huapi lake, the fluctuations in the water surface leave large areas of the lake bed uncovered and exposed to erosion by the strong winds characteristic of this area (Montes et al. 2017). In the case of the San Julian's Great Depression, dry lakes contribute to the dust budget (Mazzonia and Vazquez 2009; Li et al. 2010). Patagonian dust from these sources can cover at least 420.000 km^2 over the Atlantic coast (Crespi-Abril et al. 2018); nevertheless, Saharan dust also contributes to increase the nitrogen (N) and phosphorus (P) concentrations (Brahney et al. 2015) due to the influence of easterly winds from Africa (Baker et al. 2006). Another major source of aerosol inputs to the ocean is the volcano and fire ash and soot (i.e., black carbon). Total deposition rates of volcano and fire ash are similar in some areas to those of mineral dust, and soot delivers more P into the ocean than mineral dust (Mahowald et al. 2008; Weinbauer et al. 2017). These types of depositions additionally deliver carbon (C), N, S and bioactive trace metals in significant amounts (Weinbauer et al. 2017).

7.3 Spatio-temporal Variations of Terrestrial Inputs to Patagonian Waters

The amount of terrigenous material reaching the coastal areas through riverine runoff depends on the river discharges and its variation through time. The main discharges into the coastal Patagonian area come from rivers that originate close to the Andes mountains range and flow across Patagonia (Fig. 7.1, Depetris et al. 2005). The historical discharges of the main Patagonian Rivers (from north to south: Colorado, Negro, Chubut and Santa Cruz) exhibit high-inter-annual variability (Fig. 7.2). The Negro River has the greatest discharge, with an historical (1928–2018) mean annual value of 814.9 $m^3 s^{-1}$ (Fig. 7.2b), followed by the Santa Cruz (1956–2018) and the Colorado (1919–2014) rivers with mean annual discharges of 714.3 $m^3 s^{-1}$ (Fig. 7.2d) and 127.5 $m^3 s^{-1}$ (Fig. 7.2a), respectively; the Chubut River has the lowest mean annual discharge (1994–2018) of 34.2 $m^3 s^{-1}$ (Fig. 7.2c). The discharges from the Colorado and Negro Rivers had significant negative trends, with an overall decrease over the historical period considered (Fig. 7.2a, b). In the case of the Chubut River, a decrease in the discharges is observed but the trend is not significant (Fig. 7.2c). In contrast, the Santa Cruz River had a significantly increasing trend of discharges, probably related to the ongoing melt of the Southern Patagonia ice field due to global warming (Millan et al. 2019).

The variations in the rivers' discharges are due to three main factors. First, anthropogenic influence through dam constructions (black arrows with numbers in Fig. 7.2), for example, in the Colorado (1—Casa de Piedra) and Negro Rivers (2—El Chocón, 3—Alicurá and 4—Piedra del Águila). In the case of the Colorado River (Fig. 7.2a), the construction of the dam resulted in a reduction of the discharges for 3 years, from 1989 to 1991, while in the Negro River, they lasted

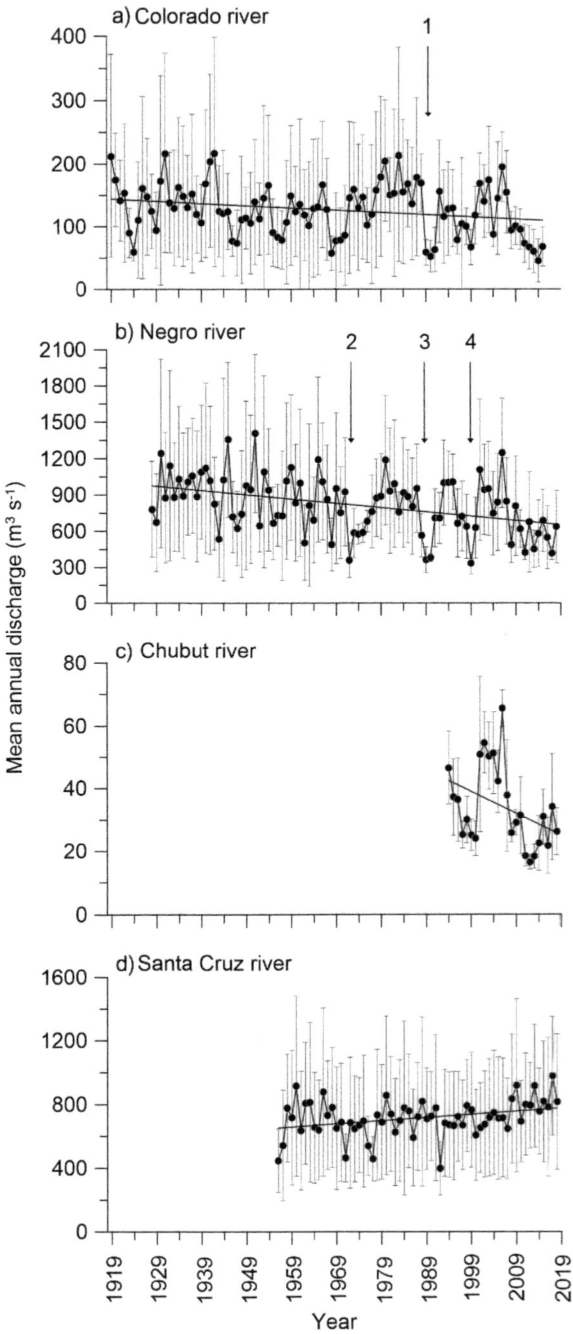

Fig. 7.2 Historical mean annual discharges of the main rivers ending in the SWAO. **a** Colorado river (1919–2014). **b** Negro river (1928–2018). **c** Chubut river (1994–2018). **d** Santa Cruz river (1956–2018). The arrows and the numbers in the Colorado and in the Negro rivers data sets indicate when the dams were constructed: 1—Casa de Piedra; 2—El Chocón; 3—Alicurá; 4—Piedra del Águila. Note the different scales on the y-axis of the panels. The vertical lines around the symbols indicate ± one standard deviation, whereas the solid lines indicate the linear regression fit of discharge versus time. Data from Sistema Nacional de Información Hídrica (http://www.snih.hidricosargentina.gob.ar)

1–3 years. Hence, the river discharges in most of the Patagonian Rivers are strongly controlled by the generation of energy (Pasquini and Depetris 2007). Second, Patagonian Rivers have variable catchment areas and are exposed to natural variations due to snow melt. Thus, higher discharges are observed in the Colorado, Negro and Santa Cruz Rivers as compared with the Chubut River (due to their larger catchment areas in the Andean mountains). In contrast, the Chubut River has a smaller upper catchment area east of the Andes, and hence, its discharges are influenced mainly by rainfall and by the Florentino Ameghino dam (number 5 in Fig. 7.1) regulation in its lower course (Pasquini and Depetris 2007). Variations along the year in the discharges are also evident, being generally higher in the Northern Patagonian coastal zone during the austral summer (Fig. 7.3a, Colorado River) or winter (Fig. 7.3b, Negro River). In the case of the Chubut River (Central Patagonia), the flux is more stable along the year (Fig. 7.3c), however, autumn and spring appear to have higher fluxes. In Southern Patagonia, i.e., the Santa Cruz River (Fig. 7.3d) has registered so far higher discharges during autumn, but this pattern may change due to the ongoing construction of dams (i.e., Cóndor Cliff, La Barrancosa). Third, rainfall directly supplies significant amounts of water through the catchment area that feed the river. Rainfall in Patagonia depends on the interaction between the westerly winds and the Andean mountains (Coronato et al. 2017). South of 40°S, the Andes do not exceed 3 km height; therefore, the Pacific air masses dominate in the Patagonia region (Labraga and Villalba 2009). The interception between the Andes and westerly winds enhances rainfall on their western and decreases on their eastern side (Depetris et al. 2005). This results in higher rainfall near the Andes ($\sim$1500 mm yr^{-1}) and lower on the Atlantic coast ($\sim$150 mm yr^{-1}; Coronato et al. 2017). In Patagonia, the temporal trends of cumulative rainfall in the coast over the last two decades show a latitudinal gradient i.e., from high values in the North (maximum of ca. 500 mm) to low in Southern Patagonia (maximum $\sim$ 100 mm) (Fig. 7.4a). Minimal cumulative rainfall was detected during the years 2006-2008, especially in northern and central latitudes (Fig. 7.4a). This inter-annual variability is also accompanied by a seasonal pattern of rainfall in the three main zones. Thus, North Patagonia presents some summer rainfall (Fig. 7.4b, Pasquini and Depetris 2007), while in Central and South Patagonia, rainfall occurs mainly during early winter (Fig. 7.4b, c) as also observed by Jobbágy et al. (1995) and Labraga and Villalba (2009).

The typical westerly winds have a daily mean velocity of 40 km h^{-1}, and the maxima are recorded in spring, reaching velocities >100 km h^{-1} (Gassó and Stein 2007). The yearly cumulative aerosol index (AI, Fig. 7.5a) used as a proxy of the atmospheric eolian dust deposition (Cabrerizo et al. 2018) is higher in Central Patagonia and with noticeable peaks in the three areas after Andean volcanic eruptions, e.g., Hudson (1991), Copahue (2000), Chaitén (2008) and Puyehue (2011) (Fig. 7.5a) due to significantly increasing amounts of ash deposition (Achterberg et al. 2013). The patterns agree with the number of deposition events (frequency) that occurred within a year, being higher during the eruptions, with maximum values during 1992 (Hudson) in Central (185 events yr^{-1}) followed by North (168 events yr^{-1}) and South Patagonia (131 events yr^{-1}). In addition, during

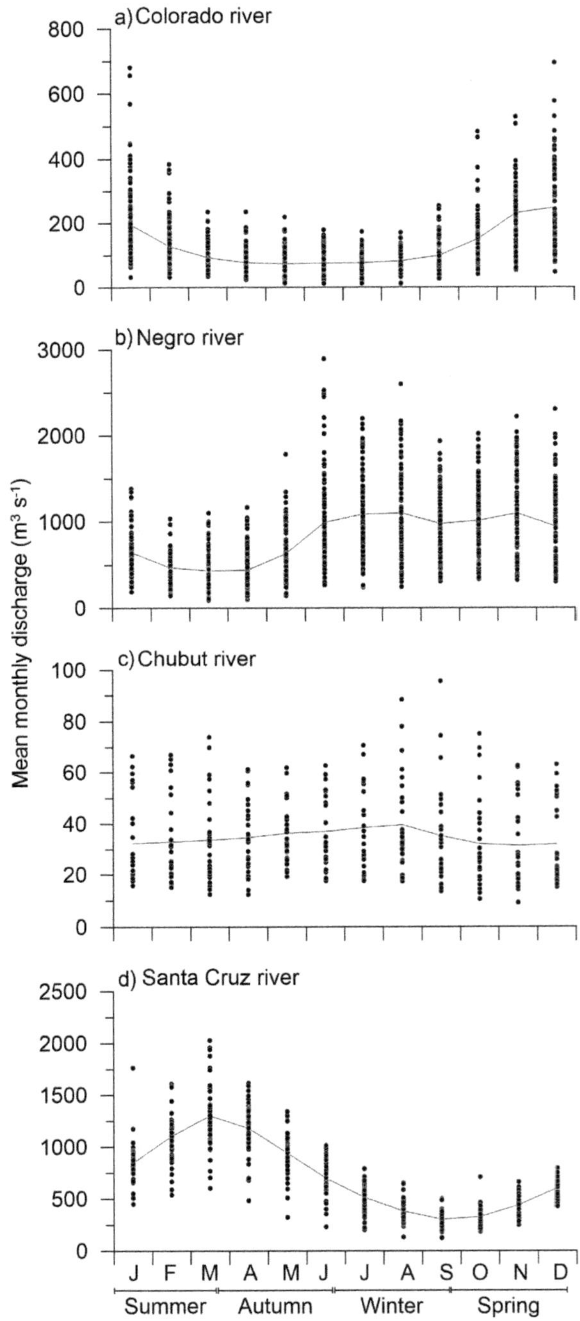

Fig. 7.3 Historical mean monthly discharges of the main rivers ending in the SWAO. **a** Colorado river (1919–2014). **b** Negro river (1928–2018). **c** Chubut river (1994–2018). **d** Santa Cruz river (1956–2018). The lines represent the historical mean monthly discharge. Note the different scales on the y-axis of the panels. Data from Sistema Nacional de Información Hídrica (http://www.snih.hidricosargentina.gob.ar)

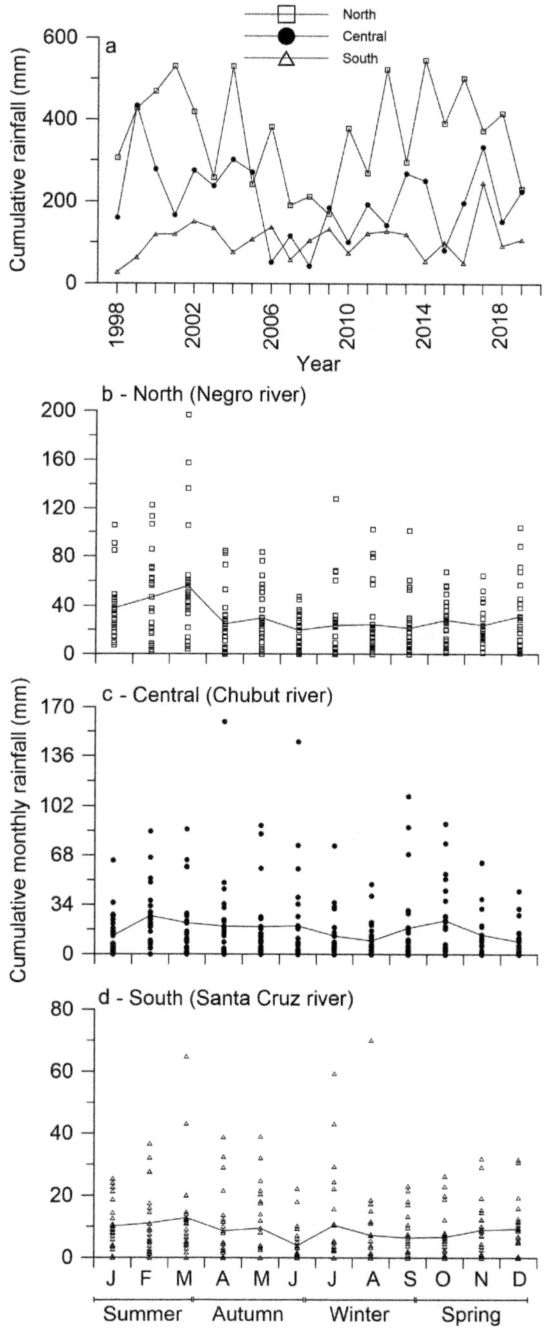

Fig. 7.4 **a** Historical (1998–2019) cumulative annual rainfall in North (squares), Central (circles) and South (triangles) areas of Patagonia. Monthly cumulative rainfall (1998–2019) in: **b** North, **c** Central and **d** South Patagonia. Data obtained from Giovanni NASA database v. 4.34 using daily area-averaged precipitation rates

the Hudson's volcanic eruption, the estimated deposition of trace elements (e.g., Al, Fe) was several thousand-fold higher than that deposited yearly by the atmospheric Patagonian dust (Gaiero et al. 2003). Finally, the AI values are higher during wintertime in North and Central Patagonia and lower during this season in South Patagonia (Fig. 7.5b). This can be related to the lower wind speeds during this season that prevents the aerosol-dust dispersion (mainly small particles, i.e., <2.5 μm) which remain in suspension in the troposphere during more time. This is consistent with recent results by Yang et al. (2017) in the Gobi desert who showed that weak wind conditions enhanced stagnation of small particles.

Changes in land use by human activities are also an important erosion factor in Patagonia (Gaiero et al. 2002) that cause an additional source of dust to the

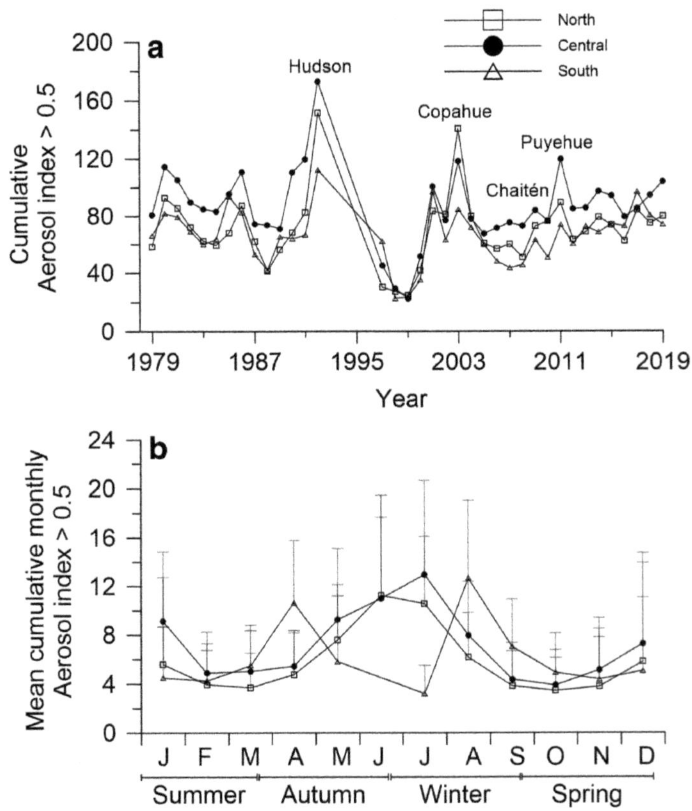

Fig. 7.5 Eolian dust over coastal Patagonian areas evaluated as aerosol index (AI) > 0.5 (1979–2019). **a** Annual cumulative events with an AI > 0.5 in north (squares), central (circles) and south (triangles) Patagonia. **b** Mean cumulative monthly events with AI > 0.5 in north (squares), central (circles) and south (triangles) Patagonia. The vertical lines on top of the symbols indicate one standard deviation. Data obtained from Giovanni NASA database v. 4.34 using the total ozone mapping spectrometer (TOMS-Nimbus-7), TOMS earth-probe and ozone monitoring instrument (OMI) satellites

atmosphere due to desertification (Muhs et al. 2014) and livestock (sheep). Using remote-sensing data, Cabrerizo et al. (2018) showed a consistent increase of AI (both in intensity and frequency) from 1995 to 2015. These increases reported in recent years, which are related with anthropogenic activities and global change, are in agreement with others variables such as wind and temperature, as well as in rainfall regimes, with the concomitant increases of nutrients and terrigenous material as observed in the Chubut River estuary (Bermejo et al. 2018).

7.4 Impacts of Riverine and Eolian Dust Inputs on Coastal Ecosystems

Riverine input of terrigenous material had been associated with the stimulation of phytoplankton biomass despite the decrease in light penetration caused by dissolved organic carbon—DOC (Klug 2002; Traving et al. 2017) as well as with increased bacterioplankton abundances with the consequent oxygen depletion (Hitchcock et al. 2010). Cotovicz Jr. et al. (2017) found that the input of terrigenous material (both DOC and particulate organic carbon—POC) caused eutrophication in an embayment of the SWAO (Brazil), increasing planktonic biomass and consequently, detrital organic carbon, which could be degraded by microorganisms and/or buried in sediments. Indeed, and although nutrient enrichment will in general favor the growth of autotrophs and increase the overall primary productivity, their excess may lead to coastal eutrophication (Cloern 2001; Rabalais et al. 2009). Eutrophication has been observed in several sites of the SWAO, and in the Patagonian coast, higher nutrient concentrations led to increases in abundance and diversity of macroalgae (Gil et al. 2019; Martinetto et al. 2010). In the Río de la Plata estuary, eutrophication together with organic pollution increased the small-sized fraction of plankton i.e., bacterioplankton, picophytoplankton and microzooplankton (Garcia and Bonel 2014).

Rainfall enhances significantly the amount of dissolved organic matter (DOM) carried by the river to the ocean, supplying half of the carbon deposited in marine sediments each year (Willey et al. 2000). Moreover, these authors found significantly higher DOM due to rainfall in comparison with other processes such as storms and hurricanes. Indeed, rainfall effects (and to some extent wind) are translated to a higher input of nutrients through riverine runoff, and several studies determined an enrichment (as nitrates and ammonium) in surface layers after rainfall (Valiela et al. 2012; Fong et al. 2020). However, the main impact of rainfall through riverine runoff is due to human activities (e.g., by the use of pesticides, fertilizers), mainly in the form of N and P. Another effect of rainfall on coastal ecosystems is the decrease of salinity after an extreme event, which can modify the community structure favoring those species with higher tolerance to changing salinities (Valiela et al. 2012).

The eolian input of materials on coastal aquatic ecosystems of Patagonia includes Fe (characteristic of the Patagonian dust) which is an essential micronutrient for phytoplankton (Jickells and Moore 2015). The addition of dust also brings macronutrients (N, P; Crespi-Abril et al. 2018) which favored phytoplankton growth (Johnson et al. 2011), and the C-sink capacity in unproductive waters (Cabrerizo et al. 2016), whereas it reduced such capacity by 30% in highly productive ecosystems (Cabrerizo et al. 2018). However, in other coastal waters (or species/systems) worldwide other responses had been observed. For example, eolian input increased the relative abundance of Haptophyceae as compared to that of Bacillariophyceae, Dinophyceae and Cryptophyceae in the East China Sea (Meng et al. 2016) and the relative abundance of toxic species, i.e., *Vibrio* (Westrich et al. 2016). Dust inputs also increased the bacterial and picoplankton primary production as well as egg production of copepods in surface waters of the Mediterranean Sea (Tsagaraki et al. 2017). Other reports have shown that dust deposition differentially favored growth of *Prochlorococcus* as compared to that of nanoplankton, in a low-nutrient/low-Chl-*a* ocean off Barbados, due to its high efficient P acquisition mechanisms, together with the low P requirements (Chien et al. 2016). However, recent results obtained by Zhang et al. (2020) in the Northwest Pacific evidenced changes toward larger cell sizes that increased their abundance when dust, containing the limiting nutrients, where added to the samples. Other reports have shown that the degree of oligotrophy controlled plankton responses to riverine or dust inputs. For instance, Marañón et al. (2010) showed that increased bacterial production was the dominant response to dust inputs in ultra-oligotrophic open-ocean environments, whereas primary productivity increased when the oligotrophy decreased. This result agrees with studies carried out in subtropical gyres of the Atlantic Ocean that also showed an accentuation of heterotrophy (i.e. production/respiration ratio < 1) in microbial food webs communities after amendments with organic and organic and inorganic nutrient inputs (Martínez-García et al. 2012). Also, in coastal phytoplankton communities, responses to riverine and eolian dust nutrient inputs were greatly varied among seasons and were positively correlated with dissolved inorganic nitrogen concentrations. Specifically, plankton responses (as Chl-*a* and primary productivity) to nutrient inputs were stronger when in situ concentrations were low (i.e., summer) than when communities were growing under nutrient-replete concentrations (i.e., spring). No effect or negative responses to nutrient input were observed during autumn (Teixeira et al. 2018). Therefore, the variability in responses toward dust input reflects a range of factors that can also condition them, such as the source of the material, initial conditions of the community and pre-existing nutrient availability (Jickells and Moore 2015). Not only dust but also ash from volcanic eruptions with considerable amounts of nutrients may have significant effects on the ocean (Achterberg et al. 2013). For instance, volcanic ash aerosols promoted massive phytoplankton blooms, suggesting that the production to respiration ratio was shifted toward autotrophy (Browning et al. 2014; Mélançon et al. 2014). Black carbon aerosols, however, seem to stimulate bacterial but not primary production, hinting at a shift toward heterotrophy after such events (Malits et al. 2015).

Browning et al. (2014) found in the Southern Ocean water a stimulation of the photochemistry performance (F_v/F_m) of phytoplankton communities and increases in Chl-*a* concentration after additions of volcano ash, in comparison with the addition of only Fe. The impact of volcano ash varies according to the species considered and the ash components, i.e., trace metals—Fe, Cu, Pb. For example, with additions of volcanic ash, there was a general increase in the growth of the diatom *Thalassiosira pseudonana* due to Fe fertilization, and either no changes or decrease of growth of the coccolithophorid *Emiliania huxleyi* was observed due to toxic metals such as Cu (Hoffmann et al. 2012); there are also reports of Cu toxicity on phytoplankton growth due to both, dust (Paytan et al. 2009) and ash (Jordi et al. 2012).

7.5 The Chubut River Estuary: A Reference Ecosystem in Patagonia

The Chubut River is one of the very few in the Patagonian region of Argentina; it originates in the Andean mountains and flows into the coastal SWAO water (Fig. 7.1; Piccolo and Perillo 1999). Several cities—Dolavon, Gaiman, Trelew and Rawson, are located downstream of the Florentino Ameghino dam (120 km from the mouth of the Chubut River, Fig. 7.1). In its lower course, the river is diverted into various irrigation channels that supply water for agricultural and cattle rising activities (Fig. 7.6; lower inset). The total length of the channel system is ca. 360 km and 22,000 ha of land are under irrigation when the channels are open (Fig. 7.6, inset) receiving water from the Chubut River from spring to early autumn and being closed during the rest of the year (FAO 2015).

There are reports of increases in the number of extreme events and frequency of rainfall (Castañeda and González 2008; IPCC 2013), and historical records in the area (from 1972 to 2019) show a 30% increase in the intensity of rainfall events (Vizzo et al. 2021). The increased input of terrigenous material, associated to rainfall, has not only altered several biogeochemical and physiological processes (see below) but also slowed down the water purification process, limiting its consumption (Pessacg et al. 2020). Indeed, the amount of terrigenous material transported by Chubut River (estimated as DOC) during rainfall conditions and when irrigation channels are open (September–March) was circa double that when they are closed and with no rainfall (Fig. 7.7). These DOC concentrations became even higher, rising by fivefold after extreme rainfall events with respect to normal conditions and resulted in a PAR attenuation coefficient (kd_{PAR}) variable between ca. 1.3 and >8 m^{-1} in the outer regime (seawater side) of the Chubut River estuary, for the two contrasting conditions, i.e., closed channels with no rainfall versus extreme rainfall, respectively, (Fig. 7.7). Other human-derived activities (i.e., agricultural, industrial and urban) together with the above conditions resulted in changes in rainfall and wind patterns which have considerably affected the coastal

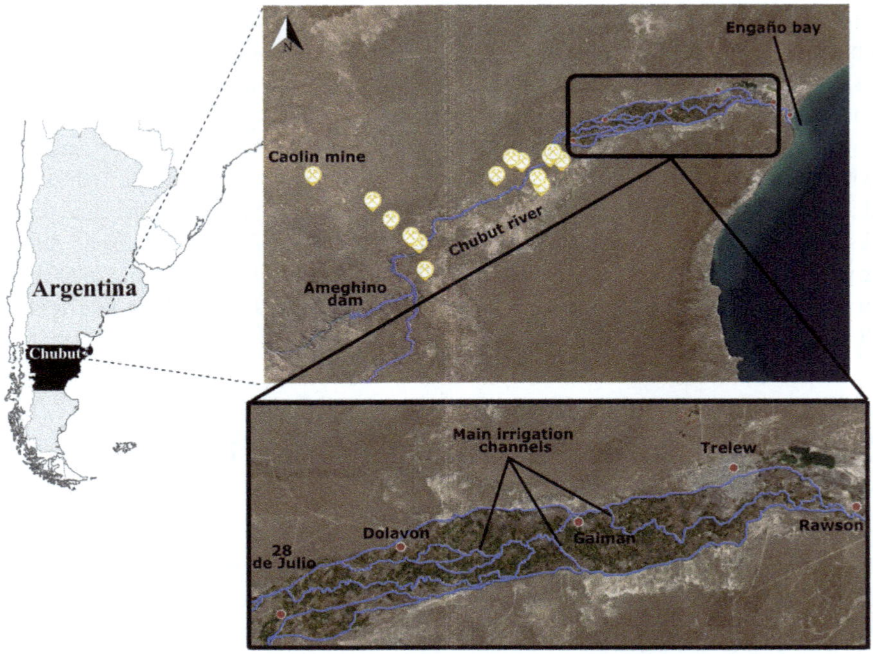

Fig. 7.6 Map of Argentina showing the location of the Chubut province. Enlarged are modified satellite images (downloaded from Instituto Geográfico Nacional, https://mapa.ign.gob.ar) showing the lower course of Chubut river and the location of the Florentino Ameghino dam (120 km upstream from the mouth of the river). The blue lines indicate the main irrigation channels diverted from Chubut river which are used for irrigation from September to March each year. Caolin mines next to the river are indicated with yellow symbols

ecosystem adjacent to the mouth of the Chubut River. Moreover, extreme rainfall increases the discharge of the Chico River, a temporary tributary water stream of the Chubut River (Kaless et al. 2019) with the consequent increase of sediments that are carried into the coastal area. The amount of terrigenous material entering the coastal aquatic system close to the Chubut River estuary during the year is a function of the intensity of transport of particles. Also, increases in wind speeds should increase the transport of material toward coastal waters. We evaluated the strength of these transport variables during the year (Fig. 7.8), and found that wind speed was higher in spring-summer than in winter, while rainfall increases in late summer and mid-spring. The river discharge, however, remained with similar values throughout the year. Therefore, the amount of terrigenous material entering in the Chubut River estuary is continuous along the year being influenced by a balance between river discharge, the dominant transport agent, rainfall and wind.

The type of materials carried to the coastal water either by the river or through eolian transport is variable, and few measurements/experimental determinations have been carried out in the outer regime of the Chubut River estuary. Under rainfall events, the presence of high amounts of caolin is evident, a clay mineral,

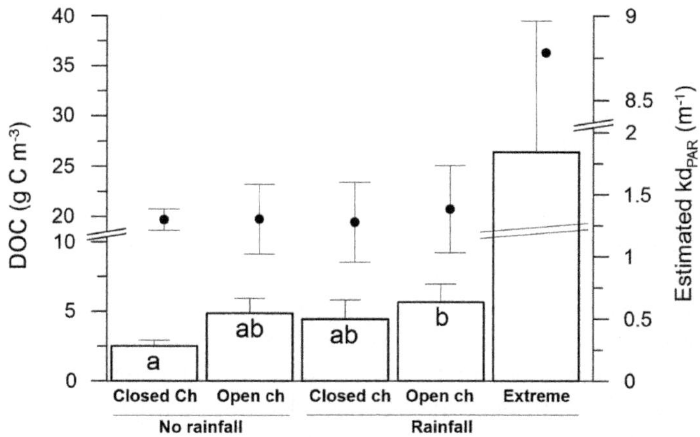

Fig. 7.7 Terrigenous material concentrations in the Chubut river estuary (estimated as dissolved organic carbon, DOC using spectrophotometric techniques) when the irrigation channels are open or closed under the absence, normal and extreme rainfall conditions. The lines on top of the bars indicate one standard deviation, whereas the letters indicate significant differences after a Tukey post hoc test, following one-way ANOVA. The black circles indicate the estimated PAR attenuation coefficient (kd_{PAR}) for each condition using the equation of Morris et al. (1995); the lines around the symbols indicate the standard deviation

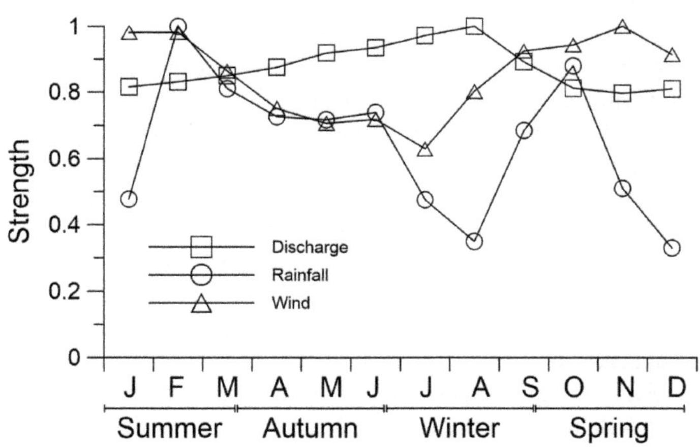

Fig. 7.8 Changes in the main drivers of transport (as strength of the transport, in relative units) of terrigenous material in the outer Chubut river estuary as a function of the time of the year. The value 1 indicates a mean strength of transport of 39.7 m^3 s^{-1} for riverine discharge, 26.4 mm of rainfall or 5.39 m s^{-1} of wind speed

from the mines located in the upper course of the river (Fig. 7.6) and ending up in the mouth of the Chubut River estuary (Cravero et al. 1991; Domínguez et al. 2013). There is also a spatial variability with regard to the origin of the materials present in the Chubut River. Commendatore and Esteves (2004) reported low hydrocarbon values with a biogenic origin in stations upstream from the river mouth; however, they found high hydrocarbon concentrations of predominantly anthropogenic origin at the mouth of the river, likely due to the Rawson's harbor and its related fishing vessel activities. In fact, Bermejo et al. (2018) determined higher amounts of nutrients (mainly nitrogen and phosphorus) at the mouth of the river during 2015, as compared to a historical reference period (2001–2014), and they were associated with agricultural and industrial activities which resulted in a transport of byproducts via riverine runoff. Furthermore, Scapini et al. (2010) compared the humic substances in the Chubut River and in the estuary and found that they were mainly from biogenic origin (autochtonous organic carbon produced by phytoplankton), but in the river, they had an additional pedogenic origin (al-lochtonous organic carbon from runoff). Also, several pollutants (i.e., organic pesticides and polychlorinated biphenyls) are clearly increasing in the lower part of the river due to agricultural activities. The increase of some heavy metal concentrations in the sediments (Cd, Zn, Cu and Hg) is suggesting other man-triggered activities (Isla et al. 2015) as also seen in other Patagonian rivers (e.g., Negro river, Gaiero et al. 2002). Finally, the presence of oil from fishery ships at the mouth of the Chubut River have recently called for the need of treatment of these wastes (González Zevallos et al. 2020). It is clear thus that the Chubut River evidences important anthropogenic disturbance, which was confirmed through analysis of diatoms assemblages in sediment cores showing changes toward species adapted to eutrophic and polluted conditions (Espinosa et al. 2020).

The interplay between some of the environmental change drivers of local interest in the area of the Chubut River estuary is resumed in Fig. 7.9. The coastal area receives important inputs of terrigenous material (both from eolian transport and riverine runoff) which finally end up in the coastal waters. These inputs are further enhanced when the irrigation channels, used for agriculture purposes, are open. Extreme rainfall events that brings more terrigenous material to the system not only provide an extra input of nutrients but also significantly alter the optical conditions (Fig. 7.7) of this coastal ecosystem (e.g., during the heavy rainfalls of the years 1998, 2014 and 2017, Bilmes et al. 2016). Volcanic eruptions occurring in the Andes mountains transport significantly higher amounts of ash through wind, also altering the underwater radiation field. In addition, byproducts from urban activities, e.g., industry or dumping of crude sewage, end up in the river mouth and reach the coastal ecosystem. These inputs of materials, together with the ongoing global change occurring in the area, e.g., warming (Gil et al. 2019) and acidification (Orselli et al. 2018) will affect the physiology and structure of phytoplankton communities. In the following section, we will present the available data for the study area obtained through experimental work, considering the input of terrigenous material, of different origin, in combination with other drivers of global change (e.g., solar UVR, acidification and mixing), and we will describe their effects on coastal phytoplankton.

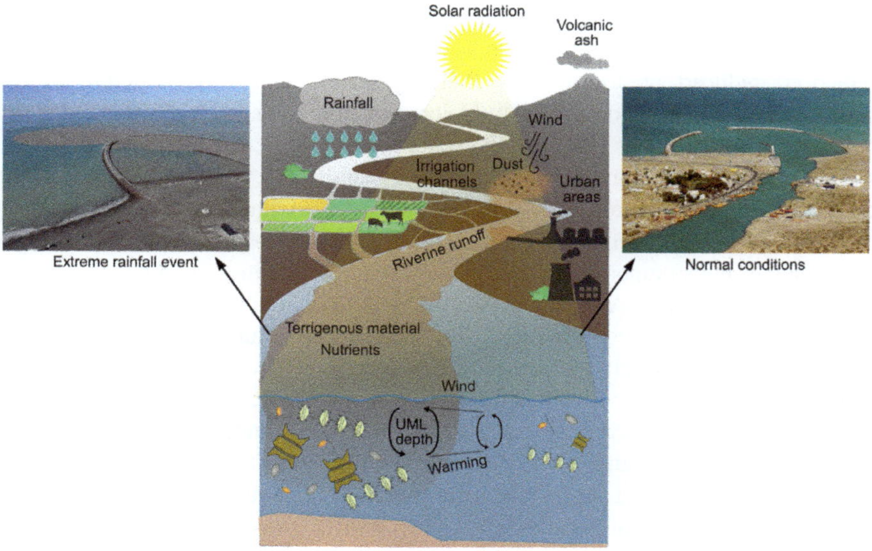

Fig. 7.9 Graphical scheme indicating the sources of terrigenous material and other environmental drivers influenced by global change affecting physico-chemical properties and functioning of the Chubut river estuary. The photograph show the visual effects of an extreme rainfall event (left) and of normal conditions (right). Photograph from Diario Jornada (www.diariojornada.com.ar)

7.6 Impacts of Terrestrial Material on Phytoplankton from the Chubut River Estuary

In general, the addition of inorganic nutrients, mimicking riverine inputs by the Chubut River, resulted in higher growth and better photosynthesis performance of phytoplankton communities (as compared to samples under ambient nutrient conditions), and these inputs acted antagonistically with ultraviolet radiation (UVR), i.e., reducing its negative effects (Helbling et al. 2005). Similar results were determined in a comparative study carried out with phytoplankton communities from the Chubut River estuary and from a tropical site (Babitonga bay, Southern Brazil, Villafañe et al. 2017). In both cases, nutrient inputs improved photosynthetic performance; however, they produced a change in the taxonomic composition of the tropical community, i.e., from diatoms to flagellates, while in the Chubut River, diatoms dominated regardless of the treatments (Villafañe et al. 2017). Also, the effects of two different sources of nutrients inputs, riverine and eolian were analyzed by Cabrerizo et al. (2017, 2018), and they found that there was a stimulatory effect of UVR on the net primary productivity (NPP) and respiration under inputs of nutrients from eolian origin, and an inhibition under the riverine inputs, with a consequent reduction in the C-sink capacity. Despite this differential effect reported at a metabolic level, both inputs prompted a similar shift in the community structure toward a dominance of nanoflagellates.

The effects of DOM were studied in post-bloom communities (Villafañe et al. 2018), finding that its addition (either as a source of nutrients or as absorber of solar radiation) reduced photosynthetic efficiency and changed the structure of the communities initially dominated by microplankton toward those dominated by nanoplanktonic species. Likewise, the effects of gradual and extreme rainfall on summer and winter communities were recently studied by Vizzo et al. (2021). The authors found that increases of terrigenous material and nutrients, mimicking normal and extreme rainfall, increased the growth of nanoplanktonic diatoms and flagellates in the summer while only nanoplanktonic diatoms significantly grew in the winter community. Also, the impact of gradual rainfall was higher than that of the extreme rainfall, especially in the summer community. By contrast and regardless of the rainfall scenario and the community considered, there was an increased photosynthetic efficiency per carbohydrates unit produced. The competitive advantage of centric diatoms was also registered under inputs of inorganic nutrients, acidification and UVR or vertical mixing (Villafañe et al. 2015; Bermejo et al. 2020). However, Duran-Romero et al. (2017) reported that when flagellates dominated the community, subsequent alterations in nutrient concentrations, pH values and solar UVR, as expected in the future global change scenarios, did not alter its dominance as compared to other taxonomic groups, but increased primary productivity per cell.

Finally, Bermejo et al. (2020) found a decrease in NPP and higher photosystem II (PSII) inhibition in post-bloom communities of the Chubut River estuary under nutrient inputs and acidification. These results contrast with those obtained by Villafañe et al. (2015) under nutrients and acidification, since the community improved its photosynthetic performance. The mechanism underpinning such differences could be that the summer post-bloom community used by Bermejo et al. (2020) was subjected to a fluctuating solar radiation regime. This regime, which mimics vertical mixing events in surface waters, could increase the sensitivity/susceptibility of communities to the interacting drivers assayed, by indirectly accentuating the harmful effect of the dominant driver (i.e., UVR) on PSII photochemistry and C-fixation (Helbling et al. 2015; Hoppe et al. 2015).

7.7 Conclusions and Perspectives

Studies related with nutrient effects on coastal phytoplankton commonly show a positive effect on different levels of biological organization. However, most studies focused on riverine input, but currently, we lack of a comprehensive understanding about how other nutrient sources—such as atmospheric dust deposition exported from arid areas and deserts, or ash coming from volcanoes, impact the basis of food webs in coastal areas. In this sense, most of the available evidences come from short-term open-ocean studies where the lack of some macronutrients (e.g., N, P), trace metals (e.g., Fe) or co-factors (e.g., Co) contained in dust-eolian or ash inputs are limiting in the ocean productivity (Moore et al. 2013). These studies propose that these inputs can enhance auto- and heterotrophic biomass production (Lekunberri et al. 2010; Rahav et al. 2018) and nitrogen fixation (Guieu et al. 2019), photosynthetic efficiency (Browning et al. 2014; 2020) or carbon sequestration in deep water (Pabortsava et al. 2017); however, it remains greatly unknown how the timing of this input and/or its interaction with other global change drivers can alter phytoplankton responses over the seasonal succession, and whether the effects of these source of nutrients can be, directly or indirectly, propagated to higher trophic levels. Scarce available evidences, which come from freshwater ecosystems, show a boosted herbivorous growth after inorganic and dust-derived nutrient input due to an increased availability of phytoplankton/seston carbon. Furthermore, this nutrient input can synergistically accentuate the effects of other major global change drivers (e.g., CO_2); hence, the effects of rising CO_2 concentrations in the atmosphere on other trophic levels may be strongly influenced by eutrophication processes and/or increases in dust deposition events (Villar-Argaiz et al. 2018). Nevertheless, recent studies have also shown that the positive bottom-up effects found on one trophic level (i.e., phytoplankton) and propagated to another trophic level (i.e., zooplankton) after nutrient input do not always occur (Gusha et al. 2019). Therefore, considering that local/regional processes, as presented in this chapter, can have global implications, we stress the need of understanding how plankton organisms can adapt to present-day eolian dust and volcanic ash events to predict how they will respond to the ongoing changes that coastal ecosystems will experience in a dustier world.

Acknowledgements This work was supported by Agencia Nacional de Promoción Científica y Tecnológica - ANPCyT (PICT 2015-0462) and Fundación Playa Unión. MJC was supported by Juan de la Cierva Formación (FJCI2017-32318) and Incorporación (IJC2019-040850-I) contracts from Ministerio de Ciencia, Innovación y Universidades. We thank the Cooperativa Eléctrica y de Servicios de Rawson for providing building's infrastructure to carry out this work. This work is in partial fulfillment of the Ph.D. thesis of JIV. This is Contribution N° 186 of Estación de Fotobiología Playa Unión.

References

Achterberg EP, Moore CM, Henson SA, Steigenberger S, Stohl A, Eckhardt S, Avendano LC, Cassidy M, Hembury D, Klar JK (2013) Natural iron fertilization by the Eyjafjallajökull volcanic eruption. Geophys Res Lett 40:921–926

Baker AR, Jickells TD, Witt M, Linge KL (2006) Trends in the solubility of iron, aluminium, manganese and phosphorus in aerosol collected over the Atlantic Ocean. Mar Chem 98:43–58

Barletta M, Lima ARA, Costa MF (2019) Distribution, sources and consequences of nutrients, persistent organic pollutants, metals and microplastics in South American estuaries. Sci Total Environ 651:1199–1218

Bermejo P, Durán-Romero C, Villafañe VE, Helbling EW (2020) Influence of fluctuating irradiance on photosynthesis, growth and community structure of estuarine phytoplankton under increased nutrients and acidification. J Exp Mar Biol Ecol 526 (151348). https://doi.org/10.1016/j.jembe.2020.151348

Bermejo P, Helbling EW, Durán-Romero C, Cabrerizo MJ, Villafañe VE (2018) Abiotic control of phytoplankton blooms in temperate coastal marine ecosystems: a case study in the South Atlantic Ocean. Sci Total Environ 612:894–902

Bilmes A, Pessacg N, Alvarez MP, Brandizi L, Cuitiño JI, Kaminker S, Bouza PJ, Rostagno CM, Núñez de la Rosa D, Canizzaro A (2016) Inundaciones en Puerto Madryn: Relevamiento y diagnóstico del evento del 21 de Enero de 2016. Informe Técnico CCT CONICET-CENPAT:1–20

Brahney J, Mahowald N, Ward DS, Ballantyne AP, Neff JC (2015) Is atmospheric phosphorus pollution altering global alpine lake stoichiometry? Global Biogeochem Cy 29:1369–1383

Browning TJ, Al-Hashem AA, Hopwood MJ, Engel A, Wakefield ED, Achterberg EP (2020) Nutrient regulation of late spring phytoplankton blooms in the midlatitude North Atlantic. Limnol Oceanogr 65:1136–1148

Browning TJ, Bouman HA, Henderson GM, Mather TA, Pyle DM, Schlosser C, Woodward EMS, Moore CM (2014) Strong responses of Southern Ocean phytoplankton communities to volcanic ash. Geophys Res Lett 41:2851–2857

Cabrerizo MJ, Carrillo P, Villafañe VE, Helbling EW (2017) Differential impacts of global change variables on coastal South Atlantic phytoplankton: Role of seasonal variations. Mar Environ Res 125:63–72

Cabrerizo MJ, Carrillo P, Villafañe VE, Medina-Sánchez JM, Helbling EW (2018) Increased nutrients from aeolian-dust and riverine origin decrease the CO_2-sink capacity of coastal South Atlantic waters under UVR exposure. Limnol Oceanogr 63:1191–1203

Cabrerizo MJ, Medina-Sánchez JM, González-Olalla JM, Villar-Argaiz M, Carrillo P (2016) Saharan dust and high UVR jointly alter the metabolic balance in marine oligotrophic ecosystems. Sci Rep 6:35892. https://doi.org/10.1038/srep35892

Cassar N, Bender ML, Barnett BA, Fan S, Moxim WJ, Levy H II, Tilbrook B (2007) The Southern Ocean biological response to aeolian iron deposition. Science 317:1067–1070

Castañeda M, González M (2008) Statistical analysis of the precipitation trends in the Patagonia region in southern South America. Atmósfera 21:303–317

Chien C-T, Mackey KRM, Dutkiewicz S, Mahowald N, Prospero JM, Paytan A (2016) Effects of African dust deposition on phytoplankton in the western tropical Atlantic Ocean off Barbados. Global Biogeochem Cy 30:716–734

Cloern JE (2001) Our evolving conceptual model of the coastal eutrophication problem. Mar Ecol Prog Ser 210:223–253

Cloern JE, Foster SQ, Fleckner AE (2014) Phytoplankton primary production in the world's estuarine-coastal ecosystems. Biogeosciences 11:2477–2501

Commendatore M, Esteves JL (2004) Natural and anthropogenic hydrocarbons in sediments from the Chubut River (Patagonia, Argentina). Mar Pollut Bull 48:910–918

Coronato F, Pessacg N, Alvarez M (2017) The climate of Península Valdés within a regional frame. In: Bouza PJ, Bilmes A (eds) Late Cenozoic of Península Valdés. Springer, Patagonia, Argentina, pp 85–104

Cotovicz LC Jr, Knoppers BA, Brandini N, Poirier D, Costa Santos SJ, Cordeiro RC, Abril G (2017) Predominance of phytoplankton-derived dissolved and particulate organic carbon in a highly eutrophic tropical coastal embayment (Guanabara Bay, Rio de Janeiro, Brazil). Biogeochemistry 137:1–14

Cravero MF, Domínguez E, Murray HH (1991) Valores δO18 y δD en caolinitas, indicadores de un clima templado moderado durante el Jurásico Superior-Cretácico Inferior de la Patagonia, Argentina. Rev Asoc Geol Argentina 46:20–25

Crespi-Abril AC, Soria G, De Cian A, López-Moreno C (2018) Roaring forties: An analysis of a decadal series of data of dust in Northern Patagonia. Atmos Environ 177:111–119

Depetris PJ, Gaiero DM, Probst JL, Hartmann J, Kempe S (2005) Biogeochemical output and typology of rivers draining Patagonia's Atlantic seaboard. J Coast Res 21:835–844

Domínguez E, Iglesias C, Dondi M, Brendel M (2013) Arcillas y caolines de Chubut y Santa Cruz. Geología, propiedades cerámicas, recursos y perspectivas. X Congreso de Geología Económica

Duran-Romero C, Villafañe VE, Valiñas MS, Gonçalves RJ, Helbling EW (2017) Solar UVR sensitivity of phyto- and bacterioplankton communities from Patagonian coastal waters under increased nutrients and acidification. ICES J Mar Sci 74:1062–1073

Espinosa MA, Vélez-Agudelo C, Isla FI (2020) Diatom responses to natural and anthropogenic environmental changes in a Patagonian river, Argentina. J South Am Earth Sci 102. https://doi.org/10.1016/j.jsames.2020.102677

FAO (2015) Estudio del potencial de ampliación del riego en Argentina, 136 p

Fong CR, Gaynus CJ, Carpenter RC (2020) Extreme rainfall events pulse substantial nutrients and sediments from terrestrial to nearshore coastal communities: a case study from French Polynesia. Sci Rep 10:1–8. https://doi.org/10.1038/s41598-020-59807-5

Franco BC, Defeo O, Piola AR, Barreiro M, Yang H, Ortega L, Gianelli I, Castello JP, Vera C, Buratti C, Pájaro M, Pezzi LP, Möller OO (2020) Climate change impacts on the atmospheric circulation, ocean, and fisheries in the southwest South Atlantic Ocean: a review. Clim Change. https://doi.org/10.1007/s10584-020-02783-6

Gaiero D, Probst J, Depetris P, Lelyter L, Kempe S (2002) Riverine transfer of heavy metals from Patagonia to the southwestern Atlantic Ocean. Reg Environ Change 3:51–64

Gaiero DM, Probst JL, Depetris PJ, Bidart SM, Leleyter L (2003) Iron and other transition metals in Patagonian riverborne and windborne materials: Geochemical control and transport to the southern South Atlantic Ocean. Geochim Cosmochim Acta 67:3603–3623

Garcia MD, Bonel N (2014) Environmental modulation of the plankton community composition and size-structure along the eutrophic intertidal coast of the Río de la Plata estuary, Argentina. J Limnol 73:562–573

Garreaud R, Lopez P, Minvielle M, Rojas M (2013) Large scale control on the Patagonia climate. J Clim 26:215–230

Gassó S, Stein AF (2007) Does dust from Patagonia reach the sub-Antarctic Atlantic Ocean? Geophys Res Lett 34:1–5

Gassó S, Torres O (2019) Temporal characterization of dust activity in the Central Patagonia desert (years 1964–2017). J Geophys Res Atm 124:3417–3434

Gil MN, Giarratano E, Barros V, Bortolus A, Codignotto JO, Schenke RD, Góngora ME, Lovrich G, Monti AJ, Pascual M, Rivas AL, Tagliorette A (2019) Southern Argentina: the Patagonian continental shelf. In: Sheppard C (ed) World seas: environmental evaluation, vol 1. Academic Press, pp 783–811

Ginoux P, Prospero JM, Gill TE, Hsu NC, Zhao M (2012) Global-scale attribution of anthropogenic and natural dust sources and their emission rates based on MODIS Deep Blue aerosol products. Rev Geophys 50:RG3005. https://doi.org/10.1029/2012rg000388

González R, Caille G, Narvarte M (2007) An assessment of the sustainability of the hake *Merluccius hubbsi* fishery at San Matías gulf, Patagonia, Argentina. Fish Res 87:58–67

González Zevallos DR, Góngora ME, Durán Romero C (2020) Abordaje socioambiental con énfasis en los residuos sólidos generados por la flota pesquera de Rawson, Patagonia Argentina. Interciencia 45:142–149

Granéli W, Lindell M, Tranvik L (1996) Photo-oxidative production of dissolved inorganic carbon in lakes of different humic content. Limnol Oceanogr 41:698–706

Guieu C, Al Azhar M, Aumont O, Mahowald N, Levy M, Ethé C, Lachkar Z (2019) Major impact of dust deposition on the productivity of the Arabian Sea. Geophys Res Lett 46:6736–6744

Gusha MNC, Dalu T, Wasserman RJ, McQuaid CD (2019) Zooplankton grazing pressure is insufficient for primary producer control under elevated warming and nutrient levels. Sci Total Environ 651:410–418

Heileman S (2009) XVI-55 Patagonian Shelf: LME #14. In: The UNEP large marine ecosystem report: a perspective on changing conditions. UNEP Regional Seas Report and Studies. Nairobi, Kenya, 11 p

Helbling EW, Banaszak AT, Villafañe VE (2015) Differential responses of two phytoplankton communities from the Chubut river estuary (Patagonia, Argentina) to the combination of UVR and elevated temperature. Estuar Coast 38:1134–1146

Helbling EW, Barbieri ES, Marcoval MA, Gonçalves RJ, Villafañe VE (2005) Impact of solar ultraviolet radiation on marine phytoplankton of Patagonia, Argentina. Photochem Photobiol 81:807–818

Hitchcock JN, Mitrovic SM, Kobayashi T, Westhorpe DP (2010) Responses of estuarine bacterioplankton, phytoplankton and zooplankton to dissolved organic carbon (DOC) and inorganic nutrient additions. Estuar Coast 33:78–91

Hoffmann LJ, Breitbarth E, Ardelan MV, Duggen S, Olgun N, Hassellöv M, Wängberg S-Å (2012) Influence of trace metal release from volcanic ash on growth of *Thalassiosira pseudonana* and *Emiliania huxleyi*. Mar Chem 132:28–33

Hoppe CJM, Holtz L-M, Trimborn S, Rost B (2015) Ocean acidification decrease the light-use efficiency in an Antarctic diatom under dynamic but not constant light. New Phytol 207:159–171

IPCC (2013) Climate Change 2013. The physical science basis. Cambridge University Press, New York, USA, p 1535

Isla F, Espinosa M, Rubio B, Escandell A, Gerpe M, Miglioranza K, Rey D, Vilas F (2015) Avulsion at a drift-dominated mesotidal estuary: the Chubut River outlet, Patagonia, Argentina. J Hydrol 529:632–639

Jickells TD, Moore CM (2015) The importance of atmospheric deposition for ocean productivity. Annu Rev Ecol Evol S 46:481–501

Jobbágy EG, Paruelo JM, León RJC (1995) Estimación del régimen de precipitación a partir de la distancia a la cordillera en el noroeste de la Patagonia. Ecol Austral 5:47–53

Johnson MS, Meskhidze N, Kiliyanpilakkil VP, Gassó S (2011) Understanding the transport of Patagonian dust and its influence on marine biological activity in the South Atlantic Ocean. Atmos Chem Phys 11:2487–2502

Jordi A, Basterretxea G, Tovar-Sánchez A, Alastuey A, Querol X (2012) Copper aerosols inhibit phytoplankton growth in the Mediterranean Sea. Proc Natl Acad Sci 109:21246–21249

Kaless G, Pascual MA, Flaherty S, Liberoff AL, Garcia Asorey MI, Brandizi LD, Pessacg NL (2019) Ecos de la tormenta de Comodoro Rivadavia en el Valle Inferior del Río Chubut. Aporte de sedimentos al Río Chubut desde la cuenca del Río Chico. In: Paredes J (ed) Comodoro Rivadavia y la catástrofe de 2017. Universidad Nacional de la Patagonia San Juan Bosco, pp 290–303

Kirk JTO (1994) Light and photosynthesis in aquatic ecosystems. Cambridge University Press, Cambridge, p 509

Klug JL (2002) Positive and negative effects of allochthonous dissolved organic matter and inorganic nutrients on phytoplankton growth. Can J Fish Aquat Sci 59:85–95

Labraga JC, Villalba R (2009) Climate in the Monte Desert: past trends, present conditions, and future projections. J Arid Environ 73:154–163

Lekunberri I, Leport T, Romero E, Vázquez-Domínguez E, Romera-Castillo C, Marrasé C, Peters F, Weinbauer M, Gasol JM (2010) Effects of a dust deposition event on coastal marine microbial abundance and activity, bacterial community structure and ecosystem function. J Plankton Res 32:381–396

Li F, Ginoux P, Ramaswamy V (2010) Transport of Patagonian dust to Antarctica. J Geophys Res 115 (D18). https://doi.org/10.1029/2009jd012356

Longhurst A, Sathyendranath S, Platt T, Caverhill C (1995) An estimate of global primary production in the ocean from satellite radiometer data. J Plankton Res 17:1245–1271

Mahowald N, Jickells TD, Baker AR, Artaxo P, Benitez-Nelson CR, Bergametti G, Bond TC, Chen Y, Cohen DD, Herut B, Kubilay N, Losno R, Luo C, Maenhaut W, McGee KA, Okin GS, Siefert RL, Tsukuda S (2008) Global distribution of atmospheric phosphorus sources, concentrations and deposition rates, and anthropogenic impacts. Global Biochem Cy 22. https://doi.org/10.1029/2008GB003240

Malits A, Cattaneo R, Sintes E, Gasol JM, Herndl GJ, Weinbauer MG (2015) Potential impacts of black carbon on the marine microbial community. Aquat Microb Ecol 75:27–42

Marañón E, Fernández A, Mouriño-Carballido B, Martínez-García S, Teira E, Cermeño P, Chouciño P, Huete-Ortega M, Fernández E, Calvo-Díaz A, Morán XAG, Bode A,

Moreno-Ostos E, Varela MM, Patey MD, Achterbergf EP (2010) Degree of oligotrophy controls the response of microbial plankton to Saharan dust. Limnol Oceanogr 55:2339–2352

Martinetto P, Alemany D, Botto F, Mastrángelo M, Falabella V, Acha EM, Antón G, Bianchi A, Campagna C, Cañete G, Filippo P, Iribarne O, Laterra P, Martínez P, Negri R, Piola AR, Romero SI, Santos D, Saraceno M (2020) Linking the scientific knowledge on marine frontal systems with ecosystem services. Ambio 49:541–556

Martinetto P, Daleo P, Escapa M, Alberti J, Isacch JP, Fanjul E, Botto F, Piriz ML, Ponce G, Casas G, Iribarne O (2010) High abundance and diversity of consumers associated with eutrophic areas in a semi-desert macrotidal coastal ecosystem in Patagonia, Argentina. Est Coast Shelf Sci 88:357–364

Martínez-García S, Teira E, Fernández E (2012) Changes in the metabolic balance of contrasting microbial food webs after nutrient enrichment. Mar Ecol Prog Ser 462:9–19

Mazzonia E, Vazquez M (2009) Desertification in Patagonia. Develop Earth Surf Proc 13:351–377

Mélançon J, Levasseur M, Lizotte M, Delmelle P, Cullen J, Hamme RC, Peña A, Simpson KG, Scarratt M, Tremblay J-E, Zhou J, Johnson K, Sutherland N, Arychuk M, Nemcek N, Robert M (2014) Early response of the northeast subarctic Pacific plankton assemblage to volcanic ash fertilization. Limnol Oceanogr 59:55–67

Meng X, Chen Y, Wang B, Ma QW, Wang FJ (2016) Responses of phytoplankton community to the input of different aerosols in the East China Sea. Geophys Res Lett 43:7081–7088

Millan R, Rignot E, Rivera A, Martineau V, Mouginot J, Zamora R, Uribe J, Lenzano G, De Fleurian B, Li X, Gim Y, Kirchner D (2019) Ice thickness and bed elevation of the Northern and Southern Patagonian icefields. Geophys Res Lett 46:6626–6635

Montes A, Rodríguez SS, Domínguez CE (2017) Geomorphology context and characterization of dunefields developed by the southern westerlies at drying Colhué Huapi shallow lake, Patagonia Argentina. Aeol Res 28:58–70

Moore CM, Mills MM, Arrigo KR, Berman-Frank I, Bopp L, Boyd PW, Galbraith ED, Geider RJ, Guieu C, Jaccard SL, Jickells TD, La Roche J, Lenton TM, Mahowald NM, Marañón E, Marinov I, Moore JK, Nakatsuka T, Oschlies A, Saito MA, Thingstad TF, Tsuda A, Ulloa O (2013) Processes and patterns of oceanic nutrient limitation. Nature Geosci 6:701–710

Morris DP, Zagarese HE, Williamson CE, Balseiro EG, Hargreaves BR, Modenutti BE, Moeller R, Quemaliños CP (1995) The attenuation of solar UV radiation in lakes and the role of dissolved organic carbon. Limnol Oceanogr 40:1381–1391

Muhs DR, Prospero JM, Baddock MC, Gill TE (2014) Identifying sources of aeolian mineral dust: Present and past. In: Mineral dust: a key player in the Earth system. Springer, pp 51–74

Orselli IBM, Kerr R, Ito RG, Tavano VM, Mendes CRB, Garcia CAE (2018) How fast is the Patagonian shelf-break acidifying? J Mar Syst 178:1–14

Osburn CL, Morris DP (2003) Photochemistry of chromophoric dissolved organic matter in natural waters. In: Helbling EW, Zagarese H (eds) UV effects in aquatic organisms and ecosystems. Comprehensive Series in Photochesmitry & Photobiology, vol 1. The Royal Society of Chemistry, Cambridge, pp 185–217

Pabortsava K, Lampitt RS, Benson J, Crowe C, McLachlan R, Le Moigne FAC, Moore CM, Pebody C, Provost P, Rees AP, Tilstone GH, Woodward EMS (2017) Carbon sequestration in the deep Atlantic enhanced by Saharan dust. Nat Geosci 10:189–194

Paparazzo FE, Crespi-Abril AC, Gonçalves RJ, Barbieri ES, Gracia Villalobos LL, Solís ME, Soria G (2018) Patagonian dust as a source of macronutrients in the Southwest Atlantic Ocean. Oceanography 31:33–39

Pasquini AI, Depetris PJ (2007) Discharge trends and flow dynamics of South American rivers draining the southern Atlantic seaboard: an overview. J Hydrol 333:385–399

Paytan A, Mackey KRM, Chen Y, Lima ID, Doney SC, Mahowald N, Labiosa R, Post AF (2009) Toxicity of atmospheric aerosols on marine phytoplankton. Proc Natl Acad Sci 106:4601–4605

Pessacg NL, Flaherty S, Solman SA, Pascual MA (2020) Climate change in northern Patagonia: critical decrease in water resources. Theor Appl Climatol 140:807–822

Piccolo MC, Perillo GME (1999) Estuaries of Argentina: A review. In: Perillo GME, Piccolo MC, Pino Quivira M (eds) Estuaries of South America: their geomorphology and dynamics. Environmental science series. Springer, Berlin, pp 101–132

Piola AR, Palma ED, Bianchi AA, Castro BM, Dottori M, Guerrero RA, Marrari M, Matano RP, Möller Jr OO, Saraceno M (2018) Physical oceanography of the SW Atlantic Shelf: a review. In: Hoffmeyer MS, Sabatini ME, Brandini FP, Calliari DL, Santinelli NH (eds) Plankton ecology of the Southwestern Ocean—from the subtropical to the subantarctic realm. Springer, pp 37–56

Prospero JM, Ginoux P, Torres O, Nicholson SE, Gill TE (2002) Environmental characterization of global sources of atmospheric soil dust identified with the NIMBUS-7 Total Ozone Mapping Spectrometer (TOMS) absorbing aerosol product. Rev Geophys 40:2-1–2-31

Rabalais NN, Turner RE, Díaz RJ, Justic D (2009) Global change and eutrophication of coastal waters. ICES J Mar Sci 66:1528–1537

Rahav E, Belkin N, Paytan A, Herut B (2018) Phytoplankton and bacterial response to desert dust deposition in the coastal waters of the Southeastern Mediterranean Sea: a four-year in situ survey. Atmosphere 9:305. https://doi.org/10.3390/atmos9080305

Scapini MC, Conzonno VH, Balzaretti VT, Fernández Cirelli A (2010) Comparison of marine and river water humic substances in a Patagonian environment (Argentina). Aquat Sci 72:1–12

Stockwell JD, Doubek JP, Adrian R, Anneville O, Carey CC, Carvalho L, De Senerpont Domis LN, Dur G, Frassi MA, Grossart H-P, Ibelings BW, Lajeunesse MJ, Lewandowska AM, Llames ME, Matsuzaki S-I, Nodine ER, Nõges P, Patil VP, Pomati F, Rinke K, Rudstam LG, Rusak JA, Salmaso N, Seltmann CT, Straile D, Thackeray SJ, Thiery W, Urrutia-Cordero P, Venail P, Verburg P, Woolway RI, Zohary T, Andersen MR, Bhattacharya R, Hejzlar J, Janatian N, Kpodonu ATNK, Williamson TJ, Wilson HL (2020) Storm impacts on phytoplankton community dynamics in lakes. Global Change Biol 26:2756–2784

Teixeira IG, Arbones B, Froján M, Nieto-Cid M, Álvarez-Salgado XA, Castro CG, Fernández E, Sobrino C, Teira E, Figueiras FG (2018) Response of phytoplankton to enhanced atmospheric and riverine nutrient inputs in a coastal upwelling embayment. Estuar Coast Shelf Sci 210:132–141

Thornton DCO (2014) Dissolved organic matter (DOM) release by phytoplankton in the contemporary and future ocean. Eur J Phycol 49:20–46

Traving SJ, Rowe O, Jakobsen NM, Sørensen H, Dinasquet J, Stedmon CA, Andersson A, Riemann L (2017) The effect of increased loads of dissolved organic matter on estuarine microbial community composition and function. Front Microbiol 8. https://doi.org/10.3389/fmicb.2017.00351

Tsagaraki TM, Herut B, Rahav E, Berman Frank IR, Tsiola A, Tsapakis M, Giannakourou A, Gogou A, Panagiotopoulos C, Violaki K (2017) Atmospheric deposition effects on plankton communities in the Eastern Mediterranean: a mesocosm experimental approach. Front Mar Sci 4. https://doi.org/10.3389/fmars.2017.00210

Valiela I, Camilli L, Stone T, Giblin A, Crusius J, Fox S, Barth-Jensen C, Monteiro RO, Tucker J, Martinetto P (2012) Increased rainfall remarkably freshens estuarine and coastal waters on the Pacific coast of Panama: magnitude and likely effects on upwelling and nutrient supply. Global Planet Change 92:130–137

Villafañe VE, Cabrerizo MJ, Erzinger GS, Bermejo P, Strauch SM, Valiñas MS, Helbling EW (2017) Photosynthesis and growth of temperate and sub-tropical estuarine phytoplankton in a scenario of nutrient enrichment under solar ultraviolet radiation exposure. Estuar Coasts 40:842–855

Villafañe VE, Paczkowska J, Andersson A, Durán Romero C, Valiñas MS, Helbling EW (2018) Dual role of DOM in a scenario of global change on photosynthesis and structure of coastal phytoplankton from the South Atlantic Ocean. Sci Total Environ 634:1352–1361

Villafañe VE, Valiñas MS, Cabrerizo MJ, Helbling EW (2015) Physio-ecological responses of Patagonian coastal marine phytoplankton in a scenario of global change: role of acidification, nutrients and solar UVR. Mar Chem 177:411–420

Villar-Argaiz M, Cabrerizo MJ, González-Olalla JM, Valiñas MS, Rajic S, Carrillo P (2018) Growth impacts of Saharan dust, mineral nutrients, and CO_2 on a planktonic herbivore in southern Mediterranean lakes. Sci Total Environ 639:118–128

Vizzo JI, Cabrerizo MJ, Helbling EW, Villafañe VE (2021) Extreme and gradual rainfall condition growth, taxonomy and photosynthesis of temperate estuarine phytoplankton communities (Patagonia, Argentina). Mar Environ Res. 163:1052535. https://doi.org/10.1016/j.marenvres.2020.105235

Weinbauer MG, Guinot B, Migon C, Malfatti F, Mari X (2017) Skyfall—neglected roles of volcano ash and black carbon rich aerosols for microbial plankton in the ocean. J Plankton Res 39:187–198

Westrich JR, Ebling AM, Landing WM, Joyner JL, Kemp KM, Griffin DW, Lipp EK (2016) Saharan dust nutrients promote *Vibrio* bloom formation in marine surface waters. Proc Natl Acad Sci 113:5964–5969

Willey JD, Kieber RJ, Eyman MS, Avery GB Jr (2000) Rainwater dissolved organic carbon: concentrations and global flux. Global Biogeochem Cycles 14:139–148

Yang Y, Russell LM, Lou S, Liao H, Guo J, Liu Y, Singh B, Ghan SJ (2017) Dust-wind interactions can intensify aerosol pollution over eastern China. Nat Comm 8. https://doi.org/10.1038/ncomms15333

Zhang C, He J, Yao X, Mu Y, Guo X, Ding X, Yu Y, Shi J, Gao H (2020) Dynamics of phytoplankton and nutrient uptake following dust additions in the northwest Pacific. Sci Total Environ 739. https://doi.org/10.1016/j.scitotenv.2020.139999

Chapter 8
Marine Eutrophication: Overview from Now to the Future

Paulo Antunes Horta, Leonardo Rubi Rörig, Giulia Burle Costa, José Bonomi Baruffi, Eduardo Bastos, Lyllyan Santos Rocha, Giovanna Destri, and Alessandra Larissa Fonseca

Abstract Marine pollution caused a growing trend of aesthetic and functional disturbances in coastal ecosystems with environmental, social and economic impairment, with abrupt acceleration in the 1970s. Within these last 50 years, the over-enrichment of continental runoff by nutrients has emerged as one of the leading causes of water quality impairment, with tremendous losses for ecosystem services, the foundations to sustain the fishery, aquaculture, recreation and tourism. Most of these scenarios are anthropogenically driven eutrophication (ADE). Algal blooms and oxygen depletion are among the symptoms of a problem that characterizes the anthropocene together with other global stressors. Worldwide, one of the most iconic cases of ADE on beaches is the green tides of *Ulva* spp., such as those occurred in Qingdao beach, China, in 2008 and the most recent *Sargassum* bloom in central tropical Atlantic, with floating material observed for 8000 km, from Africa to Central America, highlighting the problem to the global audience. In this chapter, we present the magnitude of ocean eutrophication, their causes and consequences to human beings and marine biodiversity, and potential solutions to increase the resilience of marine environments and coastal communities.

Keywords Marine pollution · Algal bloom · Coastal management · Ocean health · Dead zone · Biodiversity loss

---

P. A. Horta (✉) · L. R. Rörig · G. B. Costa · J. B. Baruffi · E. Bastos · L. S. Rocha · G. Destri
Phycology Laboratory, Botany Department, Federal University of Santa Catarina, 88040900 Florianópolis, Santa Catarina, Brazil
e-mail: paulo.horta@ufsc.br

P. A. Horta · L. R. Rörig · G. B. Costa · J. B. Baruffi · L. S. Rocha · G. Destri · A. L. Fonseca
Center of Physical and Mathematical Sciences, Federal University of Santa Catarina, 88040900 Florianópolis, Santa Catarina, Brazil

A. L. Fonseca
Marine Biogeochemistry Laboratory, Oceanography Department, Federal University of Santa Catarina, 88040900 Florianópolis, Santa Catarina, Brazil

© The Author(s), under exclusive license to Springer Nature Switzerland AG 2021
D.-P. Häder et al. (eds.), *Anthropogenic Pollution of Aquatic Ecosystems*,
https://doi.org/10.1007/978-3-030-75602-4_8

8.1 Introduction

Marine pollution causes a growing trend of aesthetic and functional disturbances in coastal ecosystems with environmental, social and economic impairment, with abrupt acceleration in the last 50 years. The observed over-enrichment of continental runoff of nutrients from human activities has emerged as one of the leading causes of seawater quality impairment, inducing significant losses of ecosystem services that sustain the fishery, aquaculture, recreation and tourism. The main consequence of this anthropogenic impact is the phenomenon denominated as cultural eutrophication and results from the lack of control over anthropic activities, especially related to agriculture, livestock and urbanization in continental areas (Malone and Newton 2020). Increasing in frequency and severity of harmful algal blooms (HAB) and oxygen depletion in coastal areas are among the significant symptoms of this environmental problem, characterizing a critical aspect of the proposed new geological Era—the Capitalocene. It is interacting with other global stressors, compromising the systemic functionality and ecosystem services necessary for human well-being, as well as biodiversity conservation (Altvater et al. 2016).

Ecosystems were altered since the early years of civilization organization, by human uses of resources and landscape modification, where agricultural and cattle raising were essential to establishing societies and development of culture. Considering the marine environment, fisheries' uses of resources were fundamental for the establishment of human communities in coastal areas. Unfortunately, the seawaters were also considered by society as the sinking of discharges, contributing historically to pollution and alterations in seawater physical-chemical characteristics worldwide.

Additionally, the expansion of hypoxic and culturally eutrophicated sites in coastal areas was registered through last 170 years (Fig. 8.1a), reinforcing a positive and strong correlation with the changes resulting from the intensification of land uses by human food production, disposal of wastewaters, and ecosystem modification. Historical records indicated that during the XIX century, anthropogenic damages—as hypoxia and cultural eutrophication—could be observed in three sites (Fig. 8.1b), evolving for dozens during Second World War (Fig. 8.1c) and widely distributed through all continental coastal areas during the last century (Fig. 8.1d). This scenario was significantly intensified during the 1980s, justifying the main concern among environmental activists and researchers with anthropogenically driven eutrophication (ADE) consequences to human and marine ecosystem future, as the disruption in nutrients balance produces changes in community structure by the elevated mortality of species of canopy-seaweeds, associated invertebrates, and fishes (www.wri.org). Observing the modifications in the microbial loop, due to the substitution of microorganisms that cycle nutrients for other, e.g., sulfur bacteria, the spread of hypoxic, or even the dead zones, where the levels of oxygen are lower than 2 mg L^{-1} (Diaz and Rosenberg 2008), can be understood. Worldwide, the major consequence of anthropic eutrophication is the occurrence of a phenomena knowledge as "green tides" by the massive and extensive reproduction of species from *Ulva* genus, such as in the event recorded in Qingdao beach, China in 2008 (Smetacek and Zingone 2013), and in a coastal area of France in 2012 (Diaz et al. 2013). The most recent *Sargassum* bloom in the central tropical Atlantic, with floating material observed for 8000 km, from Africa to Central America, stands out as a problem to the global audience (Wang et al. 2019). Despite the advances in mapping and understanding ADE's mechanisms, we still have extensive and important areas, such as the Africa coast or other regions in the Indo-Pacific and Atlantic, where the available information is not enough to identify the phenomenon, characterizes its causes and environmental consequences. Another essential aspect is ADE's interaction with multiple stressors related to the local and global scales. Therefore, the interactions among ADE, global warming, ocean acidification and the impact of emergent pollutants should be evaluated considering particularities we observed with the biogeographic provinces. The observed trends and uncertainties reinforce the needs for a global orchestration to solve ADE's causes and consequences.

However, this recognition did not promote the implementation of solutions we have, or we need to develop, mitigate or avoid this growing global warning phenomenon. Therefore, to restore foundations for the concept and trend of

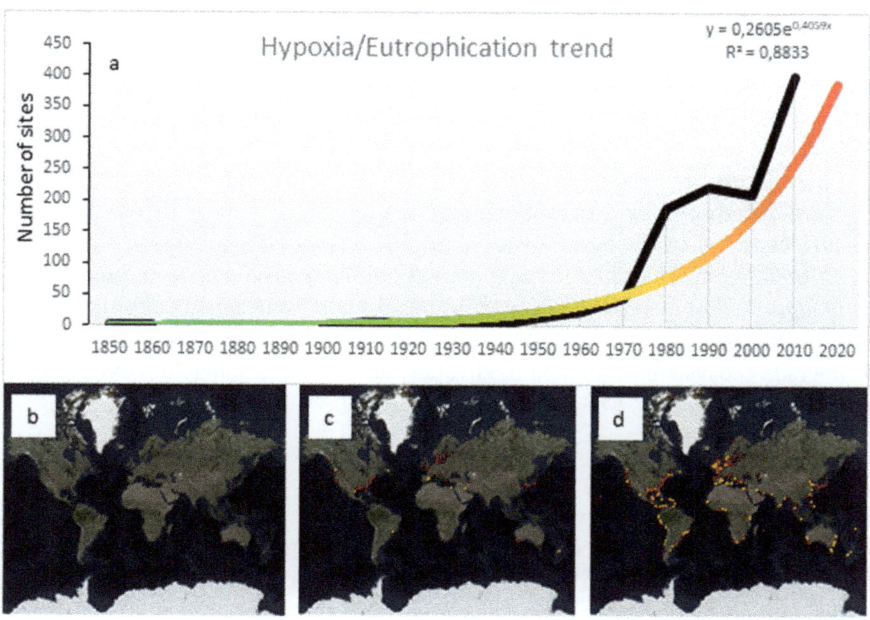

Fig. 8.1 Evolution of eutrophic (yellow spots) and hypoxic sites (red spots) through time and around the world oceans after the 1850, highlighting the periods of **a** 1850–1910, **b** 1910–1960; **c** 1960–2000 and the positive and strongly correlated ($R^2 = 0.8833$) exponential trend ($y = 0.2811e^{0.3942x}$). *Source* https://www.wri.org/ and Diaz and Rosemberg (2008); uncertainties related with scarce source of data from sites less our without monitoring programs

eutrophication, we should reinforce the necessity of institutional mechanisms to implement solutions (Duarte and Krause-Jensen 2018).

Fifty years ago, the first studies describing marine eutrophication phenomena were published, reinforcing the discussions started by limnologists, freshwater ecologists, and managers at the beginning of that century. In this context, eutrophication has been conceptualized as the process of water nutrient enrichment by an increase in dissolved forms of phosphorus (P) and nitrogen (N). This concept incorporates the consequent abundant plant and macrophyte growth or the bloom of phytoplankton in freshwater; consequently, the similar process was used to describe the marine phytoplankton and seaweed blooms, among other marine plants in seawater (Kouwe 1979). Since these first efforts, the concept was related to direct or indirect human influence in increasing the total concentration of phosphate and nitrogen nutrients in environment (Cuwvo-Wlgroep VI 1976; see the Box 8.1). In some cases, besides the key organisms and main driver are related to photosynthetic groups and inorganic nutrients, we should also consider the saprobiontic species due to the supply of organic pollutants and the abundance of groups as Euglenophyta (Schmidt-Van Dorp 1978). Since the pioneer studies in earlier 1970s in marine environments, the ADE in coastal waters has been correlated with the continental runoff enriched by organic matter and labile N and P (Nixon 1995). These main drivers interfere in biogeochemical cycles and modify the chemical dynamics of nutrients in marine water, reinforcing the interdependence and equilibrium among land and sea processes to the environmental regulation in a biological balance.

Box 8.1

The European Union describing good environmental status, defines **Human-induced or Cultural eutrophication** *as a process driven by the enrichment of water by nutrients, especially compounds of nitrogen and/or phosphorus, leading to increased growth, primary production, and biomass of algae, changes in the balance of organisms and water quality degradation.* The consequences of eutrophication are undesirable if they appreciably degrade ecosystem health and biodiversity and/or the sustainable provision of goods and services (Malone and Newton 2020).

The current and future scenarios of climate change can enrich all these basic knowledge and theoretical discussion. Ocean warming should increase metabolic rates (Martens et al. 2015) and higher availability of CO_2. The related ocean acidification provides an additional resource to the primary production, resulting in harmful algal blooms and green, golden and red tides, and all negative socio-environmental-economic consequences of ADE (Gobler 2020). Opportunistic organisms that can produce harmful algal blooms (HABs) can intensify ADE and its environmental and socioeconomic effects (Riebesell et al. 2018). Therefore, the eutrophication current concept* should evolve and be circumscribed as the anthropogenically driven nutrient-enhanced primary production and biomass accumulation of algae and macrophytes. These blooms thrive in warmer and acidified waters, resulting in bottom oxygen consumption and hypoxia or anoxia, leading to dead zone formation in different marine ecosystems and world regions (Fig. 8.1). The integrated concept involving a complex interaction among local and global multi-stressors, considering nutrient availability (N and P among others), ocean warming and acidification, primary producer blooms and the oxygen depletion, reinforces the necessity of integrated coastal management, fostering the adoption of multiple mechanisms to improve continental runoff quality, and mitigation of historical trophic disturbances (Fig. 8.2).

This chapter maps the global academic efforts regarding ADE, describing the causes and consequences of losses of biodiversity and environmental goods and services. To clarify likely future scenarios, discussing some cases, we present potential solutions regarding the vocations of different environments and regions. Moreover, describing the causes and consequences of ADE can predict the possible consequences for environmental goods and services provided to human well-being

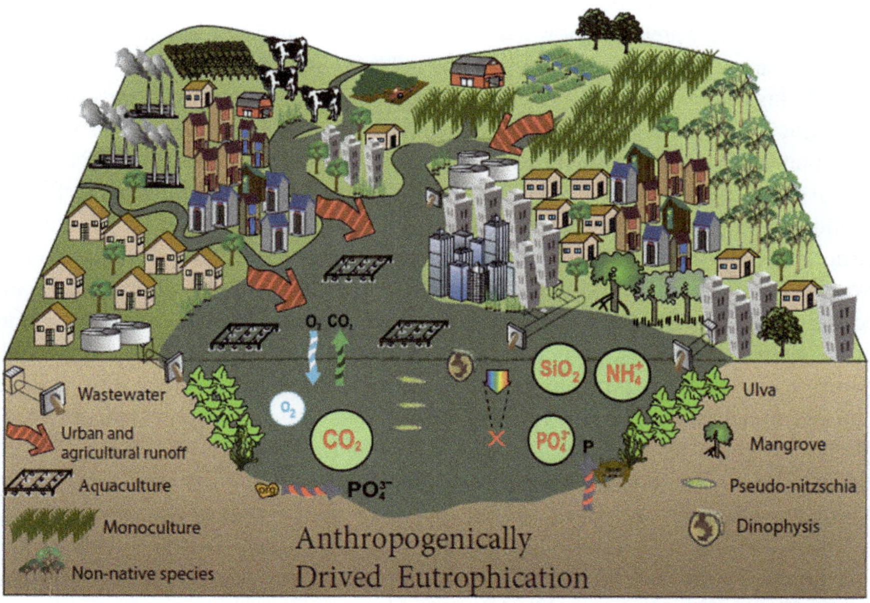

Fig. 8.2 Multi-stressor scenario related with anthropogenically driven eutrophication (ADE), where urban, industrial and agricultural (monoculture, deforestation, cultivation of non-native species) runoff combined and without wastewater treatment, enhance the organic matter and dissolved inorganic nutrients (represented by SiO_2, PO_4^{3-} and NH_4^+) facilitating the bloom of opportunistic seaweeds (i.e. *Ulva* spp.) and microalgae as diatoms and dinoflagellates (i.e., *Pseudo-nitzschia* and *Dinophysis*, respectively) in a scenario of loss of mangroves among other coastal marine forests. Because of these blooms, algae death and organic matter mineralization, consumption of O_2 and release of CO_2 are observed (adapted from Brauko et al. 2020)

and systemic conservation (Fonseca et al. 2021). The coming UN ocean decade should foster the intensification of ADE areas and the application of available techniques to mitigate or solve coastal pollution to attend the SDG 14, increasing the resilience of biodiversity and reducing the human being's vulnerabilities.

8.2 Conceptualizing the Anthropogenic Eutrophication

8.2.1 Causes of Eutrophication in Coastal Areas and Marine Systems

Coastal pollution is a global environmental problem of concern. Organic matter and nutrient pollution released to different coastal marine environments and rivers of different categories and distances of the littoral zone have diverse origins, including agriculture, aquaculture, septic tanks, urban wastewater, urban stormwater runoff, industry and fossil fuel combustion. These materials are a source of different forms of dissolved nitrogen and phosphorous, main nutrients that enter aquatic ecosystems via surface water, groundwater or even through water-atmosphere interactions.

There are significant variations in the relative importance of nutrient sources considering the historical moment, world region or country. For thousands of years, agriculture, the related deforestation and land use represent the main source of environmental nutrient enrichment in coastal marine ecosystems. In the Northern Hemisphere, in countries as those of the European Union and the United States, chemical fertilizers and animal manure are the primary sources of nutrient enrichment of marine and freshwater (Malone and Newton 2020). The savanna and tropical forest deforestation, the chemical fertilization for monocultures, the inadequate land management, the massive livestock rearing, the expansion of mechanized monoculture and animal protein production, mainly in Asia and South America, resulted in significant growth of the agricultural-related sources of aquatic nutrient enrichment. These processes raised ADE as an environmental threat of global concern, as worrying as global warming (Mvungi and Pillay 2019). Besides US (13 million tonnes per year) and Europe (8 million tonnes per year), China (32 million tonnes per year) and India (17 million tonnes per year) standout as main consumers, followed by countries as Brazil and Australia with annual consumption rates around 4 and 1.5 million tonnes per year, respectively (FAO 2014). These amounts grow with global growing consumption of commodities with no robust regional or global regulation that considers the impacts the related P and N applied in excess would be lost from the soil profile through leaching and surface runoff, polluting water bodies and respective coastal ecosystems (Hart et al. 2004). This scenario has with the distribution, frequency and intensity of ADE events, mostly concentrated in the Northern Hemisphere and river mouths where continental runoff concentrates the nutrient erosion from main global corn belts. Besides the American and European corn belt, the South America region that drains to the Plata River mouth must receive major attention. The agricultural and grazing areas of La Plata Basin contributes to altering the N cycle in the South Atlantic. Soybean production plays a critical role in La Plata runoff. It contributes an annual input of about 1800 Gg, due to biological N fixation, and the export roughly 1000 Gg per year (Bustamante et al. 2015).

After the Second World War, a period recognized as the "green revolution," N and P fertilizers' global use increased nine and fourfold, respectively. The pollution related to these activities is related to the crop, agricultural technique and soil characteristics.

Studies have shown that fertilizers are often applied more than crops need (Pellegrini and Fernández 2018). The soil exposition is associated with a 50% increase in the erosion rates, that take 20% of fertilizers in the surface runoff and leaching to groundwater. In addition, nutrients are lost, nitrogen vaporizes in the atmosphere in the form of ammonia, a process of volatilization. Under some conditions, 40–60% of the N applied to crops can be lost by volatilization (University of Delaware Cooperative Extension 2009). A portion of this volatilized ammonia pollutes rivers and marine environments through atmospheric deposition. Phosphorus, which binds to the soil, is generally lost through soil erosion from agricultural lands and dispersion of sediments in plumes distributed by riverine transport (Ekholm and Lehtoranta 2012).

This scenario of environmental modification and over natural resources uses incurs in ADE events more frequent and intense. In the North Hemisphere, river mouths concentrate elevated charges of nutrients related to superficial leaching. Especially in the USA, the elevated concentration of nutrients from main global corn belts is the principal cause of eutrophication in this country and the dead zone in the Gulf of Mexico (Breitburg et al. 2018). For the Southern Hemisphere, the Plata River Plume reaches the Brazilian Continental Shelf carrying red tides (HAB) that ban the commercialization of the aquaculture products, like mussel and oyster and promote losses in the regional economy (Proença et al. 2017; Fonseca et al. 2021).

Other important sources of nutrients that contribute to anthropic eutrophication are the urbanization process, driving marine and coastal ecosystems' eutrophication. Sewage is estimated to contribute between 12 and 68% of nitrogen enrichment of superficial and groundwaters. Regional variability is due to absence or differences in sewage treatment levels. In developing countries, fewer than 35% of cities have any sewage treatment (UNEP and WHRC 2007); the sewage treatment objectives remove solids or mineralize organic matter, but not nutrients (Selman et al. 2009). The global population growth is followed by a general and consolidated urbanization trend in coastal regions. Considering the higher prevalence of urbanization centers in coastal areas, special attention to the development of sewage management strategies to nutrient decrease is necessary to avoid intensification of ADE and losses to marine ecosystems (Lapointe et al. 2019). It is predicted that the human population grows from the current 7.8 billion to nearly 9.7 billion in 2050 and 10.9 billion by 2100 (United Nations 2019). Coastal areas are expected to grow from 1.2 billion people (ca. 1990) up to 5.2 billion by the 2080s (Rabalais et al. 2009). Population growth is associated with a 2.2% annual growth per capita demand for food, land, energy and other natural resources that will increase the sewage production with negative impacts (Häder et al. 2020).

Considering the main population growth will occur in less developed countries, where the fragility or the absence of the sewage treatment system plays a role (UNPD 2008), the rainfall and runoff, eventually combined with sewer overflow, can contribute to a higher input of organic matter and nutrients to hydrographic basins and then, polluting coastal marine systems. Related to the climatic changes expected to this century, the severity of monsoons and continental runoff will produce significant variability in the wastewater characteristics, associated with the erosive potential of the continental areas (Jarmalavicius et al. 2016). These extreme events can exceed the capacity of the sewage treatment system and even the household's septic systems. As a result, the excess wastewater, including raw sewage, can be discharged directly into nearby ecosystems (Wetz and Yoskowitz 2013). The urban runoff can also be contaminated with industrial sewage. Food processing plants among others agroindustrial activities (Koyer et al. 1995) and vessel discharges are important sources of additional nutrient pollution (Chen et al. 2016; Wilewska-Bien et al. 2016).

The demand for protein production fosters shrimp and fish farm growing and the consequent nutrient pollution related to these animals excretion, and excessive or inadequate feeding promotes the eutrophication process (Garlock et al. 2020). Global aquaculture has grown at an average annual rate of 2%, reaching 114.5 million tonnes in live weight by 2018 (FAO 2020a). Currently, 56% of all aquaculture production is within marine or brackish environments (FAO 2020b). Marine shrimp and fish farming often occur in estuarine ponds, cages, or net pens situated in enclosed bays. These farms generate concentrated amounts of organic waste and related dissolved nitrogen and phosphorus from excrement and uneaten food. Improper management and operations can result in severe aquaculture impacts on aquatic ecosystems, as nutrient-enriched waters are discharged directly into the surrounding ecosystems increasing oxygen demand and reducing pH. For every ton of fish, aquaculture operations produce between 42 and 66 kg of nitrogen waste and between 7.2 and 10.5 kg of phosphorus waste (Strain and Hargrave 2005).

All fossil fuels, from power plants to our cars, are another source of nitrogen pollution since the burn process releases NOx into the atmosphere. NOx can settle out after a dry by wet deposition, polluting land and waters. Fossil fuel combustion contributes approximately 22 teragrams of nitrogen pollution globally every year, approximately one-fifth of the contribution of synthetic nitrogen fertilizers. Atmospheric deposition can represent 25–30% of N inputs in marine environments and can exceed riverine contribution (Paerl et al. 2002; Spokes and Jickells 2005).

Climate changes can exacerbate the consequences of nutrient fertilization and eutrophication impacts of coastal water (Nazari-Sharabian et al. 2018). Oceanic and atmospheric warming increase water demand due to rising temperatures and evaporation while decreasing water availability in coastal environments. On the other hand, extreme events can increase coastal erosion, surface runoff and flooding, deteriorating coastal water quality.

High frequencies of extreme rain events and the sea level rise can overload the sewage system, increasing the frequency and intensity of coastal ecosystem contamination. Also, the increase in temperature and ocean acidification can accelerate the opportunistic species growth, increase the environmental stress of calcified bioengineer groups (hermatypic corals and coralline algae, among others), intensifying the negative impact of ADE (O'Neil et al. 2012; Snickars et al. 2015).

8.3 Algal Blooms: The Main Effect of Anthropic Eutrophication in Marine Systems

8.3.1 Conceptualizing Harmful Algal Bloom

The ADE events have as first consequence the increased productivity and algal biomass in receiving waters such as lakes and rivers (Smith 2016). The deterioration of the environmental health of these waters should be considered as evidence that the respective locality is part of the contributors of the global phenomenon of marine/coastal ADE (Smith 2003). The observed shift and losses of marine forests, structured by fauna and flora foundation species as corals and seaweeds (Gorman et al. 2020), represent the first movements of increased amount of opportunistic species that grow as epiphytes, as floating seaweed and microalgae blooms (Grall and Chauvaud 2002). These organisms and ecophysiological strategies compromise the capacity of different marine ecosystems to keep O_2 levels, favouring the dominance of anaerobic bacteria and the spread of anoxia. This process is followed by a change in the energy flow through the food web, from the benthic primary producers to higher-level predators. Under normoxia, the efficiency in carbon conversion can reach 75% by macrobenthic organisms, as example of canopy-forming species of seaweeds; but after mild or periodic hypoxia, there can be a pulse of benthic energy to other trophic levels (Fig. 8.2). With the intensification of hypoxia, the proportion of benthic energy transferred to the microbiome increases, as well as the abundance of anaerobic bacteria and resulting in persistent dead zones (Diaz and Rosemberg 2008). The losses of benthic primary production, depletion of O_2, and acidification transform autotrophic in heterotrophic communities and later into an anaerobic system. In biogenic reefs, corals and coralline red algae lose fitness and are overgrowth by opportunistic species. This is an unequivocal trend worldwide.

Estuarine and coastal marine ecosystems are among those that first receive nutrient-enriched waters and are more severely affected by ADE. In these environments, the most common and related biological process is the algal bloom, including micro and macroalgae species. These massive proliferation results in biomass accumulation, toxins production by HAB species (Box 8.2), and adverse effects associated with thallus death and decomposition. This combination of factors results in a favorable condition for the development of nuisance, toxic or more frequent and intense harmful algal blooms (Heisler et al. 2008).

Box 8.2
The algal toxins are metabolic derivatives, synthesized under stress situations and massive proliferation of algae. There are different classes of algal toxins, to be cited; the major four are: neurotoxins, hepatotoxins, cytotoxins and dermatotoxins. Those metabolites, when ingested or aspired by other organisms, can induce effects of lethargy, gastrointestinal disturbances and poisoning, and in extreme cases, the mortality of organisms including humans. Because of this toxic effect, these massive proliferations are also described as "harmful algal blooms" (HABs).

Among sites and cases reported worldwide, one of the clearest examples of the direct development of a toxic species in response to increased nutrient loading is the harmful blooms of the diatom *Pseudo-nitzschia* spp. on the Gulf of Mexico under the influence of Mississippi River plume. Blooms of *Pseudo-nitzschia* spp. develop in high abundances during the spring, when nutrients reach the highest concentrations. Historical data and frustules preserved in the sand bottom indicate a significative increase in *Pseudo-nitzschia* spp. abundance since the 1950s, concomitant with increases in nutrient loading (Anderson et al. 2002). The field observation was corroborated by mesocosm experiments when nutrient pulsing effect was evaluated in *Pseudo-nitzschia* spp. dynamics (Dortch et al. 2000), with a similar effect in populational growth.

A Similar positive correlation has been described among the other cases of macroalgae blooms, or the golden and green tides. In the specific case of *Sargassum* blooms occuring in the Central Atlantic, high surface temperatures and nutrient input from the Amazon (Djakouré et al. 2017), as a result of deforestation and intensified agricultural (Wang et al. 2019), resulted in the historical accumulation of this seaweed in the region, with drastic consequences to coastal environments as the Caribbean coral reef systems (Gouvea et al. 2020).

The ADE can produce high amounts of biomass, attaching the tonnes of algae free-floating or deposited in coastal areas. The *Sargassum* spp. algal biomass is recognized by the ability to adsorb and absorb elements as heavy and trace metals, e.g., the metalloid arsenic, cadmium, copper, mercury, zinc and other potentially toxic elements. These elements can accumulate in organisms, with effect to populational health and dynamic disruption; especially, mercury deserves attention by the effect of biomagnification, when the metal is transferred by trophic chain and impacts the higher levels of predators, as sharks and marine mammals (Souza-Araujo et al. 2020; Merly et al. 2019). These well recognized potentially toxic elements associated with the emergent properties of other contaminants, as recently observed in microplastic impacts, can interact with the poisoning effect of ADE. With this, it is important to reinforce the future correlations among nutrients inputs and synergic negative effects with other anthropic activities (e.g., mining, fisheries, industrial manufactures, deforestation and land uses, agriculture in rain-forest areas) to marine fauna and flora. The example of extensive golden tide, formed by *Sargassum* spp. in South and Central Atlantic, can be indicated as a result of multiple divers acting to allow the HABs and, consequently, damaging the ecosystem health in wide scale (Sissini et al. 2017).

8.3.2 Impacts of HABs on Benthic Organisms

The fertilization of these coastal ecosystems with N and P negatively affects coral calcification, likely both a direct physiological response to nutrients and an indirect response to a shifting pH natural variability pattern. Therefore, ADE could make reefs more vulnerable to ocean warming and acidification, compromising net community calcification and accelerating the predicted shift from net accretion to net erosion (Silbiger et al. 2018). In addition, ADE can increase the severity of coral diseases compromising the entire ecosystem health (Bruno et al. 2003).

Despite to be recognized worldwide, the water quality is not considered as a factor affecting the resilience of some reef systems, as the Great Barrier Reef (Hughes et al. 2017), where the dispersal and the low anthropogenic pressure reduce the significance of any generalization. However, growing the perspective reinforces that these local stressors are relevant and should interact with global stressors (as warming and acidification), compromising the fitness of reef builders and their environmental roles (Hall et al. 2018).

The losses of coral reef functioning considering calcification and the compromise of coasts goods and services can be even worse when considering that ADE indirectly can reduce carbon sequestration in tropical seagrass beds. Considering that these environments can store 10% of the annual buried organic carbon in the oceans, measures should be taken to reduce nutrient input into these coastal ecosystems to preserve or enhance its carbon sink potential (Jiang et al. 2018). The ADE-related algal blooms decreases water column transparency and can result in additional biotic stress associated with the decline of critical foundation species of seagrass. It is essential to highlight that degradation of these foundation species can cause cascading effects, including losses of higher trophic levels as their habitat, food source or both can be compromised or even disappear (Bittick et al. 2018). High nutrient levels can cause seagrass shoot loss of about 60% when combined with other stressors, as the burial of fronds related to extreme events such as storms (Ceccherelli et al. 2018). The resilience of these systems is directly related to their abundance and diversity. Sites with reduced abundance and diversity were less resilient to nutrient enrichment (Gladstone-Gallagher et al. 2018).

Aquaculture and fishery represent important socioeconomic segment threatened by ADE that enhances concern about their complex global consequences.

Intensification of ADE and hypoxia inshore and offshore systems affects mobile wild fishes and invertebrates, compromising the yield of these important economic sectors. Reduced O_2 concentration, H_2S release from organic matter decomposition, associated with HABs, resulted in mass mortality of fishes (Buslé 1995) reduced shrimp and bivalve production (Guyoneaud et al. 1998; O'Connor and Whitall 2007). The hypoxic event of New York—New Jersey in 1976, affected an area of about 1000 km^2, displacing demersal fishes and migration of pelagic bluefish compromising fishery (Azarovitz et al. 1979). Therefore, ADE can alter fisheries stocks' spatial distribution compromising food availability and security (Selberg et al. 2001; Craig et al. 2005). Considering aquaculture, the frequent HABs compromise the yield and commercialization due to the high concentration of different algal toxins and associated reduced fitness (Díaz 2010).

The socioeconomic impacts of ADE and the correlated algal blooms and mass mortality can also impact segments dependent on the coastal ecosystems' aesthetic values, such as tourism. Besides the unpleasant odour and the high biomass covering beaches and coastal waters, massive biomass production related to ADE affects human health due to the gases produced during decomposition. Sulfuric acid and ammonia released during this process can be toxic under chronic exposure. The exposure to these gasses can irritate the upper airways, induce headache, nausea, confusion and, in extreme cases, result in hypoxic pulmonary, neurological and cardiovascular lesions (Chávez et al. 2020). Therefore, the consequences of ocean health loss caused by ADE can impose negative economic impacts, including decreased property values and reduced recreational uses.

8.4 Systematic Review—Evidencing the Eutrophication in Coastal Systems

A review of the last 10 years regarding Ocean eutrophication revealed studies focused on evaluating the effects of eutrophication in the field, through sampling and/or monitoring, showed the highest proportions among the three evaluated categories (54.95%). For seagrass, field experiments (15.38%) were representative to understand the effects in community structure and herbivory influence in seagrass beds. The major efforts were addressed in temperate areas (45 studies), followed by tropical regions (19 studies), possibly correlated to the higher prevalence of scientific research institutes in these areas, and because the specific interest in coastal systems prevalent in these areas, with efforts especially related with coral reefs (see details in the following discussion). Tropical regions received more attention in the last decade (Fig. 8.3) what indicate the necessity to attention to tropical ecosystems considering their potential vulnerability to eutrophication (Wiedenmann et al. 2013; Lapoint et al. 2019).

Major efforts identified as not informed (NI) region because included multiple data from different regions sites or are just generic descriptions, not focused in any

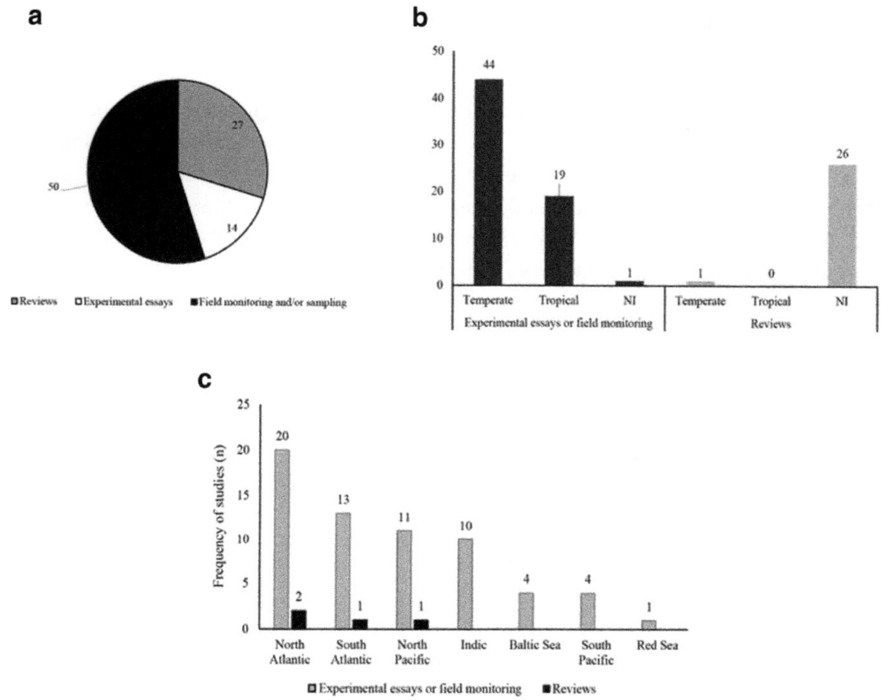

Fig. 8.3 Efforts distributed among categories (**a**) and biogeographic zones (**b**) and regions (**c**). The abbreviation NI represents studies that have no information specific data for biogeographic provinces, including modeling studies in the experimental group and generic reviews

specific example. Interesting to observe that for both groups, any record was obtained for polar or colder areas in higher latitudes, deserving attention, considering that submerged ecosystems, as kelp forests, can be in threat of being endangered by increasing nutrients distribution by oceanographic currents. Few records were obtained for seas or oceans in review studies, and only four articles focused on a specific coastal area. Even the Baltic Sea and Red Sea are in Northern Hemisphere, these areas are presented individually due to the economic and cultural importance of the coastline, receiving the efforts of field monitoring in four and one study, respectively. The marine area that received more attention was the North Atlantic, concentrating 21.97% studies (Fig. 8.3).

The analyses by ecosystems showed differences among the group of experimental essays or field monitoring and the reviews categories (Fig. 8.4). In the first group, the major focus was addressed on estuarine and lagoon systems (42.19%). This higher frequency is represented by the efforts realized in the United States of America, Brazil, and China, where the estuarine systems are very representative among the analyzed articles. Moreover, the mangroves (18.75%) were widely investigated in countries as India, South Africa, Brazil and small island countries in the Pacific.

It is important to highlight the low efforts to analyze eutrophication effects in rhodolith beds and maërl, with two studies for each (6.25% altogether). These systems are very delicate considering different sources of pollution (Bjork 1995) and are under pressure by the interactive effects of multiple stressors, such those related to global warming, marine heatwaves and ocean acidification (Schubert et al. 2019). In a recent literature review by Koerich et al (2021), the author highlights the urgency to better understand the associative effects of diverse drivers and stressors over rhodolith beds. This review indicated the absence or few literature descriptions for the impacts of eutrophication interactively to global stressors, e.g., marine heatwaves, global warming, ocean acidification or chemical pollution. With this, specific efforts must be addressed to understand the interactive and

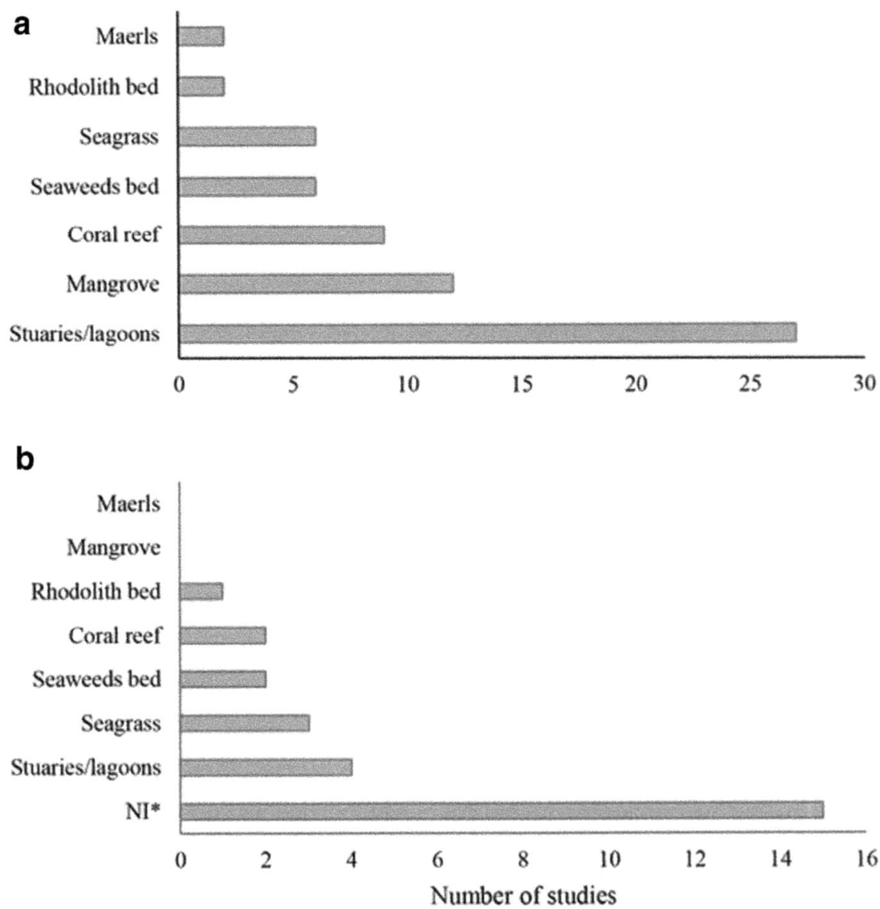

Fig. 8.4 Last ten years of scientific efforts with the evaluation of causes and consequences of ocean eutrophication considering different ecosystems. **a** Experimental essays and field monitoring and **b** reviews

collapsed interactions among those drivers to the conservation of coralline crustose algae worldwide.

All these drivers present relevant impact over calcareous algae physiology, with recognized consequences for the calcification process and photosynthetic performance (Schubert et al. 2019). In recent years, the interaction between global changes and local impacts deserved attention, considering the collapse that interactions among different stressors can exercise on calcareous algae. Special attention has been directed to the eutrophication, considering the higher sensibility of calcareous algae to higher nutrient concentrations. The production of monitoring data for these systems is imperative and urgent and must be stimulated in different countries as a strategy to conserve a system unique and able to act as a carbon sink, contributing to ameliorate the effects of climatic changes.

The data produced by this type of review is relevant to discuss the main drivers correlated to the eutrophication process and the expected effect that can be input for these systems. However, it is also important to highlight that even though there are gaps in knowledge, to determine exactly the effects of eutrophication on marine organisms, especially seaweeds, reviews that bring only compilation of well-established data are not quite relevant to improve the solution strategies to protect and restore the coastal systems. This kind of literature must be avoided in a closer future, or at least a special effort can be stimulated to complementary analyses such as metanalytical reviews and application to specific coastal management strategy.

The review categories evidenced the estuarine systems as the most evaluated ecosystem (14.81%). Presently, were not identified reviews focused on the eutrophication impacts in mangroves and maërls. Considering the high effort to monitor mangroves worldwide, a specific evaluation of actions to monitor the eutrophication impacts on mangroves can contribute to focus main efforts in producing data to infer this system health. Additionally, the absence of reviews applied to maërls deserve attention as an important marine system with poor knowledge about the impacts of the eutrophication process and interaction with other global changes.

8.5 Solutions: How is It Possible to Prevent the ADE Effects on the Ocean

The 14th UN goal for sustainable development (SDG-14) fosters the development of actions and solutions related with aquatic environments (Griggs, 2013), establishing management responses that reduce the flux of nutrients and organic matter to the environment and promoting the conservation and restoration of marine ecosystems.

The drives of ADE involve complex and interrelated socioeconomic factors in association with ecological-structural functions of the environment (Selman and

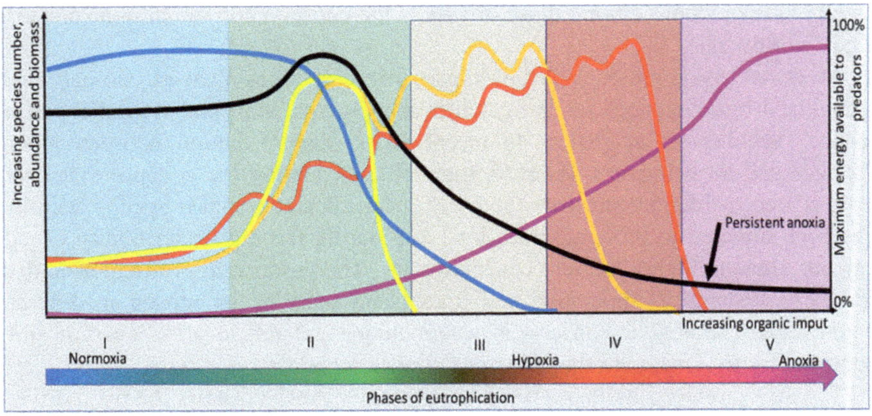

Fig. 8.5 Conceptual model of the changes observed in euphotic environments dominated by marine forests, like corals, seaweeds and seagrasses, during increasing phases of eutrophication and growing anoxia where: blue line: marine forest abundance, yellow line: opportunistic species abundance, brown line: free-floating seaweed abundance, red line: Phytoplankton abundance, purple line: anaerobic bacteria abundance. Black line maximum energy available to predators, black arrow indicate persistent anoxia in the timeline (adapted from Grall and Chauvaud 2002; Diaz and Rosenberg 2008)

Greenhalgh 2009; Elliott et al. 2017). In addition, propose solutions to complex problems requires strategies, including institutional and civil society actions, in a combinatory perspective that respects environmental characteristics, from local to global scales, and vice versa (Fig. 8.5). Mitigation programs and management strategies must consider other pressures that act in synergy with the eutrophication (Malone and Newton 2020). Overfishing, climatic changes and extreme events (heatwaves, cold spells and monsoons), ocean acidification and other chemical sources of pollution can exacerbate the impacts of the eutrophication symptoms in the ecosystem services (Brauko et al. 2020; Fonseca et al. 2021). All scenarios must be considered to properly implement management strategies to mitigate the effects of eutrophication, as well reduces the harms from other environmental stressors and anthropic disturbances.

8.6 The Role of Governance Initiative in Mitigate ADE Effects

It is important to reinforce the role and strategic position of environmental managers and politics to the decisions and response plans to mitigate the effects of eutrophication. Despite the perception that human activities interfere in the quality of seawater and coastal ecosystems, managers and civil society members often do not completely understand origins, causes, consequences and initiatives to avoid socio-environmental losses imposed by eutrophication. (Selman and Greenhalgh 2009). In this context, strengthening cooperation among scientists and researchers with environmental managers can improve decision making in restoration or conservation actions. Additionally, the comprehension of eutrophication processes and correlated causes and consequences to human well-being and ecosystem health needs to be popularized to improve the public engagement with this issue (Boesch 2019). The step forward is to improve the environmental education and the access to scientific information, communicating and dialoguing the problematic to a collective decision about the better strategies to provide accountable governance (Ghilardi-Lopes 2015; Gough 2017).

8.7 Sludge and Sewage Treatment: A Key Initiative to Reduce Nutrients Input into the Environment

The improvement of continental runoff involves complex actions at different scales. Its management requires planning, a broad spatial understanding of the causes and consequences and the articulation of multiple actors: from rural to industrial owners; from municipal to federal governments; including teaching and research institutions; the entire population residing in the coverage areas. These solutions can be divided into (1) preventive or management actions and (2) corrective or remediation actions. In the first case, the actions are focused on the causes and in the second case, the consequences are acted upon (Fig. 8.6).

Regarding preventive or management actions, the starting point is the overview understanding of the process, the causes and susceptibility of the ecosystem for the ADE. With this mapping, a diagnosis of the land and water use is carried out to feed integrated planning, which guides integrated management.

The most common management actions involve the implantation of wastewater treatment systems (tertiary plants), the ordination and change of land use, implementation of marine protected areas and the restoration of buffer ecosystems in the watershed (e.g., riparian forests, wetlands) and on the coastal fringe (e.g., mangroves, salt marshes, seagrass and seaweed beds). The restoration actions mentioned here can also be considered remediation actions. However, they differ from the informal classification adopted here as they are closer to the origins of the problem than to the place affected by eutrophication symptoms. There is a broad literature and many successful examples of management actions aimed at reducing the nutrient load, involving land (Nair and Graetz 2004), sea (Bergström et al. 2015; Duarte et al. 2018) or integrated land/sea approaches (Halpern et al. 2019).

Corrective or remediation actions involve emergency procedures, that is, when the environment is already compromised and such that actions at the level of causes do not produce measurable effects in a timely manner and need to be accompanied by extra measures. This is the case for most sites prone to coastal eutrophication. Some remediation measures include the restoration and/or creation of coastal

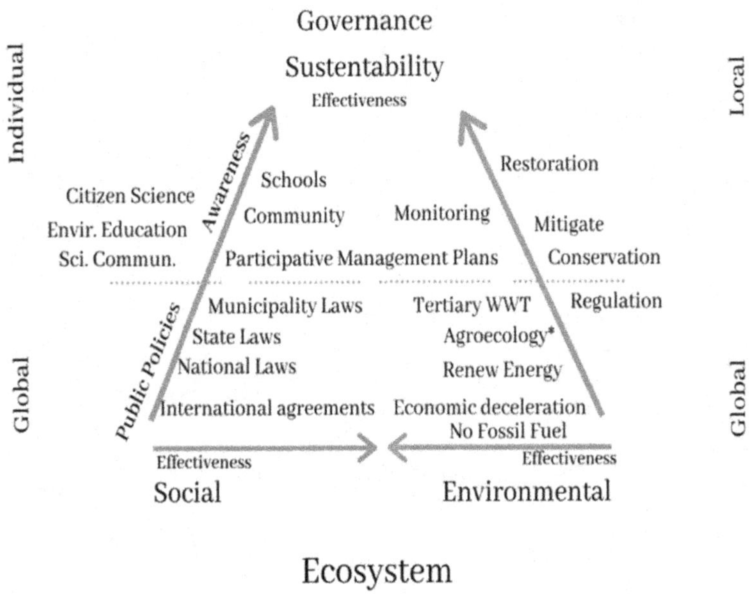

Fig. 8.6 Conceptual framework necessary to achieve resilient solution involving governance for sustainability

ecosystems in situ, in the affected location to retain or absorb pollutants, the installation of macroalgae cultivation systems or integrated multitrophic aquaculture, the construction of algal turf scrubber systems and the management of blooms that occur in the affected places because of the advanced eutrophication process.

The efficiency of the implementation of practices, as reduce fertilizing uses and operation of urban wastewater treatment plants (WWT plants), to control the eutrophication processes depends not only on legal directives, but also on the individual engagement. For example, by individual decision, many residential units did not have their sewage connected to the WWT plants in a coastal lagoon watershed of South Brazil, contributing to the maintenance of the eutrophication in the system (Cabral et al. 2019). On the other side, stakeholders and educators' active involvement in the environmental monitoring program of Chesapeake Bay promoted more effective environmental policy (Zint et al. 2002; Bell et al. 2003). The stakeholders' relevance in participating actively to achieve the goals to reduce nutrient pollution and improve conservation is explicit in complex systems, like the Baltic Sea in Northern Europe (Grimvall et al. 2017) and the coastal zone of the United States (Ardoin and Merrick 2013).

With this, the better solutions to be proposed are the ecosystem-based management, focusing in restore and use the natural potential of ecosystems to cycle and transform the nutrients input (Borja et al. 2016; Malone and Newton, 2020). For example, Algae-based solutions (AbS) can represent an important alternative to reduce coastal eutrophication with the production of biomass and socio-environmental and economic development. Restoration of mangroves, seagrasses, salt marshes and other coastal environments, combined with multitrophic aquaculture, complement conjunction of actions that will sink CO_2, produce O_2, filter dissolved inorganic nutrients and reduce socio-environmental vulnerabilities, increasing biodiversity and health of our ocean (Fig. 8.7).

Fig. 8.7 Coastal scenario resulting from science-based management, where coastal runoff and wastewater receive treatment (removing contaminants and dissolved nutrients). Algae-based solutions (AbS) can represent an important alternative to reducing coastal eutrophication with biomass production and socio-environmental and economic development. Restoration of mangroves, seagrasses and salt marshes among other coastal environments, combined with multitrophic aquaculture complement conjunction of actions that will sink CO_2, produce O_2, and reduce socio-environmental vulnerabilities, increasing biodiversity and health of our oceans (adapted from Brauko et al. 2020)

8.8 The Future We Want and Need

The future of the ocean and all coastal systems depends on a multidisciplinary approach to mitigate and ameliorate the impacts of anthropogenic eutrophication. As highlighted, there are many literature gaps, describing the effects of eutrophication in fragile systems, like coral reefs, rhodolith and maërls beds. This scenario reinforces the importance of consolidating and expanding scientific efforts to better understand ADE's consequences in marine systems and the synergic effects with other local to global stressors that can induce severe impacts in systemic functionality.

Additionally, it is important to improve initiatives of environmental education and develop a green economy, based on good and sustainable practices in production. Those initiatives can contribute to individual actions environmentally friendly and the conscience of good practices to preserve fresh and marine water health. This perception is urgent and relevant to develop society and guarantee the human well-being for the next generations.

Finally, the eutrophication requires institutional and governmental efforts to reduce, mitigate and prevent anthropogenic inputs of nutrients. As proposed by SDG 14, a global concern must be addressed to conserve marine environments and biodiversity, restoring the systemic balance. The acceleration of an eventual virtuous cycle demands a re-evaluation of the economic model (Altvater et al. 2016). It can be induced and further optimized by legislation and instruments to foster the adoption of solutions and restoration against eutrophication impacts. The nature-based solutions could provide good options and must be constantly evaluated through monitoring programs. Therefore, coastal management and the integration of traditional and academic knowledge can produce efficient solutions to reduce vulnerabilities and foster socio-environmental resilience. Marine scientists and managers, stakeholders and policymakers, with the discussion, development and implementation of these science-based techniques and actions, should involve communities in a wave of recovery of ocean health, preserving the environment to the well-being of the future generations.

Beyond a nature-based solution, cultural eutrophication is driven for food production (by intensive fertilization and land use) and for expansion of urban areas and infrastructure (Malone and Newton, 2020), which is related to the historical and usual business model for prosperity. Absolute decoupling (AD) was considered as a way when raw resources use declines in absolute terms while the gross domestic product (GDP) grows, a step for environmental sustainability (Otero et al. 2020). However, the AD as an alternative to modern society has not occurred so far (Alexandre et al. 2015). It is necessary to put in place a truly new paradigm shift, reducing the use for natural resources and the cultural consumerism (Acosta and Brand 2018). According to these authors, to target this change, society needs to change its consumerism values and policies need to promote economic slowdown and economic equality. Our society must focus on biodiversity and well-being (Otero et al. 2020). The change of human behavior, and its relationship with nature, is a challenge for our society, and it may start from knowledge, and the awareness of environmental problems is vital (Reid et al. 2010). In this way, environmental education, science communication and citizen science connected with human sustainability are powerful strategies and need to be developed in formal and informal educational systems.

The future of the ocean and all coastal systems depends on a multidisciplinary approach to mitigate and ameliorate anthropogenic eutrophication impacts. As highlighted, there are many literature gaps, describing the effects of eutrophication in fragile systems, like coral reefs, rhodolith beds and maerls. This scenario reinforces the importance to consolidate and expand scientific efforts to better understand the consequences of ADE in marine systems, as well as the synergic effects with other local to global stressors that can induce severe impacts in systemic functionality.

Additionally, it is important to improve environmental education initiatives and develop a green economy based on good and sustainable practices in production. Those initiatives can contribute for individual actions environmentally friendly and the conscience of good practices to conserve fresh and marine water. This perception is urgent and relevant to develop society and guarantee the human well-being for the next generations.

Finally, the eutrophication derivative from anthropic inputs of nutrients requires institutional and governmental efforts to be reduced, mitigated and prevented. As proposed by SDG 14, a global concern must be addressed to conserve marine environments and biodiversity, restoring the systemic balance.

References

Altvater E, Crist EC, Haraway DJ, Hartley D, Parenti C, McBrien J (2016) Anthropocene or capitalocene? Nature, history, and the crisis of capitalism. Oakland Press, CA, PM Press

Ardoin N, Merrick C (2013) Environmental education: a brief guide for US grantmakers. Retrieved from NAAEE website: https://naaee.org/eepro/resources/environmental-education-brief-guide-us

Azarovitz TR, Byrne CJ, Silverman MJ, Freeman BL, Smith WG, Turner SC, Halgren BA, Festa PJ (1979) Effects on finfish and lobster. In: Swanson RL, Sindermann CJ (eds) Oxygen depletion and associated benthic moralities in New York Bight, 1976. NOAA Professional Paper 11, pp 295–314

Bell WH, Fowler EM, Stein JA (2003) Coastal seas as a context for science teaching: a lesson from Chesapeake Bay. Mar Pollut Bull 47:253–259. https://doi.org/10.1016/S0025-326X(02)00468-X

Bergström P, Lindegarth S, Lindegarth M (2015) Modeling and predicting the growth of the mussel, *Mytilus edulis*: implications for planning of aquaculture and eutrophication mitigation. Ecol Evol 5(24):5920–5933

Bittick SJ, Sutula M, Fong P (2018) A tale of two algal blooms: Negative and predictable effects of two common bloom-forming macroalgae on seagrass and epiphytes. Mar Environ Res 140:1–9

Boesch DF (2019) Barriers and bridges in abating coastal eutrophication. Front Mar Sci 6:1–25. https://doi.org/10.3389/fmars.2019.00123

Bruno JF, Petes LE, Harvell CD, Hettinger A (2003) Nutrient enrichment can increase the severity of coral diseases. Ecol Lett 6(12):1056–1061

Truslé J (1995) The impact of harmful algal blooms on finfish. Mortality, pathology and toxicology. 6th Conférence Internationale Sur Le Phytoplancton Marin, Nantes, 18-22 Octobre 1993. IFREMER, Plouzane. The impact of harmful algal blooms on finfish: occurrence of fish kills, pathology, toxicological mechanisms, ecological and economic impacts. A review. 75 p

Cabral A, Bercovich MV, Fonseca A (2019) Implications of poor-regulated wastewater treatment systems in the water quality and nutrient fluxes of a subtropical coastal lagoon. Reg Stud Mar Sci 29:100672 https://doi.org/10.1016/j.rsma.2019.100672

Ceccherelli G, Oliva S, Pinna S, Piazzi L, Procaccini G, Marin-Guirao L, Dattolo E, Gallia R, La Manna G, Gennaro P, Costa MM, Barrote I, Silva J, Bulleri F (2018) Seagrass collapse due to synergistic stressors is not anticipated by phenological changes. Oecologia 186(4):1137–1152

Chávez V, Uribe-Martínez A, Cuevas E, Rodríguez-Martínez RE, van Tussenbroek BI, Francisco V, Estevez M, Celis LB, Monroy-Velázquez LV, Leal-Bautista R, Álvarez-Filip L, García-Sánchez M, Masia L, Silva R (2020) Massive Influx of Pelagic *Sargassum* spp. on the Coasts of the Mexican Caribbean 2014–2020: challenges and Opportunities. Water 12 (10):2908

Chen CW, Ju YR, Chen CF, Dong CD (2016) Evaluation of organic pollution and eutrophication status of Kaohsiung Harbor, Taiwan. Int Biodeterior Biodegradation 113:318–324

Craig JK, Crowder LB, Henwood TA (2005) Spatial distribution of brown shrimp (*Farfantepanaeus aztecus*) on the northwestern Gulf of Mexico shelf: effects of abundance and hypoxia. Can J Fish Aquat Sci 62:1295–1308

de Barros Marangoni LF, Ferrier-Pagès C, RottierC, Bianchini A, GroverR (2020) Unravelling the different causes of nitrate and ammonium effects on coral bleaching. Sci Rep 10(1):1–14

Diaz M, Darnhofer I, Darrot C, Beuret JE (2013) Green tides in Brittany: what can we learn about niche–regime interactions? Environ Innov Soc Transitions 8:62–75

Díaz RJ (2010) Agriculture's impact on aquaculture: hypoxia and eutrophication in marine waters 275–318

Dortch QML, Parsons GJ, Doucetre GA, Frell, A Mamr A, Thzssmg CL Powf A, SONIAT M (2000) *Pseudo-nitzschia* spp. in the northern Cuff of Mexico: overview and response to increasing eutrophication, p. 27. Ta Symposium on Harmful Marine Algae in the U.S., December 4-9, 2000. Marine Biological Laboratory, Woods Hole, Massachusetts

Duarte CM, Krause-Jensen D (2018) Intervention options to accelerate ecosystem recovery from coastal eutrophication. Front Mar Sci 5:470

Elliott M, Burdon D, Atkins JP, Borja A, Cormier R, de Jonge VN, Turner RK (2017) "And DPSIR begat DAPSI(W)R(M)!"—a unifying framework for marine environmental management. Mar Pollut Bull 118:27–40. https://doi.org/10.1016/j.marpolbul.2017.03.049

FAO (2020a) The state of world fisheries and aquaculture 2020. FAO, Rome, p 2020

FAO (2020b) Global aquaculture production. Online query. http://www.fao.org/fishery/statistics/global-aquaculture-production/query/en. Accessed 09 Dec 2020

Garlock T, Asche F, Anderson J, Bjorndal T, Kumar G, Lorenzen K, Ropicki A, Smith MD, Tveteras R (2020) A global blue revolution: aquaculture growth across regions, species, and countries. Rev Fish Sci Aquac 28(1):107–116

Ghilardi-Lopes NP (2015) Citizen science combined with environmental education can be a powerful tool for coastal-marine management. J Coast Zo Manag 18(3):407

Gladstone-Gallagher RV, Hughes RW, Douglas EJ, Pilditch CA (2018) Biomass-dependent seagrass resilience to sediment eutrophication. J Exp Mar Biol Ecol 501:54–64

Gough A (2017) Educating for the marine environment: challenges for schools and scientists. Mar Pollut Bull 124:633–638. https://doi.org/10.1016/j.marpolbul.2017.06.069

Grimvall A, Sundblad EL, Sonesten L (2017) Mitigating marine eutrophication in the presence of strong societal driving forces (88)

Häder D-P, Banaszak AT, Villafañe VE, Narvarte MA, González RA, Helbling EW (2020) Anthropogenic pollution of aquatic ecosystems: emerging problems with global implications. Sci Total Environ 713:136586

Hall ER, Muller EM, Goulet T, Bellworthy J, Ritchie KB, Fine M (2018) Eutrophication may compromise the resilience of the Red Sea coral *Stylophora pistillata* to global change. Mar Pollut Bull 131:701–711

Halpern BS, Cottrell RS, Blanchard JL, Bouwman L, Froehlich HE, Gephart JA, Jacobsen NS, Kuempel CD, McIntyre PB, Metian M, Moran DD, Nash KL, Többen J, Williams DR (2019) Opinion: putting all foods on the same table: achieving sustainable food systems requires full accounting. Proc Natl Acad Sci 116(37):18152–18156

Hart MR, Quin BF, Nguyen ML (2004) Phosphorus runoff from agricultural land and direct fertilizer effects: a review. J Environ Qual 33(6):1954–1972

Heisler J, Glibert PM, Burkholder JM, Anderson DM, Cochlan W, Dennison WC, Dortch Q, Gobler CJ, Heil CA, Humphries E, Lewitus A, Magnien R, Marshall HG, Sellner K, Stockwell DA, Stoecker DK, Suddleson M (2008) Eutrophication and harmful algal blooms: a scientific consensus. Harmful Algae 8(1):3–13

Hughes TP, Kerry JT, Álvarez-Noriega M, Álvarez-Romero JG, Anderson KD, Baird AH, Babcock RC, Beger M, Bellwood DR, Berkelmans R, Bridge TC, Butler IR, Byrne M, Cantin NE, Comeau S, Connolly SR, Cumming GS, Dalton SJ, Diaz-Pulido G, Eakin CM, Figueira WF, Gilmour JP, Harrison HB, Heron SF, Hoey AS, Hobbs JPA, Hoogenboom MO, Kennedy EV, Kuo CY, Lough JM, Lowe RJ, Liu G, McCulloch MT, Malcolm HA, McWilliam MJ, Pandolfi JM, Pears RJ, Pratchett MS, Schoepf V, Simpson T, Skirving WJ, Sommer B, Torda G, Wachenfeld DR, Willis BL, Wilson SK (2017) Global warming and recurrent mass bleaching of corals. Nature 543(7645):373–377

Jarmalavičius D, Šmatas V, Stankūnavičius G, Pupienis D, Žilinskas G (2016) Factors controlling coastal erosion during storm events. J Coast Res 75:1112–1116

Jiang Z, Liu S, Zhang J, Wu Y, Zhao C, Lian Z, Huang X (2018) Eutrophication indirectly reduced carbon sequestration in a tropical seagrass bed. Plant Soil 426(1–2):135–152

Koerich G, Costa GB, Sissini MN, Ortiz CL, Canever BF, Oliveira W, Tonkin JD, Horta PA (2021) Physiology, Niche characteristics and extreme events: current and future habitat suitability of a Rhodolith-forming species in the southwestern Atlantic. Mar Environ Res x:x–x

Lapointe BE, Brewton RA, Herren LW, Porter JW, Hu C (2019) Nitrogen enrichment, altered stoichiometry, and coral reef decline at Looe Key, Florida Keys, USA: a 3-decade study. Mar Biol 166:108

Malone TC, Newton A (2020) The globalization of cultural eutrophication in the coastal ocean: causes and consequences. Front Mar Sci 7:670

Merly L, Lange L, Meÿe M, Hewitt AM, Koen P, Fischer C, Muller J, Schilack V, Wentzel M, Hammerschlag N (2019) Blood plasma levels of heavy metals and trace elements in white sharks (*Carcharodon carcharias*) and potential health consequences. Mar Pollut Bull 142:85–92

Mvungi EF, Pillay D (2019) Eutrophication overrides warming as a stressor for a temperate African seagrass (*Zostera capensis*). PLoS ONE 14(4):

Nair VD, Graetz DA (2004) Agroforestry as an approach to minimizing nutrient loss from heavily fertilized soils: the Florida experience. Agrofor Syst 61(1–3):269–279

Nazari-Sharabian M, Ahmad S, Karakouzian M (2018) Climate change and eutrophication: a short review. Eng Technol Appl Sci Res 8(6):3668

O'Connor T, Whitall D (2007) Linking hypoxia to shrimp catch in the northern Gulf of Mexico. Mar Pollut Bull (54):460–463

O'Neil JM, Davis TW, Burford MA, Gobler CJ (2012) The rise of harmful cyanobacteria blooms: the potential roles of eutrophication and climate change. Harmful Algae 14:313–334

Paerl HW, Dennis RL, Whitall DR (2002) Atmospheric deposition of nitrogen: implications for nutrient over-enrichment of coastal waters. Estuaries 25(4):677–693

Rabalais NN, Turner RE, Díaz RJ, Justic D (2009) Global change and eutrophication of coastal waters. ICES J Mar Sci 66(7):1528–1537

Riebesell U, Aberle N, Achterberg E, Algueró-Muñiz M, Alvarez-Fernandez S, Aristegui J, Bach L, Boersma M, Boxhammer T, Guan W, Haunost M, Horn H, Löscher C, Ludwig A,

Spisla C, Sswat M, Stange P, Taucher J (2018) Toxic algal bloom induced by ocean acidification disrupts the pelagic food web. Nat Clim Change 8(12):1082–1086

Selberg CD, Eby LA, Crowder LB (2001) Hypoxia in the Neuse River Estuary: responses of blue crabs and crabbers. North Am J Fish Manag 21:358–366

Selman M, Greenhalgh S (2009) Eutrophication: policies, actions, and strategies to address nutrient pollution. WRI Policy note—Water Qual Eutrophication hypoxia 16

Silbiger NJ, Nelson CE, Remple K, Sevilla JK, Quinlan ZA, Putnam HM, Fox MD, Donahue MJ (2018) Nutrient pollution disrupts key ecosystem functions on coral reefs. Proc R Soc B: Biol Sci 285(1880): 20172718

Smetacek V, Zingone A (2013) Green and golden seaweed tides on the rise. Nature 504:84–88

Smith VH (2003) Eutrophication of freshwater and coastal marine ecosystems a global problem. Environ Sci Pollut Res 10(2):126–139

Smith VH (2016) Effects of eutrophication on maximum algal biomass in lake and river ecosystems. Inland Waters 6(2):147–154

Snickars M, Weigel B, Bonsdorff E (2015) Impact of eutrophication and climate change on fish and zoobenthos in coastal waters of the Baltic Sea. Mar Biol 162(1):141–151

Souza-Araujo J, Souza-Junior OG, Guimarães-Costa A, Hussey NE, Lima MO, Giarrizzo T (2020) The consumption of shark meat in the Amazon region and its implications for human health and the marine ecosystem. Chemosphere 129132

United Nations (2019) World population prospects 2019: Highlights (ST/ESA/SER.A/423). Department of Economic and Social Affairs, Population Division

Wang M, Hu C, Barnes BB, Mitchum G, Lapointe B, Montoya JP (2019) The great Atlantic *Sargassum* belt. Science 365(6448):83–87

Wetz MS, Yoskowitz DW (2013) An "extreme" future for estuaries? Effects of extreme climatic events on estuarine water quality and ecology. Mar Pollut Bull 69(1–2):7–18

Wiedenmann J, D'Angelo C, Smith EG, Hunt AN, Legiret FE, Postle AD, Achterberg EP (2013) Nutrient enrichment can increase the susceptibility of reef corals to bleaching. Nat Clim Change 3:160–164

Wilewska-Bien M, Granhag L, Andersson K (2016) The nutrient load from food waste generated onboard ships in the Baltic Sea. Mar Pollut Bull 105(1):359–366

Zint M, Kraemer A, Northway H, Lim M (2002) Evaluation of the Chesapeake Bay Foundation's conservation education programs. Conserv Biol 16:641–649. https://doi.org/10.1046/j.1523-1739.2002.00546.x

Chapter 9
Anthropogenic Pollution of Coastal Ecosystems in Brazil

Sebastian M. Strauch and Gilmar S. Erzinger

Abstract Brazil is the 5th largest country with the 6th largest population in the world; 25% of its population live along the coast of the Atlantic Ocean. Urban and industrial development was historically not met by the implementation of appropriate sanitation infrastructure and is therefore insufficient today. Additionally, only minimal treatment is required to dispose of sewage via submarine outfalls into the sea. Marine pollution is diverse: domestic sewage carries feces, which can contain potentially harmful microorganisms. Feces are fertilizers, causing eutrophication, and they contain pharmaceuticals and their metabolites. Industrial sewage contains trace metals in large quantities and organic pollutants. Huge amounts of mine tailings are accidentally spilled into rivers far from the coast but reach the sea eventually. Careless handling of products in harbors and during transport leads to unintended release of organic and petrochemical pollutants, fertilizers and microplastic pellets. Heavy rainfall flushes litter from landfills or from unattended trash collection sites into rivers and the sea. This chapter gives an overview over types of pollution occurring in Brazil and a more detailed picture of the pollution situation in two of the areas under most anthropogenic pressure, the Santos Bay Region in São Paulo State and Guanabara Bay in Rio de Janeiro State.

Keywords Marine pollution · Brazil · Guanabara Bay · Santos Bay

9.1 Introduction

Brazil's area of 8,510,295.914 km^2 makes it the 5th largest country in the world and the largest in South America. It ranks 6th place worldwide with a population of 211,755,692 inhabitants (IBGE 2020). Its coastline spans 7491 km (CIA 2020). Many bays and estuaries of mighty rivers offered safe harbors during the time of

S. M. Strauch (✉) · G. S. Erzinger
Postgraduate Program in Health and Environment, University of the Region of Joinville—UNIVILLE, Rua Paulo Malschitzki, 10, Joinville, Santa Catarina, Brazil
e-mail: sebastian.m.strauch@gmx.de; sebastian.michael@univille.br

© The Author(s), under exclusive license to Springer Nature Switzerland AG 2021

D.-P. Häder et al. (eds.), *Anthropogenic Pollution of Aquatic Ecosystems*,
https://doi.org/10.1007/978-3-030-75602-4_9

discovery and colonization, so some of the oldest and biggest Brazilian cities are located on bays, like Salvador (Bay of All Saints) and Rio de Janeiro (Guanabara Bay). Today, around 25% of the total population live in coastal municipalities (IBGE 2019). 14 of the 26 cities with more than 750,000 inhabitants are coastal or estuarine cities, which leads to even higher local concentrations of the population and hence to a more severe anthropogenic impact.

The landscapes of the coastline are diverse (Fernandez et al. 2019): Mangroves dominate the northern coast, but stretch down to Santa Catarina State in a more patchy pattern, covering roughly between 9,600 km^2 (Giri et al. 2011) and 11,000 km^2 (Magris and Barreto 2010). They play an ecologically important role especially as breeding ground for many species of fish, birds, crustaceans and mollusks. Consequently, this ecosystem has an indirect economical value, which is unfortunately often overlooked. Mangroves form a protective belt in the tidal zone, diminishing the force of the sea. By slowing water flow, they promote sedimentation of particles, building up a finely grained soil, in which excess nutrients (Ferreira and Lacerda 2016) and pollutants from the water are sequestered. These pollutants can then accumulate in the mangrove trees and in the animals living there. Because fish, crabs and oysters from the mangrove forests are an important part of the diet of the poorer part of the local population, this poses a human health risk. Besides the mangroves, there are more than 4000 beaches along the coast, attracting tourists from all over the world and playing a central part in coastal Brazilian culture (Da Klein and Short 2016). From Rio de Janeiro State to Santa Catarina State, the mountain range Serra do Mar parallels the coast, leaving only narrow coastal plains (Fernandez et al. 2019). Serra do Mar forms the watersheds of south-western Brazil in a peculiar way: Big cities like Curitiba and São Paulo lie within 50 km from the sea, but their rivers turn to the west, discharging into the Rio Paraná basin and finally the Rio de la Plata between Argentina and Uruguay, traveling more than 4000 km.

Brazil's population grew tenfold in the twentieth century from ca. 17 to 170 million (IBGE 2000). Consequently, the necessary infrastructure could not keep pace, a problem that persists to the present: In 2018, 53.2% of the population had access to sanitation, and of the collected wastewater, only 46.3% was treated (SNIS 2018). There are notable differences in the five regions of Brazil (see Fig. 9.1) in regard to access to sanitation, from 10.5% in the North to 79.2% in the Southeast, and treatment of the collected wastewater, from 21.70% in the North to 53.88% in the Central-West (SNIS 2018).

The Brazilian legislature allows for the discharge of untreated wastewater directly into a receiving water body if a number of conditions are met (Ministério de Meio Ambiente, Conselho Nacional do Meio Ambiente 2011). These conditions are mostly physicochemical parameters like pH, temperature, concentration of metals, nutrients, oils, organic chemicals and settling particles and the absence of floating material. Biological oxygen demand is the only biological parameter, no microbiological parameter is defined. Hospital effluents must be treated separately or in the public system if it is capable to do so. The concentration limits for the pollutants mentioned above in effluents from treatment facilities are almost the same as for

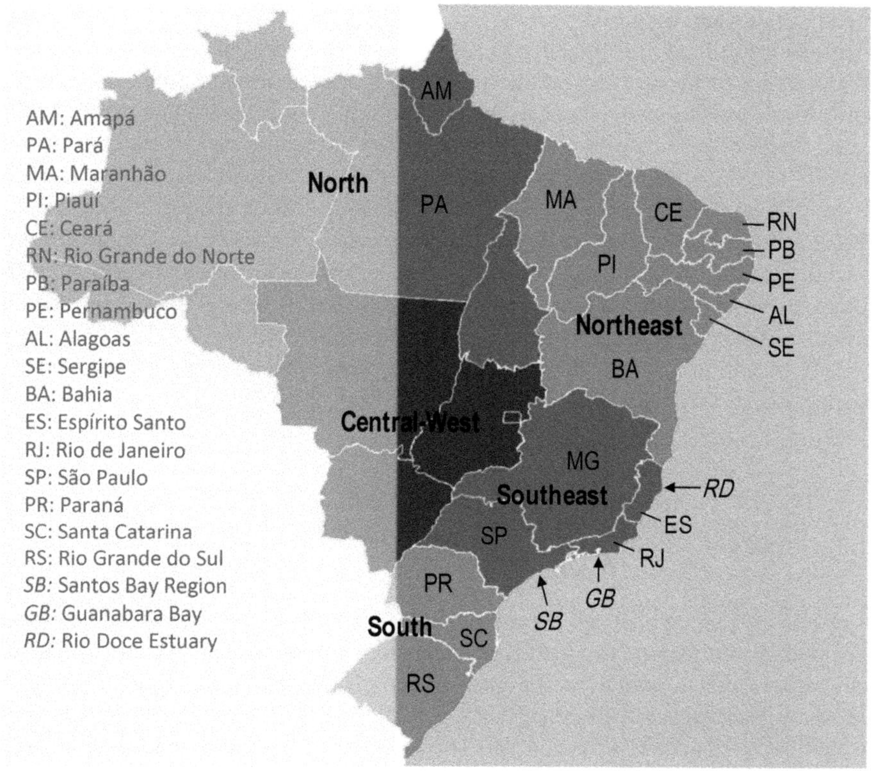

Fig. 9.1 Map of Brazil with names of the states mentioned in the chapter and location of Guanabara Bay, Santos Bay Region and Rio Doce Estuary. Shades of gray indicate the five regions. Map redrawn from free online sources

untreated sewage (a higher concentration of oils and fats are permitted). For dispatching via submarine outfall, the criteria appear to be even more permissive and define only limits for pH and temperature; the effluent must be free of floating and settling material.

Many sanitation systems in Brazil do not separate sewage collection from stormwater runoff. Heavy rainfall, which occurs frequently in the tropics and subtropics, easily overwhelm these systems.

Submarine outfalls are a common practice around the world to remove wastewater from coastal settlements, and deemed especially suitable for South America (Salas 2000). Effluents, both domestic and industrial, are collected and minimally treated (solids are removed and settling of sediments is allowed (Jordão and Rosman 2016), and sometimes chlorination is done to disinfect the water). If the local limits for physicochemical parameters are met, the liquid waste is pumped into a pipeline that goes out into the ocean. The length of the tubing can be as short as 50 m, but also reach tens of kilometers. At the end, the waste is mixed with the surrounding water by means of diffusors, which can be simply holes in the pipeline. The idea is to dilute the effluent effectively in a distance far enough from the shore, so the return of the waste is unlikely. It is also expected that the ocean water has some kind of disinfecting effect on freshwater-based wastewater (Yang et al. 2000). Exact knowledge of the hydrology is necessary to ensure proper function of the outfall: The directions of the winds and currents can have adverse effects and can occur differently depending on the season. A study found indications that one of Brazil largest submarine outfall servicing the city of Salvador de Bahia does not work as intended (Roth et al. 2016): During summertime, traces of effluents were found on beaches around the city. The sewage plume is clearly visible on satellite imagery (Google Earth, e.g., 11/2014, 13° 01′ 55.6″ S, 38° 29′ 22.7″ W), suggesting that dilution and mixture with ocean water fails. The only treatment the wastewater in this case is subjected to is removal of 10-30% of suspended solids (Roth et al. 2016).

9.2 Feces

A characteristic component of domestic sewage is feces. Typical fecal indicators are fecal sterols and potentially pathogenic microorganisms, such as bacteria, protozoa and viruses. Fecal coliforms (or thermotolerant coliforms) are a small group of intestinal bacteria, used to detect fecal contamination in water, even though their usefulness is disputed (Erickson and Doyle 2006). While most intestinal bacteria are nonpathogenic, pathogenic strains exist and can cause outbreaks of severe disease, such as enterohaemorrhagic *Escherichia coli*, EHEC. Apart from feces, domestic sewage contains other potential nutrients, like detergents. Molecular indicators for the presence of detergents are linear alkylbenzenes, LABs (Cabral and Martins 2018). LABs can also indicate oil pollution; the ratio between LABs of different sizes is indicative of the origin. A study of sterols and LABs in Babitonga Bay, Santa Catarina State, found a clear correlation between proximity of urban areas and sterol and LAB concentrations (Martins et al. 2014).

Bacteria from feces are not restricted to the receiving water body. Beach sand can be a substrate for microorganisms, posing a health risk for visitors. A study on three beaches in Fortaleza, Ceará State, collected wet and dry sand samples over 10 months (Vieira et al. 2001). 31% of the samples contained the indicator for fecal matter *E. coli* and the pathogens *Salmonella* (2%), *Vibrio parahaemolyticus* (5%) and *Candida albicans* (8%). Beaches are highly variable interfaces between sea and land, receiving and releasing all sorts of organisms from and to the sea; the presence of fecal matter indicators and pathogens demonstrates the health risk for beach visitors.

Viruses in wastewater are a particularly complex problem: On the one hand, patients suffering from gastroenteric viral diseases can shed up to 10^{11} viral particles/gram stool (Bosch 1998). On the other hand, wastewater treatment plants remove only between 50 and 90% of the viruses from the water (Okoh et al. 2010), and some viruses are highly virulent, meaning not many viable virus particles are needed to cause an infection. Once set free into the environment, viruses remain infectious for much more time than enteric bacteria (Bosch 1998) and can accumulate in the sediments, from where they can be mobilized later (Bosch et al. 1988). Because bacteria are easier to detect than viruses, standard techniques for water quality evaluation rely on bacteriological techniques. As there is not necessarily a correlation between the occurrence of bacteria and viruses, water that is considered to be safe can cause outbreaks of viral diseases, as documented in (Bosch et al. 1991). Finally, viruses can accumulate in shellfish that is growing in contaminated waters and lead to gastrointestinal infections (Abad et al. 1997): Arraial do Cabo in Rio de Janeiro state is a tourist hotspot, famous for its pristine waters ideally suited for scuba diving. During summer, the city of approximately 28,000 inhabitants accommodates more than 300,000 visitors (Sarmento et al. 2020); appropriate sanitation is not possible under these conditions. A study investigated the viral load of human norovirus GI and GII in locally produced shellfish (Sarmento et al. 2020). 41.5% of the mussel *Perna perna* and the oyster *Pseudochama cristella* were found to be contaminated. This poses a considerable human health risk, as shellfish tends to be consumed raw or undercooked, and local healthcare conditions during the main season are not suited to cope with an outbreak. Another study analyzed activated sludge and wastewater after sludge separation in a wastewater treatment plant in Florianopolis and found human adenovirus (HAdV), hepatitis A virus, rotavirus, and poliovirus (the latter in sludge only, Schlindwein et al. 2010). All viruses were present in viable form, with HAdV being the most prevalent (in all sludge samples and 75% of the water samples). However, the water samples do not indicate how many viruses were set free into the environment from this particular wastewater treatment plant, as chlorination was following before the actual release. It is noteworthy that this step is not always included and deemed both adverse and unnecessary if a submarine outfall is used (Yang et al. 2000).

The danger from pathogenic bacteria is aggravated by the fact that some pathogenic strains are resistant against (sometimes more than one) antibiotic(s), a treat acquired due to the widespread and improper use of these classes of drugs in the past (Davies and Davies 2010). A study in Rio de Janeiro State found bacterial communities tolerating ampicillin in concentrations 600 times higher than in clinical use (Coutinho et al. 2014). In a lagoon in an urban area, the authors found a strain of *Vibrio cholerae*, the pathogen causing cholera, to be resistant against six of the twelve tested antibiotics.

Coprostanol is the result of digestion of cholesterol in humans (Leeming et al. 1996). Because other animals modify cholesterol into slightly different substances, it can serve as a specific indicator for human feces. In sediments, coprostanol is rather persistent (Bartlett 1987), while it is quickly decomposed in the aerobic water column. Therefore, it is both an acute indicator of fecal pollution in the water and a historic record of fecal contamination in sediments (Carreira et al. 2004). Usually, ratios between coprostanol and other sterols are used to assess possible anthropogenic contamination in sediments (Martins et al. 2014). However, there is a certain consensus that the ratios need to be recalibrated for use in tropical and subtropical regions, as they were developed in studies in temperate climates and therefore might underestimate contamination in hotter environments.

9.3 Pharmaceutical Pollutants

Domestic sewage is the gateway for another class of pollutants: pharmaceutical compounds (PhCs). While hospital effluents are subjected to special treatment (at least in theory), many drugs are administered at home, like antihypertensive, antidiabetics, contraceptives, antibiotics or antidepressants. Many drugs are only partially absorbed by the human body, so that considerable amounts simply pass through the digestive tract and enter the sanitation system. Another important source is animal husbandry. Even with a full wastewater treatment, some drugs are only partially retained (Pivetta et al. 2020; Verlicchi et al. 2012). Once set free into the environment, PhCs can accumulate in the sediments, especially in those with a fine grain size. Some PhCs can act in the environment as endocrine disruptors (although this group contains also other organic pollutants): They have a quasi-hormonal effect on exposed organisms (Oberdörster and Cheek 2001). A study in All Saints Bay, Bahia, collected sediment samples from 17 sites both within the bay and at the adjacent ocean coast (Beretta et al. 2014); among the analyzed PhCs were Carbamazepine, Ibuprofen, Diclofenac, Atenolol and Diazepam, which were found in 41.2%, 100%, 94.1%, 100% and 70.6% of the samples, respectively, demonstrating the ubiquity of some PhCs. The concentration of PhCs was significantly correlated to the clay content of the sediment, being lowest at the ocean coast, which is mostly composed of coarse sand.

9.4 Metals

Trace elements are elements which only occur in very small quantities in organisms. They can be divided into essential, possibly beneficial and others (which includes both irrelevant and harmful elements). This categorization is organism specific. Most of the trace elements are metals (apart from halogens). In seawater, they are highly diluted (Loganathan et al. 2017); their concentration in industrial wastewater can be considerably higher. As some of the trace metals exert a negative effect on the metabolism of an organism exposed to them already in very small quantities, their ecological impact is potentially high. Some trace metals bio-accumulate: Once ingested, the organism is unable to excrete them and they accumulate in the tissue, sometimes very specifically. This leads to increasing contamination upwards in the food web and poses a health risk for humans as final consumers. One of the most harmful trace elements is mercury (Hg), feared since the acute methyl mercury poisoning in Minamata, Japan, in 1953 (Ekino et al.

2007). A study investigated mercury concentrations in the dusky grouper *Epinephelus marginatus* at the mouth of Patos Lagoon, Rio Grande do Sul State (Condini et al. 2017) and compared them to *E. marginatus* caught off an island 16 nautical miles offshore. Hg concentration correlated with the size and hence the age of the fish, indicating bioaccumulation. All fishes from the littoral site showed Hg concentration below the threshold considered safe for the animals (0.3 mg/kg wet weight), as did individuals <650 mm from the neritic site. However, 2/3 of the individuals >650 mm (only found at the neritic site) exceeded 0.3 mg/kg body weight, the limit above which adverse effects on the reproduction of fish are to be expected. This is especially concerning as fish of that size (and age, consequently) are the reproductively active group. Around 20% of these large specimens contained above 0.5 mg/kg, which is the limit considered safe for human consumption, suggesting that consumption of fish of that size class should be avoided. The counterintuitively higher Hg concentration in fish from habitats distant from the anthropogenic influence in the lagoon might be explained by the fact that many prey species of *E. marginatus* use the lagoon as feeding and growing grounds, but migrate to the ocean to reproduce, thereby vectoring high Hg concentrations from one habitat to another. While there are limits for metals in wastewater, there are none for marine or estuarine sediments (Andrade et al. 2020), e.g., to prevent fishing or rearing of mussels for human consumption.

In the past decade, Brazil suffered two major mining disasters, in 2015 the Mariana dam disaster (Carmo et al. 2017) and in 2019 the Brumadinho dam disaster (Vergilio et al. 2020). In both cases, dams containing mine tailings broke and released toxic mud into the adjacent rivers. The mud from the Mariana dam accident entered Rio Doce and reached the coast 670 km and 17 days later, forming a brown opaque plume (Gomes et al. 2017) and leading to up to fivefold concentrations of trace metals. Two years later, a study evaluated the long-term effects by analysis of trace metals in fish muscles and livers and found bioaccumulation of Cr, Zn, Mn, As, Cu, Pb and Cd up to a level above Brazilian and international guidelines (Gabriel et al. 2020). While there are notable differences between fish species, it indicates a potential health hazard for the communities around the Rio Doce estuary. The Brumadinho dam disaster occurred within the Rio São Francisco watershed; up to now, there are no studies about the impact of the disaster on the river's estuary.

As Brazil is still a country with a growing economy, many baseline studies can be found that will help evaluate environmental impact of current infrastructure projects, or to determine general trends in environmental impacts. Di Beneditto et al. (2019) established a baseline of trace metals (Al, Fe, Zn, Cu, Mn, Pb, V, Ni, Cr, Cd) in two species of commercially exploited shrimps off the coast of Rio de Janeiro state, where port activities are expected to increase. While this baseline waits to be compared to future findings, values of Pb and Cr already exceed the maximum permissible limits.

The waters far away from the coast are less accessible to monitoring. One way to estimate the presence of metals in remote areas is to analyze migratory animals. Sea turtles, for example, usually lay their eggs on the beaches where they themselves hatched but roam the oceans in between reproduction cycles. According to Sakai et al. (1995), the heavy metal concentration in sea turtle eggs is an appropriate proxy for the assessment of the heavy metal concentrations in the adults. An advantage of this approach is that no adults have to be sacrificed or even stressed, as non-developed eggs can be recovered from post-hatching clutches. However, for a given area on the globe, baselines must be established in order to monitor future changes in exposure to heavy metals. For the green turtle *Chelonia mydas*, baselines of Al, As, Ba, Cd, Cr, Cu, Fe, Mn, Ni, Pb, V and Zn were established for populations nesting at the Rocas Atoll, a pristine location in a marine biodiversity reserve 266 km off the NE coast of Brazil (Agostinho et al. 2020). Compared to metals found in eggs on a similarly protected site on the Trindade/Martim Vaz islands (Souza et al. 2018), 1140 km off the SE coast, this baseline showed similar values in Ba, Cd, Cr, Fe, Mn, Sr and Zn; values for Cu and Pb, however, were 1.7 and 19 fold higher, respectively, than the baseline. This indicates that the female turtles must have experienced exposure to anthropogenic influence somewhere else. In the same study, Souza et al. compared trace metals in *Chelonia mydas* to those found in *Caretta caretta* from a continental nesting site in Espiritu Santo State. *C. caretta* showed higher values, which is commonly attributed to the fact that it is a carnivorous species which accumulates pollutants because of its higher trophic level. Cu and Zn influence the hatching success of both species. Another baseline study (de Macêdo et al. 2015) established values (for As, Al, Ba, Ca, Cd, Co, Cr, Cu, Fe, Hg, K, Mg, Mn, Mo, Na, Ni, Pb, Sb, Se, Sr, V, Zn) in *C. mydas* and the hawksbill turtle *Eretmochelys imbricata* on the Northern coast of Brazil, analyzing different tissues of specimens found dead on beaches in the region. However, due to different methodologies (analyzed metals and tissues), it is impossible to compare many studies.

9.5 Microplastics

While plastic litter in the ocean and the problem it causes is known since the 1970s (e.g., Carpenter et al. 1972; Colton et al. 1974, cf. Chap. 17, this volume), science started to pay attention to microscopic plastic fragments only after the turn of the millennium (Castro et al. 2018; Thompson et al. 2004). Once knowing what to look for, Thompson et al. detected microplastic particles (MPs) in historic plankton samples from the 1960s onwards and showed a significant increase in abundance over the following decades. They also showed that MPs are ingested by marine organisms. MPs can either be generated by the decay of larger plastic pieces by various degradation mechanisms (Andrady 2011) or be introduced in its characteristic small size; for example, one item of clothing can generate more than 1900 microfibers in a machine wash (Browne et al. 2011). Wastewater treatment does not necessarily retain all particles of that size, therefore something as trivial as a washing machine can produce substantial amounts of MPs. Larger items from landfills can be flushed into rivers and bays due to heavy rainfall and then degrade to MPs. Nonregular or missing trash collection in coastal regions can aggravate the problem (Possatto et al. 2015). For the Brazilian equatorial margin, Garcia et al. (2020) established a baseline, sampling 18 different locations along ca. 650 km coast from Maranhão State to Ceará State. Overall, particle density was 0.14 ± 0.11 particles/m^3 and 0.02 ± 0.01 particles/m^3 in 120 μm and 300 μm mesh samples, respectively. Fibers and filaments were present in every sample regardless of mesh size. Hotspots in density could be correlated to urbanization (close to the city of Fortaleza, a metropolitan region with ca. 4 million inhabitants) or coastal dynamics (a permanent gyre, probably accumulating particles).

9.6 Organic Pollutants

The main toxicological impact from MPs does not derive from them per se, but from the load they carry: persistent organic pollutants (POPs). POPs are typically halogenated organic compounds, which are either intentionally produced or are byproducts of the production of other chemicals (including plastics). Naturally occurring quantities (if they exist) are negligible compared to the amount of manmade output. Due to their stability, they are spread across the globe (Xu et al. 2013). Because of their chemical nature, POPs are adsorbed onto MPs in the ocean (Mato et al. 2001; Rios et al. 2007), accumulating to much higher concentrations than those found in the surroundings. Upon ingestion, they can become bio-available and bio-accumulate in fatty tissues, especially in top members of the food web, as shown in a classical paper from 1970 (Prestt et al. 1970). Adverse effects of POPs include cellular damage, disruption of feeding behavior and chemical communication and possibly endocrine disruption (Besseling et al. 2013; Browne et al. 2013; Rochman et al. 2014; Trotter et al. 2019). Initiated by the United Nations Environmental Program, in 2001, the Stockholm Convention named 12 POPs to be banned from production and use (or at least restricted to specific use), among them PCBs (polychlorinated biphenyls) and DDT (dichlorodiphenyltrichloroethane, Secretariat of the Stockholm Convention 2019). Meanwhile, two amendments expanded the list. In recent years, the analysis of biliary metabolites became a standard method for detection of exposure to POPs in wildlife.

Polycyclic aromatic hydrocarbons (PAHs) originate from incomplete combustion or pyrolysis of fossil fuels (pyrogenic fraction), oil spills (petrogenic fraction) or domestic or industrial sewage (Bigus et al. 2014). Their molecular weight is correlated to the process of their origin. They accumulate in the sediment and give an historic record of pollutions in the past (Martins et al. 2011). Aliphatic hydrocarbons (AHs), come either from biogenic sources, such as vascular plants, phyto- or zooplankton, or are anthropogenic, from oil and its derivatives (de Abreu-Mota et al. 2014). Groups of AHs can be distinguished from each other and therefore be used to quantify the biogenic and petrogenic contributions in the study area (de Abreu-Mota et al. 2014). A recent study quantified AHs and PAHs in mangrove sediments in six states, covering the entire Brazilian coastline (Araújo et al. 2020). Pollution levels were low to moderate, with pronounced biogenic input from the mangroves themselves. A clear correlation was found between the degree of pollution and proximity to urbanized areas. Effects on the biota are unlikely considering the present concentrations. This study will be a valuable baseline for future investigations.

Pollution with oil and its derivatives is a consequence of exploration activities off the Brazilian shore and mishaps during handling of petrochemicals during industrial processing, including transport. In late August 2019, lumps of crude oil appeared on beaches along the northeast coast of Brazil, along almost 2500 km of coastline (Disner and Torres 2020; Lourenço et al. 2020). The response to the crisis from the government was widely criticized and deemed too slow (Brum et al. 2020; Soares et al. Soares et al. 2020a). A forensic analysis indicated that the oil was Venezuelan in origin, suggesting it spilled during transport (Oliveira et al. 2020). Later, this was confirmed by the Brazilian Federal Police, who identified a Greek tanker to be the likely cause for the spill, which was denied by the shipping company (Escobar 2019). A study identified 27 threatened animal species in the polluted area and estimated 870,000 people to be affected (Magris and Giarrizzo 2020). This event is considered the most severe environmental disaster in tropical coastal areas ever (Soares et al. 2020b), and also had a huge impact on tourism. While not uncommon, oil spills on the Brazilian coast are usually much smaller in volume and impacted area (Maggi et al. 2013), but there are occasional large-scale events (Disner and Torres 2020).

9.7 Nutrients

Heavy rainfalls are common in the rainy season in Brazil. They produce runoff of agricultural areas, which can contain substantial amounts of nutrients in the form of nitrogen species (nitrate, ammonia) and phosphorus. If the nutrient input exceeds the assimilative capacity of the rivers which receive the runoff and occurs sufficiently close to the coast, the nutrients are introduced into bays and estuaries, and finally, the sea. Another source for nutrient input are harbors, where fertilizers or agricultural products are loaded or unloaded (Mizerkowski et al. 2012). Domestic sewage also contains substantial amounts of nitrogen and phosphorus in the form of feces. Phosphorus is a bottleneck nutrient in tropical and subtropical coastal areas (Feller 1995), and its elevated levels can lead to (toxic) algal blooms, and further to oxygen depletion.

9.8 Litter

Finally, pollution occurs in the form of litter, predominately of plastic, but also cigarette buds, coal, metal, porcelain, rubber and Styrofoam. It ends up on beaches or in gyres or decomposes over time, adding to the pool of MPs or other kinds of small particles. It is also mistaken for prey and/or accidentally ingested by animals: A study of 86 turtles of five species found on beaches in Rio Grande do Sul State revealed that 47 of them had ingested plastic, rubber, wood, paper or cloth items (Rizzi et al. 2019), which in some cases caused obstructions. In a tropical country with rainy seasons characterized by torrential rain, runoff of landfills carries all sorts of particles into nearby rivers; lacking communal waste collection, and disposal infrastructure also play a role (Possatto et al. 2015). Consequently, the closer a beach is located to an estuary, the more litter is likely to be found (Andrades et al. 2020). There is also beach litter produced in situ: Beaches do not always provide trash bins and their number is negatively correlated with certain kinds of residues left by beachgoers (Andrades et al. 2020).

9.9 Case Study 1: Guanabara Bay, Rio de Janeiro

Guanabara Bay is the second largest bay in Brazil. It is located in the heart of the Rio de Janeiro Metropolitan region, home to approximately 13,000,000 people (for an extensive overview, see Fistarol et al. 2015; Soares-Gomes et al. 2016). Its waters are considerably polluted by urban, industrial and petrochemical effluents: The bay receives around 18 m^3/s of untreated sewage (Fistarol et al. 2015), around 813 t/d of solid waste (Coelho 2007), and around 18 t/d of petroleum hydrocarbons (Ferreira 1995). There is a sharp contrast between highly urbanized zones and huge protected areas on the north and northeast shore of the bay, totaling around 90 km^2 of mangrove forests (Kjerfve et al. 2001). The Olympic Games in Rio de Janeiro in 2016 sparked hopes of a restoration of the bay and investments in modern sanitation systems (Fistarol et al. 2015); many papers from after 2016 comment on the fact that these hopes did not come true.

Rodrigues et al. (2020) investigated the mercury content in swimming crabs 1 year before, during and 1 year after dredging activities in Guanabara Bay. Dredging is necessary to counteract sediment accumulation in ports (in this case, the port of Rio de Janeiro). These operations resuspend great amounts of sediments and increase bioavailability of the substances previously buried in the ground. To account for the complex water movement within the bay, crabs were sampled at five locations: the harbor area itself, two sites toward the internal part of the bay and two sites toward the mouth. Results showed that total Hg (THg) concentration in crab muscle tissue increased during dredging in the internal and external sampling sites (no crabs were found at the harbor during the works). Afterward, THg concentration decreased, but did not reach the pre-dredging levels. Due to the importance in the local diet, the consumption of swimming crabs may lead to health risks, even though THg concentrations in the animals remained below legal limits.

A study analyzed the water of 8 of the 12 main rivers discharging into Guanabara Bay for fecal sterols and thermotolerant coliforms, three in the less impacted northeastern (NE) sector, and five in the intensely urbanized western (W) sector (Assis Costa et al. 2018). Samples were taken in spring, summer and fall. Spatial differences of coprostanol were very pronounced with low levels in the NE sector and one to three orders of magnitude higher values on the W sector; temporal variations did not occur in the NE sector and were inconclusive in the W sector. All but one sample was contaminated with coliforms; again, the highest contamination rate with 1.2×10^{12} most probable number/100 mL (legal limit for singular sample is 2×10^3 MPN/100 mL) occurred in the W sector.

There are various studies on the efficiency of wastewater treatment against enteric viruses in the area of Guanabara Bay, for example, human polyomavirus JC (JCPyV, Fumian et al. 2010), human adenovirus (HAdV) and JCPyV (Fumian et al. 2013), and norovirus GI and GII (Fumian et al. 2019). The reduction rate by wastewater treatment agrees with previously published studies from other locations, meaning that in every study there was still a considerable number of viral particles discharged into the bay. Consequently, the papers conclude that virological markers are needed for water quality control. Dias et al. (2018) investigated three viruses (HAdV, JCPyV and hepatitis A virus (HAV)) in the waters of Jurujuba sound within Guanabara Bay. Biweekly sampling on four locations (three within the bay, one oceanside) over 6 months rendered 48 samples. HAdV and JCPyV were present at all sampling sites, but not in all samples. HAV was only found in one sample. Out of 48 samples, 41 were considered suitable for recreational use based on bacteriological standards, but 22 of them contained at least one of the viruses. No correlation was observed, which poses a potential human health risk as supposedly safe conditions could be related to outbreaks of viral diseases.

Also in Jurujuba sound, a study investigated pollution parameters, acute toxicity and estrogenic activity of the water (do Nascimento et al. 2018). Two sets of samples were collected 11 months apart at 7 locations selected for their proximity to potential pollution input. During the first campaign, only dissolved organic carbon (DOC) exceeded legal limits, while during the second one, DOC, ammoniacal nitrogen and turbidity were above the legal limits. All acute toxicity tests with *Aliivibrio fischeri* were negative. If estrogenic activity was detected, it was in the moderate range. The pollution pattern remained inconclusive.

Studies of MPs in Guanabara Bay found the abundance among the highest values worldwide for particles >300 μm: Olivatto et al. (2019) sampled three sites, finding between 1.4 and 21.3 particles/m^3 in the water. Sampling was repeated in the same season and showed a higher abundance after rainfall in the region, indicating that the fragments were brought in by the rivers. Castro et al. (2016) sampled three sites in a cove inside Guanabara Bay with mussel farming as an important economy and found on average 16.4 particles/m^3. Figueiredo and Vianna (2018) found on average 4.8 particles/m^3 on three sites. Interestingly, they found that even this high concentration is so low as compared to the zooplankton density that ingestion by fish larvae and chaetognaths is very unlikely. A study of MPs in the sediment of Guanabara Bay found between 160 and 1000 particles/kg sediment (mean 528 ± 30), a value among the highest worldwide (Alves and Figueiredo 2019). While fibers were the least abundant type of MP in the water of the bay, it was the most common in the sediment, indicating that fibers tend to sink. The influence of rainfall to MP abundance in the water column was not observed in the sediment. Finally, Carvalho and Baptista Neto (2016) analyzed MP abundance in 34 locations on 17 beaches in and around Guanabara Bay in summer and winter. Sampling at the high tide line they found between 12 and 1300 particles/m^2 in summer, and 3–743 particles/m^2 in winter, values in the mid-range compared to other beaches worldwide. Fibers were the least abundant MP, which again suggests that fibers tend to sediment. The authors show some correlation between the occurrence of MPs and the quality of the sample location, but do not discuss the seasonal difference. The group also investigated the continental shelf adjacent to the bay's mouth (Baptista Neto et al. 2019). Here, sediments removed from Rio de Janeiro harbor are dumped in five designated areas. The authors collected 68 samples in a wide grid, including the sediment disposal sites. About half of the MPs found were fibers, which is in good agreement with the sediment samples from within the bay mentioned above. However, MPs were not limited to the proximity of the sediment disposal sites, but omnipresent in the sampled grid: Every single sample contained MPs.

A comparison of PCB and DDT metabolites in muscle tissue of two valuable fish species (*Mugil liza* (mullet) and *Micropogonias furnieri* (croaker)) caught in Guanabara Bay and Paraty Bay, a relatively unpolluted area, was performed (Da Silva et al. 2016). There were notable differences in the frequency of the 29 chemicals in the fish species, probably due to different feeding habits; this shows that not only top-level predators can accumulate pollutants, but also bottom feeding fish in close contact to the sediments. While it is safe to eat these fish for the average consumer, fishermen and their families with their elevated consumption are exposed to a potential health risk: In mullets from Guanabara Bay, the estimated daily intake of DDT metabolites reached 131% of the acceptable daily intake as recommended by the WHO.

9.10 Case Study 2: Santos Bay Region, São Paulo

The Santos Bay region consists of the Santos Estuary, with the Island of São Vicente in its center. Santos Bay is located on the southern side of São Vicente Island, toward the ocean. The largest commercial harbor of South America and more than 1100 companies of the metallurgical, fertilizer, and petrochemical industry make it one of Brazil's most important economical and metropolitan areas, which consequently suffers from massive pollution (Martins et al. 2011). Industrial pollution can be tracked in sediments of the estuary (Luiz-Silva et al. 2008; Martins et al. 2011) and attributed to related branches of industry and historic events, like the time of local industrial growth, or the world oil crisis. In 1978, a submarine outfall was installed to discharge into Santos Bay, 4 km offshore (Abessa et al. 2005), and to date, there are at least three additional outfalls (Rodrigues et al. 2016). However, values for discharge volumes in the literature are conflicting, and the sewage of slum areas enters the bay directly in a ubiquitous and unquantifiable manner.

Pinheiro et al. (2012) investigated the concentration of six metals (Cd, Cr, Cu, Hg, Mn, Pb) in the mangrove crab *Ucides cordatus* and in its food, the red mangrove tree *Rhizophora mangle* in two locations in Santos Estuary. Buds, green or senescent leaves of the trees were analyzed individually, as were muscle, hepatopancreas and gills of the crabs. Pb and Hg were not detected in any sample, while Cd was missing in the plant, but not in the animal. Cu was mostly present in buds, Mn in green leaves, and Cr in senescent leaves. In the crabs' tissues, Cu, Cd, Cr and Mn accumulated in patterns related to age of the animal and function of the tissues. While the concentrations of Cu, Cd and Mn remained below the limit considered safe for human consumption (Mn does not have a limit according to Brazilian law), Cr concentration exceeded it, posing a serious health risk, as *U. cordatus* is an important protein source for locals. Banci et al. (2017) analyzed mangrove sediments and water samples in the Santos bay region for Cd, Pb, Cu, Cr and Hg and compared their findings to the genetic diversity of *U. cordatus*. They found a gradient of the metals (except for Hg, which was always below the detection limit) in the sediments according to the distance of the three sample sites to industrial complexes; however, concentration remained always well below threshold effect levels. In the water samples, all metals were below detection limits except for Hg in one site, where it exceeded legal limits 2200-fold. No correlation between genetic variability and metal exposure was found. Brown mussels *Perna perna* were analyzed for metals (Al, Cd, Cr, Cu, Fe, Mn, Ni, Pb and Zn). The study (Campolim et al. 2017) was performed in each season and divided the collected mussels in two sex and three size classes. All metals were detected, with differences mainly due to seasonality, as Cd, Cu, Pb and Zn were highest in summer, the season with the highest rainfall. Cu was the only metal found in higher concentrations in females, all the others showed no difference between the sexes. Differences in metal concentration over size classes might reflect changes in the life cycle. No information was given about the actual concentrations of the investigated elements in the water.

Andrade et al. (2020) sampled sediment and water to analyze it for carbapenem-resistant bacteria. Santos Bay served as an example for a polluted area and was compared to Araçã Bay, São Sebastian, SP, an area adjacent to two environmental protection areas. Three of eleven resistant isolates came from Araça Bay. Two-thirds of the bacteria came from the sediment, which is a critical factor, as Brazilian legislation defines limits for pollutants only for water, not for sediment. In addition to antibiotic resistance, the strains also showed a much higher tolerance to metals (Cd, Co, Cu, Ni, Zn) as compared to a control bacterium (*E. coli* ATCC 25922). As sediments can accumulate both metals and antibiotics, this environment favors the emergence of such co-occurrence of resistances.

Rodrigues et al. (2016) evaluated the presence of diarrheagenic *Escherichia coli* at three recreational beach areas in São Paulo State. The beaches are in differently impacted areas: In the vicinity of Santos Bay, there are four submarine outfalls releasing partly treated domestic sewage, in São Sebastião three for domestic and one for industrial sewage and none in Ubatuba. Populations in the areas at the time were around 750,000, 100,000 and 80,000, respectively. The authors found few

E. coli and even less pathogenic strains in Ubatuba and increasing numbers in São Sebastião and Santos Bay. However, the microbial risk assessment showed the risk of getting ill due to exposition on the beach in São Sebastião one order of magnitude lower than in Santos Bay. The values for Santos Bay were partly overlapping with acceptable risk as by Brazilian legislature (Conama Resolution n°274/2000). However, the overall quality of the waters might well be unacceptable, as only a small fraction of potential pathogens was analyzed.

Pereira et al. (2016) screened for 33 pharmaceuticals in the water of Santos Bay. Samples were taken in proximity of a submarine sewage outfall at five locations and two depths (surface and bottom). They found nine compounds: the antihypertensives valsartan, losartan and atenolol, the analgesic paracetamol, the non-steroidal anti-inflammatories ibuprofen and diclofenac, caffeine, cocaine and its main human metabolite benzoylecgonine (BE). Not all compounds were found in all samples, and sometimes only below the level of quantification. Cocaine and BE were found in concentrations consistent with other literature values for freshwater systems. Compared to other marine studies, the concentration in this study was much higher (12.6–537.0 ng/L). This might be related to the fact that sampling took place one day after carnival, a very important holiday in Brazil, and therefore input (as well as the population) might have been considerably higher than normal. Ibuprofen was found most abundantly in concentrations of up to 2 μg/L. For the other compounds, there was no clear correlation with values in studies from other, environmentally comparable places elsewhere in the world; this might reflect preferences of application practice in other countries. In a follow-up study, the group investigated the presence of cocaine and BE over the course of one year, sampling at similar locations close to the outfall and a more remote, supposedly uninfluenced point (Fontes et al. 2019). Samples were taken once at the end of every season. Cocaine levels were highest in spring (68.62–203.60 ng/L), while in the rest of the year they were mostly below level of quantification or detection. BE levels were highest in autumn, when cocaine was below the quantification level in ten out of twelve samples.

14 surface sediment samples around São Vicente Island were analyzed for levels of three estrogens (Pusceddu et al. 2019). This area is not only subjected to four sewage outfalls, but also to ubiquitous input of untreated and uncontrolled sewage discharge from irregular housing areas. Estrogen concentrations found were always higher than those reported from other countries. One estrogen was present in concentrations above its known effect concentration, which not only indicates a potential danger to exposed wildlife, but also raises concerns regarding local consumption of fish, mussels, and crabs.

MPs have a profound influence on filter feeders and enter the food chain on their level. Brown mussels *Perna perna* were sampled at Santos Bay, close to Santos Harbor (Santana et al. 2016). Among the goods passing through the harbor are plastic pellets of various sizes. Additionally, the bay receives huge amounts of insufficiently treated domestic wastewater and is adjacent to landfills. Of all the collected mussels, 75% showed contamination with microplastic. The study established the suitability of the mussel as a sentinel for MP contamination.

Albergaria-Barbosa et al. (2018) established a baseline for POP biliary metabolites in four species of fish in Santos Bay. Values indicated a light to moderate contamination with differences between fish species. Recently, this group performed a baseline study on POP biliary metabolites in *Spheniscus magellanicus*, the Magellanic Penguin, a frequent winter guest on Brazil's shores coming up from Patagonia (Barreto et al. 2020). Naphthalene was the dominant metabolite in accordance with studies on other penguin species. Overall, the biliary metabolite concentrations in the penguins' bile were lower than in fish bile and differed in composition, which is no surprise given the differences in metabolism and biochemistry in fish and birds.

Cascaes et al. (2014) found POPs in the liver of *Rhizoprionodon lalandii*, the Brazilian sharpnose shark, an economically important species. The PCB levels were in accordance with findings in other sharks in the Mediterranean. They were one or two magnitudes higher than what was found in blue shark elsewhere on the Brazilian coast (Cascaes 2009), which might be due to the different feeding behavior of the two sharks: The sharpnose is a bottom feeder, more exposed to sediments than the blue shark, a pelagic hunter. The levels were also higher than in the muscle tissue of three sharks (smooth hammerhead (*Sphyrna zygaena*), shortfin mako (*Isurus oxyrinchus*) and bigeye thresher shark (*Alopias superciliosus*)) caught off the coast of Espirito Santo State (Azevedo e Silva et al. 2009). This is to be expected, as the liver as a fatty organ can store more organochlorines than muscles. Concentrations of POPs in the muscles were well below acceptable daily intake, so consumption did not pose any human health risk.

9.11 Final Remarks

The environmental problems Brazil is facing are very difficult to resolve: The densely populated coastal cities are often surrounded by protected areas, where construction of wastewater treatment plants would sacrifice landscape forms worth protecting, like mangroves. Slums are a common feature of big cities in Brazil; they usually do not have a sewer system, and it is virtually impossible to build one underneath. While some pollutants might be reduced by educating the population (e.g., proper disposal of drugs and garbage, proper use of detergents, etc.), others are much more difficult to resolve. They require political action on the federal level by passing appropriate laws (e.g., pollutant limits in industrial sewage or sewage treatment obligations), enforcing them and by providing financial relief. After Jair Bolsonaro was elected President of Brazil in 2018, his legislation began to relax restrictions for exploration of natural resources, including oil, to advance economic growth. Many publications cited in this chapter lament the situation, "which privileges large enterprises and disqualifies the authority of the environmental agencies, in favor of accelerated economic development" (Araújo et al. 2020). Therefore, an improvement of the environmental situation in the near future is very unlikely.

References

Abad FX, Pintó RM, Gajardo R, Bosch A (1997) Viruses in mussels: public health implications and depuration. J Food Prot 60:677–681. https://doi.org/10.4315/0362-028X-60.6.677

Abessa DMS, Carr RS, Rachid BRF, Sousa ECPM, Hortelani MA, Sarkis JE (2005) Influence of a Brazilian sewage outfall on the toxicity and contamination of adjacent sediments. Mar Pollut Bull 50:875–885. https://doi.org/10.1016/j.marpolbul.2005.02.034

de Abreu-Mota MA, de Moura Barboza CA, Bícego MC, Martins CC (2014) Sedimentary biomarkers along a contamination gradient in a human-impacted sub-estuary in Southern Brazil: a multi-parameter approach based on spatial and seasonal variability. Chemosphere 103:156–163. https://doi.org/10.1016/j.chemosphere.2013.11.052

Agostinho KFF, Lacerda D, Tostes ECL, Baldassin P, Di Beneditto APM, de Carvalho CEV (2020) Trace elements in green turtles (*Chelonia mydas*) from Rocas Atoll, NE Brazil: baseline reference from a pristine nesting site. Mar Pollut Bull 157:111271. https://doi.org/10.1016/j.marpolbul.2020.111271

de Albergaria-Barbosa ACR, da Silva DAM, da Silva Rocha AJ, Taniguchi S, Patire VF, Dias JF, Fernandez WS, Bícego MC (2018) Evaluation of polycyclic aromatic hydrocarbons bioavailability on Santos Bay (Brazil) through levels of biliary metabolites. Mar Pollut Bull 129:822–828. https://doi.org/10.1016/j.marpolbul.2017.10.006

Alves VEN, Figueiredo GM (2019) Microplastic in the sediments of a highly eutrophic tropical estuary. Mar Pollut Bull 146:326–335. https://doi.org/10.1016/j.marpolbul.2019.06.042

Andrade VC, Caetano T, Mendo S, de Oliveira AJFC (2020) Carbapenem resistant Enterobacteriaceae from port areas in São Paulo State (Brazil): isolation and molecular characterization. Mar Pollut Bull 159:111329. https://doi.org/10.1016/j.marpolbul.2020.111329

Andrades R, Pegado T, Godoy BS, Reis-Filho JA, Nunes JLS, Grillo AC, Machado RC, Santos RG, Dalcin RH, Freitas MO, Kuhnen VV, Barbosa ND, Adelir-Alves J, Albuquerque T, Bentes B, Giarrizzo T (2020) Anthropogenic litter on Brazilian beaches: baseline, trends and recommendations for future approaches. Mar Pollut Bull 151:110842. https://doi.org/10.1016/j.marpolbul.2019.110842

Andrady AL (2011) Microplastics in the marine environment. Mar Pollut Bull 62:1596–1605. https://doi.org/10.1016/j.marpolbul.2011.05.030

Araújo MP, Hamacher C, de Oliveira Farias C, Martinho P, de Oliveira Chaves F, Gomes Soares ML (2020) Assessment of Brazilian mangroves hydrocarbon contamination from a latitudinal perspective. Mar Pollut Bull 150:110673. https://doi.org/10.1016/j.marpolbul.2019.110673

Assis Costa LA de, Mano Pessoa DM, Da Silva Carreira R (2018) Chemical and biological indicators of sewage river input to an urban tropical estuary (Guanabara Bay, Brazil). Ecological Indicators 90:513–518. https://doi.org/10.1016/j.ecolind.2018.03.046

Azevedo e Silva CE de, Azeredo A, Dias AdCL, Costa P, Lailson-Brito J, Malm O, Guimarães JRD, Torres JPM (2009) Organochlorine compounds in sharks from the Brazilian coast. Mar Pollut Bull 58:294–298. https://doi.org/10.1016/j.marpolbul.2008.11.003

Banci KRdS, Mori GM, de Oliveira MA, Paganelli FL, Pereira MR, Pinheiro MAA (2017) Can environmental pollution by metals change genetic diversity? *Ucides cordatus* (Linnaeus, 1763) as a study case in Southeastern Brazilian mangroves. Mar Pollut Bull 116:440–447. https://doi.org/10.1016/j.marpolbul.2017.01.039

Baptista Neto JA, Carvalho DG de, Medeiros K, Drabinski TL, Melo GV de, Silva RCO, Silva DCP, Sousa Batista L de, Dias GTM, da Fonseca EM, dos Santos Filho, João Regis (2019) The impact of sediment dumping sites on the concentrations of microplastic in the inner continental shelf of Rio de Janeiro/Brazil. Mar Pollut Bull 149:110558. https://doi.org/10.1016/j.marpolbul.2019.110558

Barreto IS, de Albergaria-Barbosa ACR, Patire VF, de Jesus Silva M, Baldassin P, Taniguchi S, Montone RC, Gallo H, Maranho A, Bícego MC (2020) Bioavailability of polycyclic aromatic hydrocarbons to penguins on the coast of southeastern Brazil. Mar Pollut Bull 157:111306. https://doi.org/10.1016/j.marpolbul.2020.111306

Bartlett PD (1987) Degradation of coprostanol in an experimental system. Mar Pollut Bull 18:27–29. https://doi.org/10.1016/0025-326X(87)90653-9

Beretta M, Britto V, Tavares TM, da Silva SMT, Pletsch AL (2014) Occurrence of pharmaceutical and personal care products (PPCPs) in marine sediments in the Todos os Santos Bay and the north coast of Salvador, Bahia, Brazil. J Soils Sediments 14:1278–1286. https://doi.org/10.1007/s11368-014-0884-6

Besseling E, Wegner A, Foekema EM, van den Heuvel-Greve MJ, Koelmans AA (2013) Effects of microplastic on fitness and PCB bioaccumulation by the lugworm *Arenicola marina* (L.). Environ Sci Technol 47:593–600. https://doi.org/10.1021/es302763x

Bigus P, Tobiszewski M, Namieśnik J (2014) Historical records of organic pollutants in sediment cores. Mar Pollut Bull 78:26–42. https://doi.org/10.1016/j.marpolbul.2013.11.008

Bosch A, Lucena F, Girones R, Jofre J (1988) Occurrence of enteroviruses in marine sediment along the coast of Barcelona, Spain. Can J Microbiol 34:921–924. https://doi.org/10.1139/m88-163

Bosch A (1998) Human enteric viruses in the water environment: a minireview. Int Microbiol 1:191–196

Bosch A, Lucena F, Diez JM, Gajardo R, Blasi M, Jofre J (1991) Waterborne viruses associated with hepatitis outbreak. J-Am Water Works Assoc 83:80–83. https://doi.org/10.1002/j.1551-8833.1991.tb07119.x

Browne MA, Crump P, Niven SJ, Teuten E, Tonkin A, Galloway T, Thompson R (2011) Accumulation of microplastic on shorelines woldwide: sources and sinks. Environ Sci Technol 45:9175–9179. https://doi.org/10.1021/es201811s

Browne MA, Niven SJ, Galloway TS, Rowland SJ, Thompson RC (2013) Microplastic moves pollutants and additives to worms, reducing functions linked to health and biodiversity. Curr Biol 23:2388–2392. https://doi.org/10.1016/j.cub.2013.10.012

Brum HD, Campos-Silva JV, Oliveira EG, Sills J (2020) Brazil oil spill response: government inaction. Science 367:155–156. https://doi.org/10.1126/science.aba0369

Cabral AC, Martins CC (2018) Insights about sources, distribution, and degradation of sewage and biogenic molecular markers in surficial sediments and suspended particulate matter from a human-impacted subtropical estuary. Environ Pollut 241:1071–1081. https://doi.org/10.1016/j.envpol.2018.06.032

Campolim MB, Henriques MB, Petesse ML, Rezende KFO, Barbieri E (2017) Metal trace elements in mussels in Urubuqueçaba Island, Santos Bay, Brazil. Pesquisa Agropec Bras 52 (12):1131–1139. https://doi.org/10.1590/S0100-204X2017001200001

Carmo FFd, Kamino LHY, Junior RT, de Campos IC, Carmo FFd, Silvino G, Da Castro KJSXd, Mauro ML, Rodrigues NUA, Miranda Marcos Paulo, de Souza Pinto CEF (2017) Fundão tailings dam failures: the environment tragedy of the largest technological disaster of Brazilian mining in global context. Perspect Ecol Conserv 15:145–151. https://doi.org/10.1016/j.pecon.2017.06.002

Carpenter EJ, Anderson SJ, Harvey GR, Miklas HP, Peck BB (1972) Polystyrene spherules in coastal waters. Science 178:749–750. https://doi.org/10.1126/science.178.4062.749

Carreira RS, Wagener AL, Readman JW (2004) Sterols as markers of sewage contamination in a tropical urban estuary (Guanabara Bay, Brazil): space–time variations. Estuar Coast Shelf Sci 60:587–598. https://doi.org/10.1016/j.ecss.2004.02.014

de Carvalho DG, Baptista Neto JA (2016) Microplastic pollution of the beaches of Guanabara Bay, Southeast Brazil. Ocean Coast Manag 128:10–17. https://doi.org/10.1016/j.ocecoaman.2016.04.009

Cascaes MJ, Oliveira RT, Ubarana MM, Sato RM, Baldassin P, Colabuono FI, Leonel J, Taniguchi S, Weber RR (2014) Persistent organic pollutants in liver of Brazilian sharpnose shark (*Rhizoprionodon lalandii*) from southeastern coast of Brazil. Mar Pollut Bull 86:591–593. https://doi.org/10.1016/j.marpolbul.2014.05.032

Cascaes MJ (2009) Ocorrência de PCBs, PBDEs e pesticidas organiclorados em *Prionace glauca* da costa sul brasileira. Master Thesis, Universidade de São Paulo

Castro RO, Silva ML, Marques MRC, de Araújo FV (2016) Evaluation of microplastics in Jurujuba Cove, Niterói, RJ, Brazil, an area of mussels farming. Mar Pollut Bull 110:555–558. https://doi.org/10.1016/j.marpolbul.2016.05.037

Castro RO, Silva MLd, de Araújo FV (2018) Review on microplastic studies in Brazilian aquatic ecosystems. Ocean Coast Manag 165:385–400. https://doi.org/10.1016/j.ocecoaman.2018.09.013

CIA (2020) The World Factbook 2020. https://www.cia.gov/library/publications/resources/the-world-factbook/index.html

Coelho VMB (2007) Baía de Guanabara: Uma história de agressão ambiental. Casa de Palavra, Rio de Janeiro

Colton JB, Burns BR, Knapp FD (1974) Plastic particles in surface waters of the northwestern atlantic. Science 185:491–497. https://doi.org/10.1126/science.185.4150.491

Condini MV, Hoeinghaus DJ, Roberts AP, Soulen BK, Garcia AM (2017) Mercury concentrations in dusky grouper *Epinephelus marginatus* in littoral and neritic habitats along the Southern Brazilian coast. Mar Pollut Bull 115:266–272. https://doi.org/10.1016/j.marpolbul.2016.12.006

Coutinho FH, Silveira CB, Pinto LH, Salloto GRB, Cardoso AM, Martins OB, Vieira RP, Clementino MM (2014) Antibiotic resistance is widespread in urban aquatic environments of Rio de Janeiro, Brazil. Microb Ecol 68:441–452. https://doi.org/10.1007/s00248-014-0422-5

Da Klein AHF, Short AD (2016) Brazilian beach systems: introduction. Brazilian beach systems, vol 17. Springer, Cham, pp 1–35

Da Silva AM, Pavesi T, Rosa ACS, Santos TPd, de Medeiros Tabalipa M, Lemes VRR, Alves SR, de Novaes Sarcinelli P (2016) Organochlorines and polychlorinated biphenyl environmental pollution in south coast of Rio De Janeiro state. Mar Pollut Bull 108:325–331. https://doi.org/10.1016/j.marpolbul.2016.03.067

Davies J, Davies D (2010) Origins and evolution of antibiotic resistance. Microbiol Mol Biol Rev 74:417–433. https://doi.org/10.1128/MMBR.00016-10

Dias J, Pinto RN, Vieira CB, de Abreu Corrêa A (2018) Detection and quantification of human adenovirus (HAdV), JC polyomavirus (JCPyV) and hepatitis A virus (HAV) in recreational waters of Niterói, Rio de Janeiro, Brazil. Mar Pollut Bull 133:240–245. https://doi.org/10.1016/j.marpolbul.2018.05.031

Disner GR, Torres M (2020) The environmental impacts of 2019 oil spill on the Brazilian coast: overview. Rev Bras Gest Amb Sustent 7:241–255. https://doi.org/10.21438/rbgas(2020)071518

do Nascimento MTL, Santos ADdO, Felix LC, Gomes G, Oliveira E Sá M de, da Cunha DL, Vieira N, Hauser-Davis RA, Baptista Neto JA, Bila DM (2018) Determination of water quality, toxicity and estrogenic activity in a nearshore marine environment in Rio de Janeiro, Southeastern Brazil. Ecotoxicol Environ Saf 149:197–202. https://doi.org/10.1016/j.ecoenv.2017.11.045

Ekino S, Susa M, Ninomiya T, Imamura K, Kitamura T (2007) Minamata disease revisited: an update on the acute and chronic manifestations of methyl mercury poisoning. J Neurol Sci 262:131–144. https://doi.org/10.1016/j.jns.2007.06.036

Erickson MC, Doyle MP (2006) Closing the door on the fecal coliform assay. Microbe Magazine 1:162–163. https://doi.org/10.1128/microbe.1.162.1

Escobar H (2019) Mystery oil spill threatens marine sanctuary in Brazil. Science 366:672. https://doi.org/10.1126/science.366.6466.672

Feller IC (1995) Effects of nutrient enrichment on growth and herbivory of dwarf red mangrove (*Rhizophora mangle*). Ecol Monographs 65:477–505. https://doi.org/10.2307/2963499

Fernandez GB, da Rocha TB, Barboza EG, Dillenburg SR, da Camara Rosa MLC, Angulo RJ, de Souza MC, de Oliveira LHS, Dominguez JML (2019) Natural Landscapes Along Brazilian Coastline. In: Salgado AAR, Santos LJC, Paisani JC (eds) The physical geography of Brazil: environment, vegetation and landscape. Springer, Cham, pp 199–218

Ferreira AC, Lacerda LD (2016) Degradation and conservation of Brazilian mangroves, status and perspectives. Ocean Coast Manag 125:38–46. https://doi.org/10.1016/j.ocecoaman.2016.03.011

Ferreira HO (1995) Aporte de hidrocarbonetos de petróleo para a Baia de Guanabara. MSc Thesis, Universidade Federal Fluminense

Figueiredo GM, Vianna TMP (2018) Suspended microplastics in a highly polluted bay: abundance, size, and availability for mesozooplankton. Mar Pollut Bull 135:256–265. https://doi.org/10.1016/j.marpolbul.2018.07.020

Fistarol GO, Coutinho FH, Moreira APB, Venas T, Cánovas A, Paula SEM de, Coutinho R, Moura RL de, Valentin JL, Tenenbaum DR, Paranhos R, do Valle RdAB, Vicente ACP, Amado Filho GM, Pereira RC, Kruger R, Rezende CE, Thompson CC, Salomon PS, Thompson FL (2015) Environmental and sanitary conditions of Guanabara Bay, Rio de Janeiro. Front Microbiol 6:1232. https://doi.org/10.3389/fmicb.2015.01232

Fontes MK, de Campos BG, Cortez FS, Pusceddu FH, Moreno BB, Maranho LA, Lebre DT, Guimarães LL, Pereira CDS (2019) Seasonal monitoring of cocaine and benzoylecgonine in a subtropical coastal zone (Santos Bay, Brazil). Mar Pollut Bull 149: https://doi.org/10.1016/j.marpolbul.2019.110545

Fumian TM, Fioretti JM, Lun JH, Dos Santos IAL, White PA, Miagostovich MP (2019) Detection of norovirus epidemic genotypes in raw sewage using next generation sequencing. Environ Int 123:282–291. https://doi.org/10.1016/j.envint.2018.11.054

Fumian TM, Guimarães FR, Pereira Vaz BJ, Da Silva MTT, Muylaert FF, Bofill-Mas S, Gironés R, Leite JPG, Miagostovich MP (2010) Molecular detection, quantification and characterization of human polyomavirus JC from waste water in Rio De Janeiro, Brazil. J Water Health 8:438–445. https://doi.org/10.2166/wh.2010.090

Fumian TM, Vieira CB, Leite JPG, Miagostovich MP (2013) Assessment of burden of virus agents in an urban sewage treatment plant in Rio de Janeiro, Brazil. J Water Health 11:110–119. https://doi.org/10.2166/wh.2012.123%23

Gabriel FÂ, Hauser-Davis RA, Soares LOS, Azevedo Mazzuco AC de, Rocha RCC, Saint'Pierre TD, Saggioro EM, Correia FV, Ferreira TO, Bernardino AF (2020) Metal contamination and oxidative stress biomarkers in estuarine fish following a mine tailing disaster. https://doi.org/10.1101/2020.06.29.177253

Garcia TM, Campos CC, Mota EMT, Santos NMO, Campelo RPdS, Prado LCG, Melo Junior M, Soares MdO (2020) Microplastics in subsurface waters of the western equatorial Atlantic (Brazil). Mar Pollut Bull 150: https://doi.org/10.1016/j.marpolbul.2019.110705

Giri C, Ochieng E, Tieszen LL, Zhu Z, Singh A, Loveland T, Masek J, Duke N (2011) Status and distribution of mangrove forests of the world using earth observation satellite data. Glob Ecol Biogeogr 20:154–159. https://doi.org/10.1111/j.1466-8238.2010.00584.x

Gomes LEdO, Correa LB, Sá F, Neto RR, Bernardino AF (2017) The impacts of the Samarco mine tailing spill on the Rio Doce estuary, Eastern Brazil. Mar Pollut Bull 120:28–36. https://doi.org/10.1016/j.marpolbul.2017.04.056

IBGE (2000) Evolução da População Brasileira. https://brasil500anos.ibge.gov.br/estatisticas-do-povoamento/evolucao-da-populacao-brasileira.html. Accessed 22 Nov 2020

IBGE (2019) IBGE atualiza lista dos municípios defrontantes com o mar. https://agenciadenoticias.ibge.gov.br/agencia-sala-de-imprensa/2013-agencia-de-noticias/releases/24074-ibge-atualiza-lista-dos-municipios-defrontantes-com-o-mar. Accessed 22 Nov 2020

IBGE (2020) Painel de Indicadores. https://www.ibge.gov.br/indicadores.html. Accessed 8 Nov 2020

Jordão EP, Rosman PCC (2016) Submarine outfalls are an effective solution for the disposal of properly treated sewage of coastal cities. RDAE 64:57–63. https://doi.org/10.4322/dae.2016.016

Kjerfve B, de Lacerda LD, Dias GTM (2001) Baía de Guanabara, Rio De Janeiro, Brazil. In: Seeliger U, Kjerfve B (eds) Coastal marine ecosystems of Latin America, vol 144. Springer, Berlin, Heidelberg, pp 107–117

Leeming R, Ball A, Ashbolt N, Nichols P (1996) Using faecal sterols from humans and animals to distinguish faecal pollution in receiving waters. Water Res 30:2893–2900. https://doi.org/10.1016/S0043-1354(96)00011-5

Loganathan P, Naidu G, Vigneswaran S (2017) Mining valuable minerals from seawater: a critical review. Environ Sci: Water Res Technol 3:37–53. https://doi.org/10.1039/C6EW00268D

Lourenço RA, Combi T, Alexandre MdR, Sasaki ST, Zanardi-Lamardo E, Yogui GT (2020) Mysterious oil spill along Brazil's northeast and southeast seaboard (2019–2020): Trying to find answers and filling data gaps. Mar Pollut Bull 156: https://doi.org/10.1016/j.marpolbul.2020.111219

Luiz-Silva W, Machado W, Matos RHR (2008) Multi-elemental contamination and historic record in sediments from the Santos-Cubatão Estuarine System, Brazil. J Braz Chem Soc 19:1490–1500. https://doi.org/10.1590/S0103-50532008000800008

de Macêdo GR, Tarantino TB, Barbosa IS, Pires TT, Rostan G, Goldberg DW, Pinto LFB, Korn MGA, Franke CR (2015) Trace elements distribution in hawksbill turtle (*Eretmochelys imbricata*) and green turtle (*Chelonia mydas*) tissues on the northern coast of Bahia, Brazil. Mar Pollut Bull 94:284–289. https://doi.org/10.1016/j.marpolbul.2015.02.033

Maggi P, Vaz Morgado, Claudia Do Rosario, Nóbrega De Almeida, João Carlos (2013) Offshore oil spill incidents in Brazil. In: Progress in environmental protection and processing of resource, 295–298. Trans Tech, pp 626–629

Magris RA, Barreto R (2010) Mapping and assessment of protection of mangrove habitats in Brazil. Pan-Am J Aquat Sci 5:546–556

Magris RA, Giarrizzo T (2020) Mysterious oil spill in the Atlantic Ocean threatens marine biodiversity and local people in Brazil. Mar Pollut Bull 153:110961. https://doi.org/10.1016/j.marpolbul.2020.110961

Martins CC, Bícego MC, Mahiques MM, Figueira RCL, Tessler MG, Montone RC (2011) Polycyclic aromatic hydrocarbons (PAHs) in a large South American industrial coastal area (Santos Estuary, Southeastern Brazil): sources and depositional history. Mar Pollut Bull 63:452–458. https://doi.org/10.1016/j.marpolbul.2011.03.017

Martins CC, Cabral AC, Barbosa-Cintra SCT, Dauner ALL, Souza FM (2014) An integrated evaluation of molecular marker indices and linear alkylbenzenes (LABs) to measure sewage input in a subtropical estuary (Babitonga Bay, Brazil). Environ Pollut 188:71–80. https://doi.org/10.1016/j.envpol.2014.01.022

Mato Y, Isobe T, Takada H, Kanehiro H, Ohtake C, Kaminuma T (2001) Plastic resin pellets as a transport medium for toxic chemicals in the marine environment. Environ Sci Technol 35:318–324. https://doi.org/10.1021/es0010498

Ministério de Meio Ambiente, Conselho Nacional do Meio Ambiente (2011, May 2011) Resolução No 430. http://www2.mma.gov.br/port/conama/legiabre.cfm

Mizerkowski BD, Hesse K-J, Ladwig N, da Costa Machado E, Rosa R, Araujo T, Koch D (2012) Sources, loads and dispersion of dissolved inorganic nutrients in Paranaguá Bay. Deutsche Hydrografische Zeitschrift 62:1409–1424. https://doi.org/10.1007/s10236-012-0569-x

Oberdörster E, Cheek AO (2001) Gender benders at the beach: Endocrine disruption in marine and estuarine organisms. Environ Toxicol Chem 20:23–36. https://doi.org/10.1002/etc.5620200103

Okoh AI, Sibanda T, Gusha SS (2010) Inadequately treated wastewater as a source of human enteric viruses in the environment. Int J Environ Res Public Health 7:2620–2637. https://doi.org/10.3390/ijerph7062620

Olivatto GP, Martins MCT, Montagner CC, Henry TB, Carreira RS (2019) Microplastic contamination in surface waters in Guanabara Bay, Rio de Janeiro, Brazil. Mar Pollut Bull 139:157–162. https://doi.org/10.1016/j.marpolbul.2018.12.042

Oliveira OMC de, S Queiroz AF de, Cerqueira JR, Soares SAR, Garcia KS, Filho AP, L da S Rosa M de, Suzart CM, L Pinheiro L de, Moreira ÍTA (2020) Environmental disaster in the northeast coast of Brazil: Forensic geochemistry in the identification of the source of the oily material. Mar Pollut Bull 160:111597. https://doi.org/10.1016/j.marpolbul.2020.111597

Pereira CDS, Maranho LA, Cortez FS, Pusceddu FH, Santos AR, Ribeiro DA, Cesar A, Guimarães LL (2016) Occurrence of pharmaceuticals and cocaine in a Brazilian coastal zone. Sci Total Environ 548–549:148–154. https://doi.org/10.1016/j.scitotenv.2016.01.051

Pinheiro MAA, Silva PPGE, Duarte LFdA, Almeida AA, Zanotto FP (2012) Accumulation of six metals in the mangrove crab *Ucides cordatus* (Crustacea: Ucididae) and its food source, the red mangrove *Rhizophora mangle* (Angiosperma: Rhizophoraceae). Ecotoxicol Environ Saf 81:114–121. https://doi.org/10.1016/j.ecoenv.2012.05.004

Pivetta RC, Rodrigues-Silva C, Ribeiro AR, Rath S (2020) Tracking the occurrence of psychotropic pharmaceuticals in Brazilian wastewater treatment plants and surface water, with assessment of environmental risks. Sci Total Environ 727:138661. https://doi.org/10.1016/j.scitotenv.2020.138661

Possatto FE, Spach HL, Cattani AP, Lamour MR, Santos LO, Cordeiro NMA, Broadhurst MK (2015) Marine debris in a World Heritage Listed Brazilian estuary. Mar Pollut Bull 91:548–553. https://doi.org/10.1016/j.marpolbul.2014.09.032

Prestt I, Jefferies DJ, Moore NW (1970) Polychlorinated biphenyls in wild birds in Britain and their avian toxicity. Environ Pollut 1:3–26. https://doi.org/10.1016/0013-9327(70)90003-0

Pusceddu FH, Sugauara LE, de Marchi MR, Choueri RB, Castro ÍB (2019) Estrogen levels in surface sediments from a multi-impacted Brazilian estuarine system. Mar Pollut Bull 142:576–580. https://doi.org/10.1016/j.marpolbul.2019.03.052

Rios LM, Moore C, Jones PR (2007) Persistent organic pollutants carried by synthetic polymers in the ocean environment. Mar Pollut Bull 54:1230–1237. https://doi.org/10.1016/j.marpolbul.2007.03.022

Rizzi M, Rodrigues FL, Medeiros L, Ortega I, Rodrigues L, Monteiro DS, Kessler F, Proietti MC (2019) Ingestion of plastic marine litter by sea turtles in southern Brazil: abundance, characteristics and potential selectivity. Mar Pollut Bull 140:536–548. https://doi.org/10.1016/j.marpolbul.2019.01.054

Rochman CM, Kurobe T, Flores I, Teh SJ (2014) Early warning signs of endocrine disruption in adult fish from the ingestion of polyethylene with and without sorbed chemical pollutants from the marine environment. Sci Total Environ 493:656–661. https://doi.org/10.1016/j.scitotenv.2014.06.051

Rodrigues PdA, Ferrari RG, Hauser-Davis RA, Neves Dos Santos L, Conte-Junior CA (2020) Dredging activities carried out in a Brazilian estuary affect mercury levels in swimming crabs. Int J Environ Res Pub Health 17. https://doi.org/10.3390/ijerph17124396

Rodrigues VFV, Rivera ING, Lim K-Y, Jiang SC (2016) Detection and risk assessment of diarrheagenic *E. coli* in recreational beaches of Brazil. Mar Pollut Bull 109:163–170. https://doi.org/10.1016/j.marpolbul.2016.06.007

Roth F, Lessa GC, Wild C, Kikuchi RKP, Naumann MS (2016) Impacts of a high-discharge submarine sewage outfall on water quality in the coastal zone of Salvador (Bahia, Brazil). Mar Pollut Bull 106:43–48. https://doi.org/10.1016/j.marpolbul.2016.03.048

Sakai H, Ichihashi H, Suganuma H, Tatsukawa R (1995) Heavy metal monitoring in sea turtles using eggs. Mar Pollut Bull 30:347–353. https://doi.org/10.1016/0025-326X(94)00185-C

Salas HJ (2000) Submarine outfalls, a viable alternative for sewage discharge of coastal cities in Latin America and the Caribbean. Pan American Center for Sanitary Engineering and Environmental Sciences, WHO

Santana MFM, Ascer LG, Custódio MR, Moreira FT, Turra A (2016) Microplastic contamination in natural mussel beds from a Brazilian urbanized coastal region: Rapid evaluation through bioassessment. Mar Pollut Bull 106:183–189. https://doi.org/10.1016/j.marpolbul.2016.02.074

Sarmento SK, Guerra CR, Malta FC, Coutinho R, Miagostovich MP, Fumian TM (2020) Human norovirus detection in bivalve shellfish in Brazil and evaluation of viral infectivity using PMA treatment. Mar Pollut Bull 157: https://doi.org/10.1016/j.marpolbul.2020.111315

Schlindwein AD, Rigotto C, Simões CMO, Barardi CRM (2010) Detection of enteric viruses in sewage sludge and treated wastewater effluent. Water Sci Technol 61:537–544. https://doi.org/10.2166/wst.2010.845

Secretariat of the Stockholm Convention (2019) Stockholm Convention. http://www.pops.int/. Accessed 22 Oct 2020

SNIS (2018) Esgotamento Sanitário. http://www.snis.gov.br/painel-informacoes-saneamento-brasil/web/painel-esgotamento-sanitario. Accessed 9 Nov 2020

Soares MdO, Teixeira CEP, Bezerra LEA, Paiva SV, Tavares TCL, Garcia TM, Araújo JT de, Campos CC, Ferreira SMC, Matthews-Cascon H, Frota A, Mont'Alverne TCF, Silva ST, Rabelo EF, Barroso CX, Freitas JEP de, Melo Júnior M de, Campelo RPdS, Santana CS de, Carneiro PBdM, Meirelles AJ, Santos BA, Oliveira AHB de, Horta P, Cavalcante RM (2020a) Oil spill in South Atlantic (Brazil): environmental and governmental disaster. Marine Policy 115:103879. https://doi.org/10.1016/j.marpol.2020.103879

Soares MO, Teixeira CEP, Bezerra LEA, Rossi S, Tavares T, Cavalcante RM (2020) Brazil oil spill response: time for coordination. Science 367:155. https://doi.org/10.1126/science.aaz9993

Soares-Gomes A, Da Gama B, Baptista Neto JA, Freire DG, Cordeiro RC, Machado W, Bernardes MC, Coutinho R, Thompson FL, Pereira RC (2016) An environmental overview of Guanabara Bay, Rio de Janeiro. Reg Stud Mar Sci 8:319–330. https://doi.org/10.1016/j.rsma.2016.01.009

Souza NLN, Carneiro MTWD, Pimentel EF, Frossard A, Freire JB, Endringer DC, Ferreira Júnior PD (2018) Trace elements influence the hatching success and emergence of *Caretta caretta* and *Chelonia mydas*. J Trace Elem Med Biol: Organ Soc Miner Trace Elem (GMS) 50:117–122. https://doi.org/10.1016/j.jtemb.2018.06.007

Thompson RC, Olsen Y, Mitchell RP, Davis A, Rowland SJ, John AWG, McGonigle D, Russell AE (2004) Lost at sea: where is all the plastic? Science 304:838. https://doi.org/10.1126/science.1094559

Trotter B, Ramsperger AFRM, Raab P, Haberstroh J, Laforsch C (2019) Plastic waste interferes with chemical communication in aquatic ecosystems. Sci Rep 9:5889. https://doi.org/10.1038/s41598-019-41677-1

Vergilio CDS, Lacerda D, de Oliveira BCV, Sartori E, Campos GM, Pereira ALdS, de Aguiar DB, Da Souza TS, de Almeida MG, Thompson F, de Rezende CE (2020) Metal concentrations and biological effects from one of the largest mining disasters in the world (Brumadinho, Minas Gerais, Brazil). Sci Rep 10:5936. https://doi.org/10.1038/s41598-020-62700-w

Verlicchi P, Al Aukidy M, Zambello E (2012) Occurrence of pharmaceutical compounds in urban wastewater: removal, mass load and environmental risk after a secondary treatment—a review. Sci Total Environ 429:123–155. https://doi.org/10.1016/j.scitotenv.2012.04.028

Vieira RHSdF, Rodrigues DdP, Menezes EA, Evangelista NSdS, Reis EMd, Barreto LM, Gonçalves FA (2001) Microbial contamination of sand from major beaches in Fortaleza, Ceará State, Brazil. Braz J Microbiol 32. https://doi.org/10.1590/S1517-83822001000200001

Xu W, Wang X, Cai Z (2013) Analytical chemistry of the persistent organic pollutants identified in the Stockholm Convention: a review. Anal Chim Acta 790:1–13. https://doi.org/10.1016/j.aca.2013.04.026

Yang L, Chang W-S, Huang M-N (2000) Natural disinfection of wastewater in marine outfall fields. Water Res 34:743–750. https://doi.org/10.1016/S0043-1354(99)00209-2

Chapter 10
Hydrochemical Insight and Groundwater Supply: A Case Study of Patagonia's Chubut River

Américo I. Torres, Luis F. H. Niencheski, Verena A. Campodonico, Andrea I. Pasquini, Mauricio Faleschini, and Pedro J. Depetris

Abstract The aim of this chapter is to assess the groundwater discharge component in the hydrological budget of NE Patagonia's Chubut River (CHR), using ^{222}Rn as a quantitative groundwater tracer. The main hydrological characteristics of CHR are also presented. The annual mean discharges at Valle Inferior have decreased at a mean rate of 10 m^3 per decade. Anthropogenic impacts—such as wastewater discharge—are observed in surface waters. The ^{222}Rn mass balance indicated that the groundwater inflow to the CHR ranged between 2.0 and 5.4 m^3 s^{-1}. Thus, such groundwater inputs may account for approximately 6–24% of the total water delivered by CHR to the SW Atlantic Ocean. However, a negative flux (-1.3 m^3 s^{-1}) was recorded in another sampling site suggesting that the river is losing water to the aquifer. The nutrient fluxes toward the ocean were higher in winter than in spring. The silicic acid flux (1.175×10^6 mol d^{-1}) was the highest, followed by the total inorganic nitrogen (0.08×10^6 mol d^{-1}) and then by phosphorus (0.024×10^6 mol d^{-1}). Such nitrogen flux would support a production rate of 7.42×10^6 g C d^{-1}. The results presented in this study indicate that SGD represents a relevant process, which takes place in the CHR's estuarine zone.

Keywords Hydrology · Groundwater · Tracer · Chubut River · Patagonia

---

A. I. Torres (✉) · M. Faleschini
Laboratorio de Oceanografía Química y Contaminación de Aguas, CESIMAR, CCT CONICET-CENPAT, Blvd. Brown 2915 9120, Puerto Madryn, Chubut, Argentina
e-mail: americo@cenpat-conicet.gob.ar

L. F. H. Niencheski
Instituto de Oceanografia, Fundação Universidade Federal do Rio Grande, Av. Italia km 8, Rio Grande, RS 96201-900, Brazil

V. A. Campodonico · A. I. Pasquini
Centro de Investigaciones en Ciencias de la Tierra, Consejo Nacional de Investigaciones Científicas y Técnicas y Universidad Nacional de Córdoba, Av. Vélez Sarsfield 1611, X5016CGA Córdoba, Argentina

P. J. Depetris
Academia Nacional de Ciencias, Avenida Vélez Sarsfield 229-249, X5000WAA Córdoba, Argentina

© The Author(s), under exclusive license to Springer Nature Switzerland AG 2021
D.-P. Häder et al. (eds.), *Anthropogenic Pollution of Aquatic Ecosystems*,
https://doi.org/10.1007/978-3-030-75602-4_10

10.1 Introduction

The marine coastal zone receives the discharge of both rivers and groundwater. Groundwater circulating through sediment strata can crop up in a river or in the sea, through the seabed, and this event is globally identified as submarine groundwater discharge (SGD). SGD is a complex hydrological process, which is not always evident, since it can occur along the coastline, a few meters from the coast or several kilometers offshore (Bratton 2010), depending on the hydrological characteristics of the area (Kroeger and Charette 2008). The dynamics of such a complex mechanism requires a multidisciplinary approach (Burnett and Dulaiova 2003). Groundwater contributions represent a pathway of communication and exchange of chemical components between continent and sea (Moore 1999; Burnett and Dulaiova 2003). This communication path is regulated by different processes which generate a hydraulic gradient necessary to establish the water flow (Burnett et al. 2003; Santos et al. 2012). In addition, in the last decades, SGD has been considered an important source of carbon, macro- and micronutrients, heavy metals, dissolved gases and other elements, which are transferred to the marine environment (Moore 2006; Swarzenski et al. 2006). Additionally, such chemical elements can be present in high concentrations reaching up to three orders of magnitude higher than those recorded in surface water. Globally, SGD has a net flow of 120×10^{12} m^3 y^{-1} (Kwon et al. 2014), which widely exceeds the net riverine flow (36×10^{12} m^3 y^{-1}) toward the oceans (Milliman and Farnsworth 2011). Thus, it is evident that continental nutrient contributions can affect the biogeochemical cycles in the coastal marine zone (Garcia-Solsona et al. 2010; Godoy et al. 2013). Additionally, the contribution of essential nutrients—such as N and P—in the aquatic environment is important for the development of trophic chains but also indicates the degree of eutrophication, whereas silica may suggest coastal processes related to sediment remobilization, erosion and sedimentation, evidencing the influence of freshwater.

Studies around the world have demonstrated the effective use of geochemical tracers both to detect areas with the occurrence of SGD and to study its dynamics, as well as to estimate groundwater inflow in rivers. Among the most commonly used are radioactive isotopes, such as ^{3}H, ^{219}Rn, ^{220}Rn, ^{222}Rn, ^{223}Ra, ^{224}Ra, ^{226}Ra and ^{228}Ra, and stable isotopes, like ^{2}H, ^{18}O, CFC_s, SF_6 and ^{4}He (e.g., Windom et al. 2006; Santos et al. 2009; Peterson et al. 2010; Schmidt et al. 2011; Smerdon et al. 2012; Swarzenski et al. 2017).

River waters contain low concentrations of dissolved radium and radon (Porcelli and Swarzenski 2003). Therefore, the discharge of groundwater into the deepest parts of the river channel can be detected by the characteristic radon enrichment with respect to the surface waters. Radon has a unique characteristic, because it is a noble gas and is, therefore, chemically and biologically inert. Radon is an odorless and colorless radioactive gas that occurs naturally in air, in water and in rocks and soil. Specifically, ^{222}Rn is a daughter isotope of ^{226}Ra and they both reach a radioactive equilibrium over a few weeks (Cecil and Green 2000). It has been recognized as an excellent tracer to identify areas with groundwater influx (Burnett et al. 1996; Hamada 2000; Burnett and Dulaiova 2003). Likewise, it has been proved useful for estimating the infiltration of river-fed groundwater toward other aquatic environments (Macheleidt et al. 2002). The other radon isotopes (^{219}Rn and ^{220}Rn) have a half-life of less than one minute, and for this reason they are difficult to use as tracers in environmental investigations. The use of ^{222}Rn as a proxy for the estimation of groundwater discharge stems from its significant concentrations in groundwater (i.e., 1000 times or more its concentration in surface water), its non-reactive nature and its short half-life ($t_{1/2}$ 3.82 days). These features make it useful in hydrological studies (Cecil and Green 2000).

In Argentina's Patagonia, there are eight rivers with Andean sources that discharge their waters into the Atlantic Ocean. They contribute nutrients, sediments, trace and major elements, which play an important role in sustaining the biological structure in Patagonia's coastal zone (Depetris et al. 2005). However, the available information on SGD in that coastal area is scarce. Preliminary results were obtained by Torres et al. (2018) who have identified the presence of SGD in the coastal areas from Patagonia, but the flows were not yet quantitatively measured. Currently, there is a lack of information regarding groundwater contribution to rivers and its interaction with surface waters. In addition, the role of SGD and the supply of nutrients in the biogeochemical cycles in the coastal zone are unknown. The goal of this study is using the activity of ^{222}Rn to identify the presence of groundwater discharges in Chubut River's lower reaches, estimating the groundwater flow toward the sea. In so doing, the evaluation of the groundwater chemical contribution is sought, along with the definition of the relationship between groundwater and surface waters based on chemical and hydrological characteristics.

10.1.1 Study Area

The 800-km-long Chubut River crosses the homonymous province from west to east, reaching Engaño Bay, in the SW Atlantic Ocean (Fig. 10.1). Its drainage basin exceeds ~53,200 km^2 (Subsecretaría de Recursos Hídricos 2002), with headwaters in the Cerro de las Carreras (41° S 71° 19′ W), in Argentina's Río Negro Province, at ~1800 m above sea level (a.s.l.). The main tributaries in the upper drainage are the Ñorquinco and Chico del Norte rivers, inflowing from the north, and the Tecka–Gualjaina River (i.e., drainage basin ~28,000 km^2), joining the system from the south (Fig. 10.1). Senguer-Chico del Norte River system (i.e., drainage basin of ~26,000 km^2) drains the Musters and Colhué Huapi lakes, which used to join the CHR, inflowing from the SE. Yet, due to natural obstructions that modified the original morphology, the system became endorheic.

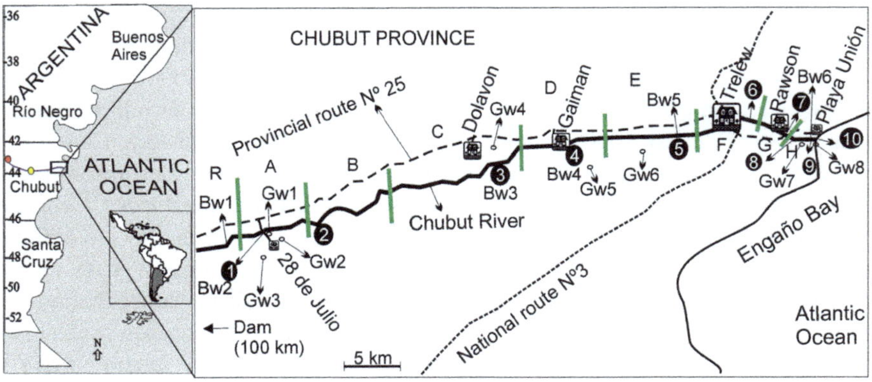

Fig. 10.1 Study area, lower reaches of Chubut River. The sampling sites are indicated as circles (i.e., samples of surface waters, from 1 to 10); Bw, bottom water samples, and Gw, groundwater samples (i.e., from wells). Green bars correspond to box boundaries used in the estimation of mass balance calculations. Red and yellow circles are indicating the location of the El Maitén and Los Altares gauging stations, respectively

The Florentino Ameghino hydroelectric dam, located 150 km upstream from the estuary, modulates Chubut's discharges. In the Chubut's lower reach, there are two cities, Trelew (~100,000 inhabitants) and Rawson (~32,000 inhabitants). The first one has a stabilization pond system and the treated liquid is naturally evaporated and infiltrated into the soil without obvious connection with the river, while the second one discharges its urban wastewater—after receiving an activated sludge treatment—into the river. In addition, other towns (<5000 inhabitants), such as Gaiman, Dolavon and 28 de Julio, which do not have a sewage network (i.e., septic tanks, instead), are a source of diffuse infiltrations that ultimately reach the river. Furthermore, in the CHR lower basin there is an area with intensive agricultural activity, irrigated with river water, which is distributed by means of an extensive channel network. Associated with this activity, there is an intensive and uncontrolled use of fertilizers and pesticides—organochlorine and organophosphate compounds (Antolini 2012) which enter the river through unused water, diffuse flows and seepage. Additionally, there is a shallow water harbor—located only 600 m from the river mouth—which is used by boats pursuing fishing activities of high economic significance for the region. As a result of the port activities, hydrocarbons (Commendatore and Esteves 2004) and heavy metals (Gil et al. 1999) are released into the river. On several occasions, dredging of the last 600 m of the navigation channel has been made to allow the transit of medium-size fishing vessels. Furthermore, there are some fish-processing plants located on the river's margins, discharging their untreated wastewaters. The salt intrusion due to tidal effects in the estuary was observed—under low river discharge—as far as 4.5 km from the mouth (Perillo et al. 1989).

The region has an arid climate due to low rainfall (<240 mm y^{-1}), a high rate of evapotranspiration (2000 mm y^{-1}) and an average annual humidity of 40% (EMC

2020). These characteristics favor the strong dryness of the soil. Spring is the windy season with prevailing winds from the west to the southwest direction with an average speed of 16.6 km h^{-1} and recorded maxima of 90 km h^{-1}. The wind-borne contribution of dust to the marine environment plays a substantial role. The addition of Patagonian dust to seawater contributes nitrate and silicate nutrients, which may affect the biological productivity depending on the dust source and on the amount added (Paparazzo et al. 2018).

10.2 Methodology

Ten surface freshwater samples, collected between June and November 2017, were taken using a Niskin bottle in the lower course of the Chubut River. The samples were stored at −20 °C preceding the measurement of the concentrations of dissolved inorganic nutrients such as ammonium ($N–NH_4^+$), nitrite and nitrate ($N–NO_3^-+NO_2^-$), phosphate (PO_4^{-3}) and silicic acid $Si(OH)_4$ using an autoanalyzer (Skalar Analytical® V.B. 2005a, b, c). The $N–NO_3^-+NO_2^-$ were determined together and are hereafter referred to as $N–NO_3^-$ concentration. Ammonium concentration was determined manually according to Strickland and Parsons (1972). Furthermore, the following parameters were measured in situ: temperature, dissolved oxygen (DO), electrical conductivity (EC), pH, redox potential, total dissolved solids (TDS) and salinity using a YSI MPS 556 sensor. Also, six bottom freshwater samples were taken in November 2017. Immediately, a volume of 4 L was placed in a glass bottle to measure the ^{222}Rn activity using a RAD-7 equipment (Durridge Co. Inc.). The RAD-7 equipment was configured to measure the ^{222}Rn activity every thirty minutes for two hours (for more details, see Campodonico et al. 2015). Another subsample was stored for the measurement of the nutrient concentrations. Additionally, groundwater samples were taken in November in seven permanent wells and in one well drilled in the beach area near the mouth of Chubut River. The physical and chemical determinations mentioned above were also carried out in groundwater samples.

The ^{222}Rn activities were measured during November 2017 at 6 sampling points along the lower stretch of the CHR and on selected groundwater samples (n = 7). ^{222}Rn measurements were performed in situ by means of a RAD-7 detector (Durridge Co. Inc.), a portable continuous radon-in-air monitor modified for radon-in-water that determines the activity of ^{222}Rn by counting its alpha-emitting daughters (^{218}Po and/or ^{214}Po). Samples were collected in 4-L plastic bottles designed to avoid gas loss (Stringer and Burnett 2004), and ^{222}Rn activity was measured in the field by means of the RAD-7, which purges ^{222}Rn in water in order to achieve a rapid equilibrium of ^{222}Rn between water and air.

10.3 Hydrological Characteristics of the Chubut River

The hydrological dynamics of the Chubut River is mainly governed by rainfall and snowfall in the upper catchments. Most of the annual discharge is delivered by the upper Chubut River (i.e., ~60–65% during the low-discharge period).

In the mountainous headwaters, atmospheric precipitations increase in May (i.e., maximum mean precipitation) and June, during the Southern Hemisphere's fall and winter, and begin a gradual decrease that reaches its minimum in November (i.e., southern spring). The uppermost Chubut River gauging station (Nacimiento, 41° 43′ S, 71° 08′ W)[1] is operative since June 1967. During a period of over 50 years (i.e., until November 2019), the maximum recorded discharge (Q_{max}) was 29.17 $m^3\ s^{-1}$, whereas the minimum (Q_{min}) was 0.41 $m^3\ s^{-1}$.

Discharge data are usually asymmetrically distributed and often adjust to a log-normal statistical distribution. Therefore, the mathematical expectation is better approached by using the geometric mean discharge (Q_g) (e.g., Davis 1986). Q_g and the corresponding geometric standard deviation (S_g) for the analyzed period were 4.8 ± 0.91 $m^3\ s^{-1}$. The highest discharges are usually recorded in July–August, due to rainfall and snowfall, and in October–November, as a result of snowmelt.

Atmospheric precipitations at the El Maitén gauging station (42° 06′ S, 71° 10′ W) follow a pattern analogous to the one recorded at Nacimiento station, although maximum mean precipitation occurs in June, with the minimum also in November. Figure 10.2 shows the significant log-normal distribution exhibited by instantaneous discharges recorded at El Maitén during the 2000–2020 record period. The station is located at ~720 m a.s.l., and the available discharge time series for the period 1953–2019 includes over 6800 instantaneous discharge measurements, with $Q_{max} \approx 240\ m^3\ s^{-1}$ and $Q_{min} \approx 1.9\ m^3\ s^{-1}$. Q_g and the corresponding S_g are 16.2 ± 2.23 $m^3\ s^{-1}$. Therefore, from a statistical point of view, 95 per cent of the assembled data are comprised within $Q_g \pm 2S_g$: $11.7\ m^3\ s^{-1} < Q_g < 20.7\ m^3\ s^{-1}$.

The Los Altares gauging station is placed in a semiarid region, about 230 km west of the city of Trelew and 323 km east of city of Esquel, on the southern margin of the Chubut River. The wild landscape is known for its high cliffs and remarkable morphology. The Los Altares station (43° 53′ 18.17″ S, 68° 23′ 57.81′ ′ W) is active since January 1943, record-keeping instantaneous discharges and other hydrometeorological variables. The linear trend of the mean annual discharge series (Fig. 10.3) shows a faintly negative slope, which lacks statistical significance. Figure 10.4 shows the mean annual hydrograph for the series of monthly mean discharges (i.e., period of 1943–2018), with maximum discharges occurring during austral spring. The average discharge (i.e., arithmetic mean) for the period was 47.04 $m^3\ s^{-1}$, whereas the annualized discharge was 14.84 km^3, the specific water yield was 2.87 $L\ s^{-1}\ km^{-2}$, and runoff exceeded 9 $mm\ y^{-1}$. In this studied time series, instantaneous Q_{max} was 524 $m^3\ s^{-1}$, and Q_{min}, 2.1 $m^3\ s^{-1}$.

[1]The hydrological data were obtained from the Argentina's Sistema Nacional de Información Hídrica (https://snih.hidricosargentina.gob.ar/Filtros.aspx).

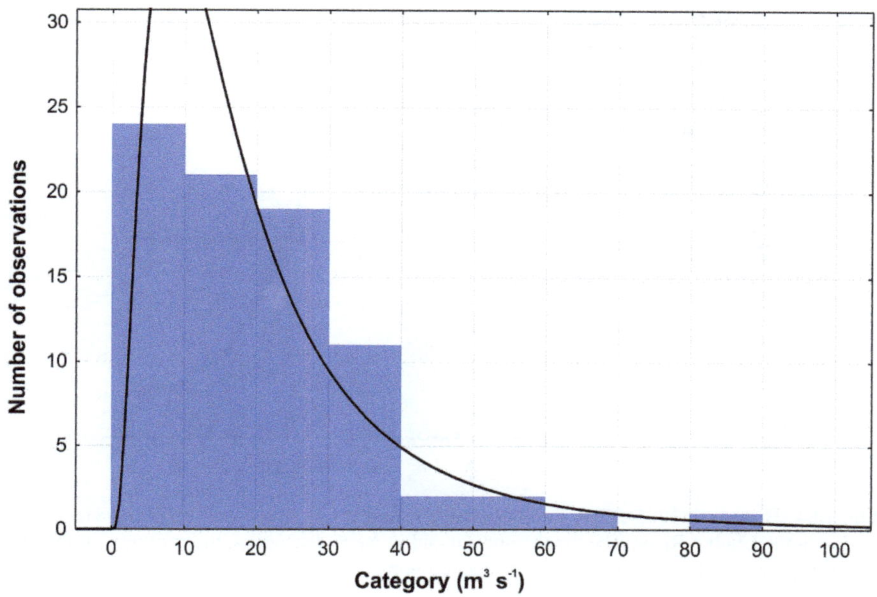

Fig. 10.2 Log-normal distribution of instantaneous discharges recorded at El Maitén gauging station (2000–2020). N = 80, chi-square = 7.01392, d.f. = 2, $p < 0.05$

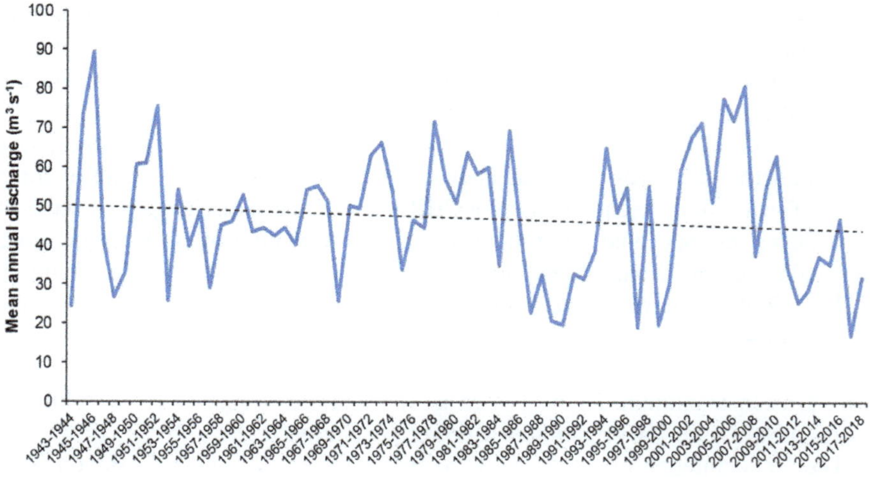

Fig. 10.3 Mean annual Chubut River discharge time series recorded at Los Altares gauging station during the 1943–2018 time period. The fitted regression line shows a nonsignificant negative trend

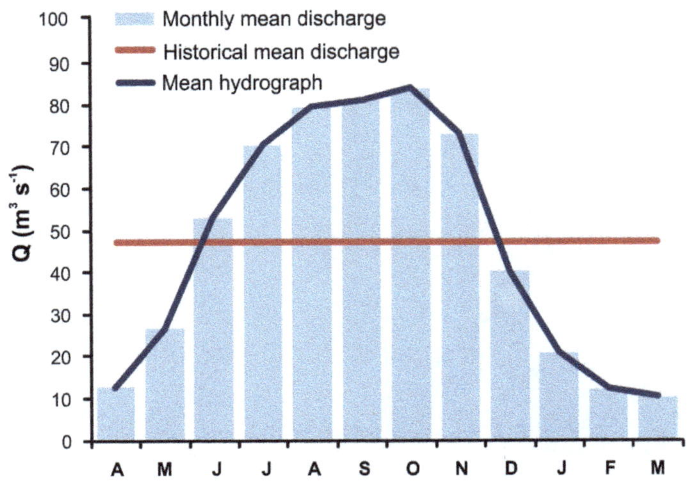

Fig. 10.4 Historical (1943–2018) mean discharge record at Los Altares, the main stage measuring station at the Chubut River. The graph shows monthly mean discharges, the resulting (arithmetic) mean discharge and the corresponding synthetic hydrograph

As it happens with the upstream discharge time series, the statistical distribution is markedly log-normal. The chi-square test performed to verify log-normality (317.14) exceeds the critical value for 9 degrees of freedom and $p < 0.001$. Hence, $Q_g \pm S_g$ at Los Altares is 35.9 ± 2.72 m^3 s^{-1}, which is lower than the arithmetic mean calculated above. The conversion to a Gaussian distribution of the discharge time series (1943–2018) allows to estimate that 95 per cent of the data in the historical discharge series fell within the range $30.5 < Q_g < 41.4$ m^3 s^{-1}.

The seasonal Kendall trend analysis (Kendall 1975; Hirsch et al. 1982) is a nonparametric statistical tool that allows establishing monotonic trends in time series. The use of the technique for the CHR monthly mean discharge time series at Los Altares station (Table 10.1) shows a statistically significant discharge decrease ($p < 0.05$) during January–March and May (i.e., the months with low-water flow).

The Chubut's lower stretch originates at the reservoir lake formed by the Florentino Ameghino dam, which was finished in 1963, 130 km west of the city of Trelew (i.e., Gaiman Department). The reservoir lake has a surface area of ~70 km^2, a mean maximum depth of ~25 m and a storage capacity of ~16 km^3 of water, mostly used for irrigation and power generation.

The Valle Inferior (43° 17′ 35.13″ S, 65°29′ 54.72″ W) gauging station is in operation since 1993, and it is located approximately 90 km downstream the Florentino Ameghino dam and less than 50 km upstream from the Chubut's estuary in the Atlantic Ocean. Contrasting with the hydrologic behavior recorded at Los Altares, the annual mean discharge time series (1993–2018 record period) shows a significant decreasing trend, a likely consequence of increased consumptive use and

Table 10.1 Seasonal Kendall test for the Chubut River monthly mean discharges at Los Altares

	N	Kendall *t*	*p**)
January	75	−2.132	**0.01651**
February	75	−2.461	**0.00693**
March	75	−2.159	**0.01541**
April	75	−0.567	0.28527
May	75	−1.935	**0.02649**
June	75	−0.814	0.20776
July	75	−0.654	0.25651
August	75	−0.174	0.43100
September	75	0.897	0.18497
October	75	−0.851	0.19743
November	75	−1.505	0.06617
December	75	−1.606	0.05418
Total	900	−1.819	0.03445

*) Statistically significant parameters in bold ($p < 0.05$)

discharge modulation imposed by the dam (Fig. 10.5). In 25 years, annual mean discharges have decreased at a mean rate of ~10 m^3 per decade. The annual mean discharge (i.e., arithmetic mean) at Valle Inferior is 34.9 ± 14 $m^3\ s^{-1}$, whereas the mean annual flow is 1.1 km^3. The specific water yield is only 1.1 $L\ s^{-1}\ km^{-2}$, and runoff is slightly less than 3.5 mm y^{-1}. In spite of likely groundwater contributions, the lower reach exhibits a mean discharge which is considerably lower than the one determined in the middle stretch. These hydrological parameters are mostly the after-effect of significant evapotranspiration and the intensive consumptive use of water.

10.4 Physical and Chemical Parameters of Surface Water and Groundwater from the Lower Basin of Chubut River

The values of physical and chemical parameters obtained in surface water from CHR in June and November are shown in Fig. 10.6. The temperature ranged from 15.6 to 19.5 °C in spring and from 6.8 to 8.2 °C in winter according to the season. In general, the DO level was close to saturation in both sampled seasons, and the values obtained in winter (98.1–114.3%) were higher than those obtained in spring (81.7–120.2%). The pH was alkaline ranging from 8.0 to 9.0 in both samplings. The EC presented low values (330 $\mu S\ cm^{-1}$) in the sites far from the mouth, but it reached a maximum value (44,730 $\mu S\ cm^{-1}$) in the estuarine zone evidencing the presence of seawater. Similarly, the TDS showed an increase toward the marine zone, varying from 0.2 to 29.1 g L^{-1}. The redox potential presented positive values in all sampling sites and ranged from 169 to 294 mV.

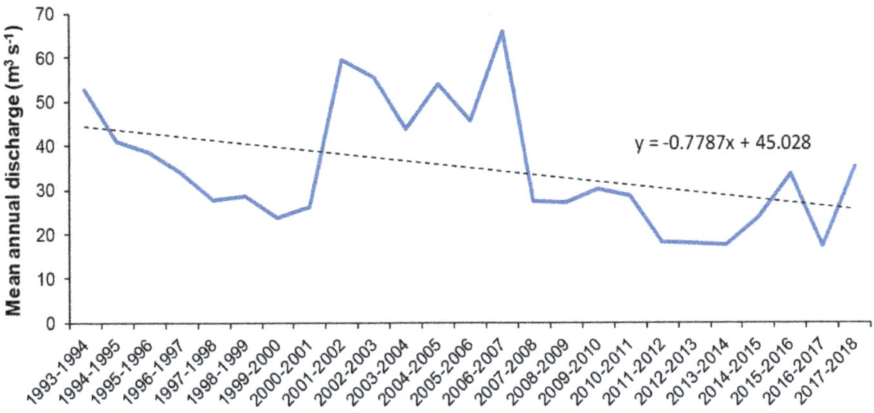

Fig. 10.5 Mean annual Chubut River discharge time series recorded at Valle Inferior gauging station. The fitted regression line exhibits a statistically significant ($p < 0.05$) trend which shows a mean discharge decrease of 25 m^3 during the 1993–2018 time period

As expected, groundwater samples exhibited for the same parameters lower values than those found in surface waters (Table 10.2). These parameters varied for DO (42.3–84.3%), temperature (13.7–16.0 °C), EC (638–3351 μS cm^{-1}), TDS (0.4–2.1 g L^{-1}), salinity (0.3–1.7 g L^{-1}), pH (6.4–7.6) and redox potential (135–406 mV). Groundwater from the well in the beach (Gw8) and close to the mouth (Gw7) presented lower temperatures, DO and pH, but showed higher EC, salinity and TDS in comparison with CHR surface waters. These values are typical of marine environments suggesting the presence of seawater. Moreover, groundwater samples presented the highest ^{222}Rn activities. At some sampling sites, ^{222}Rn activity increased tenfold. Specifically, the Gw7 site presented high ^{222}Rn activity typical of an underground aquifer, but also presented values of salinity and EC typical of the marine environment. These characteristics are suggesting the presence of a mixing zone between groundwater and seawater.

10.5 Dissolved Inorganic Nutrients in Water

In general, CHR surface waters presented lower concentrations of dissolved inorganic nutrients than groundwater (Fig. 10.7 and Table 10.3). In winter, nutrient concentrations were higher than those found in spring but they showed a similar pattern despite the higher water flow in November. Nutrient concentrations varied slightly throughout the studied area, with the exception of the sampling sites 7, 8, 9 and 10, located in the estuarine zone, which exhibited the highest range of variation (Fig. 10.7). In November, the ammonium concentration ranged from 0.35 to 4.75 μM, and the highest value was found at sampling site 7, close to the city of Rawson. This could be associated with the discharge of domestic wastewater into

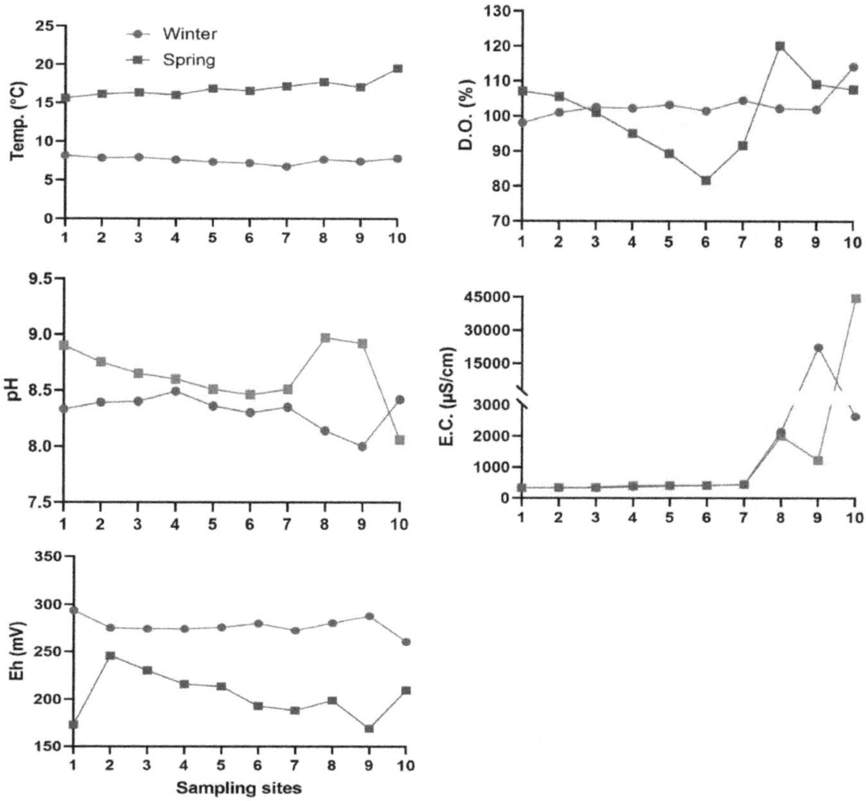

Fig. 10.6 Physical and chemical parameters measured in situ in surface waters of ten sampling sites located in the lower basin of Chubut River

the CHR. Before being delivered into the river, the residual domestic waters receive a biological treatment which consists of activated sludge and represent an aerobic process. This treatment is efficient to achieve an increase in the clarification of wastewater, but it is difficult to obtain a decrease in nutrient concentrations, mainly of nitrogen and phosphorus. Ammonium through oxidation processes produces nitrate, and this fact could be associated with the maximum of nitrate that was detected upstream in sampling site 7. Considering that nitrogen is the limiting nutrient, both in rivers and in Patagonia's marine environments (Charpy-Roubaud et al. 1982; Depetris et al. 2005), these nutrients provided by the treatment plant favor the development of primary producers such as those found in the estuarine zone of CHR (Villafañe et al. 1991; Helbling et al. 2010). In addition, nitrate concentrations ranged from 10.2 to 21.9 µM in winter, and the highest value was found at sampling site 8, which is located south of Rawson city, closer to the river mouth. At the same site, the highest phosphate concentration was found, and it varied from 0.5 to 3.6 µM. Finally, the silicic acid concentration ranged from 18.2

Table 10.2 Physical and chemical parameters measured in Chubut River's surface and bottom waters and in groundwater wells located in the studied area

Site	Temp.	D.O.	E.C.	T.D.S.	Sal.	pH	Eh	^{222}Rn	2-Sigma uncert.
	°C	%	μS cm^{-1}	g L^{-1}	g L^{-1}		mV	Dpm L^{-1}	Dpm L^{-1}
Bottom water (Bw) from Chubut River:									
Bw1	16.4	89.9	313	0.20	0.16	7.8	267	6.0	4.3
Bw2	17.9	107.7	322	0.21	0.14	7.8	283	19.9	6.9
Bw3	17.2	92.4	351	0.23	0.15	7.9	300	9.8	3.5
Bw4	16.0	91.3	409	0.27	0.19	8.6	216	177.4	15.4
Bw5	16.9	93.7	483	0.31	0.21	7.9	262	595.0	30.0
Bw6	15.4	90.3	43,930	28.50	28.57	7.8	324	0.7	1.0
Groundwater (Gw) from well (depth in m)									
Gw1-(20)	13.7	42.3	1822	1.18	0.93	7.4	238	770.4	32.3
Gw2-(30)	15.2	61.8	1038	0.70	0.54	7.4	135	577.6	27.4
Gw3-(60)	16.0	60.5	638	0.41	0.31	7.6	229	584.7	30.1
Gw4-(6)	15.5	64.4	3351	2.18	1.74	7.6	164	237.5	6.1
Gw5-(7)	13.9	47.6	807	0.66	0.51	6.4	406	444.8	33.4
Gw6-(3)	15.0	42.8	670	0.50	0.40	7.5	152	886.7	39.0
Gw7-(18)	15.1	84.3	33,686	26.10	25.70	8.6	149	237.2	19.9
Gw8-(1)	13.3	79.2	32,225	26.90	26.60	7.1	205	111.7	13.7

The number in parentheses indicates the depth of well

to 261.5 μM, and the highest values were found at sampling sites far from the mouth, while the lowest concentration was found at sampling site 10, close to the marine zone. Silicic acid presented concentrations that were two orders of magnitude higher than other nutrients in the investigated river reach. Due to its distribution pattern, it is widely used as a proxy to estimate the continental contribution to the coastal marine zone. In general, with exception of the silicic acid, nutrient concentrations increased toward the CHR's mouth.

As expected, concentrations of dissolved inorganic nutrients in groundwater were higher than in surface waters of the CHR (Table 10.3). In this sense, the ammonium and phosphate concentrations were an order of magnitude higher, and nitrate concentrations were two orders of magnitude higher than those found in surface waters. In addition, ammonium concentration reached 37.1 μM in a well (Gw7) located in a fishing plant close to the estuarine zone. This concentration is seven times higher than that found in surface waters of the CHR. The phosphate concentration reached 28.8 μM, and it was eight times higher than the concentration recorded in surface water. The nitrate concentration ranged from 1.1 to 683.3 μM, and it was fifty-eight times higher than those measured in surface waters. Similar to surface waters, groundwater presented an increase in nutrient

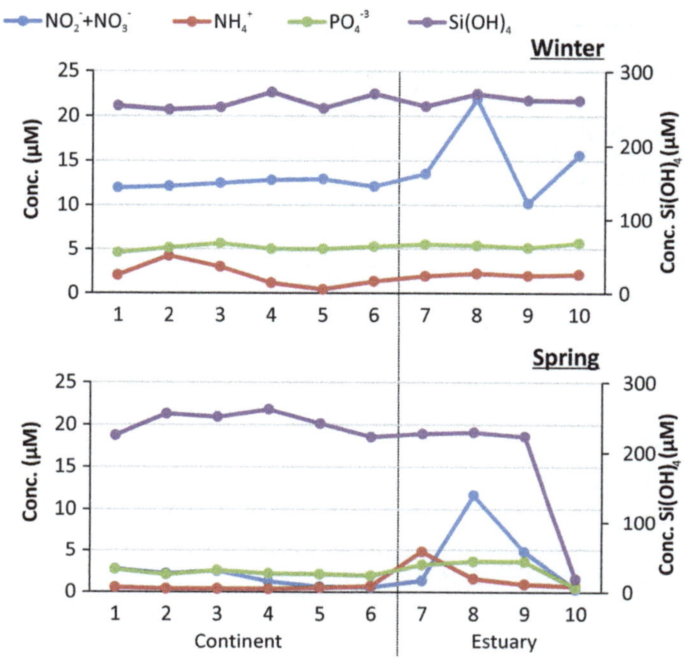

Fig. 10.7 Concentrations of dissolved inorganic nutrients in surface waters in ten sampling sites of the lower basin of the Chubut River in winter (up) and spring (down). Vertical line indicates the limit of seawater intrusion according to Perillo et al. (1989)

concentrations toward the estuarine zone. The silicic acid concentrations in groundwaters were 2.6 times higher and of the same order of magnitude than those registered in surface water. They ranged from 325 to 699 µM, with the exception of the sampling site Gw8 located close to the beach area, which presented a value of 16.1 µM, suggesting a dilution effect caused by the intrusion of the seawater.

10.5.1 Estimation of the Monthly Nutrient Flux

Considering the nutrient concentrations and river discharge, it is possible to estimate the flow of chemical substances transported by the river during the sampling time interval. Using discrete samples, an instant estimation of river contribution to the ocean was obtained. In this case, the load values for each sampling site were calculated using the nutrient concentration and the water discharge at each specific point (GESAMP 1987).

Table 10.3 Concentrations of dissolved inorganic nutrients (μM) in bottom waters from the Chubut River and groundwater from wells located in the lower basin of Chubut River. N/D: No detectable

Groundwater (Gw) from well				
Site/ nutrient	Ammonium	Nitrate	Phosphate	Silicic acid
Gw1	0.1	26.1	N/D	556.8
Gw2	0.4	17.4	0.1	624.8
Gw3	0.1	56.5	N/D	471.9
Gw4	N/D	22.9	5.0	653.9
Gw5	1.0	288.3	4.8	699.0
Gw6	0.8	362.0	17.4	550.2
Gw7	37.1	683.3	28.8	325.6
Gw8	77.3	1.1	16.0	16.8
Bottom water (Bw) from Chubut River				
Bw1	0.9	<0.1	1.2	113.3
Bw2	1.9	<0.1	0.9	119.5
Bw3	3.2	0.1	1.4	136.5
Bw4	2.6	1.2	2.1	144.8
Bw5	1.2	3.4	1.6	157.4
Bw6	1.5	3.3	1.3	27.3

The CHR-estimated nutrient flux toward the coastal marine zone was higher in June than in November (Table 10.4). Additionally, the total inorganic nitrogen (TIN) was higher than inorganic phosphorus. However, the silicic acid loads were the largest and they were an order of magnitude higher than the loads of nitrogen and phosphorus. Due to this fact, and also because this nutrient is mostly originated from a natural source, silicic acid flow is used as a tracer of riverine discharge.

In winter, the flow rate from the dam is lower than the spring discharge due to the fact that there is no snowmelt in the mountain region. In spring, there is an increase in the flow rate favored by the beginning of thawing. However, this water is less enriched in nutrient than those that drain during winter. Defrosting increases $Si(OH_4)$ concentration and, to a lesser extent, the nitrogen and phosphorous concentrations. In addition, there is a higher water demand during this season due to increased agricultural activities, which are carried out mainly in the proximities of three cities which are placed furthest from the estuarine zone (i.e., in the area of sampling sites 1–5).

At sampling site 8, there is a marked increase in the load of $N\text{–}NO_3^-$ in both sampling campaigns. This seems to be unrelated to the river discharge, and it could be caused by the discharge into the CHR of domestic wastewater from the city of Rawson. On the contrary, this nutrient load was reduced to one-half during the sampling campaign carried out in November, probably due to the dilution effect caused by higher discharge. Clearly, the strong increase of $N\text{–}NO_3^-$ load in this site is caused by anthropogenic activities from urban settlements. In winter, PO_4^{-3} loads were similar in all sampling sites. In November, the highest loads were detected at

Table 10.4 Nutrient loads (t $month^{-1}$) obtained in each sampling site along the Chubut River in June and November

June						November				
Flow rate from dam: 46.89 m^3 s^{-1}						Flow rate from dam: 51.36 m^3 s^{-1}				
Site	$N\text{–}NH_4^+$	$N\text{–}NO_3^-$	TIN	PO_4^{-3}	Si(OH)4	$N\text{–}NH_4^+$	$N\text{–}NO_3^-$	TIN	PO_4^{-3}	Si(OH)4
1	3.7	22.3	26	18.9	947.8	0.9	4.8	5.7	10.3	793.1
2	7.9	22.6	30.5	21.2	929.6	0.7	3.9	4.5	7.9	900.8
3	5.5	23.3	28.8	23.2	941.8	0.7	4.3	5	9.8	885.1
4	2.1	23.9	26	20.6	1017.3	0.6	2.2	2.8	8.3	922.2
5	0.8	24.1	24.8	20.5	936.3	0.9	1.1	2	8.1	851.2
6	2.5	22.6	25	21.6	1008.6	1.2	0.9	2.1	7.3	784.2
7	3.5	25.2	28.7	22.6	945.9	8.3	2.3	10.7	12.3	799.7
8	4.1	40.9	45	22.1	1008.7	2.7	20.5	23.2	13.9	806.5
9	3.6	19	22.6	21.1	977.4	1.6	8.3	9.9	13.7	785.7
10	3.9	29.1	33	23.3	975.3	1.1	0.6	1.7	1.9	64.2

the sampling sites located near the estuarine zone. This behavior could be similar to that described by Niencheski et al. (1999), who suggested that PO_4^{-3} is removed from the water column from Patos Lagoon estuary in the lower salinity region (less than ca. 10 g L^{-1}) and released at salinities above ca. 10 g L^{-1}. On the case of CHR, the release for the water column would occur at lower salinity region, that is, where happens the transition from the fluvial to estuarine conditions.

Silicic acid, followed by nitrogen and then phosphate were the main nutrients delivered by the CHR into Engaño Bay. These nutrient loads are consistent but lower than those reported by Helbling et al. (1992) for the CHR in the periods 1986 and 1987. Other studies have recognized that the CHR has noteworthy $N\text{–}NO_3^-$ concentrations, which is surely determined by human contribution such as the mesotrophic Florentino Ameghino dam, farmlands and the cities located in the lowermost stretch (Depetris et al. 2005). These authors have determined the nutrient fluxes from Patagonia's rivers that are discharged into the coastline. In addition, these authors have estimated for the CHR fluxes of silicic acid (1075 t $month^{-1}$), nitrate (93 t $month^{-1}$) and phosphate (28 t $month^{-1}$) which are the same order of magnitude as those obtained in this study (Table 10.4). On the other hand, the application of models demonstrated that dam-modulated rivers produce an increase of 10% in the nitrogen flux and a reduction of 25 and 30% in the $Si(OH)_4$ and PO_4^{-3} fluxes, respectively, in comparison with undammed rivers (Sferratore et al. 2008). Nitrogen is the limiting nutrient for phytoplankton production in the Patagonia′s marine environments (Charpy-Roubaud et al. 1982). In consequence, these contributions of nutrients are important because they fertilize the estuarine zone (Helbling et al. 1992), sustaining phytoplankton production (Villafañe et al. 1991), which is crucial for the marine food web.

10.6 Groundwater Contribution to the Lower Chubut River

Surface waters of CHR showed variable ^{222}Rn activities (Table 10.5). In general, lower ^{222}Rn was detected in the upstream sampling points (from 48 to 331 Bq m^{-3}), whereas at sampling sites Bw4 and Bw5, higher ^{222}Rn concentrations were recorded (from 569 to 1770 Bq m^{-3}). The sampling point located near the river outflow into the Atlantic Ocean (sampling point Bw6) showed the lowest ^{222}Rn activity ($\sim$9 Bq m^{-3}) due to marine intrusion. Based on ^{222}Rn concentrations, two groundwater end members were defined: Cg_1 = 12,752 Bq m^{-3} and Cg_2 = 7412 Bq m^{-3} (Table 10.5).

The contribution of groundwater into the lower stretch of the Chubut River—with the exception of sampling point Bw6—was estimated by means of a ^{222}Rn mass balance equation proposed by Hamada (1999):

$$Q_g = \frac{C_2Q_2 - C_1Q_1\exp(-aL)}{\frac{C_g(1-\exp(-aL))}{aL}}$$

where Qg is the groundwater inflow ($m^3\ s^{-1}$); C_1 and C_2 are the ^{222}Rn concentrations in river water at the upstream and downstream stations (Bq m^{-3}), respectively; Q_1 and Q_2 are the river discharges at the upstream and downstream stations ($m^3\ s^{-1}$), respectively; Cg is the ^{222}Rn concentration in groundwater (Bq m^{-3}); L is the distance between stations (km); and a is a parameter that can be calculated using the following equation:

$$a = (D/zhv) + (\lambda/v)$$

where D is the molecular diffusivity of ^{222}Rn, which has been defined by Peng et al. (1974) as a function of temperature $D = 10^{-(980/(T+273)+1.59)}$ ($1 \times 10^{-5}\ m^2\ s^{-1}$ at 25 °C); z is the thickness of a stagnant film (m), which is about 20 μm when the distance between stations is several kilometers (Hamada et al. 1997); h is the average stream depth (m); v is the average stream velocity (m s^{-1}); and λ is the ^{222}Rn decay constant ($2.08 \times 10^{-6}\ s^{-1}$).

The values of the parameters employed in the ^{222}Rn mass balance calculation are listed in Table 10.5, as well as the calculated groundwater contribution. Results indicate that the groundwater inflow to the river is low at sampling point Bw2 (Qg = 2.03 $m^3\ s^{-1}$), whereas a negative value was obtained at sampling point Bw3 (Qg = −1.32 $m^3\ s^{-1}$), which suggests that at this location the river is losing water to the aquifers through the streambed. Groundwater inflow slightly increases downstream at sampling point Bw4 (Qg = 2.31 $m^3\ s^{-1}$), while the highest groundwater discharge was detected at sampling point Bw5 (Qg = 5.41 $m^3\ s^{-1}$). Thus, the ^{222}Rn mass balance indicates that groundwater inputs may account for about 6–8.5% of the total water inputs to CHR upstream Gaiman city, whereas it increases downstream, reaching $\sim$24% at Trelew city.

Table 10.5 ^{222}Rn activities measured in bottom waters from Chubut River and groundwater end members, parameters used in the mass balance equation and calculated groundwater discharge to the lower Chubut River

Sample	$C^{222}Rn$	Q	h	v	L	Qg ($m^3\ s^{-1}$)	
	(Bq m^{-3})	($m^3\ s^{-1}$)	(m)	(m s^{-1})	(m)	G1	G2
Bw1	48.0	26.40	2.0	0.55			
Bw2	331.3	32.49	2.5	0.52	310	0.74	1.28
Bw3	162.5	27.65	3.0	0.42	26,980	−0.49	−0.83
Bw4	569.5	26.74	3.1	0.36	25,900	0.85	1.46
Bw5	1770.0	22.74	2.5	0.40	24,930	1.99	3.42
Bw6	8.7	–	–	–	–	–	–
G1	12,752	–	–	–	–	–	–
G2	7,412	–	–	–	–	–	–

High concentrations of ^{222}Rn are present in the waters of the CHR only in the immediate vicinity of groundwater advection points and for relatively short distances downstream from such locations. Although high activity of ^{222}Rn was observed in waters collected near the bottom, low concentrations of nutrients were detected. This was contrary to our expectations. We assume that the nutrients provided by groundwater are immediately affected by the dilution process with fresh surface waters. To confirm this hypothesis, it is necessary to use appropriate equipment to obtain the water sample at the sediment–water column interface, such as the use of benthic chambers or corers (Niencheski and Jahnke 2002). For this reason, the concentrations of nutrients associated with the groundwater that advect toward the CHR have not been sampled conveniently to indicate that the bottom water is more concentrated than the surface water. However, despite this, for the boxes (A, C, D, E) where the advection of groundwater was determined, the contributions of TIN, phosphate and silicic acid for groundwater were 3.6%, 3.4% and 6.6%, respectively, in relation to freshwater. Another factor to be considered is that the mass balance model applied in this study is considered as a site of advection, only the small area that refers to the riverbed.

As indicated by Torres et al. (2018), ^{222}Rn is a useful tracer in this Patagonian environment, showing water flowing across the sediment–water interface and into the overlying water column, regardless of its composition. It is necessary to expand these data sets to include riverine freshwater, terrestrial groundwater and saline, and recirculated seawater to better understand and quantify this important process in this particular stretch of CHR's coastal region, where the hydraulic gradient certainly is the main driving force that results in freshwater and terrestrial aquifer water discharging at the coastline.

In this particular region of CHR outflow into Engaño Bay, other driving forces that control recirculated seawater are very significant, such as tidal pumping and wave setup. We suggest that measurements of radon isotopes should continue, in both high-resolution time series and horizontal offshore transects, to allow quantifying the offshore transport rates of discharged continental groundwater. Considering that the SGD has important implications in the biogeochemical cycles of coastal environments, future studies in this region should complement isotopic measurements with determinations of inorganic nutrients, iron, carbon, rare earth elements and toxic metals. Furthermore, we suggest reducing the distance between sampling sites where bottom waters were taken in order to detect other sites with groundwater advection.

10.7 Estimation of Nutrient Mass Balance

A mass balance model was applied in the lower basin of the CHR, in order to detect the movement and inputs of nutrients in several sections. For the mass balance calculations, the CHR was divided into eight boxes (Fig. 10.1). This division was made according to the surrounding land use, the existence of urban settlements, anthropogenic impact, the location of sampling stations and the identification of the halocline inside the river. Boxes A and H represent the highest and lowest parts of the catchment, respectively. In the area encompassed around boxes A to E, strong agricultural and livestock breeding activities are developed. The largest city (Trelew) in the drainage basin is located in box F. Rawson, the capital city of Chubut Province, is located in box G. The presence of seawater has been identified strongly in box H and weakly in box G, which results in a mixing process producing modifications in the physical and chemical characteristics during the transition from the riverine environment to Engaño Bay, where marine conditions are finally reached.

For each box, the plot of the excess or deficit was integrated over the study period to estimate the net excess or deficiency of each nutrient in the estuary. The difference between the observed total content and the calculated total content of nutrients for each of the boxes in both sampling campaigns is plotted in Fig. 10.8. Only the silicic acid and TIN are presented. The excesses (values >0) and deficits (values <0) of each nutrient varied according to the box. These results were adjusted from both spatial and temporal points of view. In general, the maximum excesses were recorded in the boxes near to the mouth with exception for the $Si(OH)_4$.

Regarding the spatial variation, in winter, the mass balance of nitrogen compounds, PO_4^{-3} and $Si(OH)_4$ was positive in the estuarine zone (boxes G and H). This indicates that the CHR is discharging these nutrients into the marine coastal zone. Moreover, the high nutrient concentrations found in surface waters could suggest that the CHR receives nutrients such as nitrogen and phosphorus from external sources and silicic acid associated with sediment transport or groundwater. The evident riverine contribution of nutrients could be related mainly to the discharge of domestic wastewater from the city of Rawson. The wastewater inflow to rivers is the main problem caused by anthropogenic activities that take place in the coastline, thus stimulating the eutrophication of such aquatic environments. Soils are not the

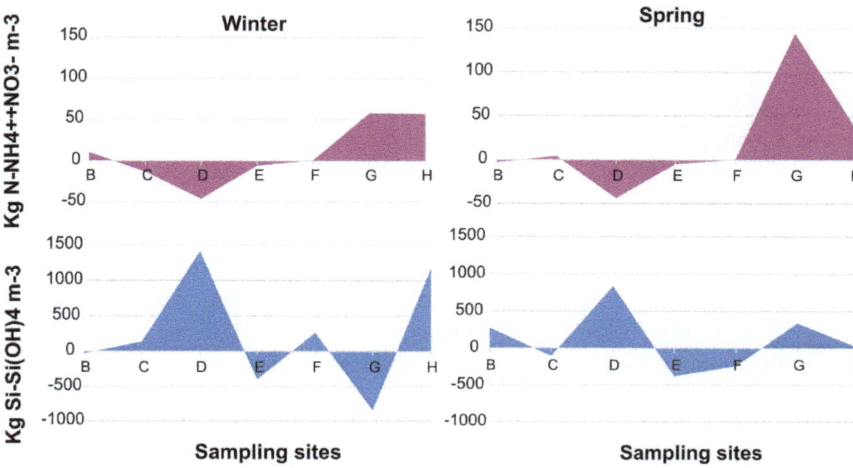

Fig. 10.8 Difference between the observed total content and the calculated total content (excess or deficit) for TIN and silicic acid in each of the seven boxes in winter (left) and spring (right)

only important source of nitrate and ammonium to the hydrosphere; additionally, nitrogen enters the river mainly through rain and porewater from the sediment (Depetris et al. 2005). Given the low rainfall prevailing in the region, their contribution is scarce, and more attention should be paid to the contribution from the sediment. In addition, the region suffers the indiscriminate use of fertilizers and pesticides in agricultural activities which are strongly developed and also represent —along with livestock—the main economic activity. Although this region is undergoing pollution originating in anthropogenic activities, its impact has not yet been measured. On the other hand, the distribution of the $Si(OH)_4$ also presented excesses in box D (Gaiman) which could be associated with the high density of human-made channels that extract freshwater from the riverbed for irrigation and the leftover is then returned downstream, thus favoring the increase of this nutrient. Since seawater presents low concentrations, the sources of $Si(OH)_4$ are the river and groundwater. Furthermore, urban pollution has no influence on its concentration.

In spring, TIN and PO_4^{-3} presented the same distribution pattern in each of the boxes. The main excesses in boxes G and H could be associated with domestic wastewater discharges from Rawson city (Fig. 10.8). Also, a marked deficit for both nutrients was found in box D (Fig. 10.8).

Regarding the observed temporal variation, all measured nutrients presented in winter and spring the same distribution pattern and, in both cases, the highest excesses were registered in the estuarine zone (boxes G and H), thus indicating a strong supply of nutrients from the CHR into the marine zone. Specifically, the CHR contributes more $N–NO_3^-$ than $N–NH_4^+$ into Engaño Bay. Contrarily, the domestic sewages are characterized by containing higher $N–NH_4^+$ than $N–NO_3^-$ concentration. However, $N–NH_4^+$ can produce $N–NO_3^-$ by a nitrification process before it reaches the sea. Furthermore, there are some fish-processing plants located on the margin of the CHR which discharge their industrial wastewater without previous or with inefficient treatment. When these nutrients reach the marine zone, they are incorporated by macroalgae which have a higher assimilation rate of N–NH_4^+ than N–NO_3^- (Torres et al. 2004; Gil et al. 2005). In addition, the $Si(OH)_4$ present in both samplings showed a marked increase in box D, which is located in the region irrigated by means of an extended channel network. In this way, they would be delivering a significant amount of this nutrient into the river. On the other hand, the residence time (not shown) was estimated for each box and significant differences were not observed between the two different sampling campaigns.

The analysis of the riverine nutrient supply to Engaño Bay was performed by taking into account only the nutrient concentrations in box H during winter because during this sampling excursion the highest nutrient concentrations were found. The total nutrient inputs are shown in Table 10.6.

During the winter season, the amount of $Si(OH)_4$ discharged into the Engaño Bay area was two orders of magnitude higher than nitrogen (as TIN) and PO_4^{-3}. The total amounts of nutrient reported in this study were lower than those noted by Helbling et al. (1992) in 1986 and 1987. In addition, in the estuarine low salinity zone the patches of diatoms *Aulacoseira granulata* (Ehr.) Simonsen and *Odontella aurita* (Lyngbye) Agardh are accompanied by extreme fluctuations in chlorophyll *a* concentrations, from less than 5 to 45 μg L^{-1} (Villafañe et al. 1991). However, although information on phytoplankton primary production levels is lacking, we can estimate the potential production rate sustained by the nitrogen flux from CHR (Table 10.6) applying a Redfield ratio of C106:Si15:N16:P1 for diatoms (Redfield et al. 1963) and the total N flux in the CHR's box H. Thus, a nitrogen flux of 0.08×10^6 mol day^{-1} would support a production rate of 7.42×10^6 g C day^{-1}. This rate is considered lower than those obtained for Patos Lagoon of 1.9×10^8 g C day^{-1} (Niencheski et al. 2007). These results are consistent if one bears in mind that the freshwater discharge into the sea at Los Patos Lagoon is two orders of magnitude higher than CHR. A comparison with the total amount of nutrients discharged by CHR into Engaño Bay shows that it is several orders of magnitude lower than other world rivers (Helbling et al. 1992).

10.8 Conclusions

The seasonal trend analysis allowed determining a historical discharge decrease during the dry season (i.e., January–March and May). The high nutrient concentrations (mainly N and P) found in surface water reflect the contributions of urban wastewater from coastal cities, hence being the more evident anthropogenic impact near the mouth of the CHR, while silicic acid is associated with natural processes such as sediment transport and advection of groundwater. The ^{222}Rn concentration recorded in bottom waters from CHR is suggesting the presence of groundwater by a molecular diffusion process. The ^{222}Rn concentrations along the CHR changed spatially between the town of 28 de Julio and Trelew city suggesting the presence

Table 10.6 Estimated nutrient fluxes to the Engaño Bay from Chubut River at box H

Nutrient	10^6 mol day^{-1}
$N–NH_4^+$	0.009
$N–NO_3^-$	0.07
N–TIN	0.08
$P–PO_4^{-3}$	0.024
$Si–Si(OH)_4$	1.175

of the groundwater. The ^{222}Rn mass balance allowed identifying areas of the river with low ^{222}Rn activities indicating that CHR is losing water to the aquifers and other areas with high ^{222}Rn activities, which are assumed to be zones where groundwater discharge into CHR is significant. Future studies should take into account performing ^{222}Rn time series in CHR throughout an entire tidal cycle to determine the effect known as "tidal pump" on SGD. Furthermore, considering that SGD interferes in the biogeochemical cycles of micro- and macro-elements we suggest performing chemical measurements in waters from the estuarine zone. Finally, we consider it relevant to perform measurements of Ra and Rn isotopes along with transects perpendicular to the coast of Engaño Bay, which will allow quantifying the transport rate of SGD. These activities must be carried out at different seasons in order to detect the variability between the wet and dry seasons.

Acknowledgements The authors wish to thank the staff of the FURG's Laboratory of Hydrochemical (Brazil); Córdoba National University (Argentina); CICTERRA-CONICET (Argentina); and Chemical Oceanography and Water Pollution Laboratory (LOQyCA) of CESIMAR—CCT CONICET—CENPAT (Argentina). This study was partially funded by the Bilateral Cooperation Project between Argentina's CONICET and Brazil's CAPES (Project BR/RED 1302) which involved the aforementioned institutions. We wish to thank D. -P. Häder for the editorial suggestions and the anonymous reviewer for the constructive comments that improved the manuscript.

References

Antolini L (2012) Percepción del riesgo y dinámica de uso de plaguicidas agrícolas en el Valle Inferior del Río Chubut. Rawson. 1. Ministerio de Salud Argentina, Technical report, 183pp

Bratton JF (2010) The three scales of submarine groundwater flow and discharge across passive continental margins. J Geol 118:565–575. https://doi.org/10.1086/655114

Burnett W, Cable J, Corbett D, Chanton J (1996) Tracing groundwater flow into surface waters using natural ^{222}Rn. In: Buddemeier R (ed) Groundwater Discharge in the Coastal Zone: Proceedings of an International Symposium. LOICZ IGBP. LOICZ/R&S/96-8, iv + 179pp. LOICZ, Texel, The Netherlands, pp 22–28

Burnett WC, Bokuniewicz H, Huettel M, Moore WS, Taniguchi M (2003) Groundwater and pore water inputs to the coastal zone. Biogeochemistry 66:3–33. https://doi.org/10.1023/B:BIOG.0000006066.21240.53

Burnett WC, Dulaiova H (2003) Estimating the dynamics of groundwater input into the coastal zone via continuous radon-222 measurements. J Environ Radioact 69:21–35. https://doi.org/10.1016/S0265-931X(03)00084-5

Campodonico VA, García MG, Pasquini AI (2015) The dissolved chemical and isotopic signature downflow the confluence of two large rivers: The case of the Parana and Paraguay rivers. J Hydrol 528:161–176. https://doi.org/10.1016/j.jhydrol.2015.06.027

Cecil LD, Green JR (2000) Radon-222. In: Cook PG (ed) Environmental tracers in subsurface hydrology. Springer US, Boston, MA, pp 175–194

Charpy-Roubaud CJ, Charpy LJ, Maestrini SY (1982) Fertilité des eaux côtières nord-patagoniques: facteurs limitant la production du phytoplancton et potentialités d'exploitation myticole. Oceanol Acta 5(2):179–188

Commendatore M, Esteves J (2004) Natural and anthropogenic hydrocarbons in sediments from the Chubut River (Patagonia, Argentina). Mar Pollut Bull 48:910–918. https://doi.org/10.1016/j.marpolbul.2003.11.015

Davis J (1986) Statistics and data analysis in geology. Wiley, New York

Depetris PJ, Gaiero DM, Probst JL, Hartmann J, Kempe S (2005) Biogeochemical output and typology of rivers draining Patagonia's Atlantic seaboard. J Coast Res 21(4):835–844. https://doi.org/10.2112/015-NIS.1

EMC. Estación Meteorológica Cenpat (2020) http://repositorio.cenpat-conicet.gob.ar:8081/xmlui/handle/123456789/400. Accessed 10 Sep 2020

Garcia-Solsona E, Garcia-Orellana J, Masqué P, Garcés E, Radakovitch O, Mayer A, Estradé E, Basterretxea G (2010) An assessment of karstic submarine groundwater and associated nutrient discharge to a Mediterranean coastal area (Balearic Islands, Spain) using radium isotopes. Biogeochemistry 97:211–229

GESAMP (1987) Land/Sea Boundary Flux of Contaminants: Contributions from Rivers. UNESCO/GESAMP Report and Studies N°. 32, 172pp

Gil MN, Harvey MA, Esteves JL (1999) Heavy metals in intertidal surface sediments from the Patagonian coast, Argentina. Bull Environ Contam Toxicol 63:52–58. https://doi.org/10.1007/s001289900947

Gil MN, Torres AI, Esteves JL (2005) Uptake of sewage derived nitrogen by *Ulva rigida* (Chlorophyceae) in Bahía Nueva (Golfo Nuevo, Patagonia, Argentine). Hydrobiologia 39–43

Godoy JM, Souza TA, Godoy MLDP, Moreira I, Carvalho ZL, Lacerda LD, Fernandes FC (2013) Groundwater and surface water quality in a coastal bay with negligible fresh groundwater discharge: Arraial do Cabo, Brazil. Mar Chem 156:85–97. https://doi.org/10.1016/j.marchem.2013.05.004

Hamada H (1999) Analysis of the interaction between surface water and groundwater using radon-222. Japan Agric Res Q 33:261–265

Hamada H (2000) Estimation of groundwater flow rate using the decay of Rn-222 in a well. J Environ Radioact 47:1–13

Hamada H, Imaizumi M, Komae T (1997) Methods of investigation and analysis on ground-water using radon as an indicator. Nat Res Inst Agric Eng 36:17–50

Helbling WE, Pérez DE, Medina CD, Lagunas MG, Villafañe VE (2010) Phytoplankton distribution and photosynthesis dynamics in the Chubut River estuary (Patagonia, Argentina) throughout tidal cycles. Limnol Oceanogr 55:55–65. https://doi.org/10.4319/lo.2010.55.1.0055

Helbling WE, Santamarina JM, Villafañe VE (1992) Chubut River Estuary (Argentina): Estuarine variability under different conditions of river discharge. Rev Biol Mar Oceanogr 27:73–90

Hirsch RM, Slack JR, Smith RA (1982) Techniques of trend analysis for monthly water quality data. Water Resour Res 18:107–121. https://doi.org/10.1029/WR018i001p00107

Kendall M (1975) Rank correlation methods. Griffin, London

Kroeger KD, Charette MA (2008) Submarine groundwater discharge: nitrogen biogeochemistry of the discharge zone. Limnol Oceanogr 53:1025–1039

Kwon EY, Kim G, Primeau F, Moore WS, Cho HM, Devries T, Sarmiento JL, Charette MA, Cho YK (2014) Global estimate of submarine groundwater discharge based on an observationally constrained radium isotope model. Geophys Res Lett 41:8438–8444. https://doi.org/10.1002/2014GL061574

Macheleidt W, Nestler W, Grischek T (2002) Determination of hydraulic boundary conditions for the interaction between surface water and groundwater. Geol Soc Publ 193(1):235–243

Milliman J, Farnsworth K (2011) River discharge to the coastal ocean: a global synthesis. Cambridge University, Cambridge

Moore WS (1999) The subterranean estuary: a reaction zone of ground water and sea water. Mar Chem 65:111–125

Moore WS (2006) Radium isotopes as tracers of submarine groundwater discharge in Sicily. Cont Shelf Res 26:852–861

Niencheski LF, Baumgarten M da G, Fillmann G, Windom HL (1999) Nutrients and suspended matter behaviour in the Patos Lagoon Estuary (Brazil). In: Perillo G, Píccolo MC, Pino-Quivira M (eds) Estuaries of South America their geomorphology and dynamics. Springer, Berlin Heidelberg, pp 67–81

Niencheski LF, Jahnke RA (2002) Benthic respiration and inorganic nutrient fluxes in the estuarine region of Patos Lagoon (Brazil). Aquat Geochemistry 8:135–152. https://doi.org/10.1023/A:1024207220266

Niencheski LFH, Windom HL, Moore WS, Jahnke RA (2007) Submarine groundwater discharge of nutrients to the ocean along a coastal lagoon barrier, Southern Brazil. Mar Chem 106:546–561. https://doi.org/10.1016/j.marchem.2007.06.004

Paparazzo F, Crespi-Abril A, Gonçalves R, Barbieri ES, Gracia Villalobos LL, Solís ME, Soria G (2018) Patagonian dust as a source of macronutrients in the Southwest Atlantic Ocean. Oceanography 31:33–39. https://doi.org/10.5670/oceanog.2018.408

Peng TH, Takahashi T, Broecker WS (1974) Surface radon measurements in the North Pacific Ocean station Papa. J Geophys Res 79:1772–1780

Perillo G, Piccolo MC, Scapini M del C, Orfila J (1989) Hydrography and circulation of the Chubut river estuary (Argentina). Estuaries 12:186–194

Peterson RN, Santos IR, Burnett WC (2010) Evaluating groundwater discharge to tidal rivers based on a Rn-222 time-series approach. Estuar Coast Shelf Sci 86:165–178. https://doi.org/10.1016/j.ecss.2009.10.022

Porcelli D, Swarzenski PW (2003) The behaviour of U- and Th series nuclides in groundwater. Rev Mineral Geochemistry 52:317–361

Redfield AC, Ketchum BH, Richard FA (1963) The influence of organisms on the composition of seawater. In: Hill MN (ed) The sea, vol 2. Wiley, New York

Santos IR, Dimova N, Peterson RN, Mwashote B, Chanton J, Burnett WC (2009) Extended time series measurements of submarine groundwater discharge tracers (^{222}Rn and CH_4) at a coastal site in Florida. Mar Chem 113:137–147. https://doi.org/10.1016/j.marchem.2009.01.009

Santos IR, Eyre BD, Huettel M (2012) The driving forces of porewater and groundwater flow in permeable coastal sediments: a review. Estuar Coast Shelf Sci 98:1–15

Schmidt C, Hanfland C, Regnier P, Van Cappellen P, Schluter M, Knauthe U, Stimac I, Geibert W (2011) ^{228}Ra, ^{226}Ra, ^{224}Ra and ^{223}Ra in potential sources and sinks of land-derived material in the German Bight of the North Sea: implications for the use of radium as a tracer. Geo-Marine Lett 31:259–269. https://doi.org/10.1007/s00367-011-0231-5

Sferratore A, Billen G, Garnier J, Smedberg E, Humborg C, Rahm L (2008) Modelling nutrient fluxes from sub-arctic basins: comparison of pristine vs. dammed rivers. J Mar Syst 73:236–249. https://doi.org/10.1016/j.jmarsys.2007.10.012

Skalar Analytical® VB (2005a) Skalar methods—analysis: Nitrate + Nitrite—Catnr. 461-031 +DIAMOND Issue 081505/MH/99235956. Breda, The Netherlands

Skalar Analytical® VB (2005b) Skalar methods—analysis: Phosphate—Catnr. 503-010w/r +DIAMOND Issue 081505/MH/99235956. Breda, The Netherlands

Skalar Analytical® VB (2005c) Skalar methods—analysis: Silicate—Catnr. 563-051 + DIAMOND Issue 081505/MH/99235956. Breda, The Netherlands

Smerdon BD, Payton Gardner W, Harrington GA, Tickell SJ (2012) Identifying the contribution of regional groundwater to the baseflow of a tropical river (Daly River, Australia). J Hydrol 464–465:107–115. https://doi.org/10.1016/j.jhydrol.2012.06.058

Strickland JDH, Parsons TR (1972) A practical handbook of the seawater analysis, 2nd edn. Fisheries Research Board of Canada Bulletin No 157, 310p

Stringer CE, Burnett WC (2004) Sample bottle design improvements for radon emanation analysis of natural waters. Heal Phys 87:642–646

Swarzenski PW, Burnett WC, Greenwood WJ, Herut B, Peterson R, Dimova N, Shalem Y, Yechieli Y, Weinstein Y (2006) Combined time-series resistivity and geochemical tracer techniques to examine submarine groundwater discharge at Dor Beach, Israel. Geophys Res Lett 33:L24405. https://doi.org/10.1029/2006GL028282

Swarzenski PW, Dulai H, Kroeger KD, Smith CG, Dimova N, Storlazzi CD, Prouty NG, Gingerich SB, Glenn CR (2017) Observations of nearshore groundwater discharge: Kahekili Beach Park submarine springs, Maui, Hawaii. J Hydrol Reg Stud 11:147–165. https://doi.org/10.1016/j.ejrh.2015.12.056

Torres AI, Andrade CF, Moore WS, Faleschini M, Esteves JL, Niencheski LFH, Depetris PJ (2018) Ra and Rn isotopes as natural tracers of submarine groundwater discharge in the Patagonian coastal zone (Argentina): an initial assessment. Environ Earth Sci 77:145–155. https://doi.org/10.1007/s12665-018-7308-7

Torres AI, Gil MN, Esteves JL (2004) Nutrient uptake rates by the alien alga *Undaria pinnatifida* (Phaeophyta) (Nuevo Gulf, Patagonia, Argentina) when exposed to diluted sewage effluent. Hydrobiologia 520:1–6

Villafañe VE, Helbling WE, Santamarina JM (1991) Phytoplankton blooms in the Chubut River estuary (Argentina): influence of stratification and salinity. Rev Biol Mar Oceanogr 26:1–20

Windom HL, Moore WS, Niencheski LFH, Jahnke RA (2006) Submarine groundwater discharge: a large, previously unrecognized source of dissolved iron to the South Atlantic Ocean. Mar Chem 102:252–266. https://doi.org/10.1016/j.marchem.2006.06.016

Chapter 11
Pharmaceutical Pollutants in Aquatic Ecosystems

Gilmar S. Erzinger, Sebastian M. Strauch, Monique Fröhlich, Carla Keite Machado, and Lineu del Ciampo

Abstract Pollutants, including synthetic organic materials and heavy metals, are known to adversely affect physiological systems in all animal species studied so far, especially in aquatic ecosystems. Although many individual chemicals can disrupt normal functions, the combined actions of various pollutants are particularly worrying as they can have effects even when each individual chemical is present at concentrations too low to be individually effective. A special approach is currently being given to anthropogenic products called emerging pollutants, including drugs, hormones and by-products of human and animal metabolism which, despite being in very low concentrations, can cause biological effects which differ greatly between species, reflecting differences in exposure pattern, uptake pathways, post-uptake metabolism, accumulation rates and target organ sensitivity. Thus, understanding the effects of pollutants on wildlife and aquatic ecosystems will require a detailed study of many different species representing a wide range of taxa. However, these studies can be substantiated by knowledge gained under more controlled conditions, which may indicate likely mechanisms of action and appropriate endpoint measures. Responses may be exacerbated by interactions between the effects of pollutants, and environmental stressors, such as malnutrition or osmotic stress, and changes in these variables associated with climate change may further exacerbate physiological responses to pollutant burden.

Keywords Emerging pollutant · Aquatic ecosystems · Drugs · Hormones · Heavy metals

---

M. Fröhlich · C. K. Machado · L. del Ciampo
Program in Health and Environment, University of Joinville Region Univille, Joinville, SC, Brazil

S. M. Strauch
Postgraduate Program in Health and Environment, University of Joinville Region, Rua Paulo Malschitzki, 10—Zona Industrial Norte, Joinville, SC CEP 89219-710, Brazil

G. S. Erzinger (✉)
Postgraduate Program in Health and Environment, University of the Region of Joinville—UNIVILLE, Rua Paulo Malschitzki, 10, Joinville, Santa Catarina, Brazil

© The Author(s), under exclusive license to Springer Nature Switzerland AG 2021
D.-P. Häder et al. (eds.), *Anthropogenic Pollution of Aquatic Ecosystems*,
https://doi.org/10.1007/978-3-030-75602-4_11

11.1 Introduction

The quality and quantity of water have a direct impact on economy, society and environment, in addition to being essential for maintaining life. Water quality is one of the biggest global concerns for protecting human health, due to its use in various sectors such as food and agriculture, which is threatened by increasing contamination with emerging pollutants (EPs) (Damstra et al. 2011). This quality depends directly on natural and man-made influences, which often result in the contribution of dissolved and undissolved substances in water bodies. One of the groups of special interest in this aspect is drugs. The present work focuses on emerging pharmaceutical pollutants since their consumption is intense in urban environments (Erzinger 2013).

Drugs present in the environment are considered emerging pollutants since they are not commonly monitored, but have the potential to cause adverse ecological or (suspected) human health effects (Geissen et al. 2015). Its continuous contribution to water bodies generates several problems in aquatic life and can consequently cause problems for humans, a fact that has been reported for years (Semenza et al. 1997; Halling-Sørensen et al. 1998; Rodger-Gray et al. 2000; von Donk et al. 2016; Grenni et al. 2018).

The term "emerging pollutant" refers to any type of contaminant which originates from previously undetected pollutants or products that were not deemed problematic to health and the environment. This group of pollutants includes pharmaceuticals for human and veterinary use, personal care products, and pesticides. Pharmaceuticals in aquatic ecosystems should be identified as they reach the environment through discharges or excretion in an unaltered form by animal organisms and humans. The greater attention given to this emerging type of pollutant is due to both an increasing incidence of these concentrations in the environment and an increase in studies which have demonstrated the health risks that these products can cause by long-term exposure. Emerging pollutants are found in soil, rivers, surface water and even groundwater, which in principle would require no specific treatment for its drinkability (Erzinger 2013).

The consumption of drugs generates an input into the environment that is composed of the excreted unmodified drug after consumption, metabolites and/or degradation products (Heberer 2002); another pathway for these substances to reach the environment is through the inappropriate disposal of unused or expired medicines in common garbage or sewage. There are still other sources, such as the use of medicines in animal husbandry, in which the excreta containing the metabolites and/or the unchanged medicine can be used as a fertilizer, also exposing soil, surface water and groundwater to pollutants (McEniff et al. 2020). As they are continuously introduced into the environment, even at low concentrations, adverse effects have been found in aquatic biota and, possibly, this can cause an impact on human health as well as on animals (Rivera-Utrilla et al. 2013; Petrie et al. 2015). Thus, a more accurate mapping of the fate of these pollutants in the environment is still a challenge to be overcome (Ong et al. 2018).

11.2 Sources of Active Pharmaceutical Ingredients (API) in Ecosystems

According to Burns et al. (2018) society is currently dependent on pharmaceutical products, both for human and animal health, being considered essential, but there has been great concern due to their waste discharged in aquatic environments. These facts are related to easier access to drugs and an increase in production and distribution. According to the Organization for Economic Cooperation and Development (OECD), there are more than 2000 active pharmaceutical ingredients that make up the majority of drugs in the world that comprise prescription drugs, over-the-counter therapeutic drugs and veterinary drugs (Burns et al. 2018).

One of the factors leading to concern is related to the annual rate of increase in the development and approval of new APIs. The report published by Transparency Market Research (TMR), entitled "Global Active Pharmaceutical Ingredients (API) - Global Industry Analysis, Size, Share, Growth, Trends, and Forecast, 2018–2028" assessed the global market for active pharmaceutical ingredients in the US at $162 billion in 2017; the cumulative annual growth rate (CAGR) of APIs is projected to be 5.4% from 2018 to 2026 (OECD 2019).

Kusturica et al. (2016) brought together several studies carried out in different geographical and demographic contexts and revealed the lack of knowledge about the proper disposal of unused drugs and the environmental risks associated with it. Results based on 830 studies published between 2005 and 2015 are shown in Table 11.1. Even in countries with high development, a large part of the drugs are disposed of directly into the garbage or in the waste water. Of the countries mentioned only Sweden and Germany showed adequate practice regarding the disposal of medicines (Kusturica et al. 2016).

There is growing concern about the contamination of water resources by "emerging pollutants" (Valcárcel et al. 2011; Kusturica et al. 2016) which include pharmaceutical products, specifically as residues of medicines and their derivatives. An important approach to be taken is on the aspects of biosafety of water resources for human or animal consumption, or even for use in agriculture. These pollutants are found in small concentrations in the environment, which, in principle, should not bring immediate complications to human health. However, environmental conditions and risks to human and animal health during chronic exposure should be assessed (Ferreira 2008). This discussion deserves special attention because there are few, if any, methods to remove these pollutants; research in this field is still emerging, but it is promising. Another important factor is related to the aspect of legislation for the disposal of these products in the environment, as few countries have specific laws (Erzinger 2013).

Several authors report different amounts and types of APIs. Figure 11.1 represents an average estimate and the main APIs found in aquatic freshwater environments.

Regarding contamination, there are two main models to justify the presence of these EPs in aquatic systems (Figs. 11.2 and 11.3). The infographic in Fig. 11.2

Table 11.1 Drug disposal methods and frequencies in which they are reported in different countries (Kusturica et al. 2016)

Countries	Return of drugs at the pharmacy (%)	Thrown in the trash (%)	Discarded directly into the water[a] (%)
USA	20–30	40–50	20–30
Germany	40	≤ 1	5–10
Sweden	40	≤ 1	≤ 4
England	30–40	50–60	10–20
Ireland	20–30	50–60	10–20
Lithuania	10–20	≥ 80	5–10
Serbia	10–20	≥ 80	5–10
Nigeria	10–20	80	5–10
Ghana	10–20	≥ 80	≤ 4
Saudi Arabia	10–20	50–60	20–30
Oman	20–30	40–50	–
India	10–20	≥ 80	5–10
Bangladesh	10–20	≥ 80	20–30
Thailand	10–20	60–80	≤ 4
New Zealand	20–30	50–60	10–20

[a]Flushed down the sink or toilet

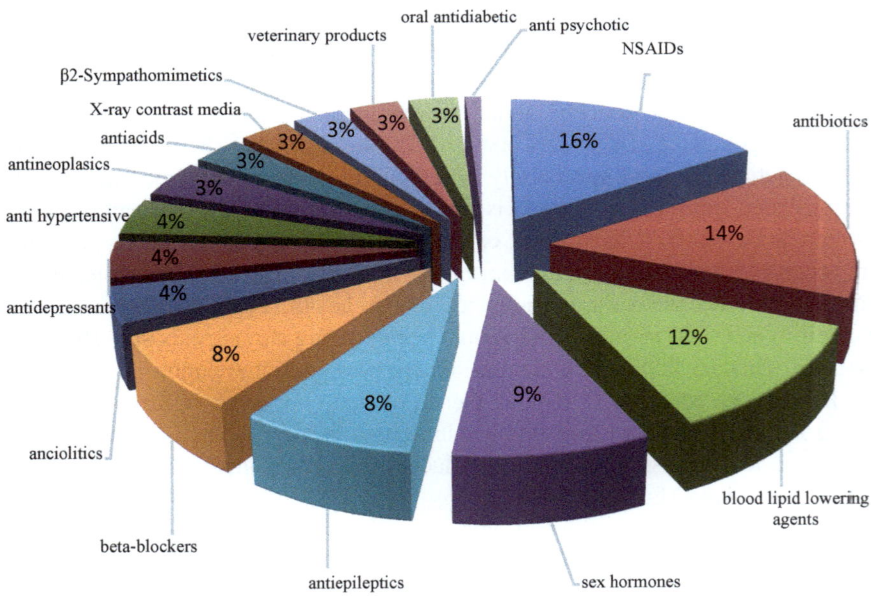

Fig. 11.1 Drug cycle in urbanized cities with effluent treatment units (Erzinger 2013; ONG et al. 2018; OECD 2019)

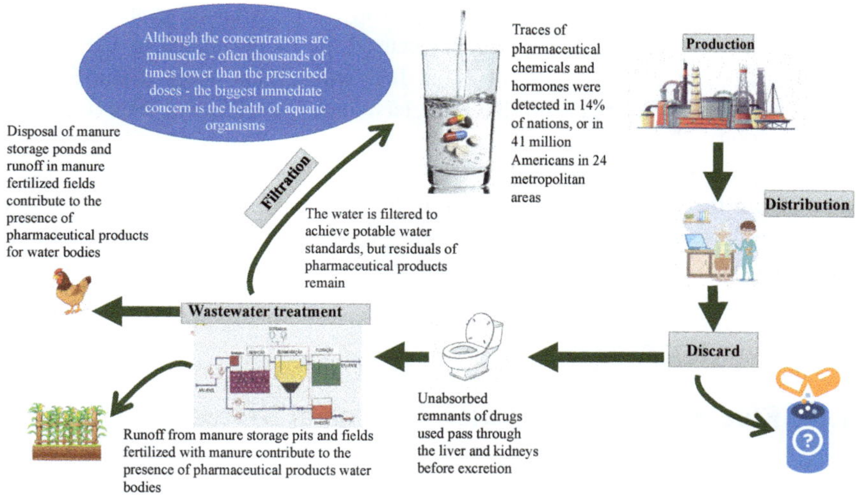

Fig. 11.2 Drug cycle in urbanized cities with effluent treatment (Aragão 2018)

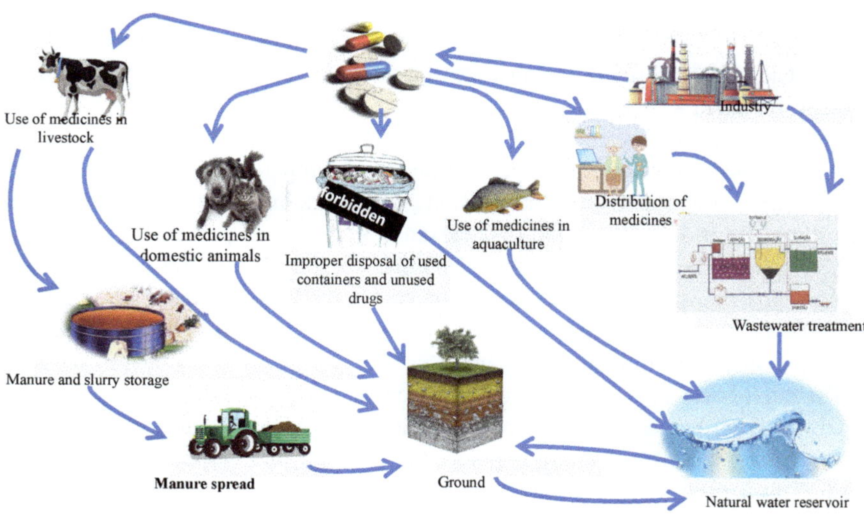

Fig. 11.3 Drug cycle in other regions without effluent treatment (Aragão 2018)

illustrates how these pharmaceutical products can end up in the water cycle when not properly disposed of. This model is used as a basis, the work described by Shea (2011), who observed that in 2009, 3.9 billion prescriptions were dispensed in the United States, and about 10–40% of the drugs were not used. This model assumes that all cities have effluent treatment recommended according to technical standards and in accordance with the legislation for water quality. According to Shea (2011),

80% of the 139 wastewater samples collected in the USA in the years 1999–2000 contained at least one drug. These data caused 22 of the states in the USA to introduce drug return policies as of 2009. It is worth mentioning that the current effluent treatment systems do not have the technology to eliminate EPs (Fig. 11.2) (Ferreira 2008).

The second model, described in Fig. 11.3, is based on the general use of medicines in the entire production system, not only taking into account the cities that have effluent treatment. This model was proposed by Boxall (2004), and notes that even for those drugs that are absorbed by the soil, the destination will always be in aquatic reservoirs and groundwater due to rain.

11.3 Efficiency of Wastewater Treatment Systems

Aus der Beek et al. (2016) created a database through a comprehensive literature review of 1016 original publications and 150 review articles on the presence of different pharmaceutical substances in wastewater treatment systems in 71 countries on all continents, excluding Antarctica. In this work, more than 100 pharmaceutical substances were found in the aquatic environment both in European countries as well as in the United States in surface waters, groundwater and/or tap/drinking water. These data further demonstrate the low degree of efficiency of current effluent treatment systems.

According to Borrely et al. (2012), both in Europe and in the United States, the drug flux in natural waters has been identified, and competent actions are being met. For example, there are limits of certain substances in some formulations: the maximum allowed concentration of triclosan in hygiene and personal care products is 0.30% (w/w), according to Union Directive 76/768/EEC European Union, and product labels must express the quantity present in the formulations. Data organized in Table 11.2 show evidence and the groups of drugs most present in sewage treatment plants, in different treatments.

The fate of drugs that reach the environment and their metabolites must be better understood, but the low volatility of the compounds indicates that their distribution occurs through transport in the aquatic environment and dispersion in the food chain. The current systems in the treatment of sewage are based on the adsorption process of these residues in suspended solids (sludge) in addition to biological degradation. The adsorption is dependent on the hydrophobicity of the molecule and the electrostatic interactions of the drug with the particles of the medium and with the microorganisms (Patel et al. 2019).

Approximately two-thirds of the Earth is covered with oceans. About 97% of the total water on the planet is salty and therefore unfit for consumption. Thus, just less than 3% is fresh water, but of this more than 2.5% is frozen in Antarctica, the Arctic and in glaciers, unavailable for immediate use. Finally, of the remaining less than 0.5%, a large part is hidden in underground aquifers. This makes Brazil the holder of about 12% of the world's surface freshwater reserves and some of the largest of

Table 11.2 Concentrations of emerging pollutants (µg/L) in affluent and effluents from sewage treatment plants found in Europe and the United States (Borrely et al. 2012)

	Affluent						Effluent							Removal rate (%)
	APIs	µg/L	dp	Average	Min.	Max.	*n*	µg/L	dp	Average	Min.	Max.	*n*	
Compounds pharmacists	Clarithromycin	0.344					2	0.15					2	56.4
	Ciprofloxacin	0.62	1.48	0.157	0.09	5>.52	13	0.234	0.649	0.021	0.007	2.378	13	62.3
	Doxycycline	0.65	0.94	0.098	0.067	2.48	10	0.42	0.426	0.227	0.038	1.09	9	35.4
	Erythromycin	0.58	0.242	0.56	0.346	0.83	3	0.297	0.237	0.2305	0.109	0.62	4	48.8
	Erythromycin H_2O	2.025					2	0.59					2	70.9
	Metronidazole	0.09					1	0.055					1	38.9
	Norfloxacin	0.115	0.056	0.0905	0.066	0.25	12	0.0526	0.098	0.0195	0.007	0.33	10	54.3
	Ofloxacin	0.482	0.884	0.156	0.007	2.275	6	0.171	0.317	0.0485	0.007	0.816	6	64.5
	Roxithromycin	0.78	0.737	0.81	0.027	1.5	3	0.472	0.435	0.54	0.008	0.87	3	39.5
	Sulfamethoxazole	0.32	0.248	0.2905	0.02	0.674	10	0.264	0.15	0.243	0.07	0.62	11	17.5
	Sulfapyridine	0.492					1	0.081					1	83.5
	Tetracycline	48					1	2.375					2	95.1
	Trimethoprim	0.43	0.401	0.251	0.0535	1.3	15	0.424	0.363	0.32	0.04	1.34	17	1.4
Antiepileptic's	Carbamazepine	0.732	0.869	0.25	0.081	1.68	6	0.774	0.789	0.37	0.042	2.1	13	–5.7
	4-aminoantipyrine	1.517					1	0.676					1	55.4
	Antipyrine	0.04					1	0.027					1	32.5
Nonsteroidal anti-inflammatory drugs (NSAIDs)	Codeine	28.605					2	1.93					2	32.5
	Diclofenac	1.039	1.283	0.232	0.16	3.1	6	0.679	0.701	0.55	0.04	2.448	11	34.6
	Ibuprofen	13.482	25.639	3.495	0.014	22.7	10	3.48	1.489	0.56	0.03	12.6	17	74.2
	Indomethacin	0.136					2	0.166	0.118	0.19	0.037	0.27	3	–22.1
	Ketoprofen	0.483	0.286	0.441	0.146	0.94	5	0.333	0.148	0.34	0.125	0.63	9	31.1
	Ketorolac	0.407					1	0.228					1	44
	Naproxen	5.077	8.251	2.363	0.206	23.21	7	0.934	0.873	0.452	0.017	2.62	13	81.6
	Clofibric acid	0.215	0.251	0.12	0.026	0.5	3	0.131	0.136	0.12	0.012	0.36	5	39.1

(continued)

Table 11.2 (continued)

	Affluent						Effluent							Removal rate (%)
	APIs	μg/L	dp	Average	Min.	Max.	*n*	μg/L	dp	Average	Min.	Max.	*n*	
Lipid regulator	Phenofibric acid	0.079					1	0.196	0.161	0.13	0.078	0.38	3	−148.1
	Bezafibrate	1.948	2.32	14.205	0.05	4.9	4	0.763	0.963	0.13	0.035	2.2	5	60.8
	Gemfibrozil	1.562	1.704	0.71	0.453	3.525	3	0.757	1.068	0.323	0.0112	2.86	6	51.5
Beta blockers	Acebutolol	0.335					1	0.14					1	58.2
	Atenolol	1.08	0.946	0.996	0.03	1.197	4	0.468	0.381	0.345	0.16	1.025	4	56.7
	Celiprolol	0.44					1	0.28					1	36.4
	Metoprolol	1.535	2.29	0.61	0.02	4.9	4	0.679	0.657	0.73	0.019	1.7	5	55.8
	Propanolol	0.198	0.269	0.005	0.036	0.51	3	0.102	0.071	0.093	0.03	0.18	5	48.5
	Sotalol	1.667					2	0.79					2	52.6
Antidepressant	Fluoxetine	5.85					1	0.112					2	98.1

n Number of concentrations recorded for affluent or effluent these underground liquid water reservoirs. On the other hand, Brazil has a technological delay in the treatment of urban effluents. In Brazil, according to the National Sanitation Information System—SNIS (2018), only 53.0% of the people have access to sewage collection, 46.0% of the country's sewage is treated and almost 100 million Brazilians do not have access to this service. The average indicator of sewage treatment in the 100 largest municipalities in 2018 was 56.0%; a small progress compared to 55.6% in 2017. According to the Institute Tata Brazil (2020), the country dumps approximately 5622 Olympic swimming pools of untreated sewage into the environment per day. Thus, the model proposed by Shea (2011) is the one that best represents the Brazilian reality in terms of the risk of the presence of EPs in water resources (Fig. 11.2) (InstitutoTata Brazil 2020).

On the other hand, according to the Association of the Pharmaceutical Research Industry 21, Brazil is the 5th largest producer of medicines in the world, behind the USA, China, Japan and Germany. Drug sales in pharmacies reached R$ 57 billion in 2018, with 162 billion doses sold and a total absence of rules or laws on the correct disposal of these drugs (Fig. 11.4).

Another important fact about pharmaceutical pollutants in aquatic ecosystems is related to the metabolism of these compounds in living beings. The APIs, when absorbed by an organism, can be excreted without any biotransformation in its unchanged form or it undergoes degradation or is metabolized by biotransformation reactions. Organic compounds follow three behaviors (Richardson and Bowron 1985): (i) they can ultimately be biodegradable, that is mineralized to CO_2 and water, such as acetylsalicylic acid; (ii) they can go through some metabolic process or be partially degraded, such as amoxicillin; or (iii) they can be persistent, such as anti-inflammatory and antilipidemic drugs (Stumpf et al. 1999).

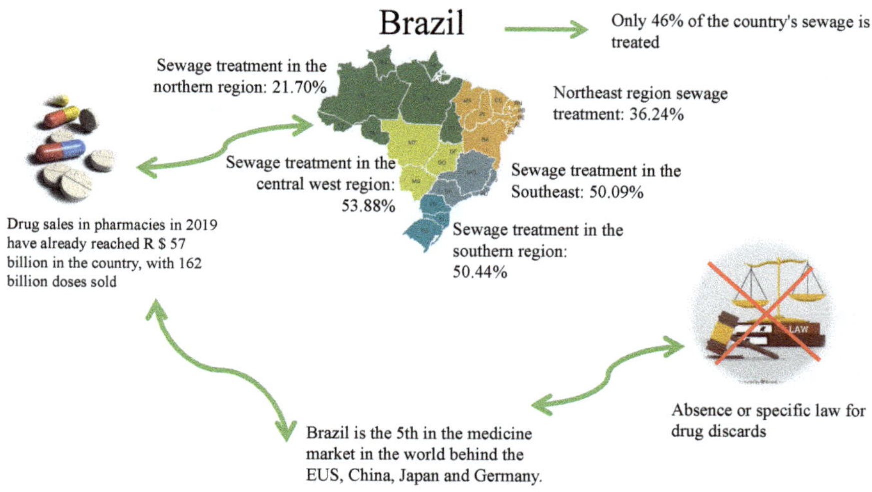

Fig. 11.4 Overview of the relationship between sewage treatment and drug production in Brazil (Aragão 2018)

Biotransformation reactions can be divided into phase I and phase II reactions. Phase I reactions are oxidation, reduction, hydrolysis and alkylation reactions. Phase II reactions are reactions such as sulfation and glucuronidation, in which more polar and hydrophilic conjugates are formed and excreted in the urine, bile, manure in the form of a metabolite or a mixture of several, which are often found in domestic sewage (Kümmerer 2009). Thus, when a drug is ingested by a living being, the organism can biotransform this drug (convert it into metabolites), degrade through environmental conditions, or even excrete without any transformation. This excretion is usually done through urine and feces, and the substances remain in the water bodies or are deposited in the sediment, when dumped directly without prior treatment. There are cases where biotransformation products are toxic, such as by-products of acetylsalicylic acid, triclosan, fluoxetine and cyclophosphamide (La Farré et al. 2008). The percentage excreted of each compound depends on the chemical interactions that the compound has with the organism. According to Watts et al. (2007), in data based on results available in the literature it was possible to establish a relationship with the estimated percentage of excretion in urine in the original form of some medications (Table 11.3).

Once excreted, drugs and their by-products can be delivered to sewage systems or, in the absence of a dedicated system, stormwater draining systems. In the case of the Metropolitan Region of São Paulo, Brazil, in 2017, only 68% of the sewage was undergoing treatment (Portal tratamento de água 2018). Drugs and their metabolites

Table 11.3 Percentage of excretion of non-metabolized active ingredients for selected drugs (Watts et al. 2007)

AIPs	Pharmaceutical group	Excreted unchanged (%)	Reference
Atenolol	Beta blocker	90	Bound and Voulvoulis (2005)
Amoxylin	Antibiotic	60	Bound and Voulvoulis (2005)
Cetirizine	Antihistamine	50	Bound and Voulvoulis (2005)
Benzafibrato	Lipid regulator	50	Bound and Voulvoulis (2005)
Felbamato	Antiepileptic	40–50	Bound and Voulvoulis (2005)
Erythromycin	Antibiotic	25	Bound and Voulvoulis (2005)
Diclofenac	Anti-inflammatory	15	Alder et al. (2006)
Ibuprofen	Analgesic	10	Bound and Voulvoulis (2005)
Clofibric acid	Active metabolite	6	Alder et al. (2006)
Carbamazepine	Antiepileptic	3	Bound and Voulvoulis (2005)

can pass through the sewage treatment without being retained; this is because they do not include adequate technology for this, eventually reaching the surface water systems and impairing the quality of drinking water. When there is no basic sanitation, the problem can be aggravated, since without treatment, contact with these waters can cause several diseases, in addition to the toxicological and ecotoxicological effects of drugs on organisms that inhabit and are associated with aquatic environments (Gagné et al. 2001, 2006; Petrie et al. 2013; Liu et al. 2015).

11.4 Prospects for Removal Processes

The most often used processes at the moment for destruction and removal of EPs are the advanced oxidative processes (AOPs). The efficiency of AOPs is widely proven, but the problem that is currently being investigated refers to compounds formed from the oxidation processes used and whether the waste generated can present a certain degree of acute or chronic environmental toxicity. The commonly used processes are effective in removing endocrine activity, but little is known about the potential risk of waste generated; studies in this field are necessary at different trophic levels (Erzinger et al. 2014; Pinto et al. 2016).

Currently, there are several processes designed to remove APIs from water and wastewater supplies (Hartung 2009; Erzinger et al. 2014; Pinto et al. 2016). While studies are being carried out under different conditions, the first step is to assess how the process occurs and the perspective of these processes under natural conditions (Hartung 2009; Erzinger et al. 2014; Pinto et al. 2016).

Some studies report the removal with the presence of several solvents grouped in the laboratory in which they create situations similar to those found in the environment. After these technology models present satisfactory results, they are applied in the environment.

AOPs are defined as the promotion of a chemical condition that generates enough hydroxyl radicals to affect a contaminating molecule in order to remove its known biological activity, promoting the purification of water. However, a hydroxyl radical attack complex can initiate a cascade of reactions that can lead to the mineralization of organic compounds (Maniero et al. 2008; Hartung 2009; Erzinger et al. 2014; Pinto et al. 2016). The most commonly used oxidizing agents are hydrogen peroxide (H_2O_2), ozone (O_3), chlorine (Cl_2) and chlorine dioxide (ClO_2). Advanced oxidation processes use H_2O_2/UV, TiO_2/UV, O_3/UV and H_2O_2/Fe^{2+} (Maniero et al. 2008; Hartung 2009; Erzinger et al. 2014; Pinto et al. 2016). They differ mainly in terms of cost, applicability in the treatment of water and sewage, and effectiveness and efficiency in removing the pollutant under study (Maniero et al. 2008; Hartung 2009; Erzinger et al. 2014; Pinto et al. 2016).

11.5 Emerging Pollutant Disposal Legislation

In Brazil, there is a water resources policy that aims to ensure the adequate availability of water for human consumption (Brazil 1997), and Ordinance 2914/12 of the Ministry of Health, which are official documents that define the standards for drinking water. Although these water quality standards exist, they do not detail and do not describe the conditions under which these drug residues are identified and, even if they are treated in water and sewage treatment plants (Erzinger and Häder 2018).

Pinto et al. (2016) demonstrated that in Brazil, drug disposal is not regulated in a specific way, and this approach does not consider environmental toxicity. Several countries and international organizations [European Union (EU), the North American Environmental Protection Agency (EPA) and the World Health Organization (WHO)] have issued different guidelines and laws in order to warn about the risks of the presence of drugs in aquatic ecosystems after studies recommended their removal, in order to establish acceptable limits for the water available for human consumption. Brazil does not have any specific legislation for this purpose (Pinto et al. 2016).

Hernando et al. (2011) reported that in Europe the environmental risk assessment (ERA) on emerging pollutants was introduced in the 1980s. This assessment was based on several protocols developed to assess the environmental risk of chemicals. The European Union introduced the regulation known as Registration, Evaluation, Authorization, and Restriction of Chemical Substances (REACH) by legislation in 2007, which is considered as the most important development in this field (Maniero et al. 2008). REACH covers materials based on nanotechnology, but the present methodologies may not be appropriate and offer an overview of approaches under water legislation in Europe, which cover ERA structures for emerging contaminants. Even though the REACH system currently has many positive aspects in the field of risk assessment (that is, in vitro tests to assess effects), the main difficulties related to the use of risk assessment is the availability of data; when data are available, they are often afflicted with uncertainty (Hernando et al. 2011).

New products are systematically evaluated in the European Union and the United States: they have to meet 61 regulations and political and social aspects. For 97% of the main chemicals in use and more than 99% of the chemicals produced by volume there are toxicity limits but they are not necessarily treated properly (Norman Network 2020). Hartung (2009) estimates that for 86% of chemicals data are missing and the REACH process aims to fill this gap. The regulation affects 27,000 companies, which are required to provide information on the toxic properties and uses of 30,000 chemicals after a pre-registration phase in 2008 (Hernando et al. 2011). The European Union has the perspective that this REACH form will be an important tool to emphasize toxicology and, mainly, to define the limits of the emerging pollutants (Hartung 2009).

Currently, most countries have laws and regulations on accepted toxicity values for the marketing of agricultural and industrial chemicals, biocides, cosmetics, food additives, medicines and other substances and use bioassays to protect human health and the environment. A current approach to regulations, political and social aspects, toxicity limits in different countries was described by Häder and Erzinger (2017).

References

Alder L, Greulich K, Kempe G, Vieth B (2006) Residue analysis of 500 high priority pesticides: better by GC–MS or LC–MS/MS? Mass Spectrom Rev 25(6):838–865

Aragão RBDA (2018) Fármacos Como Poluentes Emergentes Em Ambientes Aquáticos: Panorama De Consumo Na Região Metropolitana De São Paulo E Quadro Comparativo De Políticas Públicas Entre Países. Dissertação de mestrado, USP, Brazil, p 59

Aus der Beek T, Weber FA, Bergmann A, Hickmann S, Ebert I, Hein A, Küster A (2016) Pharmaceuticals in the environment. Global occurrences and perspectives. Environ Toxicol Chem 35(4):823–835

Borrely SI, Caminada SML, Ponezi AN, Santos DRD, Silva VHO (2012) Contaminação das águas por resíduos de medicamentos: ênfase ao cloridrato de fluoxetina. Mundo da Saúde 36(4):556–563

Bound JP, Voulvoulis N (2005) Household disposal of pharmaceuticals as a pathway for aquatic contamination in the United Kingdom. Environ Health Perspect 113(12):1705–1711

Boxall AB (2004) The environmental side effects of medication: How are human and veterinary medicines in soils and water bodies affecting human and environmental health? EMBO Rep 5 (12):1110–1116

Burns EE, Carter LJ, Snape J, Thomas-Oates J, Boxall AB (2018) Application of prioritization approaches to optimize environmental monitoring and testing of pharmaceuticals. J Toxicol Environ Health, Part B 21(3):115–141

Damstra T, Barlow S, Bergman A, Kavlock R, Van Der Kraak G (eds) (2011) Global assessment of the state-of-the-science of endocrine disruptors. Geneva: World Health Organization (WHO), International Labour Organization (ILO), United Nations Environment Programme (UNEP). Accessed at: http://www.who.int/ipcs/publications/new_issues/endocrine_disruptors/en/

Erzinger GS (2013) Emerging pollutants: environmental impact of disposal of drugs. Pharmaceut Anal Acta 4:

Erzinger GS, Häder D-P (2018) Regulations, political and societal aspects, toxicity limits. In: Häder D-P, Erzinger GS (eds) Bioassays: advanced methods and applications. Elsevier, Amsterdam, pp 51–67

Erzinger GS, Brasilino FF, Pinto LH, Häder D-P (2014) Environmental toxicity caused by derivatives of estrogen and chemical alternatives for removal. Pharm Anal Acta 5:8

Ferreira MGM (2008) Remoção da atividade estrogênica de 17ß-estradiol e de 17α-etinilestradiol pelos processos de ozonização e O3/H2O2. USP, Tese de Doutorado

Gagné F, Blaise C, André C (2006) Occurrence of pharmaceutical products in a municipal effluent and toxicity to rainbow trout (*Oncorhynchus mykiss*) hepatocytes. Ecotoxicol Environ Saf 64 (3):329–336

Gagné F, Blaise C, Salazar M, Salazar S, Hansen PD (2001) Evaluation of estrogenic effects of municipal effluents to the freshwater mussel *Elliptio complanata*. Comp Biochem Physiol Part C: Toxicol Pharmacol 128:213–225

Geissen V, Mol JGJ, Klumpp E, Umlauf G, Nadal M, Ploeg MJ, van der Zee SEATM, van der Ritsema CJ (2015) Emerging pollutants in the environment: a challenge for water resource management. Int Soil Water Conserv Res 3(1):57–65

Grenni P, Ancona V, Caracciolo AB (2018) Ecological effects of antibiotics on natural ecosystems: a review. Microchem J 136:25–39

Häder D-P, Erzinger G (eds) (2017) Bioassays: advanced methods and applications. Elsevier, Amsterdam, p 459

Halling-Sørensen BNNS, Nielsen SN, Lanzky PF, Ingerslev F, Lützhøft HH, Jørgensen SE (1998) Occurrence, fate and effects of pharmaceutical substances in the environment-a review. Chemosphere 36(2):357–393

Hartung T (2009) Toxicology for the twenty-first century. Nature 460(7252):208–212

Heberer T (2002) Occurrence, fate, and removal of pharmaceutical residues in the aquatic environment: a review of recent research data. Toxicol Lett 131(1–2):5–17

Hernando MD, Rodríguez A, Vaquero JJ, Fernández-Alba AR, García E (2011) Environmental risk assessment of emerging pollutants in water: approaches under horizontal and vertical EU legislation. Crit Rev Environ Sci Technol 41(7):699–731

Instituto Tata Brasil (2020) http://www.tratabrasil.org.br/estudos/estudos-itb/itb/ranking-do-saneamento-2020. Accessed on 30 Nov 2020

Kümmerer K (2009) Antibiotics in the aquatic environment a review–part I. Chemosphere 75 (4):417–434

Kusturica MP, Tomas A, Sabo A (2016) Disposal of unused drugs: knowledge and behavior among people around the world. In: de Voogt P (ed) Reviews of environmental contamination and toxicology, vol 240. Springer, Cham, pp 71–104

La Farre M, Pérez S, Kantiani L, Barceló D (2008) Fate and toxicity of emerging pollutants, their metabolites and transformation products in the aquatic environment. TrAC Trends Anal Chem 27(11):991–1007

Liu ZH, Lu GN, Yin H, Dang Z, Rittmann B (2015) Removal of natural estrogens and their conjugates in municipal wastewater treatment plants: a critical review. Environ Sci Technol 49 (9):5288–5300

Maniero MG, Bila DM, Dezotti M (2008) Degradation and estrogenic activity removal of 17β-estradiol and 17α-ethinylestradiol by ozonation and O_3/H_2O_2. Sci Total Environ 407 (1):105–115

McEniff G, Schmidt W, Quinn B (2020) Pharmaceuticals in the aquatic environment: a short summary of current knowledge and the potential impacts on aquatic biota and humans. EPA Research Programme 2014–2020, Published by the Environmental Protection Agency, Ireland, PG 52. Accessed on 20 Oct 2020. https://www.epa.ie/pubs/reports/research/water/Research%20142%20Report%20FINAL.pdf

Norman Network (2020) Network of reference laboratories, research centres and related organisations for monitoring of emerging environmental substance. Why do we need to address emerging substances? https://www.norman-network.net/?q=node/19. Accessed on 30 Nov 2020

OECD (2019) Pharmaceutical residues in freshwater: hazards and policy responses. https://www.oecd-ilibrary.org/sites/6a617955-en/index.html?itemId=/content/component/6a617955-en. https://doi.org/10.1787/c936f42d-en

Ong TT, Blanch EW, Jones OA (2018) Predicted environmental concentration and fate of the top 10 most dispensed Australian prescription pharmaceuticals. Environ Sci Pollut Res 25 (11):10966–10976

Patel M, Kumar R, Kishor K, Mlsna T, Pittman CU Jr, Mohan D (2019) Pharmaceuticals of emerging concern in aquatic systems: chemistry, occurrence, effects, and removal methods. Chem Rev 119(6):3510–3673

Petrie B, Barden R, Kasprzyk-Hordern B (2015) A review on emerging contaminants in wastewaters and the environment: current knowledge, understudied areas and recommendations for future monitoring. Water Res 72:3–27

Petrie B, McAdam EJ, Whelan MJ, Lester JN, Cartmell E (2013) The determination of nonylphenol and its precursors in a trickling filter wastewater treatment process. Anal Bioanal Chem 405(10):3243–3253

Pinto LH, Cardozo G, Soares JC, Erzinger GS (2016) Toxicidade ambiental de efluentes advindo de diferentes laboratórios de uma farmácia magistral. Revista Ambiente & Água 11(4):819–832

Portal tratamento de água. Segurança hídrica e abastecimento na Região Metropolitana de São Paulo (2018) https://www.tratamentodeagua.com.br/artigo/seguranca-hidrica-e-abastecimento-na-regiao-metropolitana-de-sao-paulo/. Accessed on 20 Oct 2020

Richardson ML, Bowron JM (1985) The fate of pharmaceutical chemicals in the aquatic environment. J Pharm Pharmacol 37(1):1–12

Rivera-Utrilla J, Sánchez-Polo M, Ferro-García MÁ, Prados-Joya G, Ocampo-Pérez R (2013) Pharmaceuticals as emerging contaminants and their removal from water: A review. Chemosphere 93(7):1268–1287

Rodgers-Gray TP, Jobling S, Morris S, Kelly C, Kirby S, Janbakhsh A, Harries JE, Waldock MJ, Jumpter JPO, Tyler CR (2000) Long-term temporal changes in the estrogenic composition of treated sewage effluent and its biological effects on fish. Environ Sci Technol 34(8):1521–1528

Semenza JC, Tolbert PE, Rubin CH, Guillette LJ Jr, Jackson RJ (1997) Reproductive toxins and alligator abnormalities at Lake Apopka, Florida. Environ Health Perspect 105(10):1030–1032

Shea K (2011) Infographic: unprescribed drugs in the water cycle. December 2, 2011/in North America, Pollution, Sanitation/Health, Water News, Water Policy & Politics.https://www.circleofblue.org/2011/world/infographic-unprescribed-%E2%80%94-drugs-in-the-water-cycle/. Accessed on 30 Nov 2020

SNIS—Sistema Nacional de Informações sobre Saneamento (2018) http://www.snis.gov.br/. Accessed on 30 Nov 2020

Stumpf M, Ternes TA, Wilken RD, Rodrigues SV, Baumann W (1999) Polar drug residues in sewage and natural waters in the state of Rio de Janeiro, Brazil. Sci Total Environ 225(1–2):135–141

Valcárcel Y, González Alonso S, Rodriguez-Gil JL, Romo R, Gil A, Vatala L (2011) Analysis of the presence of cardiovascular and analgesic/anti-inflammatory/antipyretic drugs in fluvial and drinking water of the Madrid Region in Spain. Chemosphere 82(7):1062–1071

Van Donk E, Peacor S, Grosser K, Domis LNDS, Lürling M (2016) Pharmaceuticals may disrupt natural chemical information flows and species interactions in aquatic systems: ideas and perspectives on a hidden global change. In: Gunther FA, de Voogt P (eds) Reviews of environmental contamination and toxicology. Springer, Cham, pp 91–105

Watts C, Maycock D, Crane M, Fawell J, Goslan E (2007) Desk based review of current knowledge on pharmaceuticals in drinking water and estimation of potential levels. Final Report. Drinking Water Inspectorate (DWI), Department for Environment, Food and Rural Affairs, 107. Accessed on 20 Oct 2020

Chapter 12
Detergents Pollution in Freshwater Ecosystems

Azizullah Azizullah, Sarzamin Khan, Sabeela Rehman, Nadia Taimur, and Donat-P. Häder

Abstract Detergents are water-soluble chemical cleaning agents which remove impurities and dust, and are made up of major chemical components i.e. surfactants, builders and additives. Detergents have emerged as major water pollutants that enter water bodies and food chain and therefore, can be hazardous to humans and other organisms. Detergent residues in water bodies come from residential areas (household detergents), runoff water of agricultural areas (herbicides and insecticides) and industrial effluents. The fundamentals on pretreatment and analysis of surfactants using various analytical approaches are discussed in detail in this chapter. In addition, levels of concentration of detergents and their degradation pathways in river, streams and lakes are also reviewed. It can be concluded that detergents are hazardous water pollutants and their components readily interfere with biological processes and therefore represents a highly toxic group of pollutants posing considerable toxicity to all aquatic organisms. It is, therefore, essential to develop accurate analytical procedures for qualitative and quantitative determination of detergents and their primary components and consequently their impact on organisms in the biosphere.

Keywords Detergents · Analytical techniques · Toxicity · Fate · Biodegradation and wastewater

A. Azizullah (✉)
Department of Biology, The University of Haripur, Haripur, Pakistan

S. Khan · S. Rehman
Department of Chemistry, University of Swabi, Swabi, Pakistan

N. Taimur
Department of Botanical and Environmental Sciences, Kohat University of Science and Technology, Kohat, Pakistan

D.-P. Häder
Department of Biology, Friedrich-Alexander Universität, 91054 Erlangen, Germany

© The Author(s), under exclusive license to Springer Nature Switzerland AG 2021

D.-P. Häder et al. (eds.), *Anthropogenic Pollution of Aquatic Ecosystems*,
https://doi.org/10.1007/978-3-030-75602-4_12

12.1 Introduction

Over the last decade, environmental problems associated with toxic chemical pollutants in aqueous environment have become a global concern due to their toxicological effects and potential risks to the ecosystem (Appannagari 2017). These pollutants are released into the environment (water bodies, sediments and soils) both from natural and anthropogenic sources. The natural sources include volcanic eruptions, leaching and weathering of rocks and anthropogenic activities such as metal mining, combustion, fossil fuels, use of agrochemicals, domestic and industrial discharges and sewage. Like other contaminants, both natural and anthropogenic sources are pathways by which detergents contaminate natural water bodies, sediments and soils (Pedrazzani et al. 2012). The detergents come from residential areas (household detergents), runoff water of agricultural areas (herbicides and insecticides) and industrial effluents. The biodegradation under the action of certain bacteria results in decomposition of about 90–97% of the detergents (Thiele and Nollet 2007). Surfactants have both polar and nonpolar characteristics, and consequently, show solubilities in various liquids; thereby reducing surface tension of aqueous and phase boundaries. These properties make these chemicals a class of special compounds to be used as washing, emulsifying, wetting and dispersing agents. Most chemical formulations of detergent, personal care products, paints, textiles, pesticides and pharmaceutical products are based on specific requirements that contain a variety of surfactants. The adverse effects of surfactants include causing foam in rivers, creating troubles in wastewater treatment and reduction of water quality by altering physical and chemical parameters of water (Prats et al. 1997). The major portion of surfactants after being used is disposed off in wastewater, where about 50% by volume of it is degraded by various chemical and biochemical processes, while the 25% as sorpted to suspended solids and the rest of 25% is dissolved. Wastewater treatment plants are the major mechanism for removal of detergents resulting in cleaning between 81 and 99.9%, although, in some instances, some types of surfactants are frequently detected in sewage effluents at concentration levels in the order of several μg L^{-1}. To eliminate the traces of surfactants after primary treatment, in most cases, secondary treatment in active sludge units is recommended (Demirbas et al. 2017). Detergents are water-soluble compounds derived from synthetic organic chemicals (mostly surfactants) that have cleaning ability. In broad terms, a detergent is a physically solid or liquid cleaning formulation chemically composed of a number of compounds, where most of which are surfactants (Aboul-Kassim and Simoneit 1993). The surfactants are a wide range of chemicals with a characteristic amphiphilic nature. Most of these molecules consist of two main parts: an hydrophilic/polar head group and hydrophobic nonpolar hydrocarbon chain. Their chemical structure grants surfactants unique characteristics like solubility in both polar and nonpolar media, micelles formation and reduction of surface tension of aqueous media. Various types of compounds have been chemically synthesized, based on the hydrophilic group of the molecule. Commercial surfactants may be anionic, cationic or nonionic, since the

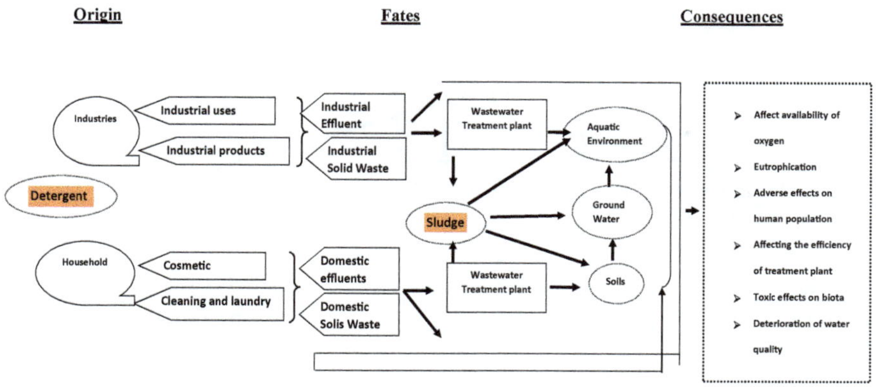

Fig. 12.1 Environmental perspective of surfactants from origin to environment and adverse effects

hydrophobic group is always nonionic in nature. Thus, the term "anionic" refers to negatively charged surfactants, "cationic" surfactants are positively charged surfactant, while "nonionic" refers to those that are uncharged in solution. Figure 12.1 shows the source, fate and effects of surfactant on the environment (Sueishi et al. 1988).

12.2 Detergents

Detergents are water-soluble chemical cleaning agents which remove impurities and dust; in general terminology, a detergent is synonymous with a surfactant, however, a detergent is made of various substances one or more of which are surfactants. Historically, the first commercially available surfactant was soap, whose definition incorporates all salts of organic carboxylates. In the first half of the last century, the development towards surface-active substances gathered pace and since its early beginning in the 1930s and the detergent science has passed through many revolutionary changes (Morelli and Szajer 2000). These are chemical substances or preparations containing surfactants, designed for water-based laundry and dish washing. Their physical form may include liquid, powder, paste, bar, etc., and in wide terms, they find their applications in domestic and industrial cleaners, cosmetic and industrial processes. Surfactants constitute primary components of a detergent and are added to achieve cleaning and rinsing or fabric softening properties. These cleaning agents belong to a group of chemical substances of high environmental relevance (Cross 1987).

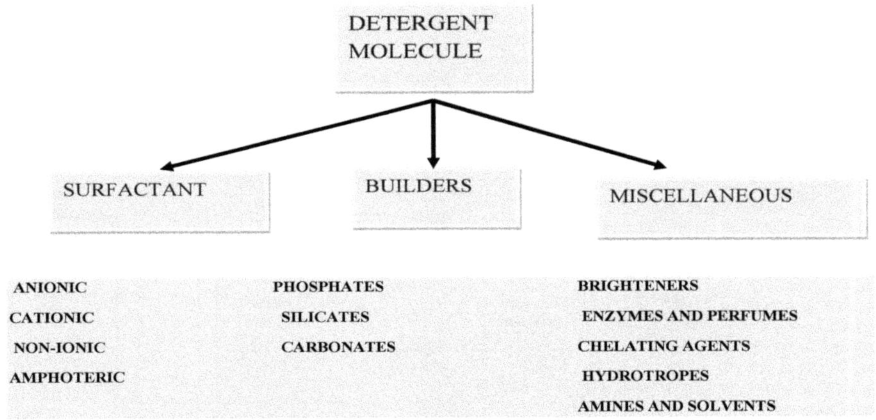

Fig. 12.2 Chemical composition of typical surfactant molecules

12.2.1 Composition and Uses of Detergents

The major chemical components of a deterrent are surfactants, builders and additives and may also contain small proportion of numerous groups of other substances (Fig. 12.2).

12.2.1.1 Surfactants

The term surface-active agents cover a heterogeneous class of compounds and long chain molecules with both hydrophilic and hydrophobic moieties. The hydrophobic group is a long aliphatic hydrocarbon (up to 20 carbon atoms), an alkyl chain of fatty acids, alkylbenzenes, carboxylates, while the hydrophilic groups are chemically sulfate, carboxylate, quaternary ammonium, polypeptide and polyoxymethylene that leads to key surface properties like emulsification, wetting, foaming, dispersing and solubilizing abilities. One of the characteristic features of surfactants is their ability to absorb at the surface/interface in an organized fashion. Therefore, surfactants have the ability to concentrate at the surface and reduce the surface tension of the medium. Based on the molecular structure, surfactants can be categorized into the following groups (Hanif et al. 2012).

Anionic Surfactants

Most detergents contain anionic surfactants. When these surfactants are dissolved in water a negatively charged anion is produced, the detergency is conferred in the anion, i.e., these surface-active agents contain an anionic group in the hydrophilic part and a small counter ion like sodium or potassium. These anionic surfactants are further classified into various groups including linear alkyl benzene sulfonates (LAS), alkyl benzene sulfonates (ABS), alkyl sulfates, alkyl ethoxylate sulfate (AES), fatty acid amide ether sulfates (FAES) and esters. The LAS contribute the major portion of the anionic surfactants, whereas secondary alkane sulfonate (SAS) is the special class of anionic surfactants used in consumer products due to their unique characteristics like fast wetting properties, high solubility, chemical stability and insensitivity to hydrolysis even at elevated pH values (Hanif et al. 2012; Painter 1992).

Cationic Surfactants

These chemical substances have a positively charged nitrogen atom and at least one hydrophobic long chain (Mata et al. 2005); in these surfactants, the detergency is in the cation. The quaternary ammonium compounds (QACs) are the common cationic surfactants. A common class of these surfactants is the alky trimethyl ammonium chloride, e.g., dodecyl trimethyl ammonium chloride (DDTMAC) and the other class contains two long chain alkyl group, e.g., dialkyl dimethyl ammonium chloride (DADMAC). The dialkyl groups are relatively less soluble in water than other classes, but they are commonly employed in detergents as fabric softener. Alkyl dimethyl ammonium chloride, also known as benzalkonium chloride, is a widely applied cationic surfactant. The quaternary ammonium cations, also known as quats, function as a common fabric softener. It also reduces the friction between the fibers and the skin, therefore, have been applied as hair conditioner. Alkyeneoxylated bisquaternary ammonium base substances also impart mildness properties to cosmetic but the problem associated with the hydrolytically stable surfactants is aquatic toxicity; therefore, these have been replaced by environmentally friendly cationic substances in surfactants like ester quats (Moulik et al. 1996).

Nonionic Surfactants

As the name indicates, these surfactants are nonionic species and the lack of charge enables them to avoid water hardness deactivation effects. They are used in various end products from many manufacturing industries like textile, detergents, paper, cosmetics and pharmaceuticals. Several classes of these surfactants have been introduced like alkyl phenol ethoxylates, fatty acid ethoxylates, sorbitan ester ethoxylates and ethylene oxide-propylene. Besides these, another widely applied class of nonionic surfactants are the polyhydroxy substances like glycol esters, glycerol esters, glucosides and other esters. However, in domestic and industrial detergents, the mostly used nonionic surfactants are alkyl ethoxylates (Lara-Martín et al. 2008; Traverso-Soto et al. 2012).

Amphoteric Surfactants

Amphoteric or Zwitterionic surfactants belong to class of surface-active agents containing both anionic and cationic functional groups, consequently, have both ionic charges (Guzmán et al. 2020). The most common amphoteric surfactants include N-alkyl betaines, for example, laurylamidopropyl dimethyl betaine, they are sometimes also known as alkyl dimethyl glycinates. The term amphoteric is generally applied to materials that show amphoteric properties, i.e. refers to those chemical species that can accept or donate a proton. In these surfactants, the molecule behaves differently in acidic and basic media, i.e. it acquires a positive charge and behaves like a cationic surfactant in an acidic medium, whereas, in alkaline pH solutions, the deprotonation takes place, and the molecule becomes negatively charged and behaves like an anionic one. The isoelectric point of these surfactants may be defined at which both ionic groups show equal ionization. The change in surface charge with the pH of the medium of amphoteric surfactants affects their properties, such as wetting, detergency and foaming. At the isoelectric point, the properties of amphoteric surfactants resemble those of nonionic surfactants. They are soluble in water, but the solubility shows a minimum at the isoelectric point. Amphoteric surfactants are compatible with other surfactants, resulting in mixed micelles. The distance between the charged groups at the isoelectric point defines the surface activity of amphoteric surfactants. The cation portion is either an amine or quaternary ammonium cation, whereas the anionic part is an organic or inorganic acid like carboxylic acid, sulfuric or phosphoric acid. The degradation of these surfactants in aqueous media indicates that most of these surfactants undergo complete biodegradation. Alkyl betaines with alky chains from 12 to 16 carbon atoms have been reported to biodegrade up to 60% and coco betaine is degraded in closed bottles more than 60% (Cha et al. 1992).

12.2.1.2 Builders

Builders are part of the chemical composition of detergents, because surfactant efficiency is reduced in hard water and even does not show good performance in soft water. Furthermore, large amounts in detergents impose heavy load on sewage water due to their environmental ecotoxicity. The builders function like a sequestering agent, because they bind hard water ions such as magnesium and calcium, enhancing the efficiency of the surfactant, and thus lower the amount of surfactants in the detergents formulations. Their other functions include attaining a desirable level of alkaline and dispersion of solid to avoid their redeposition on clothing (Yangxin et al. 2008; Cooper et al. 1988). Chemically builders may be either inorganic generally alkaline or organic materials, but the organic builders are used in relatively small amounts. Sodium triphosphate fulfill the essential requirements of a builder and therefore, has been widely applied as builder in the past; it not only removes Ca^{2+} and Mg^{2+} ions present in hard water but also enhances dissolution of detergents. The problem with phosphates is the ability to act as excellent fertilizer for algae, bacteria and other flora and fauna of rivers and lakes exhausting the oxygen supply both in the surface and in the bottom layers of water bodies and killing fish. This phenomenon is called eutrophication. Borates have also been a common constituent as builder in many types of detergents. Silicates also play an important role in the chemical formulation of detergents, having wetting and emulsifying characteristics on glass surfaces. Zeolites are also used in phosphate-free detergents in conjunction with other builders such as ethylenediaminetetraacetic acid (EDTA) or sodium carbonate. It has been claimed that zeolite is nontoxic to the environment, i.e. it does not contribute to the eutrophication of ecosystems. Organic builders such as nitrilotriacetic acid, (EDTA), disodium 3-oxapentanedioate (ODA), iminodisuccinic acid (IDA), and sodium citrate (Na–C) are good alternatives and substitutes for the traditional detergent builders, especially for phosphates builders whose application has many environmental impacts including eutrophic effect on the ecosystem (Lewis 1990; Nagarajan and Paine 1984).

12.3 Techniques for Detergent Pollution Determination in Water

Due to the specific physical and chemical properties of detergents, they are widely applied in households, industries and many other fields. Consequently, huge volumes are entering into different compartments of the ecosystems. The quantitative detection of various types of detergents have become a crucial analytical problem and the need of sensitive, accurate, new analytical methods is the demand of the day. The main issue with their quantification is due to factors like:

- Trace and ultra-trace level concentrations
- Diverse chemical structures of detergents with amphiphilic nature
- Complexity of the matrix where they reside.

For trace levels of surfactants in environmental samples with complex matrix composition, it is necessary to treat the sample at the preparation stage; it includes extraction, isolation and preconcentration to ensure appropriate selectivity and sensitivity in samples (Olkowska et al. 2014). Before processing for analysis, the proper storage of environmental samples, solid and liquid is crucial; the solid samples like soil and sediments usually require proper drying at room temperature or freeze drying followed by grinding and sieving. In the aquatic samples, most organic substances may undergo decomposition before analysis or reaching the laboratory; therefore, they need proper care, for example, to avoid biodegradation of surface-active substances. A biocide like a solution of formaldehyde is added to samples or mineral acids and then it is stored at 4 °C (Comas and Vives-Rego 1997). For analysis of surface-active agents in environmental samples that are mostly below the limit of detection of analytical protocols, commonly employed sample preparations are based on liquid-liquid extraction (LLE) and solid-phase extraction (SPE) (Marcomini et al. 1998; Cancela et al. 2017; Safarikova et al. 2005).

LLE is the first and still widely applied procedure for separation of nonionic surfactants. In LLE, the analytes (surfactants) are separated based on their relative solubilities in two different immiscible or partially immiscible liquid phases. Proper solvent systems like chloroform and other organic solvents for anion and cationic surfactants have been applied, while dichloromethane and ethyl acetate are considered suitable for extraction of nonionic surfactants from aqueous samples. The LLE is considered an efficient approach for isolation of anionic, cationic and nonionic surfactants from water samples (Idouhar and Tazerouti 2008; Olkowska et al. 2012). However, LLE is time consuming, requires a large amount of organic solvents and consequently produces toxic wastes. The other disadvantage is that it needs large volumes of samples (more than 100 mL) and the formation of an emulsion, causing difficulties in separation steps. To overcome these limitations with LLE, solid-phase extraction (SPE) is the most popular sample pretreatment and concentration technique for quantification of surfactants from aqueous samples. Contrary to LLE, this analytical approach requires less amounts of organic solvents and it is a relatively faster technique. SPE makes use of a specific material (solid phase) called sorbent that retains analytes. Cartridges using octadecyslica are used as universal sorbent for extraction of surfactants from environmental samples, and the separation of cationic surfactants is made due to strong interaction with silanol groups of the solid phase. The comparison of two sorbents, i.e., polymeric and octadecyslica for extraction of nonionic surfactants from aqueous media showed statistically similar results for analyzed analytes. The aqueous phase or extract containing the target compounds is passed through the sorbent, after washing with a proper solvent to remove interferences, and the adsorbed surfactants are eluted with suitable organic solvents like acetonitrile (ACN) ormethanol (MeOH) (Merino et al. 2003; Cruceru et al. 2012). Table 12.1 provides detailed information on procedures used in LLE and SPE for isolation of surfactants from various samples.

In the last decade, advances in SPE resulted in modern extraction techniques such as matrix solid-phase dispersion (MSPD), which is used to extract and purify target compounds simultaneously from solid matrices (Zhao et al. 1999). The other simple and low cost extraction techniques which reduce the time needed for sample preparation and decrease or eliminate solvent consumption include dispersive liquid-liquid microextraction (DLLME), a novel method based on the migration of analytes to an opaque solution, produced by mixing the extracting solvent with lower water solubility solvents (like chloroform) with a dispersant (soluble in water, e.g., acetone) in the aqueous sample. Finally, the dispersed organic phase of fine particles of the extraction phase containing analytes settles down in the lower portion of the extracting vessel by centrifugation (Deng et al. 2013). During the last years, several analytical methods have been proposed for the determination of detergents or their individual compounds such as tensiometry, potentiometric titration, chromatography and spectrophotometry (Ali et al. 2017). In tensiometry, the effect on the differential capacity due to the adsorption of surfactants on the

Table 12.1 Isolation of surfactants from various samples by LLE and SPE

Analyte	Type of sample	Volume of sample (mL)	Extraction technique	Condition of isolation	Clean-up	References
Quaternary ammonium compounds (QACs)	River water	10	SPE	Type of cartridge: strata-X 1. Conditioning: ACN, water 2. Washing: water/AA 3. Elution: ACN/AA/water	LLE(isooctane)	Bassarab et al. (2011)
Fluorescent whitening agents (FWAs)	Surface water	100	SPE	Type of cartridge: PS-DVB Solvent: methanol Ion-pair reagent: TBA	Acetonitrile solution	Shu and Ding (2005)
Alkyl benzyl dimethyl ammonium chlorides		200	SPE	Type of sorbent: Strata-X, 500 mg/6 mL 1. Conditioning: 5 mL ACN; 10 mL H2O 2. Washing: 10 mL H2O + CH3COOH (10%) 3. Elution: 8.5 mL A: B (9:1); A: ACN + CH3COOH (10%), B: $H_2O + CH_3COOH$ (10%)	Ultrapure water	Olkowska et al. (2013)
Total anionic surfactants	River water	5–50	SPE	Solvent: chloroform Ion-pair reagent: MB	LLE(water)	Roslan et al. (2010)
Benzylalkyldimethylammonium compounds (BAC)	–	50	SPE	Type of cartridge: polymeric, i.e., PLRP-s or Hysphere	The analytes from the SPE and HPLC columns consisted of acetonitrile and 10 mM ammonium formate buffer	Ferrer and Furlong (2001)

(continued)

Table 12.1 (continued)

Analyte	Type of sample	Volume of sample (mL)	Extraction technique	Condition of isolation	Clean-up	References
Nonyl phenol ethoxylates (NPEO), Octylphenol ethoxylate (OPEO) nonylphenol (NP), octylphenol (OP) Nonyl phenol ethoxylates (NPEC), Octylphenol ethoxylate (OPEC)	Sludge, wastewater	100–250	SPE	Type of cartridge: C18 1. Conditioning:MeOH, water 2. Washing: water/MeOH 3. Water/Methanol	LLE(isooctane)	Vega-Morales et al. (2010)
Total of cationic surfactant	River water	100	LLE	Solvent: chloroform (15 mL) Ion-pair reagent: patent blue V	LLE(water)	Sun et al. (2003)
Sodium linear-dodecylbenzenesulfonate (DBS)	–	2–200	SPE	Type of cartridge: C2, CH, PH	50% methanol	Tsukamoto et al. (2004)
Sodium dodecyl sulfate (SDS)	–	2–200	SPE	Type of cartridge: C2, CH, PH	50% methanol	Tsukamoto et al. (2004)
Cationic surfactant	Wastewater	100	Solvent extraction	Solvent: chloroform Ion-pair reagent: patent blue V	Pure water	Idouhar and Tazerouti (2008)
Linear alkylbenzenes (LAB)	Water	4	SPME	Type of fiber: polydimethylsiloxane	Solvent free	Penteado et al. (2006)
1,4-dioxane	Household detergents and cleaners	100	SPE	Type of cartridge: C18	2 mL acetone 5 mL dichloromethane	Tanabe and Kawata (2008)

electrode is measured, but this technique is restricted to the determination of anionic and nonionic surfactants. The potentiometric titration is based on the changes of cell electromotive force (EMF) after the addition of the variable aliquots of the samples and the ion-selective electrode located at the endpoint. This technique is widely applied for determination of the total sum of ionic surfactants. The main drawbacks of this analytical approach are low reproducibility and signal stability (Nezamzadeh-Ejhieh and Esmaeilian 2012). In spectrophotometric determination of surfactants, the ion associates of analytes after their reaction with ion-pair reagents are extracted into suitable organic solvents. Afterward, the absorbance of the organic phase is measured. This technique is simple and quick and has been the method of choice for routine analysis. The main drawback is the production of toxic wastes; another disadvantage is that it determines total surfactants instead of individual components (Coll et al. 2007). Both infrared (IR) spectroscopy and FTIR have also been widely applied for determination of ionic and nonionic surfactants in diverse environmental samples; especially, with FTIR the quantification has been accomplished at trace levels (Carolei and Gutz 2005). Recently. Kurrey et al. 2018 demonstrated a novel analytical approach for qualitative and quantitative selective determination of surfactants in water samples using ion-pair reagents (methyl orange) using diffuse reflectance-Fourier transform infrared (DRS-FTIR) spectral measurements. SPE, SPME and other techniques have been applied to sample preparation steps from liquid and solid matrices to achieve better analytical results (Kurrey et al. 2018). The total quantification of surfactants in environmental samples is considered as first step for screening, for deep and thorough study the separation of various forms of surfactants is required; therefore, the chromatographic techniques, gas chromatography (GC), high performance liquid chromatography (HPLC) with various types of detectors are used for analysis of complex samples of surfactants (Marcomini and Zanette 1996). GC is a less frequently used technique for determination of surfactants as most anionic and nonionic surfactants are not volatile and need to be chemically derivatized with specific chemical reagents before injecting them into the system (Voelkel et al. 1993). Attempts have been made for direct determination of alkylphenol biodegradation products, but better results can be obtained after chemical derivatization. The advantage of GC over HPLC is complete separation of homologues and isomers of substances like LAS on a capillary column of GC after derivatization. Several detectors are used in GC like flame ionization detector (FID), single quadrupole (MS) and tandem mass spectrometer (MS-MS), but GC is generally combined with MS or MS-MS to achieve a limit of detection at some ng mL^{-1}. For quantification of analytes, the detection can be made both by electron impact (EI) and chemical ionization (CI), but the later approach has been a widely applicable method for ionic and nonionic surfactants in environmental samples (Olkowska et al. 2012; Wulf et al. 2010). Table 12.2 provides information on analytical protocols for determination of surfactants using gas chromatography coupled with various detectors.

Liquid chromatography (LC) is currently the widely applied technique for separation and analysis of surfactants in environmental samples. LC is suitable for quantification of non-volatile analytes from low to higher molecular mass without

Table 12.2 Determination of surfactants applying gas chromatography coupled with different detectors

Analyte	Sample type	Sample preparation	Mobile phase	Column	Detection	MLD/LOD	Reference
1,4 diaxone	Household detergent, cleaners	SPE	Helium carrier gas	Fused silica J&W DB-WAX column (Agilent)	GC/MS	–	Tanabe and Kawata (2008)
1,4 diaxone	Cosmetic product	Solvent extraction	Helium carrier gas	Rxi-624Sil MS column with dimension 60 m, 0.53 mm I.D., 3.0 μm thickness	GC/MS	0.00065–0.00091 μg/mL	Alsohaimi et al. (2020)
Linear alkyl benzene sulphonates (LAS), sulfophenylcarboxylates	Aqueous environmental samples	Ion-pair LLE hydrolysis derivatization (CH2N2)	−0.7 mL/min	DB-5 (fused silica capillary column, 50 m, 0.25 mm I.D., 0.25 μm)	MS	<4μg/L	Akyüz (2007)
Octylphenol ethoxylate (OPEO) octylphenol (OP)	Soil	ASE, SPE derivatization (BSTFA)	Helium 1 mL/min	ZB-5 MS Zebron capillary column (30 m 0.25 mm I.D., 0.25)	EI-MS	3 g/kg 3 g/kg 3 g/kg	Jiménez-Díaz et al. (2010)
Nonyl phenol ethoxylates (NPEO) nonylphenol (NP)	River water	SPME without deriviatization	Helium	DB-5 (fused silica capillary column, 30 m, 0.25 mm ID, 0.25 m)	EI(+)-MS	0.09–0.45 g/L	Díaz et al. (2002)
Linear alkyl benzene sulfonates (LAS)	Laundry wastewater	SPME SPE	Helium carrier gas	NST-5-MS (NST, São Carlos, with 5% phenyl and 95% polydimethylsiloxane (30 m × 0.25 mm × 0.25 μm)	GC/MS	Below 1.2 mg/L	Motteran et al. (2019)

derivatization. Reverse phase is often employed for determination of cationic, anionic and nonionic surfactants and their degradation products after suitable extraction (Ying 2006; König and Strobel 1988). Widely applied mobile phases are deionized water and organic solvents like methanol or acetonitrile in some cases to achieve better separation additives, e.g., triethylamine or ammonium acetate, acetic acid and formic may be added to the mobile phase (Ruiz-Angel et al. 2006). The application of HPLC for determination of ionic and nonionic surfactants in diverse and complex samples of environmental origin has led to the replacement of many traditional techniques and the major advantage is the detection in solid and liquid samples. The HPLC can be coupled with different detectors like ultraviolet (UV), fluorescence (FL), MS and MS-MS for quantitative determination of surfactants and their degradation products (Bazel et al. 2014).

The separation of surfactants in HPLC is achieved by solid-liquid partition behaviors between the column materials and the properly chosen mobile phase. This technique is also suitable for nonionic surfactants, where a gain in sensitivity is possible when detection is made with a fluorescence detector instead of UV. The presence of an aromatic ring in the molecular structure of some classes of surfactants, e.g., LAS and NPEOs made them good candidates for quantitative determination using HPLC coupled to UV and FL detectors. The important information about the partial separation of positional isomers and data on the alkyl chain distribution of LAS in water can be obtained with HPLC coupled to an FL detector. Due to the lack of weak optical characteristics, aliphatic surfactants (AEOs and AES) cannot be detected with a UV or fluorescence detector; these kinds of surfactants are suitable for determination with high performance liquid chromatography coupled with mass spectrometry, where analytes are transferred to appropriate ions. For the identification and quantitative determination, suitable fragments of surfactants are employed with analyzers, such as single (Q) or triple quadrupole (TQ), quadrupole-ion-trap (QIT), time-of-flight (TOF), hybrids like quadrupole time-of-flight (Q-TOF) and quadrupole-ion-trap time-of-flight (QIT-TOF) (Andreu and Picó 2005). HPLC-MS enables simultaneous determination of various surfactants in environmental samples. The application of an MS-MS detector further improves selectivity of analytes, and various surfactants can be identified during one analysis (Wang and Kasperski 2018). Table 12.3 provides information of analytical techniques for determination of surfactants using liquid chromatography coupled with various detectors.

12.4 Detergents in Rivers, Streams and Lakes

The quality of surface and groundwater is quickly deteriorating as a result of contamination due to various pollutants from untreated municipal water and effluents from industries. These effluents ultimately find their way into rivers or are dumped into lakes causing an increase in water-borne diseases. Detergents and their components reach aquatic environments by two different ways, first, as a result of

Table 12.3 Determination of surfactants applying liquid chromatography coupled with different detectors

Analyte	Sample type	Sample preparation	Recovery (%)	Mobile phase	Column	Detection	MLD/LOD	References
Benzylalkyldimethylammonium compounds (BAC)	River water	SPE (MH/AB) SPE	95–106	MeOH with 50 mM AMF buffer (pH = 3.5)	Nova-Pack C8 (150 mm, 3.8 mm, 5 m)	ESI(+)-MS	0.004 g/L	Bassarab et al. (2011)
BAC, dialkyl dimethyl ammonium chloride (DADMAC)	Marine sediment	LLE, SPE	98–118 (DADMAC)	ACN/H2O, isopropanol, FA, AMAC	Luna C18 (150 mm, 2 mm, 5 μm)	ESI(+)-ToF-MS	0.1–2.6 ng/g (LOQ)	Li and Brownawell (2009)
BAC	Water	SPE	>71	Gradient of acetonitrile/10 mM aqueous ammonium formate	C18, PLRP-s, and Hysphere (10 mm × 2 mm inside diameter disposable precolumns	(LC/MS)	–	Ferrer and Furlong (2001)
Alkylphenol polyethoxylates	Wastewater	Gaseous stripping into ethyl acetate is employed to extract the analytes from wastewater samples	87	Gradient elution was carried out with a linear program from 97% A and 3% B to 37% A and 63% B	Bonded-phase aminosillica columns with irregularly shaped 10-μm material.	LC	1 μg/L	Ahel and Giger (1985)
Ethoxylated	Sewage	SS, SE	>80	Methanol, hexane and dichloromethane	C18 columns	RP-HPLC with UV detection	3.0 μg/L	Kiewiet et al. (1995)
NPEO, NP	Marine sediment, sewag	Sonication, SPE LLE	64–127 (sediment)	H2O, MeOH, sodium acetate	MSpak GF-310 4D filtration column (150 mm, 4.6 mm)	ESI-MS NPEO (ESI +) NP (ESI-)	0.78–37.3 ng/g	Shang et al. (1999)
4-nonylphenol, 4-NP and 4-tert-octylphenol,	Water	LLE	–	Acetonitrile/water 65:35	Zorbax Eclipse XDB-C8 column	HPLC-FLD	0.1 μg/L	Cruceru et al. (2012)

(continued)

Table 12.3 (continued)

Analyte	Sample type	Sample preparation	Recovery (%)	Mobile phase	Column	Detection	MLD/LOD	References
LAS, NPEO alcohol ethoxylates (AEO), PEG	Aqueous and solid environmental matrices	SPE	80	Mobile phase A was acetonitrile and mobile phase B was 10 mM formic acid/10 mM ammonium formate buffer in milli-Q water	Luna C18 (Phenomenex) 150 mm × 2 mm, 5 m of particle size column	ESI-MS AE, NPEO, PEG (ESI +) LAS, SPC, NPEC (ESI −)	Between 40 and 80μ g/L	Lara-Martín et al. (2011)
Cationic surfactants	River and wastewater	Microporous membrane liquid extraction	–	Chloroform, ethanol, ammonia, heptanoic acid	25032.1 mm I.D. cyanopropyl column (LiCh rosorb, Merck)	Uv	0.7–5 μg/l	Norberg et al. (2000)
NP, OP, APEOs, AEOs	Sewage sludge	PLE, ASE	89–94	Methanol–water gradient	Luna C18 (150 × 4.6 mm), 5 μm	ESI and APCI	0.3 to 30 μg kg^{-1}	Andreu et al. (2007)
Alkylphenol ethoxylates (APEOs)	water	SPE	–	A binary mobile phase gradient with 1.5 mM NH4Ac in milli-Q water (A) and MeOH with 1.5 mM NH4Ac (B)	5 μm Zorbax Eclipse XDB-C8 column	LC-MS/MS	0.04–12 ng l(−1)	(Jahnke et al. 2004)
LASs	Water	LLE	93.3	A (acetonitrile–water, 80:20) and solvent B (water	The stationary phase was a Nova-Pack C18 4 m column (300 mm × 3.9 mm	LC-MS-MS	0.03 and 0.07 mg/l	Lunar et al. (2004)
Nonylphenol ethoxylates (NPECs)	Domestic sewage sludge	Subcritical water	87	Methanol and water containing 0.2 mmol/L ammonium acetate	5-μm C18 reversed-phase packing	LC-MS	1.51μg/ml	Bruno et al. (2002)

discharge from wastewater treatment plants into rivers, lakes and oceans and second by direct discharge of raw sewage (Wang and Kasperski 2018). Mainly, anthropogenic activities such as runoff and direct discharge of effluents from industries and urban regions are the cause of surfactant contamination in aqueous environments. As a result of excessive use of surfactants, high concentrations reach wastewater treatment plants where removal has been reported to be above 95%, but still, due to huge consumption, some fraction is not removed and in some cases wastewater is discharged to aquatic ecosystem without passing through waste water treatment plants; consequently, surfactants have been detected in rivers and streams (Pérez-Carrera et al. 2010; Sun et al. 2003). It has been reported that the average concentration of surfactants may reach up to 10 mg L^{-1} in domestic waste, while concentrations higher than 300 mg L^{-1} from various industries have been found (Rivera-Utrilla et al. 2012). The concentration of both anionic and cationic surfactants in surface waters of rivers was determined both in dry and rainy seasons. The obtained results showed that the concentration of analytes was high in the dry season, and the concentration of cationic surfactant in the vicinity of towns was found to be 0.20 ± 0.21 μmol (rainy) and 0.54 ± 0.15 μmol, (dry) while residential areas showed the highest concentration of anionic surfactants (rainy = 0.34 ± 0.20 μmol, dry = 0.86 ± 0.40 μmol) (Hanif et al. 2012). In another research study, high concentrations of anionic surfactants, linear alkylbenzene sulfonates (LAS) and nonionic, nonylphenol ethoxylates were reported in urban and industrial wastewater, and concentration of linear alkylbenzene sulfonates ranged from 1155 to 9200 μg L^{-1}. The discharge of detergents is severely contaminating the freshwater lakes; for example, the Laguna de Bay in the Philippines is one of the primary source of drinking water but also receives discharges of effluents (domestic and industrial). The LC-MS analysis showed the presence of LAS at concentration levels of (1.2–73 μg L^{-1} and 2.2–102 μg L^{-1}) and alkylbenzenesulfonates (ABS) (1.1–75 μg L^{-1} and 1–66 μg L^{-1}) (Eichhorn et al. 2001). LAS in concentrations of 1090–1100 μg L^{-1} was also detected in wastewater from sewage and surface water the lower concentration was attributed to the fact that 73% of LAS was removed by aerobic degradation (Fountoulakis et al. 2009). Another group of anionic surfactants, called perfluorinated surfactants (PS) with special technological applications, also used in coating of paper, carpets and textiles are of environmental concern. Two members of this class, perfluorooctane sulfonate (PFOS) and perfluorooctane (PFOA) are considered as persistent organic pollutants. A study was conducted in the Yodo River basin (Japan) to find out the contamination profiles for aqueous PFOS and PFOA concentrations. The LC-MS measurements revealed the average reproducible concentrations from 0.4 to 123 ng L^{-1} for PFOS and 4.2 to 2600 ng L^{-1} for PFOA; the sewage treatment plants and the tributary streams showed highest levels. It was concluded that the wastewater from industries and domestic wastewater contributed significant loads of PFOS and PFOA to water bodies (Lein et al. 2008). Quaternary ammonium compounds (QACs) are a class of chemical substances widely applied as surfactants and found in personal care, liquid medical and cleaning products. A large volume of these compounds end up as discharge in domestic and industrials waters. A recently conducted research study revealed the presence of the average concentration of QACs up to 2.5 μg L^{-1} in wastewater effluents (Pati and Arnold 2020). The discharge of wastewater and industrial effluents containing surfactants is contaminating the world's water resources. The lakes and streams are valuable sources of water; in a recent survey, in 13 selected lobelia (Poland) lakes and 14 streams the level of surfactants was monitored. The results showed the cationic and anionic surfactants in all the 27 samples ranged from 0.05 to 0.51 mg L^{-1}, and nonionic surfactants were detected in 17 of 27 samples up to 2.43 mg L^{-1} (Markowski et al. 2017). A novel approach was adapted for the identification of alkyl dimethylbenzylammonium chloride also known as benzalkonium chloride (BAC) in surface water samples, and the concentration ranged from 1.2 to 36.6 mg L^{-1}, indicating the severity of contamination of water with surfactants. Alkyphenol polyethoxylates (APEOs) are nonionic surfactants which are widely used in domestic and industrial products and found in wastewater discharges and in sewage treatment plants. The concentrations of APEOs around 45 μg L^{-1} and up to 3970 ng g^{-1} were detected near treatment plants and in the sediments. One of the important metabolites of APEOs is 4-nonylphenol (NP), the NP and octylphenol (OP) which showed more toxicity and more persistent properties than of APEOs; therefore, concern has arisen and many studies in Spain, Japan, Germany, USA and Canada found the concentrations of NPS in river water up to 17.5 μg L^{-1} (Bennie et al. 1997; Buxton and Kolpin 2005; Isobe et al. 2001; Céspedes et al. 2008).

12.5 Fate of Detergents in Water

Detergents and their degradations products present in domestic and industrial effluents are discharged into sewage treatment plants or directly into surface waters that ultimately reach sludge water bodies. Depending on the chemical nature of surfactants, detergents and its key components exhibit different behavior and fate in the environment. It is crucial for the surfactants to lose the sufractancy and minimize the toxicity in the primary degradation. The complete mineralization of surfactants avoids the formation of persistent intermediates and ensures an effective biodegradation mechanism for their elimination from the ecosystem.

Most surfactants are degraded by microbes in the environment, although some surfactants such as LAS, DTDMAC and alkylphenol do not decompose under anaerobic conditions. The microbial activity is responsible as a primary step in transformation of surfactants in water. AS is reported to be degraded by aerobic microorganism in water and the intermediates are mono- and dicarboxylic sulfophenyl acids (SPC), as well as a variety of SPCs with alkyl chain length of 4–13 have been identified. The cleavage of benzene needs molecular oxygen; therefore, no reports are available on anaerobic degradation of LAS (Yadav et al. 2001; Krueger et al. 1998). The incomplete removal of LAS and its aerobic breakdown intermediates, SPCs, reach the sewage effluent, and it has been reported in earlier studies that the half-life of LAS in rivers is less than three days; furthermore, the results revealed that more than 99% is degraded by the natural microbial flora in river water even at low temperatures. In contrast, the degradation of LAS and its intermediates is slower in the marine environment and can be attributed to lower microbial activity and presence of calcium and magnesium (Larson and Payne 1981; González-Mazo et al. 1997) QACs are a class of cationic surfactants applied as fabric softeners, and their excessive use leads to release into wastewater. The alkyl chain length in these surfactants not only determine the physical and chemical characteristics, but also has a decisive role in the fate and toxic effects of these substances in water. Like other surface-active agents, the QACs undergo the sorption and biodegradation that affect their fate in different environmental compartments. These have high adsorption affinity onto sewage sludge, sediments, and clay which proportionally increase with alkyl chain length. It has been reported that the QACs can form complexes with anionic surfactants, resulting in improved elimination from effluents. QACs are biodegradable under aerobic conditions which is generally considered the most established mechanism for removal from aqueous environment. The parameters such as chemical structure, concentration and complexation with anionic surfactant define the extent of elimination from water. Although under aerobic conditions the biodegradation for individual surfactants of this class may vary. The degradation mechanism for alkyl trimethyl ammonium and alkyl dimethyl ammonium halides (TMAC and DMAC) is reported initially being by N-dealkylation, followed by N-demethylation (Nishiyama et al. 1995).

Another class of surfactant, alkylphenol ethoxylates (APEs) comprises the world's third group of surfactants in terms of production and extensive application in domestic and industrial products, octylphenol ethoxylates (OPEs) and nonpheno ethoxylates (NPEs) are the most common surfactants in the market place with applications in domestic detergents, pesticides and industrial products. It is believed that the biodegradation of APEs initiates with a shortening of the ethoxylate chain. The common intermediates after degradation in water were reported as alkylphenol like NP and OP, short chain alkylphenol ethoxylates and a series of ether carboxylates. However, concern has arisen due to the relative stability of their metabolites in the environment; therefore, attempts were made to understand the fate and pathways of APEs in water (Brycki et al. 2014). Sorption is one of the processes that affect the fate of APEs in water. Their association with aquatic particles is well documented in the rivers and coastal environments and analyzed, for example, NP in water of Tokyo, and the concentrations ranged between 0.051 and 1.08 $\mu g\ L^{-1}$; however, about 20% of NP was found in solid particles in the aqueous environment (Isobe et al. 2001). Ekelund et al. (1993) reported that the half-life of NP after biodegradation was 58 days in seawater and 35 days in aerobic water with sediment. Laboratory experiments showed that fatty alcohol ethoxylates (AE) are easily degradable under both aerobic and anaerobic conditions. Fatty alcohol sulfates (AS) are the most biodegradable surfactants, and the fast rates are noted in primary and ultimate biodegradation steps. The mechanism of degradation by enzymatic action involves the cleavage of the sulfate ester bonds, and the products generated are inorganic sulfate and a fatty alcohol. The degradation pathway is further verified by the identification of alkylsulphatase that is responsible for the initial desulphonation step. The β-oxidation further oxidizes the alcohol to an aldehyde and finally to fatty acid, the ultimate product of biodegradation (Thomas and White 1989).

12.6 Effects of Detergents on Freshwater Biota

The wide spread of detergents and their biodegradation products from various sources into water bodies pose considreable toxicity to the aqautic organisms. Generally, the toxicity of surfactants is influenced by both abiotic and biotic factors. The abiotic factors, i.e., physico-chemical properties of water, include pH, hardness, dissolved oxygen, also some properties of surfactants like type of surfactant and concentration, may largely influence the toxicity of surfactants. While the biotic factors that affect the extent of toxicity refer to tested species, age of organism and acclimatization at lower quantity of detergents (Ivanković and Hrenović 2010; Yamane et al. 2008). In sewage effluents varying concentrations of detergents and their components may be detected for example, for anionic surfactants (e.g., LAS) up to 1090 $\mu g\ L^{-1}$, for nonionic surfactants (e.g., AFEOs), 332 $\mu g\ L^{-1}$ and up to 62 $\mu g\ L^{-1}$ for cationic surfactants (e.g., DTDMAC) have been reported (Ying 2006). Although efficient wastewater treatment ensures discharge of very low levels of these surfactants, above a specific level the aquatic ecosystems may be exposed to potential risks. For toxicity tests, most of the organisms studied as models are aquatic and include algae, fish and bacteria.

Fungi are an important class of freshwater biota and play an important role in bioremediation of pollutants in water bodies. They are also considered as important components of ecosystems and play a vital role in the food chain. Pollutants from various sources including detergents can affect these important organisms (Kumar et al. 2019). In some cases, fungi can be acclimatized to the presence of pollutants, but the type and level of contaminant can adversely affect their growth (Santos et al. 2016). The effects of household sewage, which contains various detergents, on the biomass produced in 58 fungal species was evaluated. In fungal species belonging to different genera, it was concluded that the biomass decreases with increasing concentration of detergents (Chaturvedi and Tiwari 2013). It has been reported that the algae are more sensitive to the presence of QACs than other freshwater biota. A study under standard laboratory conditions revealed that QACs have high toxicity to algae, since these cationic surfactants have a high affinity to negatively charged algal cell walls (Liang et al. 2013). The dose of alkyl trimethyl ammonium halides and alkyl benzyl dimethyl ammonium halides (for (96 h EC_{50}) on *Chlorella vulgaris* at concentrations from 0.11 to 0.23 mg L^{-1} showed toxic effects, and toxicity was inversely related to the length of the alky chain. The inhibitory effect of CTAB toward *C. vulgaris* was attributed to the fact that with increasing concentrations of CTAB, alga cell activities were inhibited and the uptake efficiency of ammonia nitrogen and total phosphorus was affected (Liang et al. 2013). It was interesting to note that even to the same QACs different species have different sensitivities. The EC_{50} of BAC and DADMAC on algae, daphnids, rotifers and protozoans varied from 21 to 4427 μg/L (Kreuzinger et al. 2007). In another study, the response of two green alga *Chlorella pyrenoidosa* and *Scenedesmus quadricauda* against 13 QACs was tested by analyzing their EC_{50}, and the results indicated that *S. quadricauda* was more tolerant to the presence of most of 13 QACs while *C. pyrenoidosa* was relatively sensitive (Jing et al. 2012).

Aquatic plants are highly sensitive to detergents these effects are highly obvious, specifically in developing countries. The exposure of two aquatic plants, i.e., *Azolla pinnata and Hydrilla verticillate,* to three detergents and their components, sodium dodecyl benzene sulfonate (SDBS), builder sodium tripolyphosphate (STPP) and sodium lauryl sulfate (SLS) showed obvious effects. The results revealed that SLS at concentration of 2–15 mg L^{-1} enhanced the biomass, the number and area of leaves in *A. pinnata,*, but the chlorophyll content of the plant was reduced at all the concentrations. The concentration of 2–5 mg L^{-1} of SLS increased the length of *H. verticillate*. Concentration of SDBS from 1 to 10 mg L^{-1} also increased the growth of *A. pinnata*, but above this concentration the growth was retarded. It was also noted that STPP increases the biomass, the number and the area of leaves. Generally, the SLS have moderate toxicity for the plants, and the STPP and SDBS have the lowest and highest toxicities, respectively (Pandey and Gopal 2010). The exposure *of Spirodela polyrhiza, Pistia stratiotes, Hydrilla verticillata, Ceratophyllum demersum* and *Salvinia molesta* to different concentrations of LAS resulted in the loss of biomass and reduction in total protein and chlorophyll content (Chawla et al. 1989). In another study, the physiological and growth response of the aquatic plant, *Potamogeton perfoliastus* L in the presence of LAS from 0.1 to 50 mg L^{-1} were monitored. A dose of 10 mg L^{-1} of LAS significantly decreased the activities of superoxide dismutase (SOD), catalase (CAT) and the photosynthetic pigments. In addition, at concentration levels of 20–50 mg L^{-1}, a significant decrease in dry weight and fresh weight was noted, confirming the toxic damage of LAS to aquatic plants (Zhou et al. 2018). The ecotoxicological effects of Microcystin-LR (MCLR), LAS and their mixture on growth of *Lemna minor* and its physiological properties showed that the growth and chlorophyll *a* content were significantly reduced after 8 days exposure at concentrations of MCLR (≥ 3 μg mL^{-1}), LAS (≥ 20 μg mL^{-1}) and to the mixture of both was still higher (3 + 10 μg mL^{-1}). It was noted that after 2 days of exposure to MCLR, LAS and to their mixture, superoxide dismutase activity and glutathione content enhanced proportionally with increasing concentrations. As LAS promoted the accumulation of MCLR in duckweed, indicating that the contaminated water with a mixture of MCLR and LAS may have negative ecological risks and higher MCLR phytoremediation potentials (Wang et al. 2012). Another study also confirmed the inhibition and accumulation effects of sodium dodecyl sulfate (SDS) on *Lemna minor* L. The growth of this plant increased from exposure to 1 to 40 mg L^{-1} of SDS and no change was noted when the concentration reached a value of 100 mg L^{-1}, while in 110-diluted Jacob's nutrient media, it was increased up to a concentration of 40 mg L^{-1} and from 40 to 80 mg L^{-1} a sharp inhibition was noted (Forni et al. 2012).

The unique chemical characteristics of detergent components induce changes in chemical and physical parameters of natural water such as salinity, turbidity, temperature and pH. The alterations in chemical and physical parameters ultimately affect the biota including fishes. For example, affect the availability of dissolved oxygen and can also cause damage their gills. The chronic effects of LAS on the growth and survival rate of fish larvae of sea bass resulted in a toxicity level (LC_{50}, 96 h) of 1.8 mg L^{-1} but further showed no effect on the growth, biomass and survival rate of larvae; however, lower concentrations of the lethal doses were mortal but higher concentration also affected the daily growth with an increase in the mortality rate (Rejeki et al. 2008). The adverse effects after exposure of fishes to high concentration of surfactants have been visualized in various organs like gills, liver, kidney, spleen and intestines of fishes. Exposure to high surfactant concentrations also resulted in gill epithelial disruption, causing subsequent asphyxiation, whereas, lower concentrations caused epithelial hyperplasia, oxidative stress and mucus layer damage in fishes (Susmi et al. 2010).

12.7 Conclusions

Detergents are a class of compounds with application in various fields related to human activities. Wastewater from industries and domestic sewage effluents introduce different amounts of detergents into the water bodies that not only affect the wastewater treatment processes but also affect the flora and fauna in the ecosystem. It is therefore essential to develop an accurate analytical procedure for qualitative and quantitative determination of surfactants. These methods may include simple and rapid measurement protocols for total ionic and nonionic surfactant determination in various environmental samples, like spectrophotometric and titrimetric techniques. For more accurate and sensitive determination, more sophisticated methods like gas chromatography and liquids chromatography are coupled with mass spectrometer (MS), thereby, applying proper separations techniques in environmental samples with different matrix composition. It is important to note that some of these surfactants may be persistent under anaerobic conditions such as LAS, while others may be partially degraded under anaerobic conditions. Higher concentrations of surfactants and their biodegradation products may be detected in various water bodies like lakes, streams and rivers that above a certain concentration pose a risk for freshwater biota like aquatic plants, fishes and algae.

References

Aboul-Kassim TA, Simoneit BR (1993) Detergents: a review of the nature, chemistry, and behavior in the aquatic environment. Part I. Chemical composition and analytical techniques. Crit Rev Environ Sci Technol 23(4):325–376

Ahel M, Giger W (1985) Determination of nonionic surfactants of the alkylphenol polyethoxylate type by high-performance liquid chromatography. Anal Chem 57(13):2584–2590

Akyüz M (2007) Ion-pair extraction and GC–MS determination of linear alkylbenzene sulphonates in aqueous environmental samples. Talanta 71(1):471–478

Ali A, Uzair S, Farooq U (2017) Interactions of cationic, anionic and nonionic surfactants with cresol red dye in aqueous solutions: conductometric, tensiometric, and spectroscopic studies. Tenside, Surfactants, Deterg 54(4):342–352

Alsohaimi IH, Khan MR, Ali HM, Azam M, Alammari AM (2020) Solvent extraction and gas chromatography–mass spectrometric determination of probable carcinogen 1, 4-dioxane in cosmetic products. Sci Rep 10(1):1–9

Andreu V, Ferrer E, Rubio JL, Font G, Picó Y (2007) Quantitative determination of octylphenol, nonylphenol, alkylphenol ethoxylates and alcohol ethoxylates by pressurized liquid extraction and liquid chromatography–mass spectrometry in soils treated with sewage sludges. Sci Total Environ 378(1–2):124–129

Andreu V, Picó Y (2005) Liquid chromatography-ion trap-mass spectrometry and its application to determine organic contaminants in the environment and food. Curr Anal Chem 1(3):241–265

Appannagari RR (2017) Environmental pollution causes and consequences: a study. North Asian Int Res J Soc Sci Humanit 3(8):151–161

Bassarab P, Williams D, Dean J, Ludkin E, Perry J (2011) Determination of quaternary ammonium compounds in seawater samples by solid-phase extraction and liquid chromatography–mass spectrometry. J Chromatogr A 1218(5):673–677

Bazel YR, Antal I, Lavra V, Kormosh ZA (2014) Methods for the determination of anionic surfactants. J Anal Chem 69(3):211–236

Bennie DT, Sullivan C, Lee H-B, Peart T, Maguire R (1997) Occurrence of alkylphenols and alkylphenol mono-and diethoxylates in natural waters of the Laurentian Great Lakes basin and the upper St. Lawrence River. Sci Total Environ 193(3):263–275

Bruno F, Curini R, Di Corcia A, Fochi I, Nazzari M, Samperi R (2002) Determination of Surfactants and some of their metabolites in untreated and anaerobically digested sewage sludge by subcritical water extraction followed by liquid chromatography—mass spectrometry. Environ Sci Technol 36(19):4156–4161

Brycki B, Waligórska M, Szulc A (2014) The biodegradation of monomeric and dimeric alkylammonium surfactants. J Hazard Mater 280:797–815

Buxton HT, Kolpin DW (2005) Pharmaceuticals, hormones, and other organic wastewater contaminants in US streams. Water Encycl 5:605–608

Cancela A, Maceiras R, Alfonsin V, Sanchez A (2017) A study on techniques for microalgae separation and lipid extraction for *Desmodesmus subspicatus*. Int J Environ Res 11(3):387–394

Carolei L, Gutz IG (2005) Simultaneous determination of three surfactants and water in shampoo and liquid soap by ATR-FTIR. Talanta 66(1):118–124

Céspedes R, Lacorte S, Ginebreda A, Barceló D (2008) Occurrence and fate of alkylphenols and alkylphenol ethoxylates in sewage treatment plants and impact on receiving waters along the Ter River (Catalonia, NE Spain). Environ Pollut 153(2):384–392

Cha K-O, Lee H-J, Kim Y-I (1992) Synthesis and (lower case) surface activities of betaine amphoteric surfactant having ether bond. J Korean Appl Sci Technol 9(1):63–71

Chaturvedi AD, Tiwari K (2013) Effect of household detergents (surfactants) degraded through aquatic fungi. Recent Res Sci Technol 5(5):12–16

Chawla G, Misra V, Viswanathan P, Devi S (1989) Toxicity of linear alkyl benzene sulphonate on some aquatic plants. Water Air Soil Pollut 43(1):41–51

Coll C, Martínez-Máñez R, Marcos MD, Sancenon F, Soto J (2007) A simple approach for the selective and sensitive colorimetric detection of anionic surfactants in water. Angew Chem Int Ed 46(10):1675–1678

Comas J, Vives-Rego J (1997) Assessment of the effects of gramicidin, formaldehyde, and surfactants on *Escherichia coli* by flow cytometry using nucleic acid and membrane potential dyes. Cytometry. J Int Soc Anal Cytol 29(1):58–64

Cooper LA, Simon J, Wilson DA (1988) Detergent compositions using an aminocarboxylic acid as builder. Google Patents, US4786440A

Cross J (1987) Nonionic surfactants: chemical analysis. https://www.osti.gov/biblio/5354890

Cruceru I, Iancu V, Petre J, Badea IA, Vladescu L (2012) HPLC-FLD determination of 4-nonylphenol and 4-tert-octylphenol in surface water samples. Environ Monit Assess 184 (5):2783–2795

Demirbas A, Edris G, Alalayah WM (2017) Sludge production from municipal wastewater treatment in sewage treatment plant. Energy Sources, Part A 39(10):999–1006

Deng Q, Chen M, Kong L, Zhao X, Guo J, Wen X (2013) Novel coupling of surfactant assisted emulsification dispersive liquid–liquid microextraction with spectrophotometric determination for ultra trace nickel. Spectrochim Acta Part A Mol Biomol Spectrosc 104:64–69

Díaz A, Ventura F, Galceran MT (2002) Simultaneous determination of estrogenic short ethoxy chain nonylphenols and their acidic metabolites in water by an in-sample derivatization/ solid-phase microextraction method. Anal Chem 74(15):3869–3876

Eichhorn P, Flavier ME, Paje ML, Knepper TP (2001) Occurrence and fate of linear and branched alkylbenzenesulfonates and their metabolites in surface waters in the Philippines. Sci Total Environ 269(1–3):75–85

Ekelund R, Granmo A, Magnusson K, Berggren M, Bergman A (1993) Biodegradation of 4-nonylphenol in seawater and sediment. Environmental pollution 79(1):59–61

Ferrer I, Furlong ET (2001) Identification of alkyl dimethylbenzylammonium surfactants in water samples by solid-phase extraction followed by ion trap LC/MS and LC/MS/MS. Environ Sci Technol 35(12):2583–2588

Forni C, Braglia R, Harren F, Cristescu S (2012) Stress responses of duckweed (*Lemna minor* L.) and water velvet (*Azolla filiculoides* Lam.) to anionic surfactant sodium-dodecyl-sulphate (SDS). Aquat Toxicol 110:107–113

Fountoulakis M, Terzakis S, Kalogerakis N, Manios T (2009) Removal of polycyclic aromatic hydrocarbons and linear alkylbenzene sulfonates from domestic wastewater in pilot constructed wetlands and a gravel filter. Ecol Eng 35(12):1702–1709

González-Mazo E, Honing M, Barceló D, Gómez-Parra A (1997) Monitoring long-chain intermediate products from the degradation of linear alkylbenzene sulfonates in the marine environment by solid-phase extraction followed by liquid chromatography/ionspray mass spectrometry. Environ Sci Technol 31(2):504–510

Guzmán E, Llamas S, Fernández-Peña L, Léonforte F, Baghdadli N, Cazeneuve C, Ortega F, Rubio RG, Luengo GS (2020) Effect of a natural amphoteric surfactant in the bulk and adsorption behavior of polyelectrolyte-surfactant mixtures. Colloids Surf, A 585:

Hanif NM, Adnassssn S, Latif M, Zakaria A, Othman M (2012) The composition of surfactants in river water and its influence to the amount of surfactants in drinking water. World Appl Sci J 17(8):970–975

Idouhar M, Tazerouti A (2008) Spectrophotometric determination of cationic surfactants using patent blue V: application to the wastewater industry in Algiers. J Surfactants Deterg 11(4):263–267

Isobe T, Nishiyama H, Nakashima A, Takada H (2001) Distribution and behavior of nonylphenol, octylphenol, and nonylphenol monoethoxylate in Tokyo metropolitan area: their association with aquatic particles and sedimentary distributions. Environ Sci Technol 35(6):1041–1049

Ivanković T, Hrenović J (2010) Surfactants in the environment. Arch Ind Hygiene Toxicol 61 (1):95–110

Jahnke A, Gandrass J, Ruck W (2004) Simultaneous determination of alkylphenol ethoxylates and their biotransformation products by liquid chromatography/electrospray ionisation tandem mass spectrometry. J Chromatogr A 1035(1):115–122

Jiménez-Díaz I, Ballesteros O, Zafra-Gómez A, Crovetto G, Vílchez J, Navalón A, Verge C, De Ferrer J (2010) New sample treatment for the determination of alkylphenols and alkylphenol ethoxylates in agricultural soils. Chemosphere 80(3):248–255

Jing G, Zhou Z, Zhuo J (2012) Quantitative structure–activity relationship (QSAR) study of toxicity of quaternary ammonium compounds on *Chlorella pyrenoidosa* and *Scenedesmus quadricauda*. Chemosphere 86(1):76–82

Kiewiet AT, van der Steen JM, Parsons JR (1995) Trace analysis of ethoxylated nonionic surfactants in samples of influent and effluent of sewage treatment plants by high-performance liquid chromatography. Anal Chem 67(23):4409–4415

König H, Strobel W (1988) Analysis of surfactants in toothpastes by HPLC. Fresenius' Zeitschrift für analytische Chemie 331(3):435–438

Kreuzinger N, Fuerhacker M, Scharf S, Uhl M, Gans O, Grillitsch B (2007) Methodological approach towards the environmental significance of uncharacterized substances—quaternary ammonium compounds as an example. Desalination 215(1–3):209–222

Krueger CJ, Radakovich KM, Sawyer TE, Barber LB, Smith RL, Field JA (1998) Biodegradation of the surfactant linear alkylbenzenesulfonate in sewage-contaminated groundwater: a comparison of column experiments and field tracer tests. Environ Sci Technol 32(24):3954–3961

Kumar A, Chaturvedi AK, Yadav K, Arunkumar K, Malyan SK, Raja P, Kumar R, Khan SA, Yadav KK, Rana KL (2019) Fungal phytoremediation of heavy metal-contaminated resources: current scenario and future prospects. In: Yadav A, Singh S, Mishra S, Gupta A (eds) Recent advancement in white biotechnology through fungi. Springer, Switzerland, pp 437–461

Kurrey R, Deb MK, Shrivas K (2018) Methyl Orange paired microextraction and diffuse reflectance-fourier transform infrared spectral monitoring for improved signal strength of total mixed cationic surfactants. J Surfactants Deterg 21(2):197–208

Lara-Martín PA, Gómez-Parra A, González-Mazo E (2008) Sources, transport and reactivity of anionic and non-ionic surfactants in several aquatic ecosystems in SW Spain: a comparative study. Environ Pollut 156(1):36–45

Lara-Martín PA, González-Mazo E, Brownawell BJ (2011) Multi-residue method for the analysis of synthetic surfactants and their degradation metabolites in aquatic systems by liquid chromatography–time-of-flight-mass spectrometry. J Chromatogr A 1218(30):4799–4807

Larson R, Payne A (1981) Fate of the benzene ring of linear alkylbenzene sulfonate in natural waters. Appl Environ Microbiol 41(3):621–627

Lein NPH, Fujii S, Tanaka S, Nozoe M, Tanaka H (2008) Contamination of perfluorooctane sulfonate (PFOS) and perfluorooctanoate (PFOA) in surface water of the Yodo River basin (Japan). Desalination 226(1–3):338–347

Lewis M (1990) Chronic toxicities of surfactants and detergent builders to algae: a review and risk assessment. Ecotoxicol Environ Saf 20(2):123–140

Li X, Brownawell BJ (2009) Analysis of quaternary ammonium compounds in estuarine sediments by LC—ToF-MS: Very high positive mass defects of alkylamine ions as powerful diagnostic tools for identification and structural elucidation. Anal Chem 81(19):7926–7935

Liang Z, Ge F, Zeng H, Xu Y, Peng F, Wong M (2013) Influence of cetyltrimethyl ammonium bromide on nutrient uptake and cell responses of *Chlorella vulgaris*. Aquat Toxicol 138:81–87

Lunar L, Rubio S, Pérez-Bendito D (2004) Differentiation and quantification of linear alkyl benzenesulfonate isomers by liquid chromatography-ion-trap mass spectrometry. J Chromatogr A 1031(1–2):17–25

Marcomini A, Pojana G, Patrolecco L, Capri S (1998) Determination of nonionic aliphatic and aromatic polyethoxylated surfactants in environmental aqueous samples. Analysis 26(2):64–69

Marcomini A, Zanette M (1996) Chromatographic determination of non-ionic aliphatic surfactants of the alcohol polyethoxylate type in the environment. J Chromatogr A 733(1–2):193–206

Markowski M, Golus W, Polkowska Ż, Kwidzińska M (2017) Surfactants: a real threat to the aquatic geoecosystems of lobelia lakes. Limnol Rev 17(4):183–193

Mata J, Varade D, Bahadur P (2005) Aggregation behavior of quaternary salt based cationic surfactants. Thermochim Acta 428(1–2):147–155

Merino F, Rubio S, Pérez-Bendito D (2003) Solid-phase extraction of amphiphiles based on mixed hemimicelle/admicelle formation: application to the concentration of benzalkonium surfactants in sewage and river water. Anal Chem 75(24):6799–6806

Morelli JJ, Szajer G (2000) Analysis of surfactants: part I. J Surfactants Deterg 3(4):539–552

Motteran F, Nascimento RF, Nadai BM, Titato GM, dos Santos Neto ÁJ, Silva EL, Varesche MBA (2019) Identification of anionic and nonionic surfactant and recalcitrants compounds in commercial laundry wastewater by GC-MS analysis after anaerobic fluidized bed reactor treatment. Water Air Soil Pollut 230(12):1–17

Moulik SP, Haque ME, Jana PK, Das AR (1996) Micellar properties of cationic surfactants in pure and mixed states. J Phys Chem 100(2):701–708

Nagarajan M, Paine H (1984) Water hardness control by detergent builders. J Am Oil Chem Soc 61(9):1475–1478

Nezamzadeh-Ejhieh A, Esmaeilian A (2012) Application of surfactant modified zeolite carbon paste electrode (SMZ-CPE) towards potentiometric determination of sulfate. Microporous Mesoporous Mater 147(1):302–309

Nishiyama N, Toshima Y, Ikeda Y (1995) Biodegradation of alkyltrimethylammonium salts in activated sludge. Chemosphere 30(3):593–603

Norberg J, Thordarson E, Mathiasson L, Jönsson JÅ (2000) Microporous membrane liquid–liquid extraction coupled on-line with normal-phase liquid chromatography for the determination of cationic surfactants in river and waste water. J Chromatogr A 869(1–2):523–529

Olkowska E, Polkowska Ż, Namieśnik J (2012) Analytical procedures for the determination of surfactants in environmental samples. Talanta 88:1–13

Olkowska E, Polkowska Ż, Namieśnik J (2013) A solid phase extraction–ion chromatography with conductivity detection procedure for determining cationic surfactants in surface water samples. Talanta 116:210–216

Olkowska E, Ruman M, Polkowska Ż (2014) Occurrence of surface active agents in the environment. J upper case anal methods chem

Painter H (1992) Anionic surfactants. In: de Oude NT (ed) Detergents. Berlin, Heidelberg Springer, pp 1–88

Pandey P, Gopal B (2010) Effect of detergents on the growth of two aquatic plants: *Azolla pinnata* and *Hydrilla verticillata*. Environ & We: Int J Sci Technol 5:107–114

Pati SG, Arnold WA (2020) Comprehensive screening of quaternary ammonium surfactants and ionic liquids in wastewater effluents and lake sediments. Environ Sci Processes Impacts 22 (2):430–441

Pedrazzani R, Ceretti E, Zerbini I, Casale R, Gozio E, Bertanza G, Gelatti U, Donato F, Feretti D (2012) Biodegradability, toxicity and mutagenicity of detergents: Integrated experimental evaluations. Ecotoxicol Environ Saf 84:274–281

Penteado JC, Bruns RE, De Carvalho LRF (2006) Factorial design optimization of solid phase microextraction conditons for gas chromatography–mass spectrometry (GC–MS) analysis of linear alkylbenzenes (LABs) in detergents. Anal Chim Acta 562(2):152–157

Pérez-Carrera E, León VM, Lara-Martín PA, González-Mazo E (2010) Influence of the hydrophilic moiety of anionic and nonionic surfactants on their aerobic biodegradation in seawater. Sci Total Environ 408(4):922–930

Prats D, Ruiz F, Vázquez B, Rodriguez-Pastor M (1997) Removal of anionic and nonionic surfactants in a wastewater treatment plant with anaerobic digestion. A comparative study. Water Res 31(8):1925–1930

Rejeki S, Desrina D, Mulyana AR (2008) Chronic effects of detergent surfactant (linear alkylbenzene sulfonate/LAS) on the growth and survival rate of sea bass (*Lates calcalifer* Bloch) larvae. J Coast Dev

Rivera-Utrilla J, Bautista-Toledo MI, Sánchez-Polo M, Méndez-Díaz JD (2012) Removal of surfactant dodecylbenzenesulfonate by consecutive use of ozonation and biodegradation. Eng Life Sci 12(1):113–116

Roslan RN, Hanif NM, Othman MR, Azmi WNFW, Yan XX, Ali MM, Mohamed CAR, Latif MT (2010) Surfactants in the sea-surface microlayer and their contribution to atmospheric aerosols around coastal areas of the Malaysian peninsula. Mar Pollut Bull 60(9):1584–1590

Ruiz-Angel M, Carda-Broch S, Berthod A (2006) Ionic liquids versus triethylamine as mobile phase additives in the analysis of β-blockers. J Chromatogr A 1119(1–2):202–208

Safarikova M, Kibrikova I, Ptackova L, Hubka T, Komarek K, Safarik I (2005) Magnetic solid phase extraction of non-ionic surfactants from water. J Magn Magn Mater 293(1):377–381

Santos DKF, Rufino RD, Luna JM, Santos VA, Sarubbo LA (2016) Biosurfactants: multifunctional biomolecules of the 21st century. Int J Mol Sci 17(3):401

Shang D, Ikonomou M, Macdonald R (1999) Quantitative determination of nonylphenol polyethoxylate surfactants in marine sediment using normal-phase liquid chromatography–electrospray mass spectrometry. J Chromatogr A 849(2):467–482
Shu W-C, Ding W-H (2005) Determination of fluorescent whitening agents in laundry detergents and surface waters by solid-phase extraction and ion-pair high-performance liquid chromatography. J Chromatogr A 1088(1–2):218–223
Sueishi T, Morioka T, Kaneko H, Kusaka M, Yagi S, Chikami S (1988) Environmental risk assessment of surfactants: Fate and environmental effects in Lake Biwa basin. Regul Toxicol Pharmacol 8(1):4–21
Sun H, Takata A, Hata N, Kasahara I, Taguchi S (2003) Transportation and fate of cationic surfactant in river water. J Environ Monit 5(6):891–895
Susmi T, Rebello S, Jisha M, Sherief P (2010) Toxic effects of sodium dodecyl sulphate on grass carp *Ctenopharyngodon idella*. Fish Technol 47(2):145
Tanabe A, Kawata K (2008) Determination of 1, 4-dioxane in household detergents and cleaners. J AOAC Int 91(2):439–444
Thiele B, Nollet LM (2007) Analysis of surfactants in samples from the aquatic environment. Handbook of Water Analysis, CRC Press
Thomas O, White G (1989) Metabolic pathway for the biodegradation of sodium dodecyl sulfate by *Pseudomonas* sp. C12B. Biotechnol Appl Biochem 11(3):318–327
Traverso-Soto JM, González-Mazo E, Lara-Martín PA (2012) Analysis of surfactants in environmental samples by chromatographic techniques. In de Azervedo Calderon L (ed) Chromatography. The most versatile method of chemical analysis. IntechOPEN
Tsukamoto M, Murata Y, Tada M, Shimoishi Y (2004) Simple determination of trace amounts of anionic surfactants in river water by spectrophotometry combined with solid-phase extraction. Biosci Biotechnol Biochem 68(4):920–923
Vega-Morales T, Sosa-Ferrera Z, Santana-Rodríguez J (2010) Determination of alkylphenol polyethoxylates, bisphenol-A, 17α-ethynylestradiol and 17β-estradiol and its metabolites in sewage samples by SPE and LC/MS/MS. J Hazard Mater 183(1–3):701–711
Voelkel A, Janas J, Garcia-Dominguez J (1993) Inverse gas chromatography in characterization of surfactants: determination of binary parameter. J Chromatogr A 654(1):135–141
Wang X, Kasperski K (2018) Quantitative determination of multiple surfactants in water using HPLC-MS/MS. Anal Methods 10(21):2512–2525
Wang Z, Xiao B, Song L, Wang C, Zhang J (2012) Responses and toxin bioaccumulation in duckweed (*Lemna minor*) under microcystin-LR, linear alkybenzene sulfonate and their joint stress. J Hazard Mater 229:137–144
Wulf V, Wienand N, Wirtz M, Kling H-W, Gäb S, Schmitz OJ (2010) Analysis of special surfactants by comprehensive two-dimensional gas chromatography coupled to time-of-flight mass spectrometry. J Chromatogr A 1217(5):749–754
Yadav JS, Lawrence DL, Nuck BA, Federle TW, Reddy CA (2001) Biotransformation of linear alkylbenzene sulfonate (LAS) by *Phanerochaete chrysosporium*: oxidation of alkyl side-chain. Biodegradation 12(6):443–453
Yamane M, Toyo T, Inoue K, Sakai T, Kaneko Y, Nishiyama N (2008) Aquatic toxicity and biodegradability of advanced cationic surfactant APA-22 compatible with the aquatic environment. J Oleo Sci 57(10):529–538
Yangxin Y, Jin Z, Bayly AE (2008) Development of surfactants and builders in detergent formulations. Chin J Chem Eng 16(4):517–527
Ying G-G (2006) Fate, behavior and effects of surfactants and their degradation products in the environment. Environ Int 32(3):417–431
Zhao M, Van der Wielen F, De Voogt P (1999) Optimization of a matrix solid-phase dispersion method with sequential clean-up for the determination of alkylphenol ethoxylates in biological tissues. J Chromatogr A 837(1–2):129–138
Zhou J, Wu Z, Yu D, Pang Y, Cai H, Liu Y (2018) Toxicity of linear alkylbenzene sulfonate to aquatic plant *Potamogeton perfoliatus* L. Environ Sci Pollut Res 25(32):32303–32311

Chapter 13
Heavy Metals Pollution in Surface Waters of Pakistan

Azizullah Azizullah, Nadia Taimur, Sarzamin Khan, and Donat-P. Häder

Abstract Fresh water resources in Pakistan are under immense duress due to uncontrolled pollution from industries, municipalities and agriculture runoffs. Microbial and chemical pollutants released into water bodies have intensified with an increase in anthropogenic activities rendering the water unfit for drinking. Of the chemical pollutants heavy metals are the most toxic and persistent contaminants that are ubiquitously disseminated in the freshwater resources throughout the country. Fresh water lakes and rivers are the most vulnerable to contamination with metal pollutants such as arsenic, lead, cadmium, chromium and nickel, emanating from untreated industrial, municipal and domestic waste. Heavy metals in concentrations above permissible limits severely alter water quality and serve as toxins with severe implications for human health. These pollutants are of major environmental concerns due to their persistence in the environment and high potential to bioaccumulate in aquatic organisms, plants, animals and human tissues posing a serious threat to the entire biosphere. The abundant occurrence of heavy metals in aquatic ecosystems is also deleterious for the biodiversity of aquatic organisms causing a severe decline in the population and diversity of freshwater fish and other aquatic species. In aquatic organisms, particularly fish, heavy metals bioaccumulate over a million-fold in metabolically active organs such as gills, liver, skin, kidneys and muscles and consequently become part of the food chain. Frequent consumption of contaminated plant-based food, freshwater fish and other seafood has endangered lives of the consumers. Data extracted from research studies conducted in the last two decades has revealed critical increase in heavy metal pollution in freshwater ecosystems particularly in rivers largely due to unabated human inter-

A. Azizullah (✉)
Department of Biology, The University of Haripur, Haripur, Pakistan

N. Taimur
Department of Botanical and Environmental Sciences, Kohat University of Science and Technology, Kohat, Pakistan

S. Khan
Department of Chemistry, University of Swabi, Swabi, Pakistan

D.-P. Häder
Department of Biology, Friedrich-Alexander Universität, 91054 Erlangen, Germany

© The Author(s), under exclusive license to Springer Nature Switzerland AG 2021
D.-P. Häder et al. (eds.), *Anthropogenic Pollution of Aquatic Ecosystems*,
https://doi.org/10.1007/978-3-030-75602-4_13

ventions. In addition to acute water scarcity, continuous disposal of pollutants in water bodies has further worsened the water crisis in Pakistan. Consequently, the majority of the population dependent on freshwater resources for their livelihood and drinking water needs is deprived of access to reliable, safe and clean drinking water. This chapter summarises the current levels of heavy metal concentrations in freshwater resources i.e. rivers and lakes in Pakistan with special emphasis on the impact of major pollutants and subsequent effects on fish, agriculture and human health.

Keywords Water pollution · Heavy metals · Industrial and agricultural effluents · Bioaccumulation of toxic metals · Contaminated fish and aquatic plants · Polluted lakes and rivers · Heavy metal poisoning · Unsafe drinking water and health hazards

13.1 Introduction

Heavy metal pollution in various segments of the environment, such as water, air, sediments as well as in flora and fauna has become a global concern due to their toxicity and potential risks to the ecosystem. These contaminants are released into the environment via geogenic sources (leaching and weathering of rock, natural alteration of the mineralised zone and volcanic eruptions) and anthropogenic activities (intensive metal mining, smelting, combustion of fossil fuels, use of agrochemicals, domestic and industrial effluents and sewage discharge). Both natural and anthropogenic sources are pathways by which heavy metals contaminate natural water bodies, sediments and soils. Heavy metals are non-biodegradable and persistent environmental pollutants and hence accumulation of these pollutants poses a longstanding and serious threat to the entire biosphere, severely affecting aquatic and terrestrial ecosystems. Heavy metals become part of the food chain via uptake through plants and marine life including fish and aquatic plants which consequently affect consumers such as other fish, birds and mammals including humans. These pollutants subsequently travel to different ecosystems and negatively impact growth physiology of plants as well as cause chronic and epidemic human health hazards due to their persistence, toxicity and bioaccumulation (Dvorak et al. 2020; Kumar et al. 2015; Muhammad et al. 2011; Peralta-Videa et al. 2009; Bradl 2005).

Water pollution caused by increasing concentrations of heavy metals in water resources is one of the major environmental concerns worldwide. Heavy metals are released into water bodies and can exist in several different forms. These include (1) dissolved species in the water, (2) suspended insoluble chemical solids or (3) components of the suspended natural sediments. Dissolved heavy metals are distributed in water as hydrated metal ions or as aqueous metal complexes with other organic or inorganic constituents. In the case of suspended sediments and metallic chemical solids, when the river flow is insufficient, these suspended particles aggregate and form denser particles that are stored in riverbed sediments. Due to their high solubility, the dissolved toxic heavy metals are easily taken up by the organisms and therefore exhibit greater potential to cause harm, however, the dissolved phase is usually found in lower concentrations than the suspended particle phase (Garbarino et al. 1995; Gao et al. 2018; Hamad et al. 2012; Zhang et al. 2014).

According to literature the term "heavy metal" is widely used as a group name for metals and metalloids (assuming that heaviness and toxicity are interrelated) with connotations of environmental contamination, pollution, ecotoxicity and their hazardous impact on the biosphere. Several definitions have been proposed to define the term "heavy metal" (Duffus 2002). The earliest definition of the term could be traced back to Bjerrum's Inorganic Chemistry, 3rd Danish edition and is based on the density of the elemental form of the metal, where metal with an elemental density above 7 g/cm^3 is considered a "heavy metal" (Bjerrum 1936). Another modified definition of the term according to Csuros and Csuros (2016) is "a metal with a density greater than 5 g/cm^3 (i.e. specific gravity greater than 5)". A new broader definition of the term has been recently proposed by Ali and Khan (2018b) which is suggested as "naturally occurring metals having atomic numbers greater than 20 and an elemental density greater than 5 g/cm^3". The term "heavy metal" has been increasingly used interchangeably with terms such as toxic or hazardous metals, trace metals, trace elements, essential metals, non-essential metals, trace minerals or micronutrients, related to their role in relevant biological and environmental processes. This heterogeneous group of elements includes both essential and non-essential heavy metals. The group of essential heavy metals include cobalt (Co), copper (Cu), iron (Fe), manganese (Mn), molybdenum (Mo), nickel (Ni) and zinc (Zn) and are required in low concentrations in various biological processes. These micronutrients or trace metals when present at optimal levels improve the nutritional level as well as normal growth and better yield of plants. However, the excess of these micronutrients has toxic impacts on plant growth by generating oxidative stress and inhibiting enzyme functions thereby compromising structural and functional integrity of the cell (Arif et al. 2016; Ali et al. 2019a). The non-essential metals include lead (Pb), cadmium (Cd), mercury (Hg), arsenic (As), chromium (Cr), silver (Ag) and antimony (Sb). Although biological roles of these elements in plant metabolisms are yet to be identified, however, a number of studies have ascertained their toxicity in both eukaryotic and prokaryotic organisms. The overloading of excess concentrations of these metals in the environment often leads to severe soil and water resources pollution and is of critical global environmental concern (Azizullah et al. 2011; Di Toppi and Gabbrielli 1999).

Ground and surface water constitute important drinking water sources and unfortunately, these are heavily contaminated with heavy metals due to intense anthropogenic activities including widely expanding industrialisation, excessive usage of pesticides and fertilizers and domestic and industrial effluent discharge. Poorly treated or untreated municipal and industrial effluent discharge represents the most significant source of heavy metal pollution in aquatic environments. Hence, the quality and accessibility of safe drinking water is a critical problem in developing countries including Pakistan. The current dismal state of Pakistan's water resources emanates from acute water scarcity and severely polluted water sources, posing a serious health problem. The International Monetary Fund (IMF) in its recently published report placed Pakistan in third place among countries that are suffering from acute water scarcity and this worsening water crisis may cause drought-like situations in the next few years. With regards to water management, Pakistan sits at the bottom of the list of countries (80th out of 122 countries) with adequately managed water quality standards. The alarming water crisis in Pakistan recently gained attention due to speculation that by 2025 the country will face a severe clean water shortage if the crisis is not addressed at the earliest. Water resources in Pakistan are most vulnerable to contamination from industrial waste leading to poor water quality. Anthropogenic activities, especially discharge of untreated industrial and municipal wastewater, are the key contributing factors increasing levels of water pollution in Pakistan (Azizullah et al. 2011; Kumar et al. 2019; Martín et al. 2015; Bhuyan and Islam 2017; Shakir et al. 2016; Sleet 2019; Shukla 2018). This chapter takes into account the current status of heavy metal pollution in surface waters of Pakistan with special emphasis on the impact of major heavy metal pollutants and their concentrations in freshwater bodies, i.e. rivers and lakes and consequent effects on fish, agriculture and human health.

13.2 Heavy Metal Pollution in Pakistan Surface Waters

In Pakistan, surface and ground water (rivers, lakes and reservoirs) constitute the principle source of drinking water for the majority of the population. The quality of these water resources is fast deteriorating as a result of contamination due to discharge of untreated municipal wastewater, excessive use of chemical fertilizers and pesticides in agriculture, effluents from industries such as textile, leather, sugar processing and many more (Tahir et al. 2010). These effluents are then carried by drains and canals to rivers or dumped into lakes causing an increase in water-borne diseases. Anthropogenic activities are the main factors polluting surface water bodies with microbial and chemical pollutants making them unfit for drinking. In Pakistan, heavy metal pollutants are the most toxic and persistent contaminants that are found in increased quantities in lakes and rivers and are responsible for various public health problems. Contamination of lakes and rivers with environmentally hazardous metal pollutants, such as arsenic, chromium, cadmium, lead and nickel, is a serious environmental concern (Azizullah et al. 2011; Waseem et al. 2014; Shakir et al. 2016; Li et al. 2019).

13.2.1 Rivers

River Indus is the main surface water resource of Pakistan and originates from the western Himalayan ranges. Along with five of its major tributaries, i.e. Jhelum, Chenab, Ravi, Sutlej and Beas to the east and the Kabul River to the west, it forms one of the longest and mightiest rivers of Pakistan. Indus River constitutes one of the primary sources of fresh water and forms the backbone of agriculture and food production in Pakistan (Qureshi 2011). Unfortunately, this valuable resource is rigorously contaminated with heavy metals that enter the river through industrial and municipal wastewater discharges, agricultural runoffs as well as urban runoffs (Imran et al. 2019). It is estimated that more than 300 million gallons per day of industrial effluents and domestic sewage discharges are carried by the Lyari and Malir Rivers to the coastal areas and represents a major environmental concern (Hasnie and Qureshi 2002). Once these rivers were home to fish, freshwater and seasonal migratory birds, but unfortunately, the two rivers no longer support ecology and instead are repositories of untreated industrial and domestic effluents (Alamgir et al. 2019). One of the significant contributors of heavy metal pollution in rivers is the relentless discharge of industrial, municipal and agricultural wastewater that poses potential risks to human health. Numerous studies conducted in the last two decades have reported the heavy metal pollution status of the rivers in Pakistan. Data extracted from these findings are presented in Table 13.1.

According to the data extracted from various research studies, concentrations of heavy metals have surpassed safe limits in fresh water sources of Pakistan especially rivers and therefore, have become unfit for human consumption (Azizullah et al. 2011; Shakir et al. 2016; Waseem et al. 2014). The concentrations of these metal pollutants in fresh water sources are found in variable quantities throughout the country. The group of essential heavy metals, i.e. cobalt, copper, iron, manganese, molybdenum, nickel and zinc are required in low concentrations in various biological processes and most of them are usually reported well below the safe limits set by WHO (2011) (Table 13.1). However, heavy metals of primary environmental concern and major pollutants of rivers include cadmium, chromium, lead and arsenic. According to the extensive research conducted in the last decade, the majority of the rivers in Pakistan are heavily polluted due to enhanced heavy metal influx from the surrounding agricultural sectors and industrial estates (Alamgir et al. 2019). The acceptable concentration level of certain metals in drinking water is set to the lowest values indicating the hazardous nature of these metals even at low concentrations. The safe standards for occurrence of cadmium in clean water are set at 0.003 mg/L by the WHO (2011). Mostly, surface water resources in Pakistan are contaminated with toxic pollutants (Nazif et al. 2006; Kashif et al. 2009; Nawab et al. 2018b; Imran et al. 2019). In one of the studies conducted by Midrar-Ul-Haq et al. (2005), higher concentrations of copper, cadmium and nickel were reported in the surface water samples of NWFP and Sindh, due to dumping of untreated industrial and domestic effluents in surface water bodies. According to a recently published research, the concentration of cadmium in river waters of Indus, Chenab

Table 13.1 Heavy metals concentrations (mg/L) in surface water samples of Pakistan

S. no	Sample locations	Cd ($mg\ L^{-1}$)	Co ($mg\ L^{-1}$)	Cr ($mg\ L^{-1}$)	Cu ($mg\ L^{-1}$)	Fe ($mg\ L^{-1}$)	Mn ($mg\ L^{-1}$)	Ni ($mg\ L^{-1}$)	Pb ($mg\ L^{-1}$)	Zn ($mg\ L^{-1}$)	Reference
1*	River Ravi, Punjab	–		–	–	5.58	1.70	1.13	0.68	1.76	Javed and Mahmood (2001)
2*	Palosi drain, Peshawar	0.0–0.004		–	0.0	0.37–0.75	0.017–0.242	0.0–0.18	0.0–0.34	0.0–0.239	Ilyas and Sarwar (2003)
3*	Surface water from different areas in Khyber Pakhtunkhwa	0.002–0.09 (0.02)		0.01–0.12 (0.04)	0.01–0.77 (0.20)	0.01–1.29 (0.19)	0.01–1.11 (0.22)	0.01–1.52 (0.21)	0.02–0.38 (0.16)	0.003–0.08 (0.04)	Midrar-Ul-Haq et al.
4*	Malir River in Karachi	0.002–0.07 (0.04)		0.03–0.29 (0.10)	0.01–0.84 (0.31)	0.13–2.91 (0.78)	0.05–0.57 (0.33)	0.02–1.06 (0.59)	0.09–0.32 (0.19)	0.06–0.29 (0.16)	
5*	Canal water, Akbarpura area, Nowshera, Khyber Pakhtunkhwa	0.09–0.14		0.13–0.17	0.59–0.73	0.90–1.02	0.61–0.71	0.33–0.39	0.34–0.43	0.04–0.05	Nazif et al. (2006)
6*	Bara River, Akbarpura area, Khyber Pakhtunkhwa	0.15–0.20		0.16–0.29	0.90–1.20	1.29–1.75	0.77–0.85	0.53–0.72	0.43–0.62	0.02–0.06	
7*	Kalar Kahar lake, Chakwal, Punjab	0.01–0.05		–	0.01–1.20	0.20–5.46	–	0.04–0.25	0.01–0.30	0.44–2.82	Raza et al. (2007)
8*	MNVD, Sehwan, Jamshoro, Sindh	0.0001–0.018		–	0.003–0.0084	0.0116–0.0185	–	0.0002–0.0052	0.004–0.0096	0.0041–0.0072	Mastoi et al. (2008)
9*	Manchar Lake Jamshoro, Sindh	0.0001–0.002		–	0.0006–0.0196	0.0073–0.0178	–	0.0004–0.0096	0.0057–0.014	0.0046–0.0348	Mastoi et al. (2008)
10*	Hudaira drain, Lahore	0.18		0.07	0.45	7–9	0.85	0.93	0.03	1.7	Kashif et al. (2009)
11*	Manchar Lake Jamshoro, Sindh	–		–	–	–	–	–	–	–	Arain et al. (2009)

(continued)

Table 13.1 (continued)

S. no	Sample locations	Cd ($mg\ L^{-1}$)	Co ($mg\ L^{-1}$)	Cr ($mg\ L^{-1}$)	Cu ($mg\ L^{-1}$)	Fe ($mg\ L^{-1}$)	Mn ($mg\ L^{-1}$)	Ni ($mg\ L^{-1}$)	Pb ($mg\ L^{-1}$)	Zn ($mg\ L^{-1}$)	Reference
12*	River Kabul, Khyber Pakhtunkhwa	0.03		0.02	0.08	–	0.18	0.65	0.09	0.2	Khan et al. (2011)
13*	Surface water from (Jijaldubair, Bisham and Alpuri) Kohistan	0.0006 0.0008 0.00013		0.007 0.0009 0.0028	0.115 0.021 0.031	– – –	0.013 0.18 0.10	0.0046 0.0028 0.0059	0.0036 0.008 0.0034	0.040 0.040 0.021	Muhammad et al. (2011)
14	Mohmand agency, Khyber Pakhtunkhwa	0.001–0.002	0.017–0.052	0.190–0.377	0.052–0.091	0.125–0.225	0.023–0.051	0.032–0.141	0.001–0.008	0.004–0.068	Shah et al. (2012)
15	Rawal Lake, Islamabad	0.001–0.042	0.001–0.415	0.001–0.274	0.001–0.051	0.001–0.451	0.001–0.041		0.001–0.663	0.001–0.066	Iqbal et al. (2013)
16	Swat River watershed, Khyber Pakhtunkhwa	0.007–0.017		0.215–0.595	0.003–0.016		0.120–0.152	0.039–0.069		0.014–0.026	Khan et al. (2013b)
17*	Water of River Kabul at Peshawar	0.03		0.5	0.07	–	–	0.2	0.52	–	Ullah et al. (2013)
18	Khanpur Lake, Khanpur, Khyber Pakhtunkhwa	0.003–0.049	0.042–0.181	0.004–0.107	0.003–0.027	0.005–0.121	0.002–0.028		0.095–0.408	0.005–0.035	Iqbal and Shah (2013)
19	Simly Lake, Rawalpindi, Punjab	0.002–0.038	0.006–0.306	0.004–0.209	0.003–0.060	0.003–0.163	0.002–0.036		0.009–0.557	0.004–0.049	Iqbal and Shah (2013)
20	Mangla Lake, Mirpur, Azad Kashmir	<0.01–0.08	0.01–0.50	0.01–0.21	<0.01–0.06	0.02–0.33	<0.01–0.06	0.01–0.42	0.02–1.5	<0.01–0.08	Saleem et al. (2014)
21	Samples collected during pre and post-monsoon at River Soan, Punjab	0.001–0.101	0.01–0.16	0.01–0.1	0.003–0.23	0.01–0.37	0.01–0.15	0.06–0.42	0.05–1.03	0.002–0.083	Nazeer et al. (2014)
22	Spring at Peshawar District, Khyber Pakhtunkhwa	0.0283–0.0384	0.125–0.0515	0.0092–0.0126				0.0072–0.0102	<0.0015	0.032–0.1185	Khan et al. (2016)

(continued)

Table 13.1 (continued)

S. no	Sample locations	Cd (mg L^{-1})	Co (mg L^{-1})	Cr (mg L^{-1})	Cu (mg L^{-1})	Fe (mg L^{-1})	Mn (mg L^{-1})	Ni (mg L^{-1})	Pb (mg L^{-1})	Zn (mg L^{-1})	Reference
23	Taunsa barrage, Indus River, Punjab			0.032–0.072	0.149–0.271	1.273–1.98		0.126–0.229	0.082–0.251	0.194–0.293	Iqbal et al. (2017)
24	Upper River Indus	0.007–0.186 (0.058)		0.048–0.102 (0.072)	.002–0.026 (0.010)			0.086–0.476 (0.237)	0.003–0.127 (0.089)	0.082–0.301 (0.124)	Nawab et al. (2018b)
25	Lower River Indus	0.015–0.088 (0.037)		0.064–0.193 (0.116)	0.012–0.188 (0.084)			0.098–0.275 (0.159)	0.043–0.349 (0.194)	0.065–0.876 (0.303)	
26	River Chenab	0.009–0.095 (0.068)		0.078–0.342 (0.187)	0.080–0.772 (0.341)			0.054–0.521 (0.305)	0.006–1.257 (0.296)	0.084–0.698 (0.397)	
27	River Kabul	0.002–0.085 (0.032)		0.0046–0.021 (0.013)	0.005–0.028 (0.009)			0.001–0.065 (0.039)	0.008–0.143 (0.069)	0.05–0.370 (0.143)	
28	Muzaffarabad, Azad Jammu and Kashmir			0.032–0.363	0.038–0.211	0.096–0.718	0.013–0.246		0.029–0.665	0.058–0.405	Ali et al. (2019b)
29	Samples collected during pre-monsoon at Indus River, Kotri barrage, lower Sindh	0.00–0.19			1.24–4.23	0.57–3.52	0.03–0.59		0.01–0.03	0.48–4.02	Imran et al. (2019)
30	Zhob River, Zhob valley, Baluchistan	0.002–0.01 (0.005)	0.04–0.6 (0.15)	0.03–0.13 (0.08)	0.04–1.5 (0.85)	0.27–3.5 (1.88)	0.08–0.6 (0.30)	0.02–0.07 (0.05)	0.01–0.04 (0.02)	0.49–1.3 (0.99)	Ullah et al. (2019)
31	Loralai River, Loralai valley, Baluchistan	0.0003–0.003 (0.002)	0.01–0.06 (0.02)	0.02–0.05 (0.03)	0.04–0.1 (0.05)	0.06–0.3 (0.16)	0.01–0.14 (0.07)	0.01–0.03 (0.02)	0.01–0.03 (0.02)	0.1–1.2 (0.7)	

(continued)

Table 13.1 (continued)

S. no	Sample locations	Cd (mg L^{-1})	Co (mg L^{-1})	Cr (mg L^{-1})	Cu (mg L^{-1})	Fe (mg L^{-1})	Mn (mg L^{-1})	Ni (mg L^{-1})	Pb (mg L^{-1})	Zn (mg L^{-1})	Reference
32	Surface water samples collected from Chenab River and canals Faisalabad, Pakistan			0.11–0.4		0.32–1.89					Mahfooz et al. (2019)
33	Shahpur Dam, Fateh Jang				1.63	1.52		1.19	8.15		Khalil et al. (2020)
34	Hunza River and its tributaries in Wakhan Block (upstream), Gilgit–Baltistan	0.00156	0.00189	0.00444	0.00556	0.0552		0.00578	0.00333	0.0638	Muhammad and Ahmad (2020)
35	Hunza River and its tributaries in Karakoram Block (midstream), Gilgit–Baltistan	0.00102	0.00245	0.00548	0.00529	0.0669		0.00610	0.00448	0.0686	
36	Hunza River and its tributaries in Kohistan Block (downstream), Gilgit–Baltistan	0.00158	0.00253	0.00156	0.00656	0.0868		0.0217	0.00332	0.0604	
	National standards for drinking water quality, Pakistan	0.01		0.05	2.0	–	0.5	0.02	0.05	5	NSDWQ-Pak (2008)
	WHO standards	0.003		0.05	2.0	–	0.50	0.07	0.01	3	WHO (2011)

Data are extracted from various individual studies and arranged chronologically based on the year of publication of the reviewed articles. Values given represent the mean values or the range from minimum to maximum and where both range and mean are given, the value in parenthesis is the mean value. Samples marked with an asterisk have been reviewed by Shakir et al. (2016)

and Kabul exceeded permissible drinking water standards by manyfold. The highest concentrations of the metal were found in the upper Indus (0.186 mg/L) followed by Chenab (0.095 mg/L), lower Indus (0.088 mg/L) and Kabul (0.085 mg/L). The contamination of the rivers mostly originates from the release of industrial and domestic effluents and agricultural practices along the banks of these rivers (Nawab et al. 2018b). Similarly, significantly higher concentrations of cadmium (0.19 mg/L) were reported from water samples collected from Indus River at Kotri barrage during the monsoon season. The concentrations of the metal clearly violated guidelines of the World Health Organization (WHO) and posed a serious health risk for the surrounding communities (Imran et al. 2019). Chromium and lead are a duo of hazardous water pollutants and are the most frequently reported of all the heavy metals. Multiple research studies have revealed higher concentrations of the two heavy metals in the rivers of the country (Table 13.1). In River Swat, Kabul, Lower Indus, Chenab and Zhob contamination of chromium was found in the range of 0.215–0.595 mg/L, 0.5 mg/L, 0.064–0.193 mg/L, 0.078–0.342 mg/L and 0.03–0.13 mg/L, respectively (Khan et al. 2013b; Ullah et al. 2013, 2019; Nawab et al. 2018b). Elevated levels of the metal were also found in water samples collected from Chenab River and canals in Faisalabad to identify water quality for drinking and irrigation purposes. The concentration raged between 0.11 and 0.4 mg/L (Mahfooz et al. 2019). In another study, surface water samples collected from household taps and natural spring sources in Muzaffarabad, Azad Kashmir contained significantly higher concentrations of chromium ranging from 0.032 to 0.363 mg/L (Ali et al. 2019b). All these concentrations clearly surpassed the permissible limit of 0.05 mg/L set by the World Health Organisation as standard for clean drinking water. Lead is another commonly found surface water contaminant that enters the environmental compartments largely because of anthropogenic activities. A multitude of recently conducted studies have found lead in significantly higher concentrations exceeding the safe limits in almost all instances (Nazeer et al. 2014; Iqbal et al. 2017; Nawab et al. 2018b; Ali et al. 2019b). Data from a recent analysis of water quality in Shahpur Dam, Fateh Jang showed enormously high magnitudes of lead in water exhibiting mean concentration of 8.15 mg/L (Khalil et al. 2020). Nawab et al. (2018b) also reported higher concentrations of lead in river water of Indus, Chenab and Kabul exceeding permissible limits for drinking water. The highest concentrations of lead found in water of the upper Indus, lower Indus, Chenab and Kabul were 0.127, 0.349, 1.257 and 0.143mg/L, respectively. Similarly, analysis of water samples collected from River Soan and Taunsa barrage and River Indus in Punjab province exhibited the highest concentrations of 1.03 and 0.251 mg/L, respectively, thereby exceeding the recommended level for lead in drinking water (Iqbal et al. 2017; Nazeer et al. 2014). Contamination by lead is also reported from water sources such as household taps and natural springs in Muzaffarabad, Azad Kashmir with significantly higher concentrations of lead ranging from 0.029 to 0.665 mg/L (Ali et al. 2019b). Arsenic and nickel are also frequently found as major pollutants of the rivers. The maximum permissible concentration for arsenic and nickel set by (WHO 2011) in drinking water is 10 μg/L and 0.07 mg/L, respectively. Arsenic is one of the hazardous naturally found elements and poses a high risk of carcinogenicity in adults. In Pakistan, arsenic contamination is mostly reported from areas in Punjab and Sindh where much higher arsenic concentration levels—i.e. over 50 μg/L—are found in surface and ground water samples (Ahmad et al. 2004). A recent report about contamination by arsenic in Chenab River and associated canals in Faisalabad has revealed arsenic contamination of 80 μg/L which is well above the safe limits for drinking water (Mahfooz et al. 2019). Elevated levels of essential trace metals such as nickel have also been reported in water supplies. Nickel, a silvery-white lustrous metal, is an essential heavy metal required in small quantities for various biological and biochemical processes. However, the occurrence of elevated levels of nickel in the ecological compartments is posing a serious threat to organisms. Both natural and anthropogenic activities contribute to the emission of nickel in air, soil and water. Main contributors of the pollutant include fossil fuel consumption, industrial production such as mining, smelting, refining, disposal of nickel compounds, steel production, other nickel-containing alloy production and municipal incineration (IARC 2012). The metal has been frequently reported in surface water resources in amounts surpassing the safe limits for drinking water (Table 13.1). Most of the recent investigations of river water quality analysis summarised in Table 13.1 have reported higher concentrations of essential metals in rivers. A recent analysis of water quality from Chenab River, Faisalabad revealed nickel concentrations as high as 1.19 mg/L indicating the unsuitability of the water for drinking and agricultural purposes (Khalil et al. 2020). Toxic and unsafe concentrations have also been reported from other areas in Pakistan such as River Soan (0.42 mg/L) (Nazeer et al. 2014), Taunsa barrage, Indus River (0.229 mg/L) (Iqbal et al. 2017), upper River Indus (0.476 mg/L), lower River Indus (0.275 mg/L) and River Chenab (0.521 mg/L) (Nawab et al. 2018b). The alarming increase in these toxic pollutants have serious implications for aquatic and human life and is reported to have drastic impacts on the genetic variability and biodiversity of aquatic life (Araújo da Silva et al. 2019).

13.2.2 Lakes

Lakes are natural or manmade wetlands created by underground sources of water and or rainfalls. Lakes are a valuable drinking water resource that provides diverse ecosystem services such as provision of habitat for fauna and flora, important source of food supply, as well as aesthetic and economic benefits such as amusement and tourism opportunities, agriculture and industry (Chaudhry 2010; Khan and Arshad 2014). Lakes regarded as the most precious ecosystems are facing continuous threats due to natural and rigorous anthropogenic activities. Toxic metal pollutants that enter lakes can have severe implications for animal and human health. Not only in Pakistan but all over the world, heavy metal pollutants such as arsenic, lead, chromium, nickel, etc. are of major concern due to their persistence in the environment and their high potential to bioaccumulate in aquatic organisms and human tissues (Shafiq et al. 2011; Chen et al. 2019; Azizullah et al. 2011; Shakir et al. 2016; Waseem et al. 2014; Ali and Khan 2018a).

Multiple research studies have reported the common occurrence of heavy metals in lakes and rivers of Pakistan (Ali and Khan 2018a; Korai et al. 2010; Shafiq et al. 2011; Imran et al. 2020; Arain et al. 2009). Arsenic is one of the most potent pollutants found in drinking water reservoirs, lakes and rivers and is classified as a human carcinogen (WHO 1981). In many regions of Pakistan, the concentrations of arsenic in drinking water exceeding the WHO standard of 10 μg/L are well documented. A pioneering study to assess the toxicity of heavy metals was conducted by the Pakistan Council of Research in Water Resources (PCRWR) and the United Nations Children Fund (UNICEF) in 2000 that highlighted the hazard of arsenic to human health via drinking water consumption. A number of water samples collected from Punjab and Sindh were analysed by PCRWR and reported elevated concentrations of arsenic in more than 50% of the samples, exceeding WHO standards for drinking water quality (PCRWR 2008). In 2009 enormously high concentrations of arsenic were found in water samples collected from Manchar Lake, Sindh. Arsenic concentration in lake water ranged between 35 and 157 μg/L which is significantly higher than the permissible limits of WHO, whereas in the lake sediments concentrations were found in the range of 11.3–55.8 mg/kg (Arain et al. 2009). In a recent study, Keenjhar Lake water quality was investigated, and the data showed that the arsenic concentration of 38 μg/L exceeded WHO or Pakistan safe limits for unrestricted use (Imran et al. 2020). Similarly, elevated concentrations of arsenic were found in surface sediments of Namal Lake revealing that the lake is endangered by pollution from domestic waste and agricultural runoff (Javed et al. 2018).

Chromium is another environmental pollutant, commonly found in areas with high anthropogenic pressure. Chromium concentrations reported in lake water are low in the country but still above the permissible level set by WHO (2011) and NSDWQ-Pak (2008). An investigation to assess water quality of Rawal Lake revealed elevated levels of 0.274 mg/L of the heavy metal, indicating unsuitability of the water for drinking purposes (Iqbal et al. 2013). Furthermore, analysis of chromium distribution in Khanpur and Simly Lakes revealed many times higher amounts of the metal when compared with the permissible levels. Maximum concentrations of 0.107 and 0.209 mg/L were measured in water samples collected from Khanpur and Simly Lake, respectively (Iqbal and Shah 2013). In a similar study, maximum concentration of chromium (0.21 mg/L in summers and 0.19 mg/L in winters) reported in water samples collected from Mangla Lake were higher than the maximum permitted concentrations established by WHO (Saleem et al. 2014). Lead is a toxic metal pollutant that is frequently found in air, soil and water. In Pakistan, most of the water samples analysed from a variety of water sources surpass the safe standard of 0.01 mg/L set by WHO. The maximum concentration of lead is reported from 0.3 to 1.5 mg/L in water samples collected from different lakes in Pakistan and clearly exceeds the safe limits for drinking water set by WHO (Table 13.1). An analysis of cadmium concentration in drinking water from different lakes in Pakistan revealed that maximum concentrations of the toxic pollutant ranged from 0.05 to 0.049 mg/L, manyfold higher than the WHO limits (Iqbal and Shah 2013; Iqbal et al. 2013; Saleem et al. 2014). In case of the toxicity assessment of other heavy metals such as cobalt, copper, iron, zinc, nickel and manganese, the concentrations were found well below acceptable levels set by NSDWQ-Pak (2008) and WHO (2011) and therefore does not pose any serious health problem. In addition, the essential heavy metals are also important micronutrients that are required in optimum concentration by plants, animals and humans for various biochemical, biological and metabolic processes. Lakes are a source of livelihood as well as food and water for local populations and therefore regular monitoring of the metal pollutants is essential especially in Pakistan where devising and implementing environmental policies are challenging tasks.

13.3 Heavy Metals in Fish of Pakistan

Occurrence of high levels of heavy metals adversely affects the aquatic ecosystems and the associated biota and can structurally modify aquatic life, especially fish contributing to the decline in their population. Heavy metals in concentrations above permissible limits severely alter water quality and serve as toxins, bioaccumulating over a million-fold in the tissues of aquatic organisms, especially fish and negatively impact quality and quantity of fish populations (Shah et al. 2012; Chang et al. 1998; Malik and Maurya 2014; Luo et al. 2014). The accumulation of heavy metals in organism is either passive or selective. Differences in patterns of assimilation or egestion of organism as well as environmental metal concentration and time of exposure are factors that influence heavy metal accumulation in various fish organs such as gills, skin, kidney, liver and gut. The concentration of heavy metals that accumulate in fish organs is million-fold higher than the toxic concentration in their environment (Annabi et al. 2013; Egila and Daniel 2011). Fish acquire heavy metals from the aquatic environment primarily through skin and gills. Gills serve as the primary entry points of the waterborne heavy metals as these organs are constantly submersed in water dissolved with heavy metal ions and hence serve as major target of metal toxicity. The metal uptake via gills is enhanced as large volumes of the water pass over the gills at any time (Olsson et al. 1998). In retrospect, fish come in contact with heavy metals in four ways. These include consuming contaminated food, via direct water uptake through gills, consumption of non-edible particles and via skin absorption. After absorption, the heavy metals enter the bloodstream and then are carried to different fish organs for storage. The accumulation of heavy metals in fish organs varies considerably and mostly depend upon the characteristics of fish and external environmental characteristics such as physicochemical properties of water and climatic conditions (Youssef and Tayel 2004; Nussey 2000; Mansouri et al. 2012).

Generally, heavy metals accumulate in large quantities in metabolically active organs such as gills, liver and kidneys compared with skin and muscles evidently due to induction or occurrence of metal-binding proteins in these organs; especially the gills and liver are the main sites of metallothionein (MT) production and metal retention. These metallothionein proteins are used as useful biomarkers for identification of early contamination through increasing levels of these proteins in target tissues. For example, increase in production of metallothionein in target tissues such as the liver indicate the bioavailability of the toxicants in the aquatic environment and hence could be used as useful tool for toxicological studies (Mansouri et al. 2012; Nagamatsu et al. 2020; Doering et al. 2015). Fish accumulate lower levels of metals in muscle tissues compared with gills, liver, skin, kidneys, etc. (Jezierska and Witeska 2006). Mostly research investigations have focussed on assessing metal concentrations in fish muscles as these measures are indicative of metal toxicity transferred to humans via consumption of the contaminated fish and the consequent effect on human health. Furthermore, assessing the heavy metals content of fish tissues may also help identify the toxic metal pollution status of the aquatic ecosystems. Hence, fish may act as bioindicators for monitoring heavy metal pollution in aquatic environments (Widianarko et al. 2000).

Pakistan is blessed with a number of rivers and lakes which represent the main surface water resources that provide drinking water to a large population of the country. The river system in Pakistan originates from the western snow-covered Himalayan and the Karakoram ranges and comprises the longest and mightiest river, River Indus and its tributaries. Five major tributaries joining its eastern side are Jhelum, Chenab, Ravi, Beas and Sutlej, and joining it towards the western side is its tributary, the River Kabul (Kahlown and Majeed 2003; Khan and Arshad 2014). In the last decade, enormous increase in population growth, rapid urbanisation and industrialisation have resulted in the production of huge quantities of solid and liquid waste including emission of heavy metals in various ecological compartments that drastically deteriorated water quality and threatened aquatic life and human health. Unfortunately, dumping of industrial and domestic effluents ultimately contaminate these water resources with huge quantities of heavy metals and become a hazard for the biota including fish (Hashem et al. 2017; Spiegel 2002; Merian 1984; Mishra et al. 2019). According to reports, one of the consequences of contamination of River Indus with heavy metals is the continuous decline in the population and diversity of freshwater fish and other aquatic species. Untreated industrial effluent discharge in water bodies not only contributes to aquatic pollution but also results in the depletion of oxygen content killing huge numbers of aquatic organisms, particularly fish (Al-Ghanim et al. 2016; Qureshi et al. 2015). In the recent past, several researchers in Pakistan have reported alarmingly high levels of heavy metal contaminations in fish species found in lakes and rivers (Table 13.2). In a study, a detailed analysis was carried out to quantify the seasonal variation in the concentrations of the heavy metals in *Cyprinus carpio*, a sediment-dwelling omnivore from Rawal Lake, Pakistan. In the summer season, the highest lead (8.796 μg/g) contamination was found in muscle tissue of the common carp, whereas relatively higher concentrations of cadmium, copper, chromium, cobalt, iron and zinc were found in stomach samples suggesting direct dietary exposure (Iqbal and Shah 2014). In a similar study by Yousafzai et al. (2012), heavy metal residues were measured in *C. carpio* body tissues from River Kabul,

Nowshera. The analysis revealed that elevated levels of heavy metals exhibited the following descending pattern: intestine > skin > liver > gills > muscle. Minimal concentrations of the pollutants were found in muscle tissues whereas highest metal accumulation occurred in the intestine suggesting dietary sources as main contributors of higher levels of the metal pollutants. Despite lower levels in muscle tissue, the concentrations of nickel, cadmium and lead surpassed the US recommended dietary allowance (RDA) limits, posing health risk to fish consumers (Yousafzai et al. 2012). Heavy metal concentrations were also reported in muscle tissue of *Wallago attu* fish species collected from two sites from River Kabul. Mean concentrations of nickel (91.3 and 112.7 μg/g), chromium (446.0 and 539.3 μg/g), cadmium (60.3 and 74.7 μg/g), lead (509.7 and 605.0 μg/g) and mercury (74.0 and 94.0 μg/g) exceeded US recommended dietary allowance (RDA) limits rendering the fish unsafe for human consumption (Siraj et al. 2016). In a similar study by Ahmed et al. (2016), heavy metal concentrations were evaluated in well-known and most consumed fish species from Tarbela Lake and River Swat in the Khyber Pakhtunkhwa region (Table 13.2). Exposure to metal pollutants in the surrounding water resulted in metal accumulation in fish species that exceeded the threshold level as described in international standards (WHO 1996). Rapid industrialisation has accelerated contamination of the rivers in Pakistan. Barandu River in the district Buner is another water resource that has been heavily contaminated with effluents from marble industries situated in the vicinity of the river. Increased bioaccumulation of heavy metals in the fish species *Tor putitora* above permissible limits can be attributed to the higher concentration of heavy metals in the water of Barandu River. The presence of heavy metals in various fish tissues adversely affects fish growth and consequently becomes unfit for consumption by humans (Mulk et al. 2017). In another study, heavy metals bioaccumulation was assessed in livers, gills and muscles of *C. carpio* and *Labeo rohita* sampled from Sardaryab, River Kabul. The comparison of the two fish species revealed that chromium and iron were abundantly found in gills and livers of both species. Mean concentration of chromium in gills, livers and muscles of *C. carpio* were recorded as 0.154, 0.188 and 0.024 μg/g, respectively. Similarly, chromium levels in *L. rohita* were found in the range of 0.133 μg/g in gills, 0.165 μg/g in liver and 0.019 μg/g in muscle tissue. The concentration of iron in carp was in the range of 0.086 μg/g in gills and 0.067 μg/g in muscles. *C. carpio* being omnivorous and a bottom feeder stored higher concentrations of metals as compared *to L. rohita*. The concentration values of heavy metals in both fish ranged in the recommended daily dietary allowances (RDA) limits and therefore, considered safe for human consumption (Yousafzai et al. 2017). Unfortunately, most rivers are contaminated with industrial and domestic effluents, and consequently the inhabiting biota contains higher deposits of metal pollutants in their tissues. One such example is River Chenab, where the increase in anthropogenic activities and hyperaccumulation of heavy metal pollutants in different fish organs has emerged as a serious public health hazard. According to Alamdar et al. (2017) levels of various pollutants found in fish muscles in descending order include zinc (35.5–54.4), copper (1.38–4.57), manganese (2.43–4.5), arsenic (0.23–1.21), chromium(0.21–0.67), nickel (0.14–0.34), lead (0.14–0.31), cobalt (0.09–0.12) and cadmium (0.07–0.12) (μg/g). Herbivore fish species i.e. *Cirrhinus reba* and *Catla catla* exhibited higher accumulations of toxic metal pollutants as compared to carnivorous species. The concentration of arsenic frequently exceeded the FAO/WHO expert committee on food additives permissible limits (JECFA 2011) and emerged as major pollutant posing chronic health risk to local inhabitants. A similar investigation was carried out to assess heavy metal pollution in River Kabul (Ali and Khan 2018a). Various concentrations of heavy metals i.e. chromium, nickel, cadmium and lead found in the analysed fish species ranged from 8.0 to 40.2, 19.8 to 129.0, 0.9 to 2.0 and 7.5 to 41.0 μg/g, respectively (Table 13.2). The data suggested that nickel accumulation in freshwater fish presented a potential health risk, whereas cadmium and lead concentrations were within safe limits. Major culprits contaminating the river include industrial effluents, domestic sewage and agricultural runoff, further deteriorating its water quality (Yousafzai et al. 2010). In a country-wide survey, higher concentrations of heavy metals were found in fish samples collected from major rivers (Chenab, Upper Indus, Lower Indus and Kabul) across Pakistan (Nawab et al. 2018b). Levels of pollutants in Chenab River were highest, followed by the Lower and Upper Indus and then River Kabul. Similarly, fish species collected from Chenab River contained higher concentrations of heavy metals compared with fish from other rivers. Arsenic was the major pollutant found in tissues of the fish from Indus and Chenab rivers. Urbanisation and industrialisation of cities situated along the banks of these rivers are the major factors that contribute to the heavy metal pollution of these rivers and ultimately contamination of fish. River Kabul and its tributary River Shah Alam are the main water resources for irrigation and freshwater fish species and provide a means of livelihood for communities living along the bank. These rivers receive industrial effluents, domestic sewage and agricultural run-off from the industrial cities Peshawar, Charsadda, Nowshera and the surrounding areas (Ahmad et al. 2015). Recently a comprehensive study has been reported by Ali et al. (2020) aimed at estimating the occurrence of chromium, nickel, cadmium and lead concentrations in the Shah Alam River fish species. The analysis revealed that the quantity of metal pollutants varied in muscle samples of the six fish species. Chromium concentrations in fish muscles ranged from 30.5 to 70.8 μg/g, nickel from 16.7 to 103.5 μg/g, cadmium from 1.1 to 2.5 μg/g and lead from 29.7 to 97.7 μg/g. The alarmingly high heavy metal load in muscle tissues reflect an unparalleled rise in contamination of the rivers with metal pollutants contributed mostly by anthropogenic activities. Apart from rivers, fish residing in farms and dams are also prone to heavy metal contamination. The commonly cultured and consumed *C. carpio* from Shahpur Dam, Fateh Jang, was used as an experimental model to assess metal toxicity in various tissues. Of all the heavy metals, lead was the most accumulated metal in fish tissues, i.e. gills, liver, muscles and kidneys followed by other metals in the order; lead > nickel > iron > copper > chromium. Lead was significantly high in all tissues ranging from 8.9 to 21.7 μg/g. The concentrations of copper and chromium ranged from 13.6 to 120.7 and 0.2–1.6 μg/g and were abundant in gills and liver. Iron ranged from 58.2 to 162.7 μg/g and was slightly elevated in the liver of the fish, whereas nickel was within permissible limits in almost all organs and therefore less harmful (Khalil et al. 2020). In another study reported by Muhammad and Ahmad (2020) a heavy metal load was detected in water and fish of the Hunza River and its tributaries in Gilgit-Baltistan. The analysis revealed that muscle tissue of fish *Salmo trutta* contained relatively higher concentrations of iron, nickel, zinc and cadmium compared with farm-bred fish. However, heavy metal concentrations were within safe limits set by FAO (1983). A summary of multiple studies conducted to analyse bioaccumulation of major heavy metal pollutants in fish tissues found in lakes, rivers and dams in Pakistan is presented in Table 13.2. The occurrence of high levels of metal toxicity in fish tissues is a major environmental and health concern and in hindsight calls for implementation of environmental laws and biomonitoring programs to remediate the fast-growing environmental crisis of heavy metal pollution in surface and ground water sources of Pakistan.

13.4 Sources of Heavy Metal Pollution in Pakistan

Environmental pollution is a debilitating ecological and global public health issue aggravated by rapid population growth, unplanned urban and industrial expansion, inadequate waste management and enormous increases in resource consumption. The magnitude of pollution has increased manyfold in developing countries like Pakistan where untreated industrial and domestic effluents continuously churn out toxic heavy metals in air, soil and water resources (Briggs 2003; Junaid et al. 2017; Ali et al. 2015; Tchounwou et al. 2012).

Various sources contribute to the contamination of heavy metals in the environment and can be categorised as anthropogenic and geogenic (particularly lithogenic) in origin. The principle natural sources of heavy metals in the environment include magmatic, sedimentary and metamorphic rocks, weathering of metal-bearing rocks and soil formation. Magma, a molten rock material that originates from the Earth's crust, is a source of a variety of elements including heavy metals. The cooling and solidification of magma results in the formation of crystalline primary rocks that contains heavy metals incorporated into their crystal lattices. Some of the notable examples include chromium and nickel that are abundantly found in the mineral ores chromite and forsterite, respectively (Bradl 2005). The weathering of mafic and ultramafic rocks releases heavy metals such as copper, lead, zinc, chromium, nickel and cadmium in soils. Heavy metals accumulated in these soils are ultimately taken up by the plants growing upon these soils (Shah et al. 2010). Aerosols produced during oceanic activities and forest fires are other contributors of heavy metals in soil and water bodies. Another natural source is the accumulated metal load in various plant organs that are released back into the environment via decomposition or leaching from leaves and stems (Nagajyoti et al. 2010).

Apart from geogenic sources, most environmental contamination occurs due to anthropogenic activities such as mining and smelting operations, coal burning in power plants, nuclear power stations, industries (textile, pharmaceuticals, petrochemicals, ceramics, oil mills, sugar industries, fertilizer factories), domestic effluents, sewage sludge and agricultural applications of chemicals containing metals and metal-containing compounds (Goyer 1993; He et al. 2005; Rehman et al. 2008). Of these, various industrial sectors located throughout Pakistan are the main contributors to heavy metal pollution in the environment. Untreated industrial wastes are dumped into surface water bodies causing an alarming increase in levels of water pollution. In a race towards industrialisation, new industries are being established. However, there is no proper management of the effluents, and these wastes are directly poured into water bodies. In Pakistan a total of 6634 industries have been registered and regrettably 19% (1228) of these have been declared as 'highly polluting' (Sial et al. 2006). Waste effluents from these industries carry high loads of toxic materials and therefore constitute a major source of water pollution in Pakistan (Sleet 2019). Unfortunately, the majority of these industries are located in the vicinities of water bodies and dispose waste effluents directly into the nearby drains, rivers, streams, ponds, ditches and open or agricultural land (Mulk et al. 2017; Ullah et al. 2013; Imran et al. 2019). According to Daud et al. (2017) contamination of drinking water resources has caused broad-spectrum water-borne ailments constituting about 80% of the total diseases resulting in 33% of deaths. Unfortunately, in Pakistan only 1% of the wastewater from industrial estates and 8% of the total municipal wastewater receives treatment and the rest is disposed untreated carrying along large amounts of toxic heavy metals including arsenic, chromium, mercury, cadmium, lead, iron, nickel, zinc, cobalt, etc. These practices have severely increased the threat to human health and ecosystems (Ilyas et al. 2019). Similarly, poor management and lack of facilities to treat domestic and municipal wastes are the main factors aggravating water pollution. The untreated effluents are discharged directly to a sewer system, a natural drain or water body, a nearby field or an internal septic tank (Murtaza and Zia 2012). Incineration of hospital and municipal waste generates concentrates of heavy metals in ash residues or disperse in the surrounding environment. The process of incineration constitutes a debilitating source of heavy metal pollution in big cities of Pakistan. The heavy metals dispersed in air return to water bodies and soil with precipitation increasing the quantity of metal pollutants in the environment (Tufail and Khalid 2008; Korai et al. 2016). Rapid urbanisation and industrialisation have resulted in an increase in the production of automobiles and motor vehicles, which is responsible for the toxic metal load in air. In Pakistan, the air in major cities is heavily polluted due to strong vehicular emissions. A recently conducted study by Anwar et al. (2020) assessed the distribution of metal pollution emitted from automobiles in the dust, soil and plant samples collected from the roadsides along Lahore to Faisalabad national highways. Vehicular density and smoke samples were also recorded at the selected sites. The analysis revealed higher concentrations of lead, cadmium, chromium, copper and iron in samples suggesting vehicular smoke emission as source of heavy metals in soil dust and plant tissues. In addition, smoke emitted from trucks, lorries and buses contained significantly higher cadmium and lead concentrations. The main contributor of metal pollutants in air is the continuing use of metal-loaded petroleum products which results in greater emission of heavy metals into the environment further deteriorating the air quality consequently impacting human health.

The agricultural sector is a giant source of water pollution globally. Pakistan's water resources are contaminated with toxic compounds such as ammonia, sulphates, phosphates, nitrates, nitrites and heavy metals originating from agricultural runoff. In addition, metal pollutants are deposited in agriculture soils due to extensive use of a variety of inorganic and organic fertilizers, pesticides, liming, irrigation water and sewage sludges. Similarly, extensive application of animal manure is also responsible for contamination of soil and water bodies with heavy metals. Abuse of pesticides and fertilizers to gain better yields continuously add heavy metals to the soil ecosystem which ultimately find their way into food crops and fodder leading to severe health concerns in consumers (Azizullah et al. 2011; Waseem et al. 2014; Ali et al. 2015; Khan et al. 2013a; Ullah et al. 2020). Heavy metal contamination of water sources with arsenic (2.4 mg/L) as a result of extensive use of fertilizers has been reported by Farooqi et al. (2009) in Punjab, Pakistan. In a similar study, higher concentrations of arsenic were found in ground water samples collected from Mailsi, Punjab, Pakistan. It was observed that contamination mainly occurred due to indiscriminate use of fertilizers that have been widely applied on various cash crops including maize, cotton, wheat, rice and sugarcane in the sampling area (Rasool et al. 2015). In developing countries like Pakistan, agriculture-based anthropogenic activities such as long-term use of fertilizers, pesticides, fungicides, manures and wastewater irrigation result in accumulation of heavy metals especially cadmium and lead in agricultural soils.

13.5 Effects of Heavy Metal Pollution in Pakistan

Exposures to heavy metal pollution from anthropogenic sources remain a major source of health risk throughout the world generally and in developing countries particularly. Metal pollutants are released into the environment from a diverse range of domestic and industrial sources, ultimately entering and deteriorating the soil and water supplies and finally the food chain via edible portions of vegetables causing potential health risks in the long term. Metal pollutants can also be present in canned foods as secondary food contamination due to industrial processing and storage. Presence of heavy metals in either raw plant/animal materials or processed foods at high concentrations induces genotoxicity and therefore, represents a serious health hazard (Briggs 2003; Kopp et al. 2018; Ali et al. 2015). The agriculture sector is also severely affected mainly due to usage of industrial wastewater for irrigation resulting in increased contamination with cadmium, lead, chromium, arsenic and other metals in crops. In addition, usage of fertilizers and pesticides manyfold higher than recommended doses also leads to increases in the concentrations of these potentially toxic heavy metals in agricultural soils and has become a major environmental concern affecting the health of crops, livestock and humans (Murtaza et al. 2015).

Table 13.2 Heavy metal concentrations (μg/g) determined in freshwater fish species

S. no	Site river/lake	Fish specie	Fish organ	Co	Cd	Cr	Cu	Fe	Mn	Ni	Pb	Zn	Reference
1	Rawal Lake, Islamabad	*Cyprinus carpio*	Gills	2.366	0.763	5.922	2.557	201.3	10.23		4.034	332.9	Iqbal and Shah (2014)
			Liver	1.597	0.385	8.757	10.20	220.3	1.441		2.564	308.1	
			Muscle	4.917	0.745	2.389	1.293	9.835	0.487		8.796	24.23	
			Stomach	16.81	2.179	26.96	18.32	1999.0	30.13		5.533	968.8	
			Intestine	11.91	0.834	15.93	14.01	1628	72.05		3.240	257.9	
3	River Kabul	*Wallago attu*	Skin		669.0	515.7	108.7	102.0	87.0	100.3	70.0	866.3	Siraj et al. (2016)
			Gills		411.7	105.3	555.3	94.0	80.0	116.3	62.7	824.0	
			Liver		596.3	148.0	485.0	98	84	99.3	66.3	485.7	
			Muscle		509.7	91.3	52.0	91.0	77.0	446.0	60.3	582.3	
4	Tarbela Lake	*Catla catla*	Muscle			0.17	1.93	18.67	3.94	0.26	0.16	13.83	Ahmed et al. (2016)
		Cyprinus carpio	Muscle			0.23	1.93	23.11	3.13	0.22	0.23	12.36	
		Tor putitora	Muscle			0.18	2.39	17.46	5.63	0.18	0.18	18.1	
		Catla catla	Liver			13.65	15.16	31.06	18.2	13.36	13.27	26.55	
		Cyprinus carpio	Liver			14.46	15.11	35.72	17.93	13.97	14.42	25.63	
		Tor putitora	Liver			13.52	14.64	29.29	18.23	12.93	12.94	30.53	
5	River Swat	*Oncorhynchus mykiss*	Liver			14.22	16.15	35.51	20.67	13.64	14.26	34.06	
		Oncorhynchus mykiss	Muscle			0.23	2.60	23.93	5.87	0.19	0.16	20.51	
6	River Barandu, district Buner, Khyber Pakhtunkhwa	*Tor putitora*	Gills		0.02–0.09	0.39–2.41	0.07–0.22	1.45–4.63	0.28–1.02	0.14–0.39	0.13–0.36	0.22–0.77	Mulk et al. (2017)
			Liver		0.04–0.13	0.51–3.38	0.08–0.22	2.61–8.32	0.27–1.26	0.51–3.38	0.19–0.73	0.31–1.48	
			Kidney		0.03–0.12	0.43–2.68	0.09–0.21	1.93–4.85	0.21–0.91	0.13–0.52	0.16–0.52	0.25–1.02	
			Muscle		0.02–0.09	0.28–1.66	0.08–0.20	1.36–3.56	0.17–0.51	0.09–0.32	0.11–0.27	0.16–0.59	

(continued)

Table 13.2 (continued)

S. no	Site river/lake	Fish specie	Fish organ	Co	Cd	Cr	Cu	Fe	Mn	Ni	Pb	Zn	Reference
7	River Kabul, Sar Daryab, Khyber Pakhtunkhwa	*Cyprinus carpio*	Gills			0.154	0.024	0.086			0.041	0.074	Yousafzai et al. (2017)
			Liver			0.188	0.089	0.08			0.142	0.07	
			Muscle			0.024	0.016	0.067			0	0.018	
		Labeo rohita	Gills			0.133	0.018	0.08			0.024	0.058	
			Liver			0.165	0.071	0.061			0.161	0.088	
			Muscle			0.019	0.01	0.05			0	0.02	
8	River Chenab	*Catla catla*	Muscle	0.12	0.08	0.40	2.85		4.63	0.34	0.20	54.1	Alamdar et al. (2017)
		Cirrhinus reba	Muscle	0.10	0.10	0.21	4.03		2.43	0.14	0.14	36.5	
		Sperata sarwari	Muscle	0.09	0.07	0.45	1.38		3.73	0.19	0.14	35.5	
		Wallago attu	Muscle	0.11	0.12	0.67	4.57		3.45	0.26	0.31	52.1	
9	River Kabul, Warsak Dam, Khyber Pakhtunkhwa	*Clupisoma naziri*	Muscles		1.3	15.7				31.2	7.5		Ali and Khan (2018a)
		Glyptothorax cavia	Muscles		1.7	13.3				46.0	33.7		
		Glyptothorax punjabensis	Muscles		1.3	8.0				39.0	15.5		
10	River Kabul, Sar Daryab, Khyber Pakhtunkhwa	*Barilius vagra*	Muscles		1.6	9.5				98.3	16.2		Ali and Khan (2018a)
		Clupisoma naziri	Muscles		1.1	13.7				98.8	9.5		
		Glyptothorax cavia	Muscles		1.4	15.7				116.3	16.2		
		Tor putitora	Muscles		1.2	12.0				94.5	13.7		

(continued)

Table 13.2 (continued)

S. no	Site river/lake	Fish specie	Fish organ	Co	Cd	Cr	Cu	Fe	Mn	Ni	Pb	Zn	Reference
11	River Kabul, Nowshera, Khyber Pakhtunkhwa	*Cirrhinus mrigala*	Muscles		2.0	40.2				38.4	41.0		Ali and Khan (2018a)
		Clupisoma naziri	Muscles		1.3	37.3				19.8	18.3		
		Cyprinus carpio	Muscles		1.0	38.7				45.5	34.2		
		Notopterus chitala	Muscles		1.6	15.8				29.2	25.0		
12	River Kabul, Jahangira, Khyber Pakhtunkhwa	*Cirrhinus mrigala*	Muscles		1.3	15.3				106.2	19.5		Ali and Khan (2018a)
		Clupisoma naziri	Muscles		0.9	16.2				129.0	15.2		
		Cyprinus carpio	Muscles		0.82	11.0				122.5	26.7		
		Schizothorax plagiostomus	Muscles		0.92	9.0				79.0	20.3		
13	River Indus, Sindh	*Labeo rohita*	Muscle		0.48	0.56	0.37			0.14	3.46	1.85	Nawab et al. (2018b)
		Wallago attu	Muscle		0.35	0.28	0.68			0.33	4.41	3.48	
		Cirrhinus mrigala	Muscle		0.89	0.27	0.11			0.22	4.89	2.18	
		Labeo calbasu	Muscle		0.04	0.48	0.48			0.52	0.84	1.04	
14	River Chenab, Punjab	*Mastacembalus aramtu*	Muscle		1.28	0.64	0.36			0.19	5.93	1.68	Nawab et al. (2018b)
		Cyprinus carpio	Muscle		1.48	0.54	0.52			0.27	4.79	1.27	

(continued)

Table 13.2 (continued)

S. no	Site river/lake	Fish specie	Fish organ	Co	Cd	Cr	Cu	Fe	Mn	Ni	Pb	Zn	Reference
		Cirrhinus mrigale	Muscle		0.88	0.63	0.38			0.49	2.53	0.86	
15	River Kabul, Khyber Pakhtunkhwa	*Cyprinus carpio*	Muscle		1.13	0.96	0.52			1.21	2.34	0.94	Nawab et al. (2018b)
		Cirrhinus mrigala	Muscle		0.49	0.21	0.11			0.13	1.78	1.27	
16	Hunza River and its tributaries in Gilgit–Baltistan	*Salmo trutta*	Muscle	0.00001	0.00027	0.00022	0.00117	0.0358		0.0299	0.0004	0.001	Muhammad and Ahmad (2020)
17	River Shah Alam, Khyber Pakhtunkhwa	*Barilius vagra*	Muscle		1.9	40.2				58.3	29.7		Ali et al. (2020)
		Clupisoma naziri	Muscle		2.5	31.0				103.5	62.7		
		Glyptothorax cavia	Muscle		1.5	30.5				78.0	57.2		
		Glyptothorax punjabensis	Muscle		2.0	32.5				99.7	40.5		
		Mastacembelus armatus	Muscle		1.4	57.7				72.7	97.7		
		Notopterus chitala	Muscle		1.1	70.8				16.7	51.0		
18	Shahpur Dam, Fateh Jang	*Cyprinus carpio*	Gills			1.6	35.5	106.5			13.9		Khalil et al. (2020)
			Liver			1.13	120.7	162.7			8.9		
			Muscle			0.2	13.6	58.2			21.7		
			Kidney			0.2	18.7	66.8			18.3		

13.5.1 Effects on Agriculture

Heavy metals affect agriculture in a number of ways. Generally, both natural and anthropogenic factors add heavy metals to soils and crops. Agricultural soils may contain above-normal levels of heavy metals derived from weathering, erosion of parent rocks and volcanic activities. Main anthropogenic sources involved in heavy metal pollution in soils include wastewater irrigation, metal-based pesticides or herbicides, livestock manure, phosphate-based fertilisers, sewage sludge-based land amendments, spillage of petroleum distillates, flooding of rivers which brings sewage and contaminated water to the land and accidental spillage of toxic chemicals from vehicles during transport (El-Kady and Abdel-Wahhab 2018; Ali et al. 2015; Webber 1981). Agriculture is severely affected mainly due to usage of industrial wastewater for irrigation resulting in increased contamination of cadmium, lead, chromium, arsenic other metals in crops (Khan et al. 2013a; Ilyas et al. 2019; Arain et al. 2009). In Pakistan, agriculture is the main source of livelihood and depends mainly on irrigation from rivers, canals, lakes and deep wells. Irrigation of agricultural soils with industrial wastewater and related effluents is a widespread practice in developing countries, including Pakistan and is a source of heavy metal pollution in soil that ultimately bioaccumulates in food crops (Khan et al. 2013a, c). A recently conducted investigation assessed the presence of heavy metals in vegetables irrigated with wastewater in the district Sahiwal, Pakistan. Accumulation of metal pollutants in vegetables was found manyfold higher than in ground water. The concentrations of lead surpassed permissible limits by four-fold and was the most abundant heavy metal found in spinach leaves and judged as a potentially serious health problem (Rehman et al. 2019). A similar study estimated heavy metal concentrations and associated health risks of their consumption in the area of Mangla Dam, Pakistan and reported bioaccumulation of cadmium, chromium and lead in *Triticum aestivum* and *Eruca sativa* irrigated with wastewater (Mehmood et al. 2019). Other agricultural inputs, such as application of phosphate fertilizers, fungicides and inorganic fertilizers rich in cadmium, chromium, nickel, lead and zinc can also lead to contamination in soil and crops. Consequently, these pollutants leach into groundwater and contaminate it. In order to avoid heavy metal accumulation, soil amendments and fertilizers must be carefully screened for contaminants before their applications to fields (Tóth et al. 2016; Malik et al. 2017; Dissanayake and Chandrajith 2009). Rice is an example of crop that readily uptakes heavy metals through its roots compared to other crops and accumulates it in the grains. Millions of tons of rice are discarded due to cadmium contamination in rice grains, grown in areas irrigated with mine wastewater or due to application of cadmium-rich phosphate fertilizers polluting paddy fields with surplus cadmium (Sebastian and Prasad 2014). Additionally, metal-based pesticides are also soil-polluting agents that are generally applied to control diseases of fruit and vegetables and grain crops and in the meanwhile contaminate agricultural soils and the produce with high levels of heavy metals such as copper, arsenic, lead, iron, manganese and mercury. Metal containing pesticides such as lead, calcium arsenate or copper sulphate have been frequently used in the past to combat fungal pathogens which culminated in contaminating soils with heavy metals. Metal pollutants, arsenic and lead were abundant in soils of apple orchards in Ontario, Canada, mainly due to the usage of lead arsenate pesticides for over a period of 70 years (Frank et al. 1976). Another group of major metal pollutants in agricultural soils includes sewage sludge and livestock waste manure that are frequently added to soils for better yields, however, heavily contaminating the fields. Sewage sludges are generally applied to agricultural lands because it supplies many essential nutrients to crop plants to increase biomass production and crop productivity and to improve physicochemical properties of soils (Riaz et al. 2018). However, sewage sludge contains excessively high concentrations of heavy metals that can negatively impact soil microbial biomass and the activity of nitrogen-fixing bacteria (Valsecchi et al. 1995; Khan et al. 2008; Ahmad et al. 2012). Likewise, livestock feed and manure have also been identified as significant sources of heavy metals. Animal diets are supplemented with heavy metals such as copper and zinc in small quantities for health reasons and as growth promoter. Copper and zinc are also added to chicken feed as poultry enzyme co-factors. Arsenic compounds such as arsanilic acid, 4-nitrophenyl arsonic acid and 4-hydroxy-3-nitrophenyl arsonic acid are commonly added to broiler rations to improve muscle growth and to fight disease. The majority of the heavy metals consumed in feed are excreted unchanged and become part of the manure that is subsequently applied to land leading to excessive concentrations in soil (Kumar et al. 2013; Nicholson et al. 2003; Webber 1981).

13.5.2 Effects on Food Quality

Foods of plant origin generally include fruits, vegetables, legumes, grains, nuts and seeds and the derived ingredients such as oils, sugars, herbs and spices. In densely populated developing countries plant-based food is grown on soils contaminated with heavy metals sourced through industrial and domestic effluents, city wastes, sewage sludge and toxic emissions. These sources are responsible for bioaccumulation of the toxic metal pollutants in plants above permissible limits and therefore constitute a major health concern for millions of people around the globe (Khan et al. 2018; Ahmad et al. 2018). Essential elements such as iron, copper, zinc, selenium, manganese and molybdenum are required in low concentrations for various biological functions whereas cadmium, mercury, lead, chromium are non-essential elements and are toxic even in trace amounts (Arif et al. 2016). Heavy metals in the soils enter the plant body through roots and constitute the primary organs of heavy metal storage resulting in higher metal content in underground parts than in parts above the grounds. Ultimately, the metal pollutants are internally translocated to other parts such as stem, leaf, fruit and grains and are thus introduced into the human food chain (Cheng 2003). Additionally, foliar absorption of particulate matter containing heavy metals further contributes to contaminant levels within the plant. Higher metal content has been reported in food crops, particularly fruits and vegetables, nuts, pulses and cereal grains. The major factor contributing to edible food crop contamination is not limited to soil enriched with heavy metals contributed by anthropogenic or other intensive agricultural practices but also due to deposition of heavy metal-loaded particles released from automobiles (Hussain et al. 2019; Nawab et al. 2018a; Sattar et al. 1989; Liu et al. 2013). Research studies have shown that metal uptake potential of leafy vegetables is comparatively higher than other vegetables and fruits, possibly due to higher transpiration rate to sustain moisture content and regulate plant growth. Roots and leaves of vegetables accumulate substantially higher amounts of heavy metals than stems and fruits (Mahmood and Malik 2014; Yargholi et al. 2008). Excessive accumulation of heavy metals inside the plants causes a wide range of physiological and biochemical disorders which in turn cause reduction in crop growth, biomass, productivity and nutrient quality (Sheetal et al. 2016; Manzoor et al. 2018). Furthermore, heavy metals such as lead, cadmium, arsenic and mercury taken up by plants constitute the main source of accumulation in food. Metal toxicity in vegetables causes a diverse range of physiological effects such as inhibition of seed germination, changes in photosynthetic efficiency, respiration, transpiration and alterations in trace metal uptake rates (Manzoor et al. 2018). Vegetables grown in metal contaminated soils show reduced secondary metabolite content. For example, Pascual et al. (2010) reported lower concentrations of capsaicin and dihydrocapsaicin in pepper fruits exposed to heavy metals contamination in sludge-amended soils. Exposure to metal pollution also decreases synthesis of proteins and reduction in uptake of essential elements thereby decreasing the nutrient quality of the plant-based food (Manzoor et al. 2018).

Consumption of crops containing trace amounts of arsenic, lead, mercury and cadmium contributes to substantial health-threatening human exposure. Accumulation of high arsenic levels in rice, grains and vegetables have been reported in several studies around the globe. The main contributor of higher arsenic concentration in the soil has been identified as arsenic contaminated ground water used for irrigation purposes. In areas such as Bangladesh heavily contaminated drinking water supply and consumption of rice containing higher levels of arsenic are the main causes of toxic pollutant exposure. Research studies have revealed loss of grain yield and deterioration of grain quality (i.e. higher grain-arsenic concentration) as a result of long-term use of contaminated irrigation water (Khan et al. 2010). Soils heavily contaminated with arsenic progressively decline rice yields. An estimated yield loss of 16% was recorded for a rice variety called BRRI dhan 29 grown in contaminated fields in Bangladesh. Reduction in yield was accompanied with a decrease in the numbers of grain-filled panicles. Total above-ground plant biomass production was also reduced to one-third in arsenic-affected soils (Panaullah et al. 2008). Cadmium ranks second on a scale of most abundant rice grain contaminants. Rice constitutes a major source of cadmium intake and poses severe health risks to a huge population in the developing world largely dependent on rice as staple food. Cadmium readily accumulates in rice compared with other crops exhibiting elemental concentrations manyfold higher than arsenic with augmented risk of soil-food chain transfer and with this attribute it has emerged as the most prominent toxic metal in food crops (Zhao and Wang 2020). Similarly, high levels of cadmium accumulate in edible parts of wheat and maize crops grown in cadmium polluted soils. The concentrations of the metal increased by 31.6-fold and 7.0-fold in maize and wheat grains, respectively. Consumption of wheat grown in metal-polluted areas poses a higher risk to the resident population (Yang et al. 2014). Cadmium becomes part of the food chain and is taken up by plants such as leafy vegetables, root crops, cereals and grains which are a major exposure route for humans. Plant-based foods constitute about 70%–80% of the dietary cadmium source. Cadmium intake in vegetarians is three-fold higher compared with non-vegetarians (Clemens et al. 2013; Khan et al. 2017). Analysis of metal toxicity of wheat cultivars grown in metal-polluted fields revealed manyfold higher concentrations of cadmium and lead in grains indicating substantial accumulation of the toxic metals and posing a serious health risk to the consumers and a threat to food safety (Guo et al. 2018).

Apart from heavy metal toxicity in food crops, cases of poisoning and toxicity in medicinal plants have also been reported. Medicinal plants are extensively cultivated for their therapeutic uses and health benefits. However, reports of heavy metal bioaccumulation largely emerged in plants cultivated in soils contaminated with toxic metal wastes from industrial areas (Mousavi et al. 2014; Okatch et al. 2012; Lal et al. 2013). A copious number of medicinal plants have been identified as accumulators of heavy metals such as cadmium, chromium, nickel, arsenic, lead, iron, etc. (Annan et al. 2010; Street et al. 2008; Bolan et al. 2017). Heavy metal stress in medicinal plants elicits varying degrees of secondary responses. For example, exposure to cadmium stress enhances therapeutically active components phyllanthin and hypophyllanthin in *Phyllanthus amarus* (Rai et al. 2004). Conversely, exposure to heavy metal pollution reduces the yield of essential oil in *Mentha* by up to 14% (Zheljazkov and Nielsen 1996). Toxic metal pollutants are also found in significant amounts in industrially processed food stuff and pharmaceuticals. Interestingly, contamination of canned vegetables, fruits and canned fish with heavy metals has been extensively reported in developed countries (e.g. Spain, Belgium, England and USA). One of the main reasons contributing to increased metal toxicity is the metallic food packaging which is mostly composed of tin- or chromium-coated steel or aluminium. However, any damage to the protective inner resin coating accelerates the release of toxic metals such as tin, iron, cadmium and lead, contaminating the food. In the case of acidic foodstuffs such as canned tomatoes, tin leach into the food and cause acute toxicity in consumers (González-Martín et al. 2018; Raber et al. 2012; Kassouf et al. 2013; Massadeh et al. 2018). Besides heavy metal intake from contaminated cereal grains and vegetables, dietary consumption of certain seafood is also a potential source of heavy metal intake, particularly arsenic. The toxic element is taken up by the phytoplankton from the sea water and converted to arsenosugar compounds which are taken up by algae. Consumption of arsenic accumulating algae by other marine animals magnifies the metal pollutant in these organisms up to 1000–10,000-fold. Intake of seafood with potentially toxic concentrations of the pollutants such as lead, cadmium, arsenic and mercury represents one of the main sources of exposure to metals in the general population (Djedjibegovic et al. 2020; Borak and Hosgood 2007; Rose et al. 2007).

13.5.3 Effects on Human Health

Of all the metal pollutants major threats to human health emanate from exposure to non-essential heavy metals such as arsenic, cadmium, lead and mercury. These metals exist in different environmental compartments such as air, water and food and gain entrance into the human body mainly through inhalation, ingestion and dermal adsorption. Of these, the gastrointestinal route, i.e. eating contaminated food such as fruits, vegetables, seafood, fish or drinking water and beverages are the main pathway of metal exposure in human populations. The consequences of metal toxicity in the human body are well documented and have been shown to cause a myriad of disorders such as mental disorder, effects on the blood constituents, damage to the lungs, liver, kidneys and other vital organs (Jaishankar et al. 2014). In Japan serious pollution-related illnesses appeared in the 60s, the cause being acute contamination of water and fish with cadmium and mercury. Three of the "four big pollution diseases" included *Itai-itai* disease, Minamata disease and Niigata Minamata disease. *Itai-itai* disease was caused by a severe type of cadmium poisoning arising from the contamination of rice fields due to dumping of liquid wastes of the Kamioka mine as the main source of cadmium pollution. Long-term ingestion of the cadmium led to kidney damage, soft or brittle bones and the intense shooting pains which gave the disease the name *Itai-itai* meaning ouch-ouch. The Minamata disease was the other pollution-related disease that garnered nationwide attention and was caused by severe mercury poisoning. The disease surfaced when industrial waste from the Chisso Corporation's chemical factory, Minamata, containing methyl mercury was released into nearby waters resulting in bioaccumulation of the contaminant in fish and shellfish. The consumption of the contaminated seafood later caused symptoms such as numbness of the hands and feet, loss of peripheral vision, hearing loss, paralysis, convulsions and, in quite a few cases, death (Fujigaki 2015). Whilst plant-based terrestrial food such as fruits, vegetables, etc. contain trace amounts of mercury, fish and seafood represent the predominant pathway of methyl mercury exposure for most human populations. It is a neurological toxicant that affects neurotransmitters and is poisonous to nerve tissue. Animals exposed to intoxicating levels of mercury exhibited drastic reduction of neuronal RNA and protein synthesis. It diminishes production of thyroid hormones and testosterone in the body. Other disorders such as slow mental development, blindness, cerebral palsy and other birth defects in humans are also associated with prolonged exposure to the toxic element (Coccini et al. 2000; Chang 1977; Chen et al. 2013).

Cadmium poisoning by plant-based foods and water supply is an extreme health hazard that has caused environmental catastrophes in the past. Vegetables constitute the predominant dietary source of cadmium and accounts for about 70% of the metal intake (Ryan et al. 1982). The occurrence of the *Itai-itai* disease is a notable example of cadmium poisoning (Fujigaki 2015). Although phosphate ore-based fertilizers are the main source of metal pollution, however, smoking and food intake constitutes two main routes of cadmium entry into the human body (Järup 2003; Chen et al. 2007). It is reported that cadmium is the only metal pollutant that can cause severe health problems in humans and animals when ingested at plant tissue concentrations that are not generally considered phytotoxic (Peijnenburg et al. 2000). Cadmium toxicity causes a number of severe ailments including damages to the lungs and respiratory irritation, stomach irritation, vomiting, diarrhoea, kidney damage, soft or brittle bones, defects in the female reproduction and endocrine system and cancer (Järup 2003; Nishijo et al. 2017; Thompson and Bannigan 2008).

It is a widely known fact that the major cause of arsenic intoxication is drinking arsenic-contaminated water. Higher levels of arsenic occur naturally in drinking water and surface soil in parts of West Bengal in India, Bangladesh, USA, Argentina, Taiwan, China, Hungary and Vietnam. In Pakistan, arsenic contamination is mostly reported from areas in Punjab and Sindh where higher arsenic concentration levels of over 50 μg/L are found in surface and ground water samples (Waseem et al. 2014). Apart from drinking water, exposure to the pollutant also occurs by consuming contaminated vegetables, rice and other cultivated grains irrigated with groundwater containing elevated levels of arsenic. Of the staple crops grown globally, rice is well known for its high potential to accumulate arsenic in large quantities. Irrigation of crops with contaminated water also affects plant height, crop yield and development of root growth (Abedin et al. 2002). According to reports, an estimated 150 million people around the world are affected due to consumption of arsenic-contaminated ground water. In Pakistan, estimates reveal that more than 50 million people are exposed to arsenic poisoning mostly due to usage of arsenic-polluted groundwater for drinking and other domestic uses, the exposure resulting in the incidence of arsenicosis in the population (Sanjrani et al. 2017; Podgorski et al. 2017). Population exposure to high levels of arsenic for longer duration causes a myriad of serious health problems such as cardiovascular diseases, lung disease, diabetes, oxidative stresses and various types of cancers (skin, bladder and lung cancer), a decrease in white and red blood cells production, gastrointestinal irritation, "pins and needles" sensation in hands and feet and other skin lesions, such as hyperkeratosis and pigmentation (Järup 2003; Huy et al. 2014; Abernathy et al. 2003).

Lead is a potent environmental toxicant that has been a subject of interest due to its ability to cause a broad range of physiological, biochemical and morphological defects in plants, animals and humans. It naturally occurs as lead sulfide, also called Galena, in the Earth's crust. It is released into the environment predominantly due to anthropogenic activities that include automobile emissions, industrial emissions, mining activities, usage of lead-containing agrochemicals, oil and paint, etc. Lead is readily available for plant uptake once in the soil. Nearly 95% of the lead accumulates in the root tissue after penetrating the plant root system and afterwards it may be translocated to arial plant parts (Kumar et al. 2020). Lead poisoning is a widely known environmental health hazard and has severe implications for the public exposed to the toxicant via different pathways. It can enter the human body through inhalation of lead-laden air, drinking contaminated water, ingestion of contaminated soil and food grown in contaminated soil. Consumption of livestock meat contaminated with higher proportions of lead is also a major health risk. Another notable lead source is the food canning industry. Lead in the glazed food containers can leach into the food deteriorating the nutritional value of the food product as well as posing a serious health risk (Sharpe and Livesey 2006; Kumar et al. 2020). When in the bloodstream, lead binds to haemoglobin and is thus transported to all vital organs of the body. Furthermore, lead accumulation results in lead poisoning giving rise to chronic health problems such as brain and nerve disorders, blood disorders, structural damage and changes in the excretory function of kidneys, mineralising of bones and teeth, digestive problems, cardiovascular problems, reproductive systems and hypertension (Abadin et al. 2007; Yuan et al. 2014). According to a detailed review by Shakir et al. (2016) a number of studies have reported elevated levels of lead in human samples such as hair, nails and blood. Higher concentrations in the range of 12.03–100.19 μg/L were reported in blood samples of children living around automobile and battery repair workshops. In addition, lead-related occupational workers were more prone to lead poisoning due to overexposure for longer durations. Lead-related health ailments such as adversely affected haematological, renal and hepatic functions are commonly found in lead-exposed workers. Pregnant women are more vulnerable to lead toxicity, and overexposure can result in low birth weight and miscarriage of the foetus. Due to an increased rate of lead absorption during pregnancy, it can hamper foetal growth and cause developmental delays. In children, exposure to lead may lead to behavioural disturbances in learning and concentration, memory deterioration and prolonged reaction times (Bellinger 2005; Järup 2003).

Trace elements or essential heavy metals are required in minimal quantities and are known to be nutritionally essential. For example, iron and copper are required in oxidation-reduction reactions in energy metabolism; iron is a constituent of haemoglobin and myoglobin; it also plays a vital role in the transport of oxygen; manganese regulates many enzymes in the body; chromium participates in the carbohydrate and lipid metabolism in the body; nickel is required for the active synthesis of urease in plant cells; cobalt is the main constituent of cobalamin (Vitamin B12); zinc functions as a cofactor for certain enzymes. Exposure to excessive doses of the mentioned trace elements can be toxic and may lead to many fatal diseases. For example, excessive levels of manganese lead to ailments of the nervous system and in some cases cause permanent neurological disorder with symptoms like those of Parkinson's disease (Neal and Guilarte 2013). Similarly, iron overexposure can lead to several serious health problems like cancer, diabetes, cardiovascular diseases as well as neurodegenerative disorders (Abbaspour et al. 2014). Occupational exposure to chromium (VI) compounds in chromate production industry can cause asthma and other respiratory distresses, skin diseases, lung cancers, contact allergic dermatitis, irritated and ulcerated skin and burns in sensitive workers (Wilbur et al. 2012). Likewise, exposure to nickel in the workplace results in ailments such as kidney disorders, cardiovascular diseases, lung fibrosis and respiratory tract cancer. One of the most frequently encountered ailments is the development of dermatitis known as 'nickel allergy' which arises when nickel coins or jewellery come in contact with the skin. The allergy also manifests itself in other forms such as headaches, gastrointestinal and respiratory symptoms (Chen et al. 2017).

13.6 Conclusions

Pollution of freshwater resources by toxic metals is a major environmental concern that is accelerating at an alarming speed mostly due to human interventions. In developing countries like Pakistan toxic metal pollution of freshwater resources is on the rise with severe implications for human health. Frequent use of heavy metal-contaminated drinking water and consumption of contaminated plant-based food, freshwater fish and other seafood has endangered lives of the consumers. Due to noncompliance of the environmental laws and continuous disposal of agricultural, domestic and industrial waste in rivers and lakes, Pakistan's fragile water infrastructure is barely able to provide clean water to the population largely dependent on freshwater resources for their livelihood and drinking water needs. High concentrations of heavy metals in effluents loaded in rivers and lakes are taken up by the aquatic organisms and thus enter the food chain. The majority of the rivers in Pakistan are contaminated with universally proclaimed toxic metals with a history of human disasters. For example, arsenic, cadmium and lead are the most commonly found contaminants that accumulate manyfold in staple crops and fish reflecting the state of their abundant occurrence in water bodies. Heavy metals are not only a hazard from the human health perspective but also deleterious for the biodiversity of aquatic organisms as evident from the continuous decline in the population and diversity of freshwater fish and other aquatic species. Underdeveloped wastewater treatment facilities in Pakistan have limited capacity and can treat only 8% of the country's total municipal wastewater and 1% of the industrial wastewater before they are discharged into the rivers further adding to the dire situation of heavy metal pollution. In a nutshell, the government must take it as a challenge to formulate policies to ensure sustainable surface and groundwater use, industrial and domestic wastewater management and take measures to implement laws and biomonitoring programmes to remediate the fast-growing environmental crisis of heavy metal pollution in surface and groundwater resources of Pakistan.

References

Abadin H, Ashizawa A, Stevens Y, Llados F, Diamond G, Sage G, Quinones A, Bosch S, Swarts S (2007) Toxicological profile for lead, Atlanta (GA): agency for toxic substances and disease registry (US). Lewis Publishers, Boca Raton

Abbaspour N, Hurrell R, Kelishadi R (2014) Review on iron and its importance for human health. J Res Med Sci 19(2):164–174

Abedin MJ, Cresser MS, Meharg AA, Feldmann J, Cotter-Howells J (2002) Arsenic accumulation and metabolism in rice (*Oryza sativa* L.). Environ Sci Technol 36(5):962–968. https://doi.org/10.1021/es0101678

Abernathy CO, Thomas DJ, Calderon RL (2003) Health effects and risk assessment of arsenic. J Nutr 133(5):1536S–1538S

Ahmad T, Kahlown M, Tahir A, Rashid H (2004) Arsenic an emerging issue: experiences from Pakistan. In: Proceedings of the 30th WEDC international conference 2004, Lao PDR Vientiane, pp 459–466

Ahmad E, Zaidi A, Khan MS, Oves M (2012) Heavy metal toxicity to symbiotic nitrogen-fixing microorganism and host legumes. In: Hashmi MZ, Kumar V, Varma A (eds) Xenobiotics in the soil environment: monitoring, toxicity and management. Springer International Publishing, Cham, pp 139–152. https://doi.org/10.1007/978-3-319-47744-2_11

Ahmad H, Yousafzai AM, Siraj M, Ahmad R, Ahmad I, Nadeem MS, Ahmad W, Akbar N, Muhammad K (2015) Pollution problem in River Kabul: accumulation estimates of heavy metals in native fish species. Biomed Res Int 2015: https://doi.org/10.1155/2015/537368

Ahmad K, Nawaz K, Khan ZI, Nadeem M, Wajid K, Ashfaq A, Munirl B, Memoonas H, Sanas M, Shaheen F (2018) Effect of diverse regimes of irrigation on metals accummulation in wheat crop: an assessessment-dire need of the day. Fresenius Environ Bull 27(2):846–855

Ahmed M, Ahmad T, Liaquat M, Abbasi KS, Farid IBA, Jahangir M (2016) Tissue specific metal characterization of selected fish species in Pakistan. Environ Monit Assess 188(4):212

Alamdar A, Eqani SAMAS, Hanif N, Ali SM, Fasola M, Bokhari H, Katsoyiannis IA, Shen H (2017) Human exposure to trace metals and arsenic via consumption of fish from river Chenab, Pakistan and associated health risks. Chemosphere 168:1004–1012. https://doi.org/10.1016/j.chemosphere.2016.10.110

Alamgir A, Fatima N, Khan MA, Rehman M, Shaukat SS (2019) A preliminary pollution appraisal of western backwater at Karachi Coastal area. Appl Water Sci 9(7):167. https://doi.org/10.1007/s13201-019-1049-y

Al-Ghanim K, Mahboob S, Seemab S, Sultana S, Sultana T, Al-Misned F, Ahmed Z (2016) Monitoring of trace metals in tissues of *Wallago attu* (lanchi) from the Indus River as an indicator of environmental pollution. Saudi J Biol Sci 23(1):72–78

Ali H, Khan E (2018a) Assessment of potentially toxic heavy metals and health risk in water, sediments, and different fish species of River Kabul, Pakistan. Hum Ecol Risk Assess Int J 24 (8):2101–2118. https://doi.org/10.1080/10807039.2018.1438175

Ali H, Khan E (2018b) What are heavy metals? Long-standing controversy over the scientific use of the term 'heavy metals'—proposal of a comprehensive definition. Toxicol Environ Chem 100(1):6–19. https://doi.org/10.1080/02772248.2017.1413652

Ali Z, Kazi AG, Malik RN, Naz M, Khan T, Hayat A, Kazi AM (2015) Heavy metal built-up in agricultural soils of Pakistan: sources, ecological consequences, and possible remediation measures. In: Sherameti I, Varma A (eds) Heavy metal contamination of soils. Monitoring and remediation. Springer International Publishing, Cham, pp 23–42. https://doi.org/10.1007/978-3-319-14526-6_2

Ali H, Khan E, Ilahi I (2019a) Environmental chemistry and ecotoxicology of hazardous heavy metals: environmental persistence, toxicity, and bioaccumulation. J Chem 2019:6730305. https://doi.org/10.1155/2019/6730305

Ali U, Batool A, Ghufran M, Asad-Ghufran M, Sabahat-Kazmi S, Hina-Fatimah S (2019b) Assessment of heavy metal contamination in the drinking water of Muzaffarabad, Azad Jammu and Kashmir, Pakistan. Int J Hydro 3(5):331–337

Ali H, Khan E, Nasir MJ (2020) Bioaccumulation of some potentially toxic heavy metals in freshwater fish of River Shah Alam, Khyber Pakhtunkhwa, Pakistan. Pak J Zool 52 (2):603–608

Annabi A, Said K, Messaoudi I (2013) Cadmium: bioaccumulation, histopathology and detoxifying mechanisms in fish. Am J Res Com 1(4):62

Annan K, Kojo AI, Cindy A, Samuel A-N, Tunkumgnen BM (2010) Profile of heavy metals in some medicinal plants from Ghana commonly used as components of herbal formulations. Pharmacognosy Res 2(1):41

Anwar S, Naz A, Ashraf MY, Malik A (2020) Evaluation of inorganic contaminants emitted from automobiles and dynamics in soil, dust, and vegetations from major highways in Pakistan. Environ Sci Pollut Res Int 27(26):32494–32508. https://doi.org/10.1007/s11356-020-09198-x

Arain M, Kazi T, Baig J, Jamali M, Afridi H, Shah A, Jalbani N, Sarfraz R (2009) Determination of arsenic levels in lake water, sediment, and foodstuff from selected area of Sindh, Pakistan: estimation of daily dietary intake. Food Chem Toxicol 47(1):242–248

Araújo da Silva F, Feldberg E, Moura Carvalho ND, Hernández Rangel SM, Schneider CH, Carvalho-Zilse GA, Fonsêca da Silva V, Gross MC (2019) Effects of environmental pollution on the rDNAomics of Amazonian fish. Environ Pollut 252(Pt A):180–187. https://doi.org/10.1016/j.envpol.2019.05.112

Arif N, Yadav V, Singh S, Singh S, Ahmad P, Mishra RK, Sharma S, Tripathi DK, Dubey NK, Chauhan DK (2016) Influence of high and low levels of plant-beneficial heavy metal ions on plant growth and development. Front Environ Sci 4(69). https://doi.org/10.3389/fenvs.2016.00069

Azizullah A, Khattak MNK, Richter P, Häder D-P (2011) Water pollution in Pakistan and its impact on public health—a review. Environ Int 37(2):479–497

Bellinger DC (2005) Teratogen update: lead and pregnancy. Birth Defects Res A Clin Mol Teratol 73(6):409–420

Bhuyan MS, Islam MS (2017) A critical review of heavy metal pollution and its effects in Bangladesh. Sci J Energy Eng 5(4):95

Bjerrum N (1936) Bjerrum's inorganic chemistry, 3rd Danish edn. Heinemann, London

Bolan S, Kunhikrishnan A, Seshadri B, Choppala G, Naidu R, Bolan NS, Ok YS, Zhang M, Li C-G, Li F, Noller B, Kirkham MB (2017) Sources, distribution, bioavailability, toxicity, and risk assessment of heavy metal(loid)s in complementary medicines. Environ Int 108:103–118. https://doi.org/10.1016/j.envint.2017.08.005

Borak J, Hosgood HD (2007) Seafood arsenic: implications for human risk assessment. Regul Toxicol Pharmacol 47(2):204–212. https://doi.org/10.1016/j.yrtph.2006.09.005

Bradl H (2005) Heavy metals in the environment: origin, interaction and remediation. Elsevier/Academic Press, London

Briggs D (2003) Environmental pollution and the global burden of disease. Br Med Bull 68(1):1–24. https://doi.org/10.1093/bmb/ldg019

Chang LW (1977) Neurotoxic effects of mercury—a review. Environ Res 14(3):329–373. https://doi.org/10.1016/0013-9351(77)90044-5

Chang S, Zdanowicz VS, Murchelano RA (1998) Associations between liver lesions in winter flounder (*Pleuronectes americanus*) and sediment chemical contaminants from north-east United States estuaries. ICES J Mar Sci 55(5):954–969. https://doi.org/10.1006/jmsc.1998.0354

Chaudhry AA (2010) Wetlands in Pakistan: what is happening to them. World Environment Day 5

Chen W, Chang AC, Wu L (2007) Assessing long-term environmental risks of trace elements in phosphate fertilizers. Ecotoxicol Environ Saf 67(1):48–58. https://doi.org/10.1016/j.ecoenv.2006.12.013

Chen A, Kim SS, Chung E, Dietrich KN (2013) Thyroid hormones in relation to lead, mercury, and cadmium exposure in the National Health and Nutrition Examination Survey, 2007–2008. Environ Health Perspect 121(2):181–186. https://doi.org/10.1289/ehp.1205239

Chen QY, Brocato J, Laulicht F, Costa M (2017) Mechanisms of nickel carcinogenesis. In: Mudipalli A, Zelikoff JT (eds) Essential and non-essential metals: carcinogenesis, prevention and cancer therapeutics. Springer International Publishing, Cham, pp 181–197. https://doi.org/10.1007/978-3-319-55448-8_8

Chen L, Zhou S, Wu S, Wang C, He D (2019) Concentration, fluxes, risks, and sources of heavy metals in atmospheric deposition in the Lihe River watershed, Taihu region, eastern China. Environ Pollut 255:

Cheng S (2003) Heavy metals in plants and phytoremediation. Environ Sci Pollut Res 10 (5):335–340

Clemens S, Aarts MG, Thomine S, Verbruggen N (2013) Plant science: the key to preventing slow cadmium poisoning. Trends Plant Sci 18(2):92–99

Coccini T, Randine G, Candura SM, Nappi RE, Prockop LD, Manzo L (2000) Low-level exposure to methylmercury modifies muscarinic cholinergic receptor binding characteristics in rat brain and lymphocytes: physiologic implications and new opportunities in biologic monitoring. Environ Health Perspect 108(1):29–33

Csuros M, Csuros C (2016) Environmental sampling and analysis for metals. CRC Press, Boca Raton

Daud MK, Nafees M, Ali S, Rizwan M, Bajwa RA, Shakoor MB, Arshad MU, Chatha SAS, Deeba F, Murad W, Malook I, Zhu SJ (2017) Drinking water quality status and contamination in pakistan. Biomed Res Int 2017:7908183–7908183. https://doi.org/10.1155/2017/7908183

Di Toppi LS, Gabbrielli R (1999) Response to cadmium in higher plants. Environ Exp Bot 41 (2):105–130

Dissanayake C, Chandrajith R (2009) Phosphate mineral fertilizers, trace metals and human health. J Nat Sci Found Sri Lanka 37(3):153–165

Djedjibegovic J, Marjanovic A, Tahirovic D, Caklovica K, Turalic A, Lugusic A, Omeragic E, Sober M, Caklovica F (2020) Heavy metals in commercial fish and seafood products and risk assessment in adult population in Bosnia and Herzegovina. Sci Rep 10(1):13238. https://doi.org/10.1038/s41598-020-70205-9

Doering JA, Beitel SC, Eisner BK, Heide T, Hollert H, Giesy JP, Hecker M, Wiseman SB (2015) Identification and response to metals of metallothionein in two ancient fishes: white sturgeon (*Acipenser transmontanus*) and lake sturgeon (*Acipenser fulvescens*). Comp Biochem Physiol C: Toxicol Pharmacol 171:41–48

Duffus JH (2002) "Heavy metals" a meaningless term? (IUPAC technical report). Pure Appl Chem 74(5):793–807. https://doi.org/10.1351/pac200274050793

Dvorak P, Roy K, Andreji J, Liskova ZD, Mraz J (2020) Vulnerability assessment of wild fish population to heavy metals in military training area: synthesis of a framework with example from Czech Republic. Ecol Ind 110: https://doi.org/10.1016/j.ecolind.2019.105920

Egila J, Daniel V (2011) Trace metals accumulation in freshwater and sediment insects of Liberty Dam, Plateau State Nigeria. Int J Basic Appl Sci 11(6):128–140

El-Kady AA, Abdel-Wahhab MA (2018) Occurrence of trace metals in foodstuffs and their health impact. Trends Food Sci Technol 75:36–45

FAO (1983) Compilation of legal limits for hazardous substances in fish and fishery products. FAO Fisheries Circular (FAO), 764

Farooqi A, Masuda H, Siddiqui R, Naseem M (2009) Sources of arsenic and fluoride in highly contaminated soils causing groundwater contamination in Punjab, Pakistan. Arch Environ Contam Toxicol 56(4):693–706. https://doi.org/10.1007/s00244-008-9239-x

Frank R, Braun H, Ishida K, Suda P (1976) Persistent organic and inorganic pesticide residues in orchard soils and vineyards of southern Ontario. Can J Soil Sci 56(4):463–484

Fujigaki Y (2015) Lessons from Fukushima: Japanese case studies on science, technology and society. Springer International Publishing, Cham, Switzerland

Gao L, Wang Z, Li S, Chen J (2018) Bioavailability and toxicity of trace metals (Cd, Cr, Cu, Ni, and Zn) in sediment cores from the Shima River, South China. Chemosphere 192:31–42

Garbarino JR, Hayes HC, Roth DA, Antweiler RC, Brinton TI, Taylor HE (1995) Heavy metals in the Mississippi river in contaminants in the Mississippi river. In: Meade RH (ed) US geological survey, Circular 1133. Retrieved from http://pubs.usgs.gov/circ/circ1133/heavy-metals.html

González-Martín MI, Revilla I, Betances-Salcedo EV, Vivar-Quintana AM (2018) Pesticide residues and heavy metals in commercially processed propolis. Microchem J 143:423–429. https://doi.org/10.1016/j.microc.2018.08.040

Goyer R (1993) Casarett and Doull's toxicology. Basic Sci Poisons 19:623

Guo G, Lei M, Wang Y, Song B, Yang J (2018) Accumulation of As, Cd, and Pb in sixteen wheat cultivars grown in contaminated soils and associated health risk assessment. Int J Environ Res Public Health 15(11):2601. https://doi.org/10.3390/ijerph15112601

Hamad SH, Schauer JJ, Shafer MM, Abed Al-Raheem E, Satar H (2012) The distribution between the dissolved and the particulate forms of 49 metals across the Tigris River, Baghdad, Iraq. Sci World J 2012

Hashem MA, Nur-A-Tomal MS, Mondal NR, Rahman MA (2017) Hair burning and liming in tanneries is a source of pollution by arsenic, lead, zinc, manganese and iron. Environ Chem Lett 15(3):501–506

Hasnie F, Qureshi N (2002) Quantification of pollution load in wastewater channels of village Rehri draining into Korangi Creek. Pak Steel Res J 3:47–54

He ZL, Yang XE, Stoffella PJ (2005) Trace elements in agroecosystems and impacts on the environment. J Trace Elem Med Biol 19(2–3):125–140

Hussain S, Rengel Z, Qaswar M, Amir M, Zafar-ul-Hye M (2019) Arsenic and heavy metal (cadmium, lead, mercury and nickel) contamination in plant-based foods. In: Ozturk M, Hakeem KR (eds) Plant and human health, volume 2: phytochemistry and molecular aspects. Springer International Publishing, Cham, pp 447–490. https://doi.org/10.1007/978-3-030-03344-6_20

Huy TB, Tuyet-Hanh TT, Johnston R, Nguyen-Viet H (2014) Assessing health risk due to exposure to arsenic in drinking water in Hanam Province, Vietnam. Int J Environ Res Public Health 11(8):7575–7591

IARC (2012) A review of human carcinogens: arsenic, metals, fibres, and dusts. IARC Monogr Eval Carcinog Risks Hum 100(Pt C):11–465

Ilyas A, Sarwar T (2003) Study of trace elements water in the vicinity of Palosi Drain, Peshawar. Paki J Biol Sci 6(1):86–91

Ilyas M, Ahmad W, Khan H, Yousaf S, Yasir M, Khan A (2019) Environmental and health impacts of industrial wastewater effluents in Pakistan: a review. Rev Environ Health 34 (2):171–186. https://doi.org/10.1515/reveh-2018-0078

Imran U, Ullah A, Shaikh K, Mehmood R, Saeed M (2019) Health risk assessment of the exposure of heavy metal contamination in surface water of lower Sindh, Pakistan. SN Appl Sci 1(6):589. https://doi.org/10.1007/s42452-019-0594-1

Imran U, Weidhaas J, Ullah A, Shaikh K (2020) Risk associated with spatio-temporal variations in trace metals and a metalloid in a major freshwater reservoir of Pakistan. Human Ecol Risk Assess: Int J 1–20. https://doi.org/10.1080/10807039.2020.1729090

Iqbal J, Shah MH (2013) Health risk assessment of metals in surface water from freshwater source lakes, Pakistan. Human Ecol Risk Assess: Int J 19(6):1530–1543. https://doi.org/10.1080/10807039.2012.716681

Iqbal J, Shah MH (2014) Study of seasonal variations and health risk assessment of heavy metals in *Cyprinus carpio* from Rawal Lake, Pakistan. Environ Monit Assess 186(4):2025–2037. https://doi.org/10.1007/s10661-013-3515-6

Iqbal J, Shah MH, Akhter G (2013) Characterization, source apportionment and health risk assessment of trace metals in freshwater Rawal Lake, Pakistan. J Geochem Explor 125:94–101

Iqbal A, Tabinda AB, Yasar A, Mahfooz Y (2017) Heavy metal uptake and toxicity in tissues of commercially important freshwater fish (*Labeo rohita* and *Wallago attu*) from the Indus River, Pakistan. Polish J Environ Stud 26(2):627–633

Jaishankar M, Tseten T, Anbalagan N, Mathew BB, Beeregowda KN (2014) Toxicity, mechanism and health effects of some heavy metals. Interdisc Toxicol 7(2):60–72

Järup L (2003) Hazards of heavy metal contamination. Br Med Bull 68(1):167–182. https://doi.org/10.1093/bmb/ldg032

Javed M, Mahmood G (2001) Metal toxicity of water in a stretch of river Ravi from Shahdera to Baloki headworks [Pakistan]. Pak J Agric Sci (Pak) 38(1–2):37–42

Javed T, Ahmad N, Mashiatullah A (2018) Heavy metals contamination and ecological risk assessment in surface sediments of Namal Lake, Pakistan. Polish J Environ Stud 27 (2):675–688

JECFA (2011) Evaluation of certain food additives and contaminants: seventy-third [73rd] report of the Joint Food and Agriculture Organisation/World Health Organization Expert Committee on food additives. World Health Organization, Geneva, Switzerland

Jezierska B, Witeska M (2006) The metal uptake and accumulation in fish living in polluted waters. In: Twardowska I, Allen HE, Häggblom MM, Stefaniak S (eds) Soil and water pollution monitoring, protection and remediation. Springer, The Netherlands, pp 107–114

Junaid M, Hashmi MZ, Tang Y-M, Malik RN, Pei D-S (2017) Potential health risk of heavy metals in the leather manufacturing industries in Sialkot, Pakistan. Sci Rep 7(1):8848. https://doi.org/10.1038/s41598-017-09075-7

Kahlown MA, Majeed A (2003) Water-resources situation in Pakistan: challenges and future strategies. In: COMSATS' series of publications on science and technology: water resources in the south: present scenario and future prospects, Islamabad, Pakistan, pp 21–40

Kashif S, Akram M, Yaseen M, Ali S (2009) Studies on heavy metals status and their uptake by vegetables in adjoining areas of Hudiara drain in Lahore. Soil Environ 28(1):7–12

Kassouf A, Chebib H, Lebbos N, Ouaini R (2013) Migration of iron, lead, cadmium and tin from tinplate-coated cans into chickpeas. Food Addit Contam Part A 30(11):1987–1992

Khalil A, Jamil A, Khan T (2020) Assessment of heavy metal contamination and human health risk with oxidative stress in fish (*Cyprinus carpio*) from Shahpur Dam, Fateh Jang, Pakistan. Arab J Geosci 13(18):928. https://doi.org/10.1007/s12517-020-05933-3

Khan A, Arshad S (2014) Wetlands of Pakistan: distribution, degradation and management. Pak Geographical Rev 69(1):28–45

Khan M, Zaidi A, Oves M, Wani P (2008) Heavy metal toxicity to legumes. In: Samuel EB, William CW (eds) Heavy metal pollution. Nova Science, Hauppauge, pp 197–225

Khan MA, Islam MR, Panaullah GM, Duxbury JM, Jahiruddin M, Loeppert RH (2010) Accumulation of arsenic in soil and rice under wetland condition in Bangladesh. Plant Soil 333 (1):263–274. https://doi.org/10.1007/s11104-010-0340-3

Khan T, Muhammad S, Khan B, Khan H (2011) Investigating the levels of selected heavy metals in surface water of Shah Alam River (a tributary of River Kabul, Khyber Pakhtunkhwa). J Himalayan Earth Sci 44(2):71–79

Khan K, Lu Y, Khan H, Ishtiaq M, Khan S, Waqas M, Wei L, Wang T (2013a) Heavy metals in agricultural soils and crops and their health risks in Swat District, northern Pakistan. Food Chem Toxicol 58:449–458. https://doi.org/10.1016/j.fct.2013.05.014

Khan K, Lu Y, Khan H, Zakir S, Ihsanullah Khan S, Khan AA, Wei L, Wang T (2013b) Health risks associated with heavy metals in the drinking water of Swat, northern Pakistan. J Environ Sci 25(10):2003–2013. https://doi.org/10.1016/S1001-0742(12)60275-7

Khan MU, Malik RN, Muhammad S (2013c) Human health risk from heavy metal via food crops consumption with wastewater irrigation practices in Pakistan. Chemosphere 93(10):2230–2238. https://doi.org/10.1016/j.chemosphere.2013.07.067

Khan S, Rauf R, Muhammad S, Qasim M, Din I (2016) Arsenic and heavy metals health risk assessment through drinking water consumption in the Peshawar District, Pakistan. Human Ecol Risk Assess: Int J 22(3):581–596. https://doi.org/10.1080/10807039.2015.1083845

Khan MA, Khan S, Khan A, Alam M (2017) Soil contamination with cadmium, consequences and remediation using organic amendments. Sci Total Environ 601–602:1591–1605. https://doi.org/10.1016/j.scitotenv.2017.06.030

Khan ZI, Ugulu I, Sahira S, Ahmad K, Ashfaq A, Mehmood N, Dogan Y (2018) Determination of toxic metals in fruits of *Abelmoschus esculentus* grown in contaminated soils with different irrigation sources by spectroscopic method. Int J Environ Res 12(4):503–511

Kopp B, Zalko D, Audebert M (2018) Genotoxicity of 11 heavy metals detected as food contaminants in two human cell lines. Environ Mol Mutagen 59(3):202–210. https://doi.org/10.1002/em.22157

Korai AL, Lashari KH, Sahato GA, Kazi TG (2010) Histological lesions in gills of feral cyprinids, related to the uptake of waterborne toxicants from Keenjhar Lake. Rev Fish Sci 18(2):157–176. https://doi.org/10.1080/10641260903430522

Korai MS, Mahar RB, Uqaili MA (2016) Optimization of waste to energy routes through biochemical and thermochemical treatment options of municipal solid waste in Hyderabad, Pakistan. Energy Convers Manag 124:333–343. https://doi.org/10.1016/j.enconman.2016.07.032

Kumar RR, Park BJ, Cho JY (2013) Application and environmental risks of livestock manure. J Korean Soc Appl Biol Chem 56(5):497–503

Kumar S, Dubey RS, Tripathi RD, Chakrabarty D, Trivedi PK (2015) Omics and biotechnology of arsenic stress and detoxification in plants: current updates and prospective. Environ Int 74:221–230

Kumar V, Parihar RD, Sharma A, Bakshi P, Singh Sidhu GP, Bali AS, Karaouzas I, Bhardwaj R, Thukral AK, Gyasi-Agyei Y, Rodrigo-Comino J (2019) Global evaluation of heavy metal content in surface water bodies: a meta-analysis using heavy metal pollution indices and multivariate statistical analyses. Chemosphere 236: https://doi.org/10.1016/j.chemosphere.2019.124364

Kumar A, Kumar A, MMS C-P, Chaturvedi AK, Shabnam AA, Subrahmanyam G, Mondal R, Gupta DK, Malyan SK, Kumar SS, Khan SA, Yadav KK (2020) Lead toxicity: Health hazards, influence on food chain, and sustainable remediation approaches. Int J Environ Res Public Health 17(7). https://doi.org/10.3390/ijerph17072179

Lal K, Yadav RK, Kaur R, Bundela DS, Khan MI, Chaudhary M, Meena RL, Dar SR, Singh G (2013) Productivity, essential oil yield, and heavy metal accumulation in lemon grass (*Cymbopogon flexuosus*) under varied wastewater–groundwater irrigation regimes. Ind Crops Prod 45:270–278. https://doi.org/10.1016/j.indcrop.2013.01.004

Li J, Luo G, Xu J (2019) Fate and ecological risk assessment of nutrients and metals in sewage sludge from ten wastewater treatment plants in Wuxi City, China. Bull Environ Contam Toxicol 102(2):259–267

Liu X, Song Q, Tang Y, Li W, Xu J, Wu J, Wang F, Brookes PC (2013) Human health risk assessment of heavy metals in soil-vegetable system: a multi-medium analysis. Sci Total Environ 463–464:530–540. https://doi.org/10.1016/j.scitotenv.2013.06.064

Luo J, Ye Y, Gao Z, Wang W (2014) Essential and nonessential elements in the red-crowned crane *Grus japonensis* of Zhalong Wetland, northeastern China. Toxicol Environ Chem 96 (7):1096–1105

Mahfooz Y, Yasar A, Sohail MT, Tabinda AB, Rasheed R, Irshad S, Yousaf B (2019) Investigating the drinking and surface water quality and associated health risks in a semi-arid multi-industrial metropolis (Faisalabad), Pakistan. Environ Sci Pollut Res 26(20):20853–20865. https://doi.org/10.1007/s11356-019-05367-9

Mahmood A, Malik RN (2014) Human health risk assessment of heavy metals via consumption of contaminated vegetables collected from different irrigation sources in Lahore, Pakistan. Arab J Chem 7(1):91–99. https://doi.org/10.1016/j.arabjc.2013.07.002

Malik D, Maurya PK (2014) Heavy metal concentration in water, sediment, and tissues of fish species (*Heteropneustis fossilis* and *Puntius ticto*) from Kali River, India. Toxicol Environ Chem 96(8):1195–1206

Malik Z, Ahmad M, Abassi GH, Dawood M, Hussain A, Jamil M (2017) Agrochemicals and soil microbes: interaction for soil health. In: Hashmi MZ, Kumar V, Varma A (eds) Xenobiotics in the soil environment: monitoring, toxicity and management. Springer International Publishing, Cham, pp 139–152. https://doi.org/10.1007/978-3-319-47744-2_11

Mansouri B, Ebrahimpour M, Babaei H (2012) Bioaccumulation and elimination of nickel in the organs of black fish (*Capoeta fusca*). Toxicol Ind Health 28(4):361–368

Manzoor J, Sharma M, Wani KA (2018) Heavy metals in vegetables and their impact on the nutrient quality of vegetables: a review. J Plant Nutr 41(13):1744–1763. https://doi.org/10.1080/01904167.2018.1462382

Martín JR, De Arana C, Ramos-Miras J, Gil C, Boluda R (2015) Impact of 70 years urban growth associated with heavy metal pollution. Environ Pollut 196:156–163

Massadeh AM, Allah A, Al-Massaedh T (2018) Determination of heavy metals in canned fruits and vegetables sold in Jordan market. Environ Sci Pollut Res 25(2):1914–1920

Mastoi GM, Shah SGS, Khuhawar MY (2008) Assessment of water quality of Manchar Lake in Sindh (Pakistan). Environ Monit Assess 141(1–3):287–296

Mehmood A, Aslam Mirza M, Aziz Choudhary M, Kim K-H, Raza W, Raza N, Soo Lee S, Zhang M, Lee J-H, Sarfraz M (2019) Spatial distribution of heavy metals in crops in a wastewater irrigated zone and health risk assessment. Environ Res 168:382–388. https://doi.org/10.1016/j.envres.2018.09.020

Merian E (1984) Introduction on environmental chemistry and global cycles of chromium, nickel, cobalt, beryllium, arsenic, cadmium and selenium, and their derivatives. Toxicol Environ Chem 8(1):9–38. https://doi.org/10.1080/02772248409357038

Midrar-Ul-Haq K, RA P, HK S, MS M KS (2005) Surface and ground water contamination in NWFP and Sindh provinces with respect to trace elements. Int J Agric Biol 7(2):214–217

Mishra S, Bharagava RN, More N, Yadav A, Zainith S, Mani S, Chowdhary P (2019) Heavy metal contamination: an alarming threat to environment and human health. In: Sobti RC, Arora NK, Kothari R (eds) Environmental biotechnology: for sustainable future. Springer, Singapore, pp 103–125. https://doi.org/10.1007/978-981-10-7284-0_5

Mousavi Z, Ziarati P, Esmaeli Dehaghi M, Qomi M (2014) Heavy metals (lead and cadmium) in some medicinal herbal products in Iranian market. Iran J Toxicol 8(24):1004–1010

Muhammad S, Ahmad K (2020) Heavy metal contamination in water and fish of the Hunza River and its tributaries in Gilgit–Baltistan: evaluation of potential risks and provenance. Environ Technol Innov 20: https://doi.org/10.1016/j.eti.2020.101159

Muhammad S, Shah MT, Khan S (2011) Health risk assessment of heavy metals and their source apportionment in drinking water of Kohistan region, northern Pakistan. Microchem J 98 (2):334–343. https://doi.org/10.1016/j.microc.2011.03.003

Mulk S, Korai AL, Azizullah A, Shahi L, Khattak MNK (2017) Marble industry effluents cause an increased bioaccumulation of heavy metals in Mahaseer (*Tor putitora*) in Barandu River, district Buner, Pakistan. Environ Sci Pollut Res 24(29):23039–23056. https://doi.org/10.1007/s11356-017-9921-5

Murtaza G, Zia M (2012) Wastewater production, treatment and use in Pakistan. In: Second regional workshop of the project 'safe use of wastewater in agriculture, pp 16–18

Murtaza G, Javed W, Hussain A, Wahid A, Murtaza B, Owens G (2015) Metal uptake via phosphate fertilizer and city sewage in cereal and legume crops in Pakistan. Environ Sci Pollut Res 22(12):9136–9147

Nagajyoti PC, Lee KD, Sreekanth TVM (2010) Heavy metals, occurrence and toxicity for plants: a review. Environ Chem Lett 8(3):199–216. https://doi.org/10.1007/s10311-010-0297-8

Nagamatsu PC, Vargas DÁR, Prodocimo MM, Opuskevitch I, Ferreira FCAS, Zanchin N, de Oliveira Ribeiro CA, de Souza C (2020) Synthetic fish metallothionein design as a potential tool for monitoring toxic metals in water. Environ Sci Pollut Res. https://doi.org/10.1007/s11356-020-11427-2

Nawab J, Farooqi S, Xiaoping W, Khan S, Khan A (2018a) Levels, dietary intake, and health risk of potentially toxic metals in vegetables, fruits, and cereal crops in Pakistan. Environ Sci Pollut Res 25(6):5558–5571. https://doi.org/10.1007/s11356-017-0764-x

Nawab J, Khan S, Xiaoping W (2018b) Ecological and health risk assessment of potentially toxic elements in the major rivers of Pakistan: general population vs. fishermen. Chemosphere 202:154–164. https://doi.org/10.1016/j.chemosphere.2018.03.082

Nazeer S, Hashmi MZ, Malik RN (2014) Heavy metals distribution, risk assessment and water quality characterization by water quality index of the River Soan, Pakistan. Ecol Ind 43:262–270

Nazif W, Perveen S, Shah SA (2006) Evaluation of irrigation water for heavy metals of Akbarpura area. J Agric Biol Sci 1(1):51–54

Neal AP, Guilarte TR (2013) Mechanisms of lead and manganese neurotoxicity. Toxicol Res (Camb) 2(2):99–114. https://doi.org/10.1039/c2tx20064c

Nicholson FA, Smith SR, Alloway BJ, Carlton-Smith C, Chambers BJ (2003) An inventory of heavy metals inputs to agricultural soils in England and Wales. Sci Total Environ 311(1):205–219. https://doi.org/10.1016/S0048-9697(03)00139-6

Nishijo M, Nakagawa H, Suwazono Y, Nogawa K, Kido T (2017) Causes of death in patients with Itai-itai disease suffering from severe chronic cadmium poisoning: a nested case-control analysis of a follow-up study in Japan. BMJ Open 7(7):

NSDWQ-Pak (2008) National standards for drinking water quality. Health Services Academy, Ministry of Health, Government of Pakistan

Nussey H (2000) Bioaccumulation of chromium, manganese, nickel and lead in the tissues of the moggel, *Labeo umbratus* (Cyprinidae), from Witbank Dam, Mpumalanga. Water SA 26 (2):269–284

Okatch H, Ngwenya B, Raletamo KM, Andrae-Marobela K (2012) Determination of potentially toxic heavy metals in traditionally used medicinal plants for HIV/AIDS opportunistic infections in Ngamiland District in Northern Botswana. Anal Chim Acta 730:42–48

Olsson P-E, Kling P, Hogstrand C (1998) Mechanisms of heavy metal accumulation and toxicity in fish. In: Langston WJ, Bebianno MJ (eds) Metal metabolism in aquatic environments. Springer, Boston, pp 321–350. https://doi.org/10.1007/978-1-4757-2761-6_10

Panaullah GM, Alam T, Hossain MB, Loeppert RH, Lauren JG, Meisner CA, Ahmed ZU, Duxbury JM (2008) Arsenic toxicity to rice (*Oryza sativa* L.) in Bangladesh. Plant Soil 317 (1):31. https://doi.org/10.1007/s11104-008-9786-y

Pascual I, Azcona I, Aguirreolea J, Fn Morales, Corpas FJ, Palma JM, Rellan-Alvarez R, Sanchez-Diaz M (2010) Growth, yield, and fruit quality of pepper plants amended with two sanitized sewage sludges. J Agric Food Chem 58(11):6951–6959

PCRWR (2008) Arsenic contamination in ground water of central Sindh Phase I. Islamabad, Pakistan: Pakistan council for Research in Water Resources (PCRWR), pp 19–38. Retrieved from http://www.pcrwr.gov.pk/Arsenic_CS/ACS_TOC.htm

Peijnenburg W, Baerselman R, De Groot A, Jager T, Leenders D, Posthuma L, Van Veen R (2000) Quantification of metal bioavailability for lettuce (*Lactuca sativa* L.) in field soils. Arch Environ Contamination Toxicol 39(4):420–430

Peralta-Videa JR, Lopez ML, Narayan M, Saupe G, Gardea-Torresdey J (2009) The biochemistry of environmental heavy metal uptake by plants: implications for the food chain. Int J Biochem Cell Biol 41(8–9):1665–1677

Podgorski JE, Eqani S, Khanam T, Ullah R, Shen H, Berg M (2017) Extensive arsenic contamination in high-pH unconfined aquifers in the Indus Valley. Sci Adv 3(8): https://doi.org/10.1126/sciadv.1700935

Qureshi AS (2011) Water management in the Indus basin in Pakistan: challenges and opportunities. Mt Res Dev 31(3):252–260

Qureshi AL, Mahessar AA, Leghari ME-U-H, Lashari BK, Mari FM (2015) Impact of releasing wastewater of sugar industries into drainage system of LBOD, Sindh, Pakistan. Int J Environ Sci Dev 6(5):381

Raber G, Stock N, Hanel P, Murko M, Navratilova J, Francesconi KA (2012) An improved HPLC–ICPMS method for determining inorganic arsenic in food: application to rice, wheat and tuna fish. Food Chem 134(1):524–532. https://doi.org/10.1016/j.foodchem.2012.02.113

Rai V, Vajpayee P, Singh SN, Mehrotra S (2004) Effect of chromium accumulation on photosynthetic pigments, oxidative stress defense system, nitrate reduction, proline level and eugenol content of *Ocimum tenuiflorum* L. Plant Sci 167(5):1159–1169

Rasool A, Xiao T, Baig ZT, Masood S, Mostofa KMG, Iqbal M (2015) Co-occurrence of arsenic and fluoride in the groundwater of Punjab, Pakistan: source discrimination and health risk assessment. Environ Sci Pollut Res 22(24):19729–19746. https://doi.org/10.1007/s11356-015-5159-2

Raza N, Niazi SB, Sajid M, Iqbal F, Ali M (2007) Studies on relationship between season and inorganic elements of Kallar Kahar Lake (Chakwal), Pakistan. J Res (Sci) Bahauddin Zakariya Univ 18:61–68

Rehman W, Zeb A, Noor N, Nawaz M (2008) Heavy metal pollution assessment in various industries of Pakistan. Environ Geol 55(2):353–358. https://doi.org/10.1007/s00254-007-0980-7

Rehman Ku, Bukhari SM, Andleeb S, Mahmood A, Erinle KO, Naeem MM, Imran Q (2019) Ecological risk assessment of heavy metals in vegetables irrigated with groundwater and wastewater: the particular case of Sahiwal district in Pakistan. Agric Water Manag 226: https://doi.org/10.1016/j.agwat.2019.105816

Riaz U, Murtaza G, Farooq M (2018) Influence of different sewage sludges and composts on growth, yield, and trace elements accumulation in rice and wheat. Land Degrad Dev 29 (5):1343–1352

Rose M, Lewis J, Langford N, Baxter M, Origgi S, Barber M, MacBain H, Thomas K (2007) Arsenic in seaweed—forms, concentration and dietary exposure. Food Chem Toxicol 45 (7):1263–1267. https://doi.org/10.1016/j.fct.2007.01.007

Ryan JA, Pahren HR, Lucas JB (1982) Controlling cadmium in the human food chain: a review and rationale based on health effects. Environ Res 28(2):251–302

Saleem M, Iqbal J, Shah MH (2014) Dissolved concentrations, sources, and risk evaluation of selected metals in surface water from Mangla Lake, Pakistan. Sci World J 2014: https://doi.org/10.1155/2014/948396

Sanjrani M, Mek T, Sanjrani N, Leghari S, Moryani H, Shabnam A (2017) Current situation of aqueous arsenic contamination in Pakistan, focused on Sindh and Punjab Province, Pakistan: a review. J Pollut Eff Cont 5(207):2

Sattar A, Wahid M, Durrani SK (1989) Concentration of selected heavy metals in spices, dry fruits and plant nuts. Plant Foods Hum Nutr 39(3):279–286. https://doi.org/10.1007/BF01091938

Sebastian A, Prasad MNV (2014) Cadmium minimization in rice. A review. Agronomy Sustain Dev 34(1):155–173

Shafiq HB, Ajaz M, Rasool SA (2011) Bacterial and toxic pollutants in lakes of river Indus. Pak J Bot 43(3):1765–1772

Shah MT, Begum S, Khan S (2010) Pedo and biogeochemical studies of mafic and ultramfic rocks in the Mingora and Kabal areas, Swat, Pakistan. Environ Earth Sci 60(5):1091–1102

Shah MT, Ara J, Muhammad S, Khan S, Tariq S (2012) Health risk assessment via surface water and sub-surface water consumption in the mafic and ultramafic terrain, Mohmand agency, northern Pakistan. J Geochem Explor 118:60–67. https://doi.org/10.1016/j.gexplo.2012.04.008

Shakir SK, Azizullah A, Murad W, Daud MK, Nabeela F, Rahman H, Ur Rehman S, Häder D-P (2016) Toxic metal pollution in Pakistan and its possible risks to public health. In: de Voogt P (ed) Reviews of environmental contamination and toxicology, vol 242. Springer International Publishing, Cham, pp 1–60. https://doi.org/10.1007/398_2016_9

Sharpe RT, Livesey CT (2006) Lead poisoning in cattle and its implications for food safety. Vet Rec 159(3):71–74. https://doi.org/10.1136/vr.159.3.71

Sheetal KR, Singh SD, Anand A, Prasad S (2016) Heavy metal accumulation and effects on growth, biomass and physiological processes in mustard. Indian J Plant Physiol 21(2):219–223. https://doi.org/10.1007/s40502-016-0221-8

Shukla V (2018) Pakistan's alarming water crisis: country to run out of clean water by 2025. Politics. https://www.valuewalk.com/2018/05/pakistan-water-crisis-clean-water-2025/. Accessed 8 June 2018

Sial R, Chaudhary M, Abbas S, Latif M, Khan A (2006) Quality of effluents from Hattar industrial estate. J Zhejiang Univ Sci B 7(12):974–980

Siraj M, Khisroon M, Khan A (2016) Bioaccumulation of heavy metals in different organs of *Wallago attu* from River Kabul Khyber Pakhtunkhwa, Pakistan. Biol Trace Element Res 172 (1):242–250

Sleet P (2019) Water resources in Pakistan: scarce, polluted and poorly governed. Independent strategic analysis of Australia's global interest, Nedlands, p 2

Spiegel H (2002) Trace element accumulation in selected bioindicators exposed to emissions along the industrial facilities of Danube Lowland. Turk J Chem 26(6):815–824

Street RA, Kulkarni MG, Stirk WA, Southway C, Van Staden J (2008) Variation in heavy metals and microelements in South African medicinal plants obtained from street markets. Food Addit Contaminants: Part A 25(8):953–960. https://doi.org/10.1080/02652030801993605

Tahir M, Rasheed H, Imran S (2010) Water quality status in rural areas of Pakistan. Pakistan Council of Research in Water Resources, Islamabad, Pakistan. Report

Tchounwou PB, Yedjou CG, Patlolla AK, Sutton DJ (2012) Heavy metal toxicity and the environment. Exp Suppl 101:133–164. https://doi.org/10.1007/978-3-7643-8340-4_6

Thompson J, Bannigan J (2008) Cadmium: toxic effects on the reproductive system and the embryo. Reprod Toxicol 25(3):304–315. https://doi.org/10.1016/j.reprotox.2008.02.001

Tóth G, Hermann T, Da Silva MR, Montanarella L (2016) Heavy metals in agricultural soils of the European Union with implications for food safety. Environ Int 88:299–309. https://doi.org/10.1016/j.envint.2015.12.017

Tufail M, Khalid S (2008) Heavy metal pollution from medical waste incineration at Islamabad and Rawalpindi, Pakistan. Microchem J 90(1):77–81

Ullah Z, Khan H, Waseem A, Mahmood Q, Farooq U (2013) Water quality assessment of the River Kabul at Peshawar, Pakistan: industrial and urban wastewater impacts. J Water Chem Technol 35(4):170–176

Ullah R, Muhammad S, Jadoon IAK (2019) Potentially harmful elements contamination in water and sediment: evaluation for risk assessment and provenance in the northern Sulaiman fold belt, Baluchistan, Pakistan. Microchem J 147:1155–1162. https://doi.org/10.1016/j.microc.2019.04.053

Ullah I, Ditta A, Imtiaz M, Mehmood S, Rizwan M, Rizwan MS, Jan AU, Ahmad I (2020) Assessment of health and ecological risks of heavy metal contamination: a case study of agricultural soils in Thall, Dir-Kohistan. Environ Monit Assess 192(12):1–19

Valsecchi G, Gigliotti C, Farini A (1995) Microbial biomass, activity, and organic matter accumulation in soils contaminated with heavy metals. Biol Fertil Soils 20(4):253–259. https://doi.org/10.1007/BF00336086

Waseem A, Arshad J, Iqbal F, Sajjad A, Mehmood Z, Murtaza G (2014) Pollution status of Pakistan: a retrospective review on heavy metal contamination of water, soil, and vegetables. BioMed Res Int 2014

Webber J (1981) Trace metals in agriculture. In: Lepp NW (eds) Effect of heavy metal pollution on plants: metals in the environment. Springer, Dordrecht, pp 159–184. https://doi.org/10.1007/978-94-009-8099-0_5

WHO (1981) Task group on environmental health criteria for arsenic. Environmental Health Criteria, vol 18

WHO (1996) Guidelines for drinking water quality, addendum to volume 2: health criteria and other supporting information. WHO Publications

WHO (2011) Guidelines for drinking-water quality: World Health Organization. Distribution and Sales, Geneva 27

Widianarko B, Van Gestel C, Verweij R, Van Straalen N (2000) Associations between trace metals in sediment, water, and guppy, *Poecilia reticulata* (Peters), from urban streams of Semarang, Indonesia. Ecotoxicol Environ Saf 46(1):101–107

Wilbur S, Abadin H, Fay M, Yu D, Tencza B, Ingerman L, Klotzbach J, James S (2012) Agency for toxic substances and disease registry (ATSDR) toxicological profiles. In: Toxicological profile for chromium. Agency for Toxic Substances and Disease Registry (US), Atlanta, GA

Yang Y, Nan Z, Zhao Z (2014) Bioaccumulation and translocation of cadmium in wheat (*Triticum aestivum* L.) and maize (*Zea mays* L.) from the polluted oasis soil of Northwestern China. Chem Speciat Bioavailab 26(1):43–51. https://doi.org/10.3184/095422914x13888342841789

Yargholi B, Azimi A, Baghvand A, Liaghat A, Fardi G (2008) Investigation of cadmium absorption and accumulation in different parts of some vegetables. American-Eurasian J Agric Environ Sci 3(3):357–364

Yousafzai AM, Khan AR, Shakoori AR (2010) Pollution of large, subtropical rivers-river Kabul, Khyber-Pakhtun Khwa province, Pakistan): physico-chemical indicators. Pak J Zoology 42(6):795–808

Yousafzai AM, Siraj M, Habib A, Chivers DP (2012) Bioaccumulation of heavy metals in common carp: implications for human health. Pak J Zoology 44(2):489–494

Yousafzai AM, Ullah F, Bari F, Raziq S, Riaz M, Khan K, Nishan U, Sthanadar IA, Shaheen B, Shaheen M, Ahmad H (2017) Bioaccumulation of some heavy metals: analysis and comparison of *Cyprinus carpio* and *Labeo rohita* from Sardaryab, Khyber Pakhtunkhwa. BioMed Res Int 2017:5801432. https://doi.org/10.1155/2017/5801432

Youssef DH, Tayel FT (2004) Metal accumulation by three Tilapia spp. from some Egyptian inland waters. Chem Ecol 20(1):61–71. https://doi.org/10.1080/02757540310001642689

Yuan G, Dai S, Yin Z, Lu H, Jia R, Xu J, Song X, Li L, Shu Y, Zhao X (2014) Toxicological assessment of combined lead and cadmium: acute and sub-chronic toxicity study in rats. Food Chem Toxicol 65:260–268. https://doi.org/10.1016/j.fct.2013.12.041

Zhang N, Zang S, Sun Q (2014) Health risk assessment of heavy metals in the water environment of Zhalong Wetland, China. Ecotoxicology 23(4):518–526

Zhao F-J, Wang P (2020) Arsenic and cadmium accumulation in rice and mitigation strategies. Plant Soil 446(1):1–21. https://doi.org/10.1007/s11104-019-04374-6

Zheljazkov VD, Nielsen NE (1996) Effect of heavy metals on peppermint and cornmint. Plant Soil 178(1):59–66

Chapter 14
Arsenic Pollution

Donat-P. Häder

Abstract Arsenic is a frequent element on Earth and occurs on all continents and has been detected in over 70 countries. It is found in 10% of the global tube wells providing groundwater for millions of people. Since it has no odor, taste or color it is often not detected especially in developing countries where it causes major deleterious effects on the biota and human health. Chronic exposure to arsenic (especially trivalent As) causes skin problems, brain damage, lung and cardiovascular diseases as well as various cancers. It accumulates in plants and enters aquatic food webs. Arsenic can be detected by laboratory tests as well as field tests, but the latter are often not very reliable. Other means of detection are by several bioassays. There are several strategies of removal of arsenic from water for human consumption as well as industrial and agricultural use such as precipitation, reverse osmosis, ion exchange or adsorption. For developing countries, cheap filtration units have been developed and tested in the field.

Keywords Arsenic · Human health · Aquatic ecosystems · Detection · Quantification · Remediation

14.1 Introduction

Growing human populations are accompanied by increasing pollution of air, land and water. In addition to persistent organic pollutions (POP), pesticides, fossil oil spillage and fertilizers are accumulating in aquatic ecosystems. Heavy metals are of special interest because they cause major damage to ecosystems and human health. Arsenic pollution has been increasing during the past few decades in many countries around the Earth especially in South America and Asia due to increased water usage from dug wells. High arsenic concentrations in Canada, the Western United States, Argentina, Brazil, Bolivia and 13 other Latin American countries constitute

D.-P. Häder (✉)
Department of Biology, Friedrich-Alexander Universität, 91054 Erlangen, Germany
e-mail: donat@dphaeder.de

© The Author(s), under exclusive license to Springer Nature Switzerland AG 2021
D.-P. Häder et al. (eds.), *Anthropogenic Pollution of Aquatic Ecosystems*,
https://doi.org/10.1007/978-3-030-75602-4_14

a major health risk to millions of people (Bundschuh et al. 2012; Welch et al. 1988; Wang and Mulligan 2006). High concentrations also occur in Vietnam, Bangladesh, India and China (Berg et al. 2001; Meharg and Rahman 2003; Rodríguez-Lado et al. 2013; Chakraborti et al. 2010).

Arsenic poisoning can result from skin absorption, intake of polluted food and air but the major exposure threat is from water. Geogenic arsenic accumulates in groundwater and pollutes drinking water pumped up for human and animal consumption as well as for irrigation (Shankar and Shanker 2014; McGrory et al. 2017; Rasool et al. 2017). At least 140 million people in Asia consume arsenic-polluted water. The arsenic concentrations in groundwater often exceed the WHO recommended limit of 10 μg/L by far. For example, the concentrations samples collected from cities and villages in Tehsil Mailsi, Punjab, Pakistan ranged from 12 to 4,++48.5 μg/L. Groundwater samples from small wells in Vietnam have been found to contain between 1 and 3,050 μg/L arsenic with an average concentration of 159 μg/L (Berg et al. 2001). The groundwater from an aquifer that supplies water for Hanoi showed between 37 and 320 μg/L. While the treatment for iron removal lowered the concentration to 25–81 μg/L, 50% of the samples exceeded the limit of 50 μg/L established for Vietnam. The pollutant concentration in the sediments is related to iron oxyhydroxides which leach to the groundwater after reduction of iron (Berg et al. 2001).

Canada has decreased the limit for arsenic from 50 to 25 μg/L and considers a further reduction to 5 μg/L. The U.S. reduced the limit from 50 to 10 μg/L (Kapaj et al. 2006). In India, many of the more than 18 million wells in the country exceed this level (Daigle 2016). These wells had been drilled, financed by a 125 million US $ program in 1969 aided by international groups including UNICEF and several subsequent programs in order to replace surface water for human consumption which is often polluted by pathogens, sewage and agricultural and industrial runoff. Groundwater from these small wells is used for drinking, cooking and irrigation by about 80% of the rural population and about 50% of the people in the cities in India with a total population of about 1.25 billion. Most wells reach a depth of 50–200 m which is the first water layer free of bacteria. Unfortunately, this is also the layer which contains the highest arsenic concentrations. Since arsenic has no taste, color or odor the problem is usually not noticed. Furthermore, drilling deeper requires more time and money and sturdier materials which poor people cannot afford (Daigle 2016).

The aim of this review is to state the global distribution of arsenic pollution in groundwater and other sources of water for irrigation, industry and human consumption. It analyzes the effects on human health and the environment and summarizes the methods of detection and quantification as well as remediation.

14.2 Global Distribution

Since arsenic is a frequent metal on Earth it is not amazing that it is found in many countries and on every continent. Detailed information on the concentrations is available for Europe, South East Asia and part of North America. Many countries in South America also have high arsenic concentrations which are due to anthropogenic activities such as mining and drilling for water. White areas exist in the maps of Northern Canada, Siberia most of Africa and Australia. Four main mechanisms for arsenic accumulation can be listed (Ravenscroft 2007):

- Reductive-dissolution in anoxic waters with neutral pH, high Fe, Mn, NH_4, HCO_3 concentrations, but low SO_4, NO_3
- Alkali-desorption results in alkaline oxic waters at pH ≥ 8 with low Fe concentrations
- Sulfide oxidation produces acid-sulfate waters at very low pH and high SO_4 concentrations
- Geothermal waters at high temperatures and high Cl concentrations.

In addition, evaporation of surface waters can concentrate arsenic.

High concentrations of arsenic can be found in alluvial deposits by rivers, wind and in ancient lakes as well as in alluvial volcanic deposits (Majumder et al. 2014; Ravenscroft 2007). Also from the alluvial period stem glacial and fluvio-glacial deposits over limestone (Warner 2001). In dry tropical areas laterite with high Fe concentrations often contains high arsenic concentrations (Partey et al. 2008). Tertiary intracontinental sediments and volcanic deposits often show high concentrations (Jamali et al. 2010). Furthermore, Paleozoic and Mesozoic sedimentary and metamorphic rocks have been found to be a source of arsenic as well as Precambrian-Paleozoic crystalline bedrock (Turekian and Wedepohl 1961).

14.3 Effects of Arsenic Pollution on Humans and the Environment

It should be mentioned in passing that arsenic has been a preferred poison to eradicate animal pests and people because this metalloid is tasteless, colorless and odorless and it is lethal at low concentrations. It had been speculated that the unexpected deaths of Francesco de'Medici and his wife Bianca Capello in 1587 were due to arsenic poisoning which was confirmed in 1945 when samples of the bodies were analyzed by inductive coupled plasma atomic emission spectrometry (Mari et al. 2006). The arsenic concentrations determined in several soft tissue samples were in the range found in a series of fatalities attributed to accidental or intentional arsenic poisoning. Many other records of death due to acute or chronic arsenic poisoning can be found in the literature (Poklis and Saady 1990; Bednarczyk and Matusiak 1982).

Historically arsenic was used for medical purposes such as the treatment of syphilis, asthma, rheumatism, cough, pruritus and itching (Gontijo and Bittencourt 2005; Phelps and Washburn 1930). Both arsenite and arsenate are toxic to human health (arsenicosis) and result in a plethora of ailments. However, compounds with As(III) are much more toxic than those with As(V); also inorganic arsenic is commonly assumed to be more toxic than arsenic in organic compounds (Edmonds and Francesconi 1993). Exposure to arsenic in drinking water affects humans which manifests itself by resulting in skin scarring and accumulation in hair and nails visible by hyperpigmentation and keratosis. Low arsenic concentrations in drinking water may result in higher concentrations in urine without inducing clinical symptoms. Chronic ingestion of arsenic-containing drinking water causes deleterious effects at lower concentrations than estimated before (Kapaj et al. 2006). It may result in lung diseases such as chronic bronchitis, chronic obstructive pulmonary disease and bronchiectasis as well as liver problems such as non-cirrhotic portal fibrosis (Mazumder 2008). When arsenic accumulates in the body it results in brain damage, cardiovascular disease, polyneuropathy and cancer of the skin as well as internal organs such as lung and bladder and has been linked with diabetes mellitus, non-pitting edema of hands and feet, weakness and anemia. Other deleterious effects are diseases of the skin and blood as well as the lymphatic and reproductive systems (Singh et al. 2007). Arsenic poisoning at levels below 10 μg/L can affect hormone regulation and gene transcription which may lead to fetal loss or premature delivery resulting in reduced birth weights of infants (Kapaj et al. 2006). The most effective way to reduce the effects of arsenicosis is avoiding drinking arsenic-polluted water since chelation therapy has been found to be of limited value and medication as unsatisfactory (Saha et al. 1999). Malnutrition and low social status augment the risks of long-term arsenic toxicity. The LD_{50} of As (III) is 15.86 mg/kg in rats after oral application (Maiti and Chatterjee 2001) and that of monomethylarsonous acid is 29.3 μmol/kg body weight in hamsters (Petrick et al. 2001). The LD_{50} for As(III) in adult humans is 34.5 mg/kg body weight and the fetal dose between 120 and 300 mg/kg (Bissen and Frimmel 2003). Arsenic has been found to inactivate several hundred enzymes involved in mitochondrial energy production as well as repair and synthesis of DNA (Ratnaike 2003).

Almost all marine organisms such as fish, crustaceans, mollusks and algae used for human consumption often contain arsenic. Most of this pollutant occurs in organic forms, mainly as arsenobetain, but inorganic arsenic [As(III) and As(V)] is also present (Edmonds and Francesconi 1993). The average As concentration in the oceans is ca. 1.7 μg/L which is about 100 times higher than the US Environmental Protection Agency recommends for fish consumption (Neff 1997). Mussels (*Mylilus galloprovincialis*) in the Venetian lagoon around the Island of Murano have been found to be contaminated with a number of heavy metals. The concentrations of most of them were below the recommended Italian and international guidelines for human consumption of shellfish with the exception of arsenic (Giusti and Zhang 2002).

Rice is a staple food for millions of people which often contains high concentrations of arsenic, but the uptake and bioavailability is not well understood.

Irrigation with water-containing sodium arsenate results in the accumulation of organic arsenic in the form of dimethylarsinic acid; however, only about 33% of the arsenic in rice is bioavailable (Juhasz et al. 2006). In contrast, when arsenic-free rice is cooked in water contaminated with sodium arsenate, the poison is present in its inorganic form which has a high bioavailability. Rice grown in the U.S. also contains considerable levels of arsenic, mostly in the form of arsenite and dimethyl arsenic acid (DMA) as detected in 25 samples (Zavala et al. 2008). While most of the U.S. samples mainly contain DMA, samples from Asia and Europe had higher inorganic arsenic concentrations which are considered more toxic, reflecting genetic differences.

14.4 Detection and Quantification

Because arsenic is odorless, colorless and tasteless it is mandatory to test the water in suspicious areas for potential pollution. Given the high number of 18 million tube wells in India shows the overwhelming task. There are attempts to visit village after village to apply an inexpensive field test. This is not as accurate as a laboratory test, but provides an immediate answer at minimal cost (~30 U.S. cent). While the older field test had a poor accuracy two recently introduced tests were found to be quite reliable as shown by comparison of 136 samples from western Nevada to laboratory measurements using atomic fluorescence spectroscopy (Steinmaus et al. 2006). The kits Quick Arsenic and Hach EZ were tested in the range of <10 to >500 μg/L arsenic and reliably identified all water samples with concentrations >15 μg/L, which, however, is above the WHO limit (World Health Organization 2006).

Several field test kits for arsenic in drinking water are based on the reaction of inorganic arsenic with zinc powder in a reaction vessel acidified by e.g. tartaric acid, sulfamic acid or hydrochloric acid (Spear et al. 2006). This converts the inorganic arsenic to arsine gas (AsH_3) which reacts with mercuric bromide on a test strip to form a yellow to brown mercury halogenated product AsH_2HgBr. The color is then compared with a color chart.

More recently, the quantification of the color change was automated using mobile phone cameras and basic image analysis software. This technique was tested with 376 well samples from Araihazar, Bangladesh (Haque et al. 2018). In the laboratory, the detection limit was 9.2 μg/L while in the field it was 21.9 μg/L attributed to variations in the illumination spectrum and camera response. Another option is to use anode stripping voltammetry: the arsenic is reduced and collected on a gold microelectrode, subsequently oxidized and analyzed following the procedure described in (AWWA 1998). Field kits have been used to test about 1.3 million wells in Bangladesh at a cost of several million U.S. dollars (Rahman et al. 2002). The wells were painted either green (<50 mg/L, safe) or red (>50 mg/L, unsafe). However, these results were unreliable: samples from 290 wells previously tested with the field kits were analyzed in the laboratory. False negatives were found in 68% and false positives in 35% of the cases.

Several laboratory methods are available for quantification of arsenic. Inorganic arsenic [total of As(III) and As(V)] can be determined using high-performance liquid chromatography coupled to inductively coupled plasma-tandem mass spectrometry (ICP-MS/MS) (Mlangeni et al. 2018; Balcaen et al. 2015). The two arsenic species can be separated on an anion exchange column using ammonium carbonate buffer as eluent.

Another method is based on flow injection hydride generation atomic absorption spectrometry (FI-HG-AAS) (Fiamegkos et al. 2016). The procedure also uses arsine vapor generated by reaction with $NaBH_4$ which produces nascent hydrogen. The detection limits are 0.25 μg/L for As(III), 0.22 μg/L for *i*As and 0.10 μg/L *t*As in water (Yang et al. 2016).

Globally about 10% of the wells for obtaining groundwater have been estimated to be contaminated with arsenic. In addition to using field and laboratory testing of water samples probability maps have been developed using geological and environmental parameters in geostatistical models compared to previously measured groundwater arsenic concentrations (Berg et al. 2016b). In order to develop prediction maps, the depositional environments in Southeast Asia have been assessed followed by logistic regression analysis to determine relationships between surface proxies such as sediment deposits and soil data as well as measured arsenic concentrations. The predicted risk areas agree well with known pollution patterns in several countries in South East Asia with known arsenic concentrations. Therefore prediction modeling of geogenic arsenic pollution could be a valuable and resource-saving tool for scientists and policy makers (Berg et al. 2016a).

Several bioassays have been developed for fast and reliable detection of arsenic pollution in water samples using e.g. fish, nematodes, and clams in lieu of costly and time-consuming traditional techniques (Dixon and Sprague 1981). As(III) and As(V) toxicity was evaluated in a bioassay measuring the growth rate inhibition over 72 h in the green freshwater unicellular algae *Chlorella* sp. and *Monoraphidium arcuatum* (Levy et al. 2005). However, the sensitivity was low with LC_{50} values between 15 and 25 mg/L. Also growth parameters such as frond count, frond area and dry biomass for the aquatic higher plant *Lemna gibba* have been used in a bioassay to detect the presence of arsenic (Mkandawire et al. 2006). In contrast, no effect on root length has been found.

Diesel and Schreiber described a bacteria-based bioassay using several strains of bacteria including *Escherichia coli*, *Bacillus subtilis*, *Staphylococcus aureus* and *Rhodopseudomonas palustris* which had been genetically modified to produce a protein in dependence of the arsenic concentration present in the water sample (Diesel et al. 2009). Another bioassay uses *E. coli* which was genetically modified to emit bioluminescence in the presence of arsenic (Trang et al. 2005). Atomic absorption spectroscopy and a bioassay were used to analyze 194 groundwater samples from the Red and Mekong Rivers in Vietnam. The bacterial cells had a detection limit of 7 μg/L arsenic in groundwater. The bioluminescence increased linearly between 10 and 100 μg As/L ($r^2 = 0.997$). When the results from the bioassay were compared with the analytical technique about 8.0% false negatives and 2.4% false positives were found predicting the WHO limit for arsenic 10 μg/L.

New instruments use genetically modified cells integrated into microdevices. Several studies have shown that these bioassays can be used under realistic field conditions (Merulla et al. 2013).

The microcrustacean *Daphnia* is a well-established organism for bioassays. The protocols for the tests are standardized. A fully automatic Daphniatox has been developed which evaluates 14 endpoints including percentage motility, moving velocity, precision of orientation by phototaxis and gravitaxis which are evaluated using computer-based real-time image analysis (Häder 2017; Häder and Erzinger 2017b). In addition, the size and several form parameters are determined. Statistical significance is warranted by automatic analysis using a high number of organisms (Häder and Erzinger 2017a). The results demonstrate that this bioassay can serve as a quick, predicable and low-cost method to determine arsenic concentrations in water for human consumption (Häder and Erzinger 2018). A more recent bioassay uses the unicellular photosynthetic flagellate *Euglena gracilis* which is sensitive to the exposure to arsenic (0.1–10 mg/L). Several endpoints including gravitaxis, motility and biochemical parameters including pigmentation and ROS (reactive oxygen species) markers were analyzed (Tahira et al. 2018). In addition, these organisms can be used to remove As(III) from the water since they accumulate 0.27 mg per gram cell dry weight.

14.5 Remediation

One way of avoiding arsenic in groundwater is digging deeper wells which tap water below the ancient river sediments with high As concentrations; but this requires more effort and money. As discussed in the previous section testing the water quality is tedious due to the gigantic number of tube wells. In addition, it may be necessary to test the water repeatedly since underground water movement can shift, as has been found in the Red River floodplain, Vietnam and other Southeast Asian locations (Postma et al. 2007; Fendorf et al. 2010).

Several strategies exist to eliminate arsenic from water such as precipitation with aluminum, iron, manganese, softening with lime, reverse osmosis, electrodialysis, ion exchange and adsorption on activated carbon or aluminum (Mondal et al. 2006). Especially for developing countries low-cost strategies are essential for providing water for human consumption, industrial and agricultural use. These techniques include ion exchange, filtration and adsorption along with bioremediation (Malik et al. 2009).

Several filter types have been developed to remove arsenic from drinking water. One type (SONO filter, invented in 2006 by Abul Hussam) is used in Bangladesh; it is based on surface complexation of arsenic on a composite iron matrix (Hussam et al. 2007). The method works without chemical treatment or regeneration and does not produce toxic wastes since it results in iron-arsenate cement which does not leach in rainwater. One unit produces about 25 L per hour, sufficient for the needs of one or two families; the running costs are about 40 $ for 5 years. About

30,000 of these filters have been installed in Bangladesh and the inventer has been recognized by the National Academy of Engineering by awarding him with the Grainger Challenge Prize for sustainability.

In Nepal, an arsenic biosand filter (KanchanTM Arsenic Filter, KAF) has been developed and deployed. In essence, it is a slow sand filter with added nails for arsenic removal (Noubactep et al. 2009). Arsenic is adsorbed and coprecipitated with iron oxyhydroxides. In addition, the KAF filters remove pathogens such as bacteria by sand filtration (Pokhrel et al. 2009). Recent modifications have been reported to adsorb arsenic on iron oxides and hydroxides including granular ferric hydroxide, goethite, akaganeite, magnetite or haematite. In addition, arsenic can be absorbed by several aluminum compounds (Giles et al. 2011).

A more recent approach describes the use of aluminum nanoparticles dispersed in a polyacrylamide matrix as arsenic adsorbent (Saha and Sarkar 2012). The loading capacity is high (6.56 mg/g) but depends on exposure time, concentration, pH and occurrence of other anions. Also the removal is pH dependent. These operation characteristics indicate that this approach is a valuable alternative to remove arsenic from water for human consumption.

Another approach utilizes the absoprtion of arsenic (as well as cadmium and zinc) by biochar, which is produced from biomass by pyrolysis (Beesley and Marmiroli 2011). Results of a column leaching experiment were promising for cadmium while for arsenic leachate concentrations did not decline sufficiently.

Plants are known to absorb arsenic from polluted waters. The water hyacinth *Eichhornia* had been cultivated for two days in containers with 10 ppm As, Cd, Pb and Hg (Chigbo et al. 1982). Cadmium was found at the highest concentration in the leaves (0.574 mg/g dried plant material) while for arsenic the lowest concentration was detected (0.3428 mg/g dried plant material). The Chinese brake fern *Pteris vittata* tolerates and hyperaccumulates very high levels of arsenic which would kill any other plant (Cai et al. 2019). RNA sequencing identified three genes coding for glyceraldehyde 3-phosphate dehydrogenase (GAOC1), an organic cation transporter 4 (OCT4) and glutathione s-transferase (GST), which are highly upregulated by arsenic. Each of these proteins is necessary for the arsenic tolerance as shown by RNAi which turns off the targeted gene. The GAPC uses phosphate to metabolize sugars for energy production. Arsenate replaces the phosphate and thus blocks the energy production. GAPC1 in this fern is slightly different from other organisms and capable of binding arsenate. OCT4 transports the bound arsenate into membrane-bound vesicles, where a 1-arseno-3-phosphoglycerate hydrolyses and releases the arsenate which in turn is reduced to arsenite by GST. The arsenite is stored in vesicles and protects the fern from insects feeding on the leaves. A similar mechanism with nearly identical genes was found in the arsenic-tolerant bacterium *Pseudomonas aeruginosa*, which pumps out the arsenite back into the environment rather than storing it in its body (Zhu et al. 2017). In Thailand native plant species have been tested for phytoremediation of arsenic from mine tailings (Visoottiviseth et al. 2002). Out of 36 analyzed plants two ferns (*Pityrogramma calomelanos* and *Pteris vittata*) and two higher plants (*Mimosa pudica* and *Melastoma malabrathricum*) were found to be useful because of their high tolerance and uptake potential. In addition, these plants have a wide distribution, a short life cycle and extensive subsurface biomass. In this study, the arsenic concentrations varied from 21 to 16,000 μg/g soil. The ferns were found to be the most efficient plants removing arsenic from the soil and accumulating up to 8,350 μg/g dry weight in the fronds. Thus, the removal of arsenic from the soil seems possible, however, subsequently the polluted plant material needs to be disposed of. A recent approach uses the green flagellate *E. gracilis* to remove As-III from water with an index of bioaccumulation of 27 (Tahira et al. 2018).

14.6 Conclusions

Arsenic is a major pollutant that causes health problems for millions of people in many countries of the world. Because it is odorless, tasteless and colorless it is often not detected in groundwater tube wells. Despite the gigantic task of testing millions of wells it is mandatory to detect arsenic concentrations exceeding the maximum allowed levels of e.g. 10 μg/L set by the WHO. Since underground water currents may change due to heavy consumption repetitive testing may be necessary. While remediation methods are being used in several countries, cheap filtration needs to be installed and maintained especially in developing countries. Since arsenic is taken up by plants and enters the aquatic food webs, concentrations need to be monitored in products for human and animal consumption.

References

AWWA AJAPHA, Washington DC, USA WEF (1998) Standard methods for the examination of water and wastewater

Balcaen L, Bolea-Fernandez E, Resano M, Vanhaecke FJAca (2015) Inductively coupled plasma-tandem mass spectrometry (ICP-MS/MS): a powerful and universal tool for the interference-free determination of (ultra) trace elements—a tutorial review. Anal Chim Acta 894:7–19

Bednarczyk LR, Matusiak W (1982) Case report: an arsenic murder. J Anal Toxicol 6(5):260–261

Beesley L, Marmiroli M (2011) The immobilisation and retention of soluble arsenic, cadmium and zinc by biochar. Environ Pollut 159(2):474–480

Berg M, Tran HC, Nguyen TC, Pham HV, Schertenleib R, Giger W (2001) Arsenic contamination of groundwater and drinking water in Vietnam: a human health threat. Environ Sci Technol 35(13):2621–2626

Berg M, Winkel L, Amini M, Rodriguez-Lado L, Hug S, Podgorski J, Bretzler A, de Meyer C, Trang P, Lan V (2016a) Regional to sub-continental prediction modeling of groundwater arsenic contamination. In: Arsenic research and global sustainability: proceedings of the sixth international congress on arsenic in the environment (As2016), June 19–23 2016, Stockholm, Sweden. CRC Press, p 21

Berg M, Winkel LHE, Amini M, Rodriguez-Lado L, Hug SJ, Podgorski J, Bretzler A, de Meyer C, Trang P, Lan V, Viet P (2016b) Regional to sub-continental prediction modeling of groundwater arsenic contamination. In: Bhattacharya V, Jarsjo K, Ahmad S, Bundschuh, Naidu (eds) Arsenic research and global sustainability. Taylor & Francis Group, London, pp 21–22

Bissen M, Frimmel FH (2003) Arsenic—a review. Part I: occurrence, toxicity, speciation, mobility. Acta Hydrochimica et Hydrobiologica 31(1):9–18
Bundschuh J, Litter MI, Parvez F, Román-Ross G, Nicolli HB, Jean J-S, Liu C-W, López D, Armienta MA, Guilherme LR (2012) One century of arsenic exposure in Latin America: a review of history and occurrence from 14 countries. Sci Total Environ 429:2–35
Cai C, Lanman NA, Withers KA, DeLeon AM, Wu Q, Gribskov M, Salt DE, Banks JA (2019) Three genes define a bacterial-like arsenic tolerance mechanism in the arsenic hyperaccumulating fern *Pteris vittata*. Curr Biol 29(10):1625–1633. e1623
Chakraborti D, Rahman MM, Das B, Murrill M, Dey S, Mukherjee SC, Dhar RK, Biswas BK, Chowdhury UK, Roy S (2010) Status of groundwater arsenic contamination in Bangladesh: a 14-year study report. Water Res 44(19):5789–5802
Chigbo FE, Smith RW, Shore FL (1982) Uptake of arsenic, cadmium, lead and mercury from polluted waters by the water hyacinth *Eichhornia crassipes*. Environ Pollut Ser A Ecol Biol 27(1):31–36
Daigle K (2016) Death in the water. Sci Am 314(1):42–51
Diesel E, Schreiber M, van der Meer J (2009) Development of bacteria-based bioassays for arsenic detection in natural waters. J Anal Bioanal Chem 394(3):687–693
Dixon D, Sprague J (1981) Acclimation-induced changes in toxicity of arsenic and cyanide to rainbow trout, *Salmo gairdneri* Richardson. J Fish Biol 18(5):579–589
Edmonds JS, Francesconi KA (1993) Arsenic in seafoods: human health aspects and regulations. Mar Pollut Bull 26(12):665–674
Fendorf S, Michael HA, van Geen A (2010) Spatial and temporal variations of groundwater arsenic in South and Southeast Asia. Science 328(5982):1123–1127
Fiamegkos I, Cordeiro F, Robouch P, Vélez D, Devesa V, Raber G, Sloth JJ, Rasmussen RR, Llorente-Mirandes T, Lopez-Sanchez J (2016) Accuracy of a method based on atomic absorption spectrometry to determine inorganic arsenic in food: outcome of the collaborative trial IMEP-41. Food Chem 213:169–179
Giles DE, Mohapatra M, Issa TB, Anand S, Singh P (2011) Iron and aluminium based adsorption strategies for removing arsenic from water. J Environ Manage 92(12):3011–3022
Giusti L, Zhang H (2002) Heavy metals and arsenic in sediments, mussels and marine water from Murano (Venice, Italy). Environ Geochem Health 24(1):47–65
Gontijo B, Bittencourt F (2005) Arsenic—a historical review (Arsênio-Uma revisão histórica). Brasilian Ann Dermatol 80(1):91–95
Häder D-P (2017) Image analysis for bioassays—the basics. In: Häder D-P, Erzinger GS (eds) Bioassays. Advanced methods and applications. Elsevier, Atlanta, GA, pp 69–98
Häder D-P, Erzinger GS (2017a) Arsenic pollution measured with an online monitoring system using *Daphnia*. Open J Environ Biol 2(1):23–30
Häder D-P, Erzinger GS (2017b) Daphniatox—online monitoring of aquatic pollution and toxic substances. Chemosphere 167:228–235
Häder D-P, Erzinger GS (2018) Daphniatox. In: Häder D-P, Erzinger GS (eds) Bioassays—advanced methods and applications. Elsevier, Amsterdam, pp 221–240
Haque E, Mailloux BJ, de Wolff D, Gilioli S, Kelly C, Ahmed E, Small C, Ahmed KM, van Geen A, Bostick BC (2018) Quantitative drinking water arsenic concentrations in field environments using mobile phone photometry of field kits. Sci Total Environ 618:579–585
Hussam A, Munir AK, A HP (2007) A simple and effective arsenic filter based on composite iron matrix: development and deployment studies for groundwater of Bangladesh. J Environ Sci Health Part A 42(12):1869–1878
Jamali H, Dilek Y, Daliran F, Yaghubpur A, Mehrabi B (2010) Metallogeny and tectonic evolution of the Cenozoic Ahar-Arasbaran volcanic belt, northern Iran. Int Geol Rev 52(4–6): 608–630
Juhasz AL, Smith E, Weber J, Rees M, Rofe A, Kuchel T, Sansom L, Naidu R (2006) In vivo assessment of arsenic bioavailability in rice and its significance for human health risk assessment. Environ Health Perspect 114(12):1826

Kapaj S, Peterson H, Liber K, Bhattacharya P (2006) Human health effects from chronic arsenic poisoning—a review. J Environ Sci Health Part A 41(10):2399–2428
Levy JL, Stauber JL, Adams MS, Maher WA, Kirby JK, Jolley DF (2005) Toxicity, biotransformation, and mode of action of arsenic in two freshwater microalgae (*Chlorella* sp. and *Monoraphidium arcuatum*). Environ Toxicol Chem 24(10):2630–2639
Maiti S, Chatterjee AK (2001) Effects on levels of glutathione and some related enzymes in tissues after an acute arsenic exposure in rats and their relationship to dietary protein deficiency. Arch Toxicol 75(9):531–537
Majumder S, Nath B, Sarkar S, Chatterjee D, Roman-Ross G, Hidalgo M (2014) Size-fractionation of groundwater arsenic in alluvial aquifers of West Bengal, India: the role of organic and inorganic colloids. Sci Total Environ 468:804–812
Malik AH, Khan ZM, Mahmood Q, Nasreen S, Bhatti ZA (2009) Perspectives of low cost arsenic remediation of drinking water in Pakistan and other countries. J Hazard Mater 168(1):1–12
Mari F, Polettini A, Lippi D, Bertol E (2006) The mysterious death of Francesco I de'Medici and Bianca Cappello: an arsenic murder? Br Med J 333(7582):1299–1301
Mazumder DG (2008) Chronic arsenic toxicity & human health. Indian J Med Res 128(4): 436–447
McGrory ER, Brown C, Bargary N, Williams NH, Mannix A, Zhang C, Henry T, Daly E, Nicholas S, Petrunic BM (2017) Arsenic contamination of drinking water in Ireland: a spatial analysis of occurrence and potential risk. Sci Total Environ 579:1863–1875
Meharg AA, Rahman MM (2003) Arsenic contamination of Bangladesh paddy field soils: implications for rice contribution to arsenic consumption. Environ Sci Technol 37(2):229–234
Merulla D, Buffi N, Beggah S, Truffer F, Geiser M, Renaud P, van der Meer JR (2013) Bioreporters and biosensors for arsenic detection. Biotechnological solutions for a world-wide pollution problem. Curr Opin Biotechnol 24(3):534–541
Mkandawire M, Taubert B, Dudel EG (2006) Limitations of growth-parameters in *Lemna gibba* bioassays for arsenic and uranium under variable phosphate availability. Ecotoxicol Environ Saf 65(1):118–128
Mlangeni AT, Vecchi V, Norton GJ, Raab A, Krupp EM, Feldmann J (2018) Comparison of on-site field measured inorganic arsenic in rice with laboratory measurements using a field deployable method: method validation. Food Chem 263:180–185
Mondal P, Majumder C, Mohanty B (2006) Laboratory based approaches for arsenic remediation from contaminated water: recent developments. J Hazard Mater 137(1):464–479
Neff JM (1997) Ecotoxicology of arsenic in the marine environment. Environ Toxicol Chem Int J 16(5):917–927
Noubactep C, Schöner A, Woafo P (2009) Metallic iron filters for universal access to safe drinking water. CLEAN-Soil, Air, Water 37(12):930–937
Partey F, Norman D, Ndur S, Nartey R (2008) Arsenic sorption onto laterite iron concretions: temperature effect. J Colloid Interface Sci 321(2):493–500
Petrick JS, Jagadish B, Mash EA, Aposhian HV (2001) Monomethylarsonous acid (MMAIII) and arsenite: LD_{50} in hamsters and in vitro inhibition of pyruvate dehydrogenase. Chem Res Toxicol 14(6):651–656
Phelps J, Washburn W (1930) Toxic effect of arsenical compounds employed in the treatment of syphilis in the United States navy. A statistical study of the results of 272,354 injected doses of neo-arsphenamine and other compounds of arsenic. Naval Med Bull 28:659–695
Pokhrel D, Bhandari B, Viraraghavan T (2009) Arsenic contamination of groundwater in the Terai region of Nepal: an overview of health concerns and treatment options. Environ Int 35(1): 157–161
Poklis A, Saady JJ (1990) Arsenic poisoning: acute or chronic? Suicide or murder? Am J Forensic Med Pathol 11(3):226–232
Postma D, Larsen F, Hue NTM, Duc MT, Viet PH, Nhan PQ, Jessen S (2007) Arsenic in groundwater of the Red River floodplain, Vietnam: controlling geochemical processes and reactive transport modeling. Geochim Cosmochim Acta 71(21):5054–5071

Rahman MM, Mukherjee D, Sengupta MK, Chowdhury UK, Lodh D, Chanda CR, Roy S, Selim M, Quamruzzaman Q, Milton AH (2002) Effectiveness and reliability of arsenic field testing kits: are the million dollar screening projects effective or not? Environ Sci 36(24):5385–5394

Rasool A, Xiao T, Farooqi A, Shafeeque M, Liu Y, Kamran MA, Katsoyiannis IA, Eqani SAMAS (2017) Quality of tube well water intended for irrigation and human consumption with special emphasis on arsenic contamination at the area of Punjab, Pakistan. Environ Geochem Health 39(4):847–863

Ratnaike RN (2003) Acute and chronic arsenic toxicity. Postgrad Med J 79(933):391–396

Ravenscroft P (2007) Predicting the global distribution of arsenic pollution in groundwater. In: Royal Geographical Society annual international conference, Department of Geography, 2007. Cambridge University, Cambridge

Rodríguez-Lado L, Sun G, Berg M, Zhang Q, Xue H, Zheng Q, Johnson CA (2013) Groundwater arsenic contamination throughout China. Science 341(6148):866–868

Saha J, Dikshit A, Bandyopadhyay M, Saha K (1999) A review of arsenic poisoning and its effects on human health. Crit Rev Environ Sci Technol 29(3):281–313

Saha S, Sarkar P (2012) Arsenic remediation from drinking water by synthesized nano-alumina dispersed in chitosan-grafted polyacrylamide. J Hazard Mater 227:68–78

Shankar S, Shanker U (2014) Arsenic contamination of groundwater: a review of sources, prevalence, health risks, and strategies for mitigation. Sci World J 304524 (2014)

Singh N, Kumar D, Sahu AP (2007) Arsenic in the environment: effects on human health and possible prevention. J Environ Biol 28(2):359

Spear JM, Zhou YM, Cole CA, Xie YF (2006) Evaluation of arsenic field test kits for drinking water analysis. J Am Water Works Assoc 98(12):97–105

Steinmaus CM, George CM, Kalman DA, Smith AH (2006) Evaluation of two new arsenic field test kits capable of detecting arsenic water concentrations close to 10 μg/L. Environ Sci 40 (10):3362–3366

Tahira S, Khan S, Samrana S, Shahi L, Ali I, Murad W, ur Rehman Z, Azizullah A (2018) Bio-assessment and remediation of arsenic (arsenite As-III) in water by *Euglena gracilis*. J Appl Phycol 31(1):1–11

Trang PTK, Berg M, Viet PH, Mui NV, van der Meer JR (2005) Bacterial bioassay for rapid and accurate analysis of arsenic in highly variable groundwater samples. Environ Sci 39(19):7625–7630

Turekian KK, Wedepohl KH (1961) Distribution of the elements in some major units of the earth's crust. Geol Soc Am Bull 72(2):175–192

Visoottiviseth P, Francesconi K, Sridokchan W (2002) The potential of Thai indigenous plant species for the phytoremediation of arsenic contaminated land. Environ Pollut 118(3):453–461

Wang S, Mulligan CN (2006) Occurrence of arsenic contamination in Canada: sources, behavior and distribution. Sci Total Environ 366(2–3):701–721

Warner KL (2001) Arsenic in glacial drift aquifers and the implication for drinking water—Lower Illinois River Basin. Groundwater 39(3):433–442

Welch AH, Lico MS, Hughes JL (1988) Arsenic in ground water of the western United States. Groundwater 26(3):333–347

World Health Organization (2006) Guidelines for drinking water quality. Guidelines for drinking water quality. Accessed 03 Sept 2006

Yang X-A, Lu X-P, Liu L, Chi M-B, Hu H-H, Zhang W-B (2016) Selective determination of four arsenic species in rice and water samples by modified graphite electrode-based electrolytic hydride generation coupled with atomic fluorescence spectrometry. Talanta 159:127–136

Zavala YJ, Gerads R, Gürleyük H, Duxbury JM (2008) Arsenic in rice: II. Arsenic speciation in USA grain and implications for human health. Environ Sci Technol 42(10):3861–3866

Zhu Y-G, Xue X-M, Kappler A, Rosen BP, Meharg AA (2017) Linking genes to microbial biogeochemical cycling: lessons from arsenic. Environ Sci Technol 51(13):7326–7339

Chapter 15
Petroleum Hydrocarbons in Atlantic Coastal Patagonia

Marina L. Nievas El Makte, Rosana Polifroni, Marcela A. Sepúlveda, and Ana Fazio

Abstract The Patagonian Atlantic region is known as a pristine area with more than 3000 km of coastline. It includes two productive oil basins, Golfo San Jorge and Austral from where oil production (about 14 million m^3/year) is transported by sea from five terminals. The maritime and coastal areas of the Argentine continental shelf are relevant for marine conservation, tourism and fishing, being spawning and nursery areas of many species. In this chapter, we describe the main aspects of oil production in the coastal basins and its maritime transport throughout the Atlantic Patagonia. Offshore oil exploitation and future development prospects on the Argentine continental shelf are introduced. We also present the main oil spills that occurred in Atlantic Patagonia and a review of hydrocarbon levels in coastal marine sediments and organisms. Some episodes of chronic contamination are depicted. The reported levels range from pristine uncontaminated areas to heavily polluted ones, affected by oil or fuel spills. Hydrocarbon pollution has a punctual distribution pattern near the sources, with the most polluted sites located in ports and oil exploitation areas. In addition, possible interactions of sensitive areas for marine conservation, fishing activities and maritime transport are discussed.

Keywords Hydrocarbon · Pollution · Oil spill · Harbor · Patagonia

---

M. L. Nievas El Makte (✉) · R. Polifroni · M. A. Sepúlveda · A. Fazio
Centro para el Estudio de Sistemas Marinos (CESIMAR-CONICET, CCT CONICET-CENPAT), Blvd. Brown 2915, Chubut U9120ACD Puerto Madryn, Argentina
e-mail: nievas@cenpat-conicet.gob.ar

M. L. Nievas El Makte
Facultad Regional Chubut, Universidad Tecnológica Nacional, Av. del Trabajo 1536, Puerto Madryn, Chubut, Argentina

© The Author(s), under exclusive license to Springer Nature Switzerland AG 2021

D.-P. Häder et al. (eds.), *Anthropogenic Pollution of Aquatic Ecosystems*,
https://doi.org/10.1007/978-3-030-75602-4_15

15.1 Petroleum Hydrocarbons Nature and Its Fate into the Marine Environment

Petroleum or crude oil is a natural and complex substance containing hydrocarbons as major components, nitrogen-sulfur-oxygen heteroatom compounds (NSO compounds) in minor proportions and minor quantities of nickel, vanadium and other trace elements. Hydrocarbons can be subdivided based on their structures into aliphatics (*n*-alkanes, branched alkanes, cycloalkanes) and aromatics (monoaromatic and polycyclic aromatic hydrocarbons (PAHs)), while NSO compounds are resins and asphaltenes. All these types of compounds have different biodegradation susceptibility and toxicity. The monoaromatic benzene, toluene, ethylbenzene and xylene (BTEXs), PAHs (2–6 rings compounds) and their alkylated derivatives are found in crude oil. Low molecular weight (LMW) PAHs comprise compounds with 2–3 aromatic rings, while high molecular weight (HMW) PAHs 4–6 aromatic rings compounds. Some PAHs are classified as priority pollutants, known as possible or probable human carcinogens and having also mutagenic and teratogenic properties. Aromatic compounds are responsible for most of the ecotoxicological issues caused by oil and fuel spills and thus are used as indicators of oil pollution in water, sediments and biota (Beiras 2018).

When hydrocarbon product spills occur in aquatic media, like the marine environment, a complex process called "weathering" occurs. This involves physical, chemical and biological processes that change the composition, the physical state, the toxicity and the distribution of the spilled oil into different environmental compartments: seawater, sediments, air and biota. The initial transformations are spreading, dispersion, dissolution and evaporation, where winds, tidal, marine currents, environmental variables (i.e., temperature, salinity) and the chemical properties of each hydrocarbon compound are involved. Highly energetic coasts may generate water-in-oil stable emulsions ("chocolate or oil mousse"), which have a long environmental half-life and reduced biodegradation susceptibility (Beiras 2018). Initially, light compounds evaporate, photo-oxidation occurs, and the overall toxicity of the spill is reduced. Only a small fraction of the hydrocarbons is dissolved in water, mainly LMW PAHs with high toxicity. In the next weeks and months, sedimentation and biodegradation are the predominant processes, carried out by microorganisms naturally occurring in the marine environment (Chapman et al. 2018). Finally, the heaviest hydrocarbons and droplets settle on the bottom sediments. Shorelines are more sensitive to deleterious effects to the ecosystem through oiling biota, fouling fine-grained sediments and impacting subtidal macroinfauna (Filler et al. 2014). Sediments are considered as sink for pollutant hydrocarbons and act as long-term reservoirs and secondary sources of pollution, while bivalves are filtering organisms that accumulate hydrophobic compounds. Thus, these matrices are used for evaluating ecotoxicological effects and the hydrocarbon pollution status of a certain site.

Oil spills have acute toxicity effects on wildlife, causing death by direct ingestion when the oiled animals try to clean themselves or by oil-contaminated prey ingestion and through the loss of the feathers or fur waterproof property, altering the animal thermoregulation ability (i.e., birds, particularly penguins). Chronic or long-term oil spill effects are evident at the population or community levels in an ecosystem. Changes in feeding, mobility, reproductive behaviors and malformations are often found as responses to oil pollution (Beiras 2018; Chapman et al. 2018).

15.2 Eastern Patagonia

Patagonia is the southernmost region of the South American continent (37–56° S), comprising portions of Argentina and Chile. The Argentinean Patagonia extends for more than 2200 km in a straight line with N–S direction, including on its coast the provinces of Río Negro, Chubut, and Santa Cruz, the southern portion of Buenos Aires Province and the Tierra del Fuego Island, with a total coast length of more than 3000 km (Coronato et al. 2008, Gil et al. 2019) (Fig. 15.1). It has a semi-arid temperate and cold climate, with almost permanent westerly winds with strong intensities that frequently exceed 100 km/h. Few cities on the coast concentrate the major population and industrial activities, such as oil, coal, fishing, aluminum, cement, renewable energy, textile, electronic production and tourism.

Argentina has ratified international treaties for the prevention of oil pollution of the sea by means of national regulations such as OILPOL, MARPOL and UNCLOS. The most important are summarized in Table 15.1. The Prefectura Naval Argentina (PNA) is the competent authority dealing with oil and other noxious substances pollution in marine and freshwater environments. The same standards

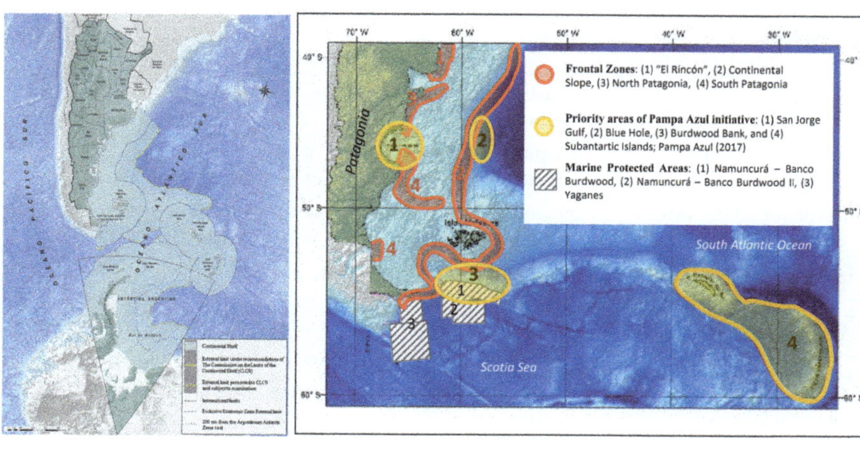

Fig. 15.1 Argentinean continental shelf. Argentinian maritime spaces (left), frontal zones, priority areas of Pampa Azul initiative and marine protected areas (right). *Source* COPLA (2017), and adapted from Allega et al. (2020), Pampa Azul (2017; 2021) and Laws 26,875/13 and 27,490/18

established by the MARPOL convention were adopted for the national order, applying them to ships with Argentine flag in international or national navigation. Two PNA rules are particularly relevant for the Atlantic Patagonian zone: (1) The declaration of special protection zones (SPZs) (Ordinance N° 12/98 PNA 1998a), in agreement with the particularly sensitive sea areas (PSSA) criteria of the International Maritime Organization (IMO), which bans the discharges of any hazardous wastes from ships, including hydrocarbons in any type; and (2) the tanker maritime route, ruled by the Ordinance N°13/98 PNA (PNA 1998b), which states that vessels transporting hazardous substances, such as oil and distillates, must take a route keeping them within 20 nautical miles of their stranded isobaths. The Argentine continental shelf (ACS) outer limit beyond 200 nautical miles was recently established by the Law N° 27,557/17, in accordance with Article 76 of UNCLOS, incorporating 1,782,500 km^2 from the 200 mile limit (COPLA 2017, Fig. 15.1). The ACS has unique characteristics worldwide where many frontal zones are generated, producing high primary productivity, turning it into one of the largest and most fertile maritime spaces in the Southern Hemisphere. In the exclusive economic zone (EEZ) (1,529,585 km^2), commercial fisheries, hydrocarbon basins and mineral deposits occur (Koutoudjian 2011; Baruj and Drucaroff 2018; MEM-SE 2021). In the Patagonian Sea, its varied fauna, especially birds and marine mammals that approach the coast, gives it an additional ecologic and economic value.

The Argentinean Patagonia economic relevance is based on natural resources exploitation. The main fishing species are red shrimp, squid and hubbsi hake. Total fishery exportation reached 2148 US million dollars in 2018 (ME-SSPME 2019).

Table 15.1 International treaties and Argentine regulations intended for the prevention of marine oil pollution

Convention	Subject	Argentine law
OILPOL 54	International convention for the prevention of pollution of the seas by oil	N° 21.353/ 1976
London Convention, 1972	Convention on the prevention of marine pollution by dumping of wastes and other matter	N° 21.947/ 1979
Intervention 1969, IMO	International convention relating to intervention on the high seas in cases of oil pollution casualties	N° 23.456/ 1987
MARPOL 73/78	International convention for the prevention of marine pollution from ships	N° 24.089/ 1992
OMI conference, 1990	International convention on oil pollution preparedness, response and co-operation (OPRC)	N° 24.292/ 1994
Convention on Biological Diversity, UN 1992	Establishes among its objectives the sustainable use of biological diversity	N° 24.375/ 1994
UNCLOS, 1992	United Nations Convention on the Law of the Sea	N° 24.543/ 1995

Source http://www.infoleg.gob.ar/

Regarding oil and gas (O&G) resources, Argentina produced in 2019 29.5 million m^3 of oil and 49,350 million m^3 of gas, being 49% of oil and 34% of gas of the total country from Easter Patagonia basins. Moreover, the economic potential of O&G offshore exploitation for the next 25 years is estimated between 40 and 60 US billion dollars (Baruj and Drucaroff 2018). On the other hand, tourism activity related to the sea on the Patagonian coast has exponentially grown in recent years. Puerto Madryn and Ushuaia are the major touristic cities, especially in relation with ecotourism with wildlife watching. For example, whale watching in Peninsula Valdés has been developed based on the observation of Southern right whales, becoming an important source of income and employment for Chubut Province (Argüelles et al. 2016).

In 2014, the Argentinean Ministry of Science, Technology and Productive Innovation promoted the "Pampa Azul" initiative, aiming to develop actions to promote the conservation and sustainable use of marine natural resources. Based on their oceanographic characteristics, five priority geographic areas were defined, four of them in Patagonia (Pampa Azul 2017, Fig. 15.1).

15.3 Petroleum Basins in Atlantic Coastal Patagonia

Argentina has two productive O&G basins in Eastern Patagonia, Golfo San Jorge (GSJ) and Austral (IAPG 2020) (Fig. 15.2). Golfo San Jorge basin (GSJB) extends over the southern part of Chubut and northern Santa Cruz provinces and to the east, over the ACS (45°–47° S and 65°–71° W), with nearly 180,000 km^2 being a third part of it situated offshore (Sylwan et al. 2011) (Fig. 15.2). It is a mature basin exploited for more than a century in onshore areas, being the most productive oil zone from Argentina in the last decade. In 2019, it reached nearly 13.3 million m^3 of oil and 4681 million m^3 of gas, with near 61% of the proven oil reserve of the country in 2018 (Fig. 15.3). The GSJB produces paraffinic medium oils (typically 24°API, TERMAP 2020): Escalante and Cañadón Seco from the north and south flanks of the basin, respectively.

The Austral Basin, also known as Austral-Magallanes basin, is located in the southernmost part of South America. It comprises the southern mainland region of Argentina and Chile, part of the Tierra del Fuego Island, and an east offshore area edging the Malvinas Basin (Fig. 15.2). It extends in NNW-SSE direction between 46°30'–54°30' S and 66°30'–73° W over 170,000 km^2 (Cuitiño et al. 2019). The basin produces mainly gas and light oil to a minor degree (~API 47°, Petrotecnia 2004). In 2019, the Austral Basin's total production reached 1.25 million m^3 of oil and 12,040 million m^3 of gas, being the gas proven reserves 29% of the total Argentine reserves in 2018 (Fig. 15.3). Thus, it is a promising area to increase gas production, considering that the country has an annual gas deficit of near 9–12 million m^3 for the last 6 years (IAPG 2020). Actually, offshore O&G production is only exploited at Austral Basin, representing 2% of the oil and 17% of the gas produced in Argentina (MEM 2017). Offshore facilities are located in the east mouth

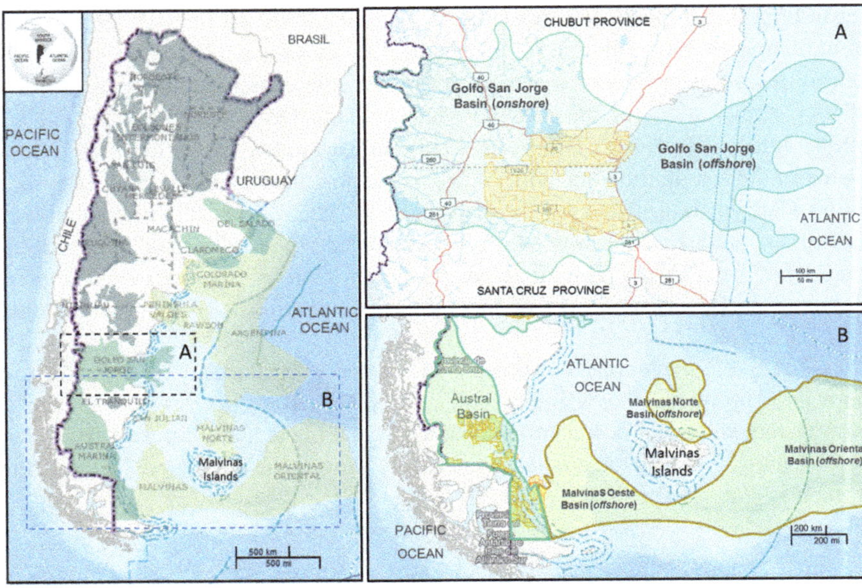

Fig. 15.2 Argentina's hydrocarbon basins (left). Productive basins in Atlantic coastal Patagonia. GSJ (top right), Austral (bottom right) basins. Adapted from https://sig.se.gob.ar/visor/visorsig.php

of the Magellan strait and at eastern Tierra del Fuego Island (52°28' S–53°40' S) near the Atlantic shore in shallow water areas (less than 100 m depth) (Figs. 15.2 and 15.4). Enap Sipetrol S.A. is the major offshore operator in Argentina with 5 platforms nearly located 20 km from shore, with 43 km of submarine collector pipelines and two submarine 22-km pipelines to transport the production to the onshore reception plant, located near Cabo Vírgenes, the end point of Argentinean mainland territory (Petrotecnia 2002; ENAP 2021). ENAP (Chile) has another 40 platforms in operation in the Chilean territory of the Magellan strait (ENAP 2021). The east offshore areas of Tierra del Fuego are operated by Total Austral S.A. or by a consortium of Total Austral S.A., Wintershall Energía S.A. and PanAmerican Energy S.A. (TAU-WS-PAE) companies. This consortium has operated the Cuenca Marina Austral-1 block since 1989, with various oilfields and a network of submarine pipelines connected to onshore plants. The most recent Vega Pléyade offshore area (20 km east from the coast in the south of Bahía San Sebastián, 50 m depth water) constitutes the southernmost offshore platform in the world starting its operation in 2016 (Petrotecnia 2016). Onshore plants on the Atlantic coast of Tierra del Fuego receive the oil and gas production from the marine platforms. The gas is injected into the main gas pipeline that runs from Tierra del Fuego to Buenos Aires ($\sim$4500 km), which crosses the Magellan strait parallel to the international limit between Argentina and Chile by means of a 36-km submarine pipeline. The oil is dispatched by maritime terminals.

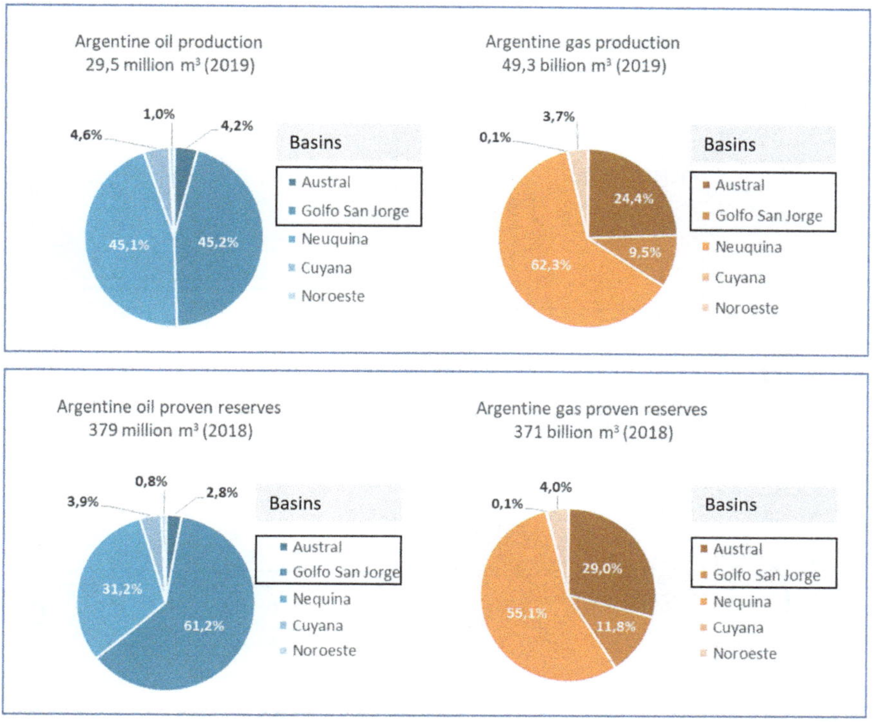

Fig. 15.3 Argentina oil and gas production in 2019 and proven reserves in 2018. *Sources* MEM-SE (2021) and SGE (2019a)

15.3.1 Prospects for Offshore O&G Development in Argentina

Argentina hydrocarbon policy promotes the development of offshore hydrocarbon activities with current regulations establishing tax deduction and a 30-years concession period for developments in the ACS (IAPG 2020). The most promising offshore areas that have concentrated the exploration efforts until now are the Colorado Marina basin at south-east of Buenos Aires province and GSJ, Austral and Malvinas Oeste basins in Atlantic Patagonia (Fig. 15.4). In 2008, the YPF Company carried out a shallow water drilling campaign in offshore GSJB, recovering hydrocarbons from wells drilled at 70–110 km from shore in the center of the gulf, but they are not actually exploited as they are considered subcommercial discoveries (YPF 2011). In addition, in 2014, the company reactivated an oil exploitation project in shallow waters in the Restinga Alí area, a maritime sector contiguous to the shore in Comodoro Rivadavia, up to 55 m depth over 167 km^2. Directed (horizontal) wells from the shore were used, up to 1600 m from the

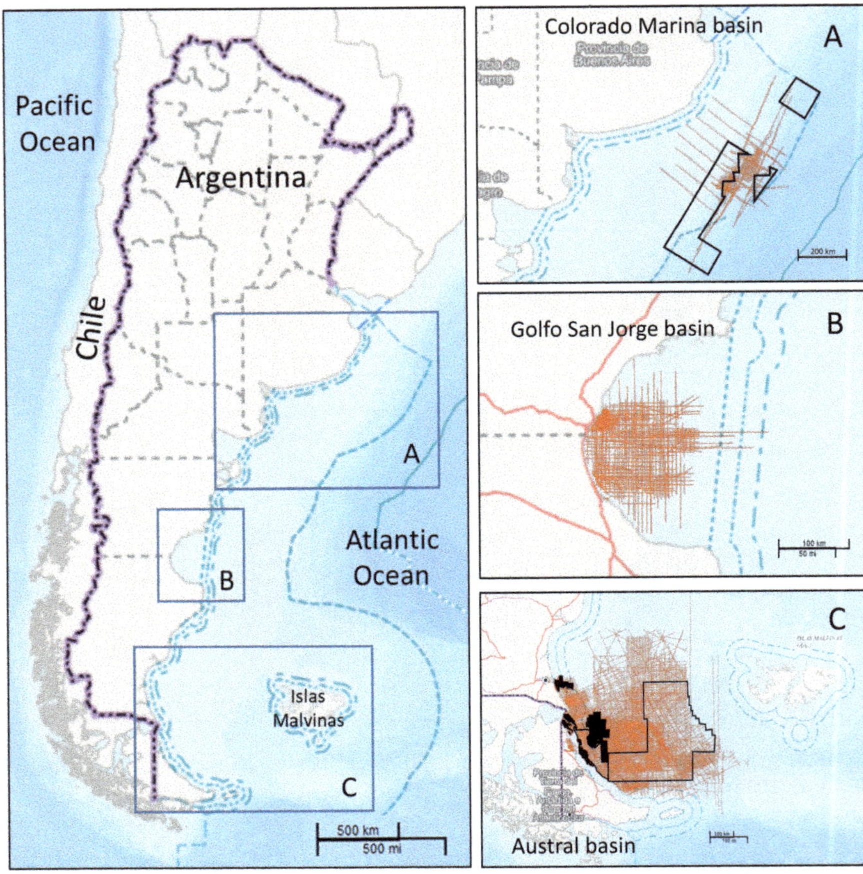

Fig. 15.4 Offshore prospects in the ACS. Main marine areas with perspectives of O&G offshore development (left). Right: 2D seismic prospection (orange lines). Offshore O&G exploration areas granted in 2018 in contour black lines (A and C), and offshore exploitation areas in Austral basin in black areas (C). Adapted from https://sig.se.gob.ar/visor/visorsig.php, and SGE (2019b)

coastline. The company plans to increment the number of wells for the Restinga Alí offshore area, being 21 wells drilled until 2019 (Van Wyk et al. 2019).

Aiming to increase the offshore O&G development the Argentinean government carried out an international bidding in 2017 to enlarge the potential production, leading to a significant increase of exploration permission in marine zones. Eighteen new marine areas were concessioned, being 94,800 km^2 granted to 13 companies in three sedimentary oceanic basins (Fig. 15.4, SGE 2019b), being 48,036 km^2 in the Cuenca Argentina Norte (CAN) Basin in deepwater (200–1300 m) and ultradeep water (1200–4000 m) blocks, 42,328 km^2 in the Cuenca Malvinas Oeste (MLO) Basin in deepwater blocks (100–700 m) and 4439 km^2 from the Cuenca Marina Austral (AUS) Basin in shallow water blocks (less than

100 m depth). The two last basins are located in the Patagonia maritime region (Fig. 15.4). Other offshore development projects have also promising perspectives of concretion. Recently, Enap Sipetrol acquired an exploitation permission in the Octans Pegaso area, located 20 km east of Santa Cruz Province. In addition, the consortium TA-WS-PAE is projecting a new development in the Fenix area in the CAM-1 block. The project includes the drilling of three wells and the construction of an underwater pipeline to Tierra del Fuego Island. The oil would be dispatched through the oil terminals in Tierra del Fuego Island.

15.3.2 Hydrocarbon Transport in Patagonia Argentina

Most of the oil production from the coastal Patagonic Basins is commercialized for local market distillation (e.g., in 2019, 78% of total production of the GSJB had this destiny). Distilleries are located near Buenos Aires and along the coast of internal rivers of the country. Thus, maritime voyages from Patagonic terminals to distilleries often take 1800–2500 km (Fig. 15.5).

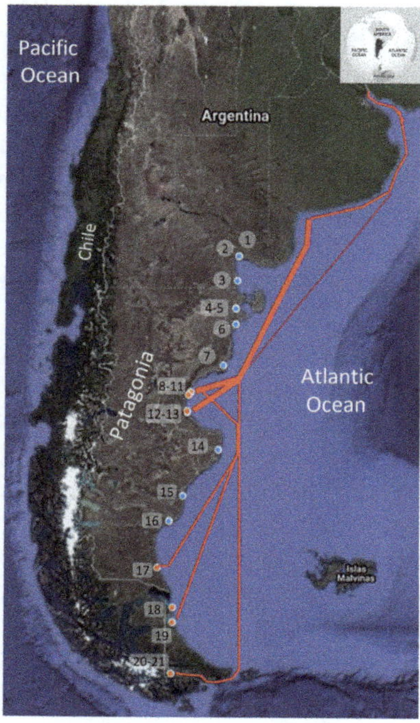

#	Harbor or Terminal	Shipping Activity
1	San Antonio Este	Fruit, trade
2	San Antonio Oeste	Fishing
3	Punta Colorada	Minning (iron)
4	Storni Pier - Puerto Madryn	Fishing, aluminium, trade
5	Piedrabuena Pier - Puerto Madryn	Turism
6	Rawson	Fishing
7	Camarones	Fishing
8	Caleta Cordova-Monobouy (CR)	Oil terminal
9	Caleta Córdova - Pier (CR)	Fishing
10	YPF Pier (CR)	Fuel
11	Comodoro Rivadavia (CR)	Trade and fishing
12	Caleta Olivia (CO)	Oil terminal
13	Caleta Paula (CO)	Fuel and Trade
14	Puerto Deseado	Fuel and trade
15	Puerto San Julián	Fishing, trade and minning
16	Punta Quilla	Fuel and trade
17	Punta Loyola	Oil terminal and coal
18	Río Cullen	Oil terminal
19	Cruz del Sur	Oil terminal
20	Ushuaia	Tourism, fishing, trade
21	Orion Plant YPF Pier	Fuel

Fig. 15.5 Harbors, oil terminals and the oil maritime route in Atlantic Patagonia. Data of oil route from https://www.shipmap.org/

The GSJB oil production is concentrated in two exit terminals, Caleta Córdova and Caleta Olivia, operated by Terminales Marítimas Patagónicas S.A. (TERMAP S.A.) (TERMAP 2020) (Fig. 5.5). The former terminal is located 20 km north of Comodoro Rivadavia city, at Caleta Córdova village, in the Chubut Province and the latter in Caleta Olivia city in Santa Cruz province in the GSJ, 70 km south of Comodoro Rivadavia. Both terminals have onshore tank farm facilities with crude storage capacities of 283,000 and 209,000 m^3, respectively, and a single point mooring (SPM) monobuoy system for tanker loading, connected to the terminal by submarine pipeline extended along the seabed. Tankers are loaded by connecting to the monobuoy by a floating hose. Both terminals admit tankers up to 160,000 DWT (TERMAP 2020).

The oil production from the Santa Cruz territory of the Austral Basin is delivered by the maritime terminal Punta Loyola (Fig. 15.5). This is an oil and coal terminal operated by YPF S.A., located in the mouth of the Gallegos River, 40 km from Río Gallegos city. It has an oil storage capacity of 180,000 m^3 and a pier for vessel loading, admitting tankers up to 60,000 DWT (Petrotecnia 2004). In the Atlantic coast of Tierra del Fuego Island two oil terminals operate: Río Cullen and Cruz del Sur. The Río Cullen terminal is part of the homonymous onshore oil and gas treatment plant located 32 km south of the international limit between Argentina and Chile. The terminal has an oil storage capacity of 92,000 m^3 and a SPM monobuoy charge system, operated by Total Austral S.A. It receives mainly the offshore production from the complex operated by the consortium TAU-WS-PAE at east Tierra del Fuego. The terminal operates with tankers up to 150,000 DWT (Petrotecnia 2004). The Cruz del Sur or San Sebastian terminal is located in the south inside the San Sebastian Bay and is operated by YPF S.A. It has a tank farm with crude storage capacity of 70,000 m^3 and a SPM monobuoy system for tanker loading. The terminal admits tankers up to 60,000 DWT (Petrotecnia 2004). Both terminals of the Tierra del Fuego Island are located in non-populated places at 140 and 75 km, respectively, away from the only city of this province at the Atlantic coast, Río Grande, with 66,000 inhabitants. The three Argentinean terminals of the Austral Basin operate with light oils.

The maritime transportation route from the Austral and GSJ Basins runs along the coast of Argentinean Patagonia from the oil terminals to the distilleries in the north of the country. The same route back is used to transport refined products from distilleries to fuel storages plants in Patagonia, from where fuels are distributed by trucks to the end user cities and industries. Three main fuel plants are located on the coast: the YPF pier in Comodoro Rivadavia (Chubut), the Caleta Paula plant in Caleta Olivia (Santa Cruz) and the Orion plant at Ushuaia (Tierra del Fuego Province). Distillate products are usually transported in 60,000 DWT tankers. The most frequent tanker's route is depicted in Fig. 15.5.

15.4 Oil Spills in Atlantic Coastal Patagonia

Eastern Patagonia has a low population density and low pollution compared with other sites worldwide. However, some oil spill episodes have taken place on its coast. Even minor but quite frequent episodes contribute to chronic pollution, particularly, in those areas around crude oil terminals and ports. Below, we depict the most relevant and documented oil or hydrocarbons spills in Atlantic Patagonia and include some low volume spill episodes that account for chronic pollution (Fig. 15.6).

Metula oil spill in 1974. The worst oil spill suffered in Patagonia up to date is the Metula accident, where 52,300 tons of Arabian light crude oil and approximately 2000 tons of Bunker C fuel oil were spilled in the Magellan Strait, in Chilean territory in August 1974. This strait connects the Atlantic with the Pacific Oceans separating the South America mainland from the Tierra del Fuego Island, with the east mouth of the strait being Argentine territory and Chilean territory the rest of the strait up to the Pacific Ocean embouchure (Fig. 15.6C). We include this spill here even though it is not strictly in the Atlantic coast, because the predominant West-East direction winds and currents spreaded the oil reaching Punta Dungeness, in the Atlantic mouth of the strait. "Oil mousse" formation occurred rapidly and around 250 km of coast were oiled, being Espora marshes (EM) heavily polluted (Fig. 15.6). Wang et al. (2001) reported 24 years after the accident, in a

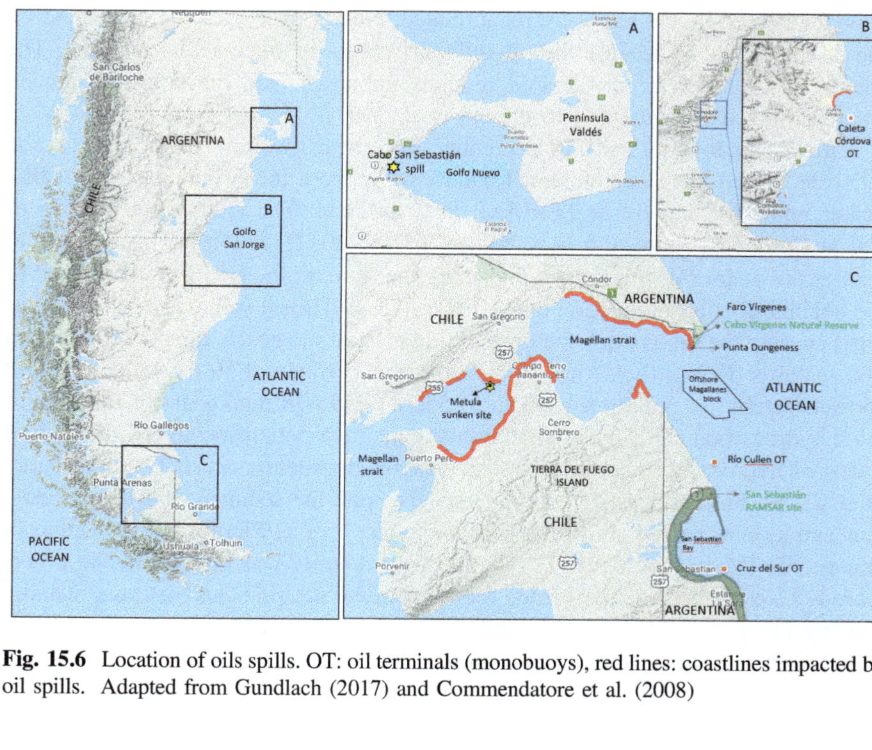

Fig. 15.6 Location of oils spills. OT: oil terminals (monobuoys), red lines: coastlines impacted by oil spills. Adapted from Gundlach (2017) and Commendatore et al. (2008)

site of east EM, oil mousse poorly weathered, showing the *n*-alkanes series from *n*-C12 and exhibiting still 27–40% of the alkylated PAHs of the reference oil, with a whole weathered percentage of only 25–32%. This confirmed a low biodegradation process. They highlighted that the PAHs and fresh oil mousse remaining in the area were expected to persist for many decades. Gundlach (2017) reported in 2015 in the same site relatively fresh oil mousse with 4–5 cm of subsurface oil mousse layers, 41 years after the spill. The high persistence of oil mousse was once again demonstrated in a low energy environment (marshes) with cold climate, even with light oil as the Arabian Light of the Metula spill. Evidence up to date makes researchers believe that predictions of 100 years of impact of the Metula spill are still likely (Gundlach 2017).

Magallanes area spills. A series of oil spills took place at the offshore Magallanes exploitation block in the Austral basin, between 2004 and 2006 (Fig. 15.6). The sources of some of the events were leakages from the submarine pipeline that connects the Enap Sipetrol S.A. platforms with the onshore storage plant near the Cabo Virgenes Natural Reserve (CVNR), where 90,000 Magellan penguins breeding pairs together with a diverse and rich marine ecosystem are protected. In 2004, the company declared a minor oil spill from a submarine pipeline. Later, tar balls appeared on the high tideline over 600 meters on the CVNR beach. In September 2005, an oil spill occurred between platforms connecting pipelines, which did not reach the shore but 50 oiled penguins died. The oil company took responsibility for both events. In April 2006, almost 400 seabirds were found oiled in CVNR and in Isla Magdalena (Chile). Around 33% of the animals died while the rest were rehabilitated. Non-impact signs were reported on the beach, probably because the oil spills drifted to the open sea to the East. The source of the 2006 event oil spill remains unknown, and the company denied responsibility (de Haro 2007; Rouppolo et al. 2007).

Caleta Córdova oil spill. In December 2007, nearly 300 m^3 of Escalante crude oil were spilled from contaminated ballast water from the Presidente Arturo Illia tanker while loading oil at TERMAP S.A. terminal. The spill affected 7 km of Caleta Córdova's shore, a fishery village (PJN 2020) (Fig. 15.6). This event constitutes the major oil spill known impacting the shore in the GSJB up to date, since the beginning of oil production in 1907 (Fig. 15.7). The oil spill occurred with East wind causing the oil to reach the highest tide zone of the beach, affecting more than 800 seabirds (FPN 2011). Analysis of sediments carried out 50 days after the spill showed patchy oil pollution with up to 55,784 and 676 mg/kg dw of total petroleum hydrocarbon (TPHs) and total aromatic hydrocarbon, respectively, the highest levels of hydrocarbons recorded in the Patagonian Atlantic coast (Commendatore et al. 2008). Furthermore, 30 months after the spill, the sediment PAHs individual concentration remained above the probable effect level (PEL) for the protection of aquatic life (Marcos et al. 2012). The local community was affected as the Caleta Córdova's coast became dangerous for recreational activities and artisanal fishing. The legal responsibility of the oil spill could not be demonstrated; nevertheless, the tanker owner company subscribed a judicial deal with the provincial state paying 5 million dollars as punishment but without assuming the spill authorship (Decreto N° 701/14). The tanker was scrapped in 2009. Other minor spills happened in the CC zone in 2017 and 2018, most of them denounced by neighbors of the village.

Undeclared oil spills event in 2010. Some hydrocarbon spills in the ocean remain unknown as they do not reach the shore. In September 2010, around 400 oiled penguins were found in Chubut coasts (FPN 2011). The largest breeding colony of Magellanic penguins in Patagonia is located at Punta Tombo. Penguins migrate annually from Punta Tombo up to southern Brazil (~2000 km), following a corridor within 250 km of shore (Stokes et al. 2014). They leave the colony from January to April returning from September to December. In the 2010 pollution event, oiled penguins found from San Lorenzo ranch (42°08.3' S, 63°58.2' W) to Punta

Fig. 15.7 Caleta Córdova (CC) oil spill in December 2007 at Golfo San Jorge, Argentina. Oiled animals in a CC rescue center (top), oil impacted beach (yellow arrows), with wastes absorbent booms and waste tank on the beach (orange arrows). *Photos* courtesy of Martin Levicoy https://martinlevicoy.com

Tombo (44°03.6’ S, 65°13.3’ W) beaches were studied (Fig. 15.8). Oiled penguin feathers were analyzed to assess hydrocarbon type and distribution, showing the chromatographic profiles of a common source of pollution with aliphatic series from *n*C12 to *n*C37 and the presence of unresolved complex mixture (UCM) compatible with oil, as shown in Fig. 15.8 (Nievas et al. 2010). The *n*–heptadecane/pristane (*n*C17/Pri) and *n*–octadecane/phytane (*n*C18/Phy) ratios, indicators of early biodegradation, were around 1.5 and 3.5, respectively, indicating fresh or slighted biodegraded oil. The oil in penguin feathers lost the lighter compounds up to *n*C15, with the exception of the northern sample site (San Lorenzo ranch) where the *n*–alkane series started from *n*C12, suggesting that penguins that reached this location were nearer to the oil spill. Escalante crude oil shows close hydrocarbons profile matching with oiled penguin, with an *n*–alkane profile from *n*C8 to *n*C37, UCM and *n*C17/Pri and *n*C18/Phy ratios of 2.9 and 3.9. Diesel fuel and bilge waste from fishing ships and Bunker C were unlikely sources of the pollution found in oiled penguins, as they had lower *n*-alkane/isoprenoids ratios than those found in these animals (Nievas 2006; Wang et al. 1997). The penguins returning route to the Punta Tombo colony matches with the oil route in Patagonia (Fig. 15.8). No shore pollution was reported associated with this event, neither declaration of tanker owner, cruises or other ships in the zone at the event time. The overall evidence let us hypothesize that the most probable situation was that an undeclared oil spill of Escalante crude oil happened in the oil route of northern Patagonia. The predominant west wind likely drove the oil slick to the open sea. The penguins returning to Punta Tombo hit the oil slick, and the most affected reached the nearest coast where they were found. No one responsible for the event was found. The oil spill is just known because of the sights of oiled penguins.

In Patagonia Argentina other events of declared or mystery oil spills affected mainly penguins (García-Borboroglu et al. 2006; 2008; FPN 2011; Ruoppolo et al. 2007). For example, from 1982 to 1991, estimates indicate that 40,000 penguins died by oil pollution along the Patagonian coast each year. In September 1991, 17,000 oiled penguins appeared in Punta Tombo by a mystery oil spill, suspected from a tanker. In 2005 and 2006, 50 and near 400 oiled penguins, respectively, appeared at Cabo Vírgenes Provincial Reserve, Santa Cruz. The cause was an oil spill from a broken submarine pipeline from offshore facilities at the Magellan strait in 2005 and by an undeclared oil spill in 2006 (see *Magallanes area spills*, in this section). In December 2007, more than 800 seabirds were registered at the Caleta Córdova spill, Chubut (see *Caleta Cordova oil spill*, in this section). The number of oiled penguins found along the coast varied among years but showed a dramatic increase in the mid-1990s, showing a close match with the increase Argentinean oil exportations from 1987 to 2002 (García-Borboroglu et al. 2006). After 1998, when the oil and hazardous substances route was moved 20 miles away from the tanker stranding depth, the number of oiled penguins arriving to the Chubut coast decreased, indicating that this was a useful measurement (García-Borboroglu et al. 2008).

Cabo San Sebastian oil spill at Puerto Madryn’s harbor. In August 2015, a storm capsized a wrecked fishing ship, Cabo San Sebastián, in the Almirante Storni harbor area at Puerto Madryn city (Figs. 15.6 and 15.9). Around 20,000 L of fuel

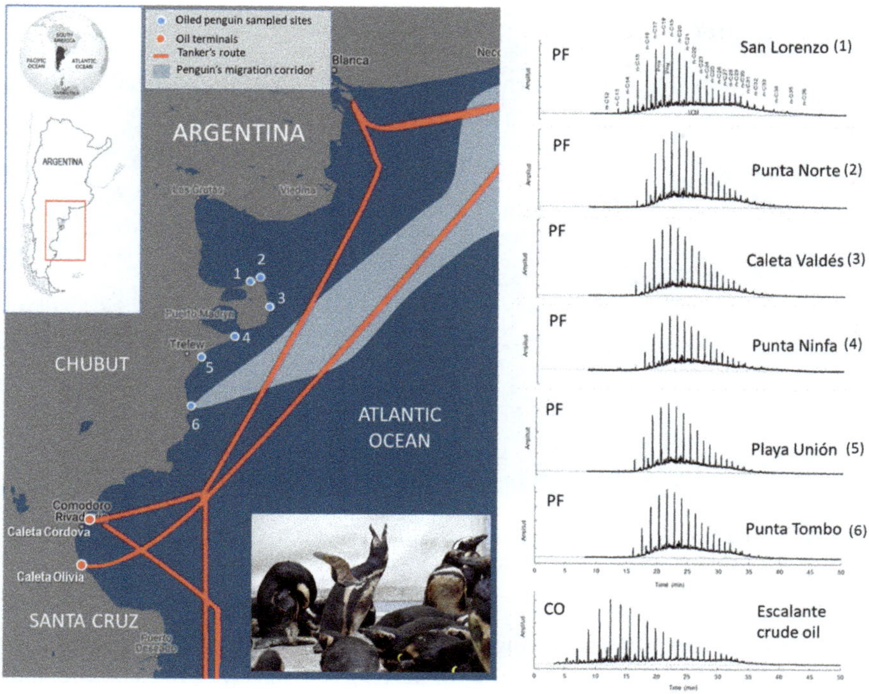

Fig. 15.8 Oiled penguins in Chubut coast in a mysterious oil spill event, September 2010. Positions of sampled oiled penguins (left), oiled penguin in recovery center in Puerto Madryn (bottom), chromatographic hydrocarbon profiles from penguin feathers (PF) and Escalante crude oil (CO) as reference adapted from Nievas et al. (2010). *Photo* Courtesy of Maxi Jonas Fotos https://www.facebook.com/maxijonasFotos/, Data of oil route from https://www.shipmap.org/

and an undetermined amount of bilge wastes were released, the slick drifting into the Golfo Nuevo and impacting the harbor beach. Pollution levels in the intertidal sediments 15 days after the event were 5–30 ppm of TPH with fresh hydrocarbon signature of medium distillates (Nievas El Makte and Sepúlveda 2016; Sepúlveda et al. 2016). This hydrocarbon spill happened in Golfo Nuevo, which is a SPZ due to its environmental sensitivity, and besides generating dock area pollution coincided with the peak of abundance of Southern right whales, which visit the gulf for their reproduction and calving every year (Fazio et al. 2015). Furthermore, the spill site (industrial pier) is less than 3 km from the busiest recreational beach in the city.

Silent hydrocarbon pollution in Puerto Madryn Harbor. In the same event that wrecked the Cabo San Sebastian ship, another seven fishing ships drifted and were hauled up on the nearby beach from the Almirante Storni harbor pier, in Puerto Madryn, in 2014. The ships increased the number of stranded vessels in a ship graveyard area of the dock (Fig. 15.10). A hydrocarbon spill was detected from one of the grounded vessels, Codepeca I, in January 2016. The ship presented hull damage after being grounded on the beach and was partially flooding inside

Fig. 15.9 Cabo San Sebastian oil spill at Puerto Madryn harbor (Chubut, Argentina) in August 2015. *Photo* courtesy of Maxi Jonas Fotos https://www.facebook.com/maxijonasFotos/

with every high tide. Water flowing out the ship percolated through the sand under the vessel and blew up in the perimeter of the ship settlement. At the spill event, the water flowing out contained a significant amount of a yellowish non-aqueous liquid phase (NALP), which formed little pools in sediments while running off down the intermareal up to the sea (Fig. 15.10). Analysis concluded that NALP was fuel from the ship assimilable to diesel oil. The nearest sediments presented high fuel pollution (up to 22% dw), being the lighter compounds quickly lost by evaporation and dissolution (Fig. 15.10). The hydrocarbon concentrations ranged from 100% for the NALP to nearly 5 ppm of total petroleum hydrocarbons for sediment away from the spill plume (Nievas El Makte and Sepúlveda 2016). The amount of spilled fuel was estimated between 10 and 20 L throughout low tide. The findings of this case indicate that the ship internally contained some amount of fuel that floated over the water column that flooded the grounded vessels at each tide cycle. The event was noticed at an extraordinary low tide day. This situation, with a longer time between flooding cycles, enabled the water to drain and also the hydrocarbons layer over it. The wrecked ship became a slow-release hydrocarbon buoy generating small fuel spills over time. As the ship was abandoned and declared unrecoverable, and no obvious hydrocarbon spills were easily appreciated, the situation remained unnoticed for 16 months, until the hydrocarbon spill event was reported. After the episode, the ship was conditioned until removed, and finally scrapped in December 2019. In Patagonia, as worldwide, harbors usually remain as hot spots regarding hydrocarbon pollution and other contaminants (Jupp et al. 2017; O'Brian 2006; NRC 2003). Thus, the management of ship graveyards at ports and programs to scrap abandoned ships are highly recommended, as they can not only enhance the visual contamination, but also avoid the invisible chronic pollution. This management became particularly relevant if the sites are in sensitive environments or nearby recreational beaches as the case of Puerto Madryn harbor.

Cruz del Sur terminal oil spill at San Sebastian Bay. A minor oil spill happened in 2019 in an episode from the terminal monobuoy at the Austral Basin (Fig. 15.6). Nearly 1500 L of released oil were declared by authorities, while only 200 L were admitted by the operator company. The spill was contained, partially recovered and did not reach the shore (Tiempo fueguino 2019). The monobuoy stopped operations due to technical problems for nearly a year, restarting in August 2020. No further information is available up to date regarding the ecological impact of the event. San Sebastian Bay, located on the Atlantic coast of Tierra del Fuego Island, is a sensitive environment declared SPZ and RAMSAR site (Fig. 15.6).

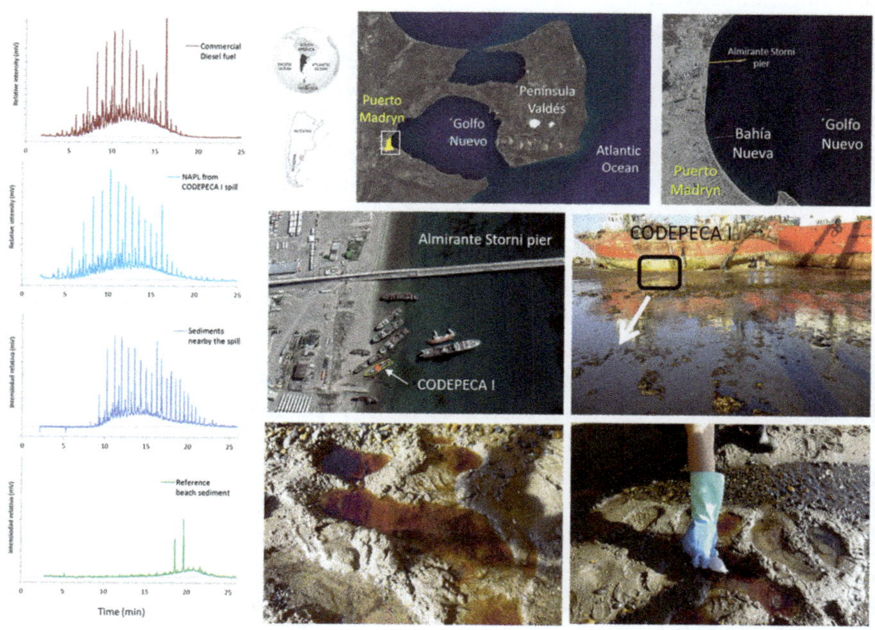

Fig. 15.10 Chronic hydrocarbon pollution by fuel spill at Puerto Madryn harbor. Chromatographic profiles (left). Center and right: Almirante Storni harbor (top), and CODEPECA I ship (middle) locations, NAPL pools in sediments in the spill runoff (bottom). Adapted from Nievas El Makte and Sepúlveda (2016)

15.5 Petroleum Hydrocarbons Levels in Atlantic Coastal Patagonia

15.5.1 Hydrocarbons Levels in Coastal Sediments

Coastal areas are usually classified according to the hydrocarbon pollution in sediments as non-polluted (hydrocarbon content <10 mg/kg dw), low to moderate polluted sediments (concentration of 10–100 mg/kg dw), moderate to high hydrocarbon levels (100–1000 mg/kg dw) and highly polluted sediments (concentration above 1000 mg/kg dw) (Commedatore et al. 2000; Commentore and Esteves 2007). While regarding PAHs levels, according to Baumard et al. (1998), the parent PAH levels (the sum of 18 compounds) can be considered as low, moderate, high and very high when they are in the ranges of 0–100, 100–1100, 1000–5000 and >5000 μg/kg, respectively.

Reports of hydrocarbon levels in coastal sediments or organisms in Atlantic Patagonia are quite scarce, even more taken into account the extension of its coastline estimated at more than 3000 km (Gil et al. 2019). Some baseline level assessments were performed along the coast (Commendatore et al. 2000; 2012; 2015; Commendatore and Esteves 2007; Esteves et al. 2006; Esteves and Commendatore 1993). Other sites description were performed in spill event reports (Iantanos et al. 2008; Commendatore et al. 2008; Nievas El Makte and Sepúlveda 2016; Sepúlveda et al. 2016), studies focused on bioaccumulation and effects of hydrocarbons on marine organisms (Massara Paletto et al. 2008; Amin et al. 2011; Primost et al. 2018; Sturla Lompré et al. 2018) or gene and enzyme prospection of microbial hydrocarbon metabolism (Guibert et al. 2012; Marcos et al. 2012). Based on this information, hydrocarbon levels in sediments can be resumed as follows: levels of hydrocarbon from low to high concentrations were found in many harbors and nearby areas in Atlantic Patagonia. Considering TPH or total aliphatics (TAl) were found: (1) in the low-moderate pollution range: coast of Comodoro Rivadavia (1995), (2) in the moderate to high range: San Antonio Oeste harbor (1995), Rawson harbor (2001), Comodoro Rivadavia harbor (1995), Caleta Córdova (1995), Ushuaia commercial port, Nautical pier and Aspirante Creek (2006) and (3) in the highly polluted levels Puerto Madryn (2016), Ushuaia combustible pier (2006) and Caleta Córdova (after the 2007 oil spill). Moreover, regarding the PAHs a similar pattern was found: low PAHs pollution in Camarones harbor (2010) as in north and south beaches of Puerto Madryn (2001); in Puerto Madryn harbors, Storni pier moderate to very high (2001 and 2005) and in Piedra Buena Pier moderate to high (2005 and 2012). In Comodoro Rivadavia harbor, the concentration was low (in 2005) and very high in Caleta Córdova after the oil spill (2007), decreasing to moderate 2.5 years after the spill in two evaluated sites. In Ushuaia Bay, moderate pollution was found in commercial harbors and nearby areas (2006), while in the Orion combustible plant, low to high PAHs pollution levels were reported in intertidal and subtidal sediments (2004–2010). Thus, the hydrocarbon pollution type seems to be associated with sources: anthropogenic activities at ports and oil industry. Typical polluted sediments from areas related to port and industrial activities are reported in the range of 650–1200 mg/kg dw of petroleum hydrocarbons (Commendatore et al. 2000). Sites where hydrocarbon spills happened recently were particularly polluted, as expected. Concentrations measured early after the spills showed the highest hydrocarbon values in sediments of the coast. The Caleta Córdova sediments showed up to 55,108 mg/kg dw of total hydrocarbons in 2008 (Commendatore et al. 2008), while in the Puerto Madryn chronic spill (see Sect. 15.4) reached a hydrocarbon content of 22% dw in the fuel spill point up to near 5 mg/kg dw away the site (Nievas El Makte and Sepúlveda 2016). These concentrations represent an acute situation which requires a following up to assess the attenuation of hydrocarbons over time, after spreading, adsorption onto the sediments and subsequent weathering. Commendatore and Esteves (2007) concluded that three types of zones could be identified in the Atlantic Patagonian coast: harbors affected by chronic pollution with concentrations in the range of 6–741 mg/kg dw of TAl and 2.5–7.7 mg/kg dw of total PAHs, an accumulation zone likely an area where an oil spill impacted the north zone of GSJ with concentrations up to 1305 mg/kg dw TAl and the rest of the assessed sites with unpolluted character with hydrocarbons concentrations in the range of 0.27–5.4 mg/kg dw TAl. Since then, more recent reports have confirmed this tendency. Sites away, even short distances, from punctual hydrocarbon pollution likely sources, have low levels of hydrocarbons and usually with biogenic character. To this category belong beaches of Golfo San Jose (1989 and 2005), Península Valdés (1989, 2009 and 2012), beaches 8 km north and 3 km south of the Puerto Madryn center in Golfo Nuevo (1989, 1995 and 2001), Chubut Atlantic coasts from Punta Ninfa to the north edge of GSJ (except Rawson harbor) (1989,1995 and 2001), most of the sites of the North Zone of GSJ in the Patagonian Austral Inter-jurisdictional Marine Coastal Park (with exception of an accumulation point in Faro Aristizabal and Camarones harbor) (1989, 1995 and 2010), Rada Tilly (1989 and 1995), Santa Cruz coast at south of GSJ (1995), the Atlantic shore of Tierra del Fuego Island (2001), the Beagle channel (with the exception of the Ushuaia bay) (2001) and Isla de los Estados (2003).

Despite available data are mostly historical and recent information is very scarce, no unexpected high levels of hydrocarbons were found in remote places away from anthropogenic activities. New assessments of hydrocarbon concentration are needed to update the available information and establish pollution trends. Moreover, some oil spills happened after the last hydrocarbon level assessments in sediments in many sites reported here, so it is expected that levels would be higher than in previous reports. The Atlantic Patagonia possesses many ports where commercial fishery activities and oil transport take place (Fig. 15.5). Minor accidental spills while bunkering ships or in loading tanker operations are frequent reasons of hydrocarbon pollution in harbors and oil terminals worldwide (O'Brian 2006; Krata and Jachowski 2021). Some spill events in harbors and in oil terminals have also contributed to chronic pollution in the Atlantic Patagonian coast.

15.5.2 Hydrocarbons Levels in Bioindicator Organisms

Bivalves are filtering organisms that accumulate hydrophobic compounds, and for this reason, they are used as pollution indicators (Sericano 2000). Many monitoring programs (i.e., Mussel Watch, NOAA) were implemented worldwide to assess persistent organic pollutants (POP) in the marine environment (Sericano 2000; Monirith et al. 2003). In addition, hydrocarbon levels in edible species are of great concern regarding the foodstuffs safety.

In Atlantic Patagonia, some authors reported hydrocarbons levels or PAHs bioaccumulation in organisms. Primost et al. (2018) compared PAHs content in the edible gastropods *Buccinanops globulosus* in two sites with different maritime influences: Puerto Madryn harbor area and a 20 km nearby beach, presumably non-polluted, finding non-detectable PAHs neither in gastropods nor in sediments of the control site. On the other hand, PAHs levels were higher in gastropod tissues (560 μg/kg dw) than in sediments (270 μg/kg dw) at the harbor area, being the PAH4 level (sum of benzo[a]anthracene, benzo[b]fluoranthene, benzo[a]pyrene and chrysene) exceeding the maximum limit established by UE for PAH4 in foodstuffs (30 μg/kg dw, EUC 2011). While the dibenzo[a,h]anthracene sediment concentration exceeded the level established by the Canadian guidelines for marine sediment quality, being lower than the PEL concentration (CCME 1999). These authors concluded that based on the PAHs levels, the consumption of these animals could be dangerous for humans according to the ingestion frequency. Hydrocarbon levels in mussel and clam were also studied in Bahía Nueva, Golfo Nuevo, Chubut, in 2001 (Massara Paletto et al. 2008). Aliphatic and aromatics compounds were assessed. The highest levels of aliphatic hydrocarbon were found for mussels at both piers, while the highest PAHs levels were also found in both piers, in an effluent discharge site and in the harbor adjacent area. Sediments of the harbor area showed the highest PAHs concentration of the study, particularly, the Piedra Buena Pier. Amin et al. (2011) studied PAHs concentration in *Mytilus edulis chilensis* (mussel) from Ushuaia Bay reporting an increment between October 1999 and October 2003. In 2003, the PAHs in mussels ranged from 2.25 to 2420 μg/g lipid having petrogenic origin with pyrogenic background, with high concentrations of naphthalenes, fluorenes, phenanthrenes and anthracenes. These facts indicated that the mussels were exposed to relatively fresh petroleum hydrocarbons. The highest levels in both years were found in mussels from the site next to the Orion combustible plant in Ushuaia Bay (Fig. 5.5), with notable high levels. The authors stated that these results reflect the environmental quality of the Ushuaia bay, which atmospheric and oceanographic conditions favors the POP accumulation and long-term permanence. These results closely agree with hydrocarbon pollution determined in the Ushuaia bay sediments (Esteves et al. 2006; Commendatore et al. 2012; Marcos et al. 2012). Evaluation of PAHs levels in the north zone of the San Jorge gulf was carried out in 2010, assessing sediments and bivalves (*Mytilus edulis*) (Commendatore et al. 2015). Generally, values found in sediments corresponded to non-polluted, except in Camarones that were slightly polluted. In addition, levels of PAHs in bivalve tissues were low and typical for locations distant from contaminant sources.

In summary, studies performed in organisms from the Atlantic coast generally reflected the hydrocarbon pollution levels of its site of origin. Thus, hydrocarbon levels in marine organisms in locations identified as hydrocarbon-polluted ones, likely shows the same trend. Consequently, closed monitoring programs are necessary to assess PAHs levels in hydrocarbon pollution hot points on the Atlantic Patagonian coast, such as near ports or in sites impacted by hydrocarbon spills where coastal artisanal fishing and mollusk gathering are economic activities, such as Caleta Córdova in Golfo San Jorge.

15.6 Resource Overlap in the Patagonian Sea: Hydrocarbon Pollution and Its Environmental Implications

The Patagonian Continental Shelf of Argentina presents a highly productive ecosystem, which provides a great richness and diversity of resources. The abundance of plankton along the fronts in Patagonian Sea promotes the reproduction of commercial fish, crustacean and squid species, which supports the development of a large-scale fishing industry (Allega et al. 2020). The Patagonian coast is also an important breeding zone for the same fishing resources since intertidal and coastal benthic communities are the diet of larvae or juveniles of commercial species or offer refuge for the deposition of eggs. Thus, benthic organisms constitute highly structured environments. GSJ and off Tierra del Fuego are hotspots for benthic species richness (Miloslavich et al. 2011). Particularly, GSJ has a relevant hubbsi hake fishery and the most economically important shrimp fishery of Argentina, with the main area of shrimp youth concentration located in the south of the gulf in Mazarredo (46°45’ S, 66°30’ W) at depths less than 40 m (Góngora et al. 2012). On the other hand, these sites are near the exit zone of oil production of Eastern Patagonia oil basins (see Sect. 15.3). Considering that oil spills are particularly harmful to benthic communities when they reach the coast, and that hydrocarbon pollution events happened in the near past in those sites (see Sect. 15.4), management actions strongly committed to prevent oil spills from hydrocarbon terminals are imperative. Any oil impact in critical areas as Mazarredo raises the risk of a fishery collapse. Adverse ecological and environmental impacts of oil spills have been well documented. For example, in the Exxon Valdez oil spill, the most notable ecological effect was the outright mortality of extensive quantities of biota at all levels, with varying degrees and rates of recovery (Gill et al. 2016). The economic and subsequent social impact of this spill was also extensive. The collapse of fishing industries, the damage of tourism along with ancillary industries, and the implications on community-based health and livelihood were consequences of the spill, together with the ecological damage (Zhang et al. 2019). Similar situations were reported in other spills as the Prestige in Spain (Martínez-Gómez et al. 2009; Penela-Arenaz et al. 2009). Special protection zones (SPZs) were created to protect sensitive environments. However, despite the rules that ban hydrocarbons discharges from vessels in normal ship operations in SPZs, accidental spills can greatly affect these areas. As an example, the oil spills described here in Caleta Córdova (2007), in Bahía San Sebastian (2019) and in Puerto Madryn (2015) all happened in SPZs. Moreover, three of the five oil terminals in Atlantic Patagonia are located in SPZs (Caleta Córdova, Caleta Olivia and Cruz del Sur), thus, any oil spill near these terminals acquires higher relevance based on the sensitivity of the zone and on the higher risk of damage.

The Patagonian coast and its adjacent waters are valuable in terms of global biodiversity, as they are used as a resting, feeding and mating site for birds and marine mammals. The predictability when those animals arrive on these coasts has given rise to ecological tourism, including tours to see penguins, sea lions, elephant seals and whales. Whale watching in particular is the main tourist activity of Chubut. It is well known the occurrence of acute and chronic toxicity in whales or in other marine mammals when these animals are in contact with hydrocarbons from oil spills or chronic pollution. Consequences of these toxicities as, for example, reduction in reproductive success or increases in respiratory infections have been reported (Hook 2020; Kellar et al. 2017). Another important tourist attraction along the Patagonian shore are the Magellanic penguin breeding colonies. As was described before, oil spill affects particularly penguins and seabirds (see Sect. 15.4). Thus, the risk of oil spills in sensitive areas such as penguin colonies or reproductive areas of marine mammals, like in the example of Golfo Nuevo, would impact on the ecology and also on tourism activities based on wild marine fauna, which is one of the socio-economical pillars of eastern Patagonia.

15.7 Concluding Remarks

The Argentinean O&G production in onshore and offshore areas is expected to increase in the next years to fulfill the country's own needs and also to improve its trade balance. This will imply an increase in platform operations and number of tankers for hydrocarbon transportation across the maritime zone. Thus, with the development of the new offshore activity, the risk of oil spill pollution will also rise. Hydrocarbon pollution in the Atlantic Patagonian coast has been mainly related to oil exploitation zones and to maritime hydrocarbon transport. The Magellan strait suffered a disaster oil spill four decades ago, which impacts still persist today. A more recent and smaller oil spill in GSJ in 2007 impacted a fishery village causing deep environmental and social consequences. The other hydrocarbon pollution focus detected in Patagonian coast is the chronic pollution related to harbors operation. Besides major oil spill, oil terminals had minor but frequent spilling affecting the surrounding area. Commercial and fishing ports are, excepting areas affected by petroleum spills, the more polluted sites in the Atlantic Patagonian coast. However, information is scarce and frequent monitoring programs would be useful to assess hydrocarbon pollution trends. Some studies have also proven the presence of hazardous PAHs in organisms near ports over safe international standard levels. The hydrocarbon pollution in the Atlantic Patagonian coast does not seem to be an extended situation, instead, it is concentrated in specific points: ports and harbors, oil terminals and nearshore or offshore oil production areas. While the impact of hydrocarbon pollution in the open sea, such as those likely from intense fishing areas near the 200 mile, is unknown.

In summary, Atlantic Patagonia has vast coast areas exempt from hydrocarbon pollution, with high biodiversity and a huge richness in biological resources. To continue this way, modern environmental management tools became indispensable, in order to have a sustainable development. Prevention of oil pollution at sea requires quick response monitoring tools covering the maritime most frequent routes as well as the hot spot fishing areas. While the pollution damage immediately begins after an oil spill, environmental crime responsibilities of its generator may take many years to be proven. Thus, legal tools and procedures of environmental authorities need to be strengthened to achieve a quick and effective response after an oil spill. Regarding chronic pollution, increasing in security requirements for oil terminals and harbor bunkering became obvious measurements. On the other hand, the implementation of programs for ship graveyard elimination, by scrapping or removing abandoned ships at harbors, would greatly enhance the recovery of harbor's nearby impacted area, not only of hydrocarbons, but also of other contaminants associated such as heavy metals. This is particularly important in sensitive environments, or where the same place is used by ports and recreational activities such as in Puerto Madryn, Rawson and Comodoro Rivadavia cities. Argentina faces the challenge to develop economic activities in its maritime continental shelf area without losing environmental quality. Marine research in pollution fields as well as in physical, chemical and biological oceanography seems to be the necessary input to take management and conservation adequate decisions.

Acknowledgements Parts of the experimental work presented here were carried out granted by the following research projects: PICT-2016-3124 from the National Agency for the Promotion of Research, Technological Development and Innovation (ANPCyT), PIO-YPF N°004/14 from YPF Foundation and CONICET, and PID-UTN 2981 from the National Technological University (UTN FRCH).

References

Allega L, Braverman MS, Cabreira AG, Campodónico S, Colonello JH, Derisio CM, Di Mauro RP, Gaitán EN, Hozbor M, Irusta G (2020) Estado del conocimiento biológico pesquero de los principales recursos vivos y su ambiente, con relación a la exploración hidrocarburífera en la Zona Económica Exclusiva Argentina y adyacencias. INIDEP, Mar del Plata, pp 1–119

Amin OA, Comoglio LI, Sericano JL (2011) Polynuclear aromatic and chlorinated hydrocarbons in mussels from the coastal zone of Ushuaia, Tierra del Fuego, Argentina. Environ Toxicol Chem 30(3):521–529

Argüelles MB, Coscarella M, Fazio A, Bertellotti M (2016) Impact of whale-watching on the short-term behavior of Southern right whales (*Eubalaena australis*) in Patagonia, Argentina. Tour Manag Perspect 18:118–124

Baruj G, Drucaroff S (2018) Estimaciones del potencial económico del océano en la Argentina. Centro Interdisciplinario de Estudios en Ciencia, Tecnología e Innovación, pp 1–83. http://www.ciecti.org.ar/wp-content/uploads/2018/04/IT10-Pampa-azul_vDigital_16-abril-2018.pdf. Last accessed 20 Jan 2021

Baumard P, Budzinski H, Garrigues P (1998) Polycyclic aromatic hydrocarbons (PAHs) in sediments and mussels of the western Mediterranean Sea. Environ Toxicol Chem 17:765–776

Beiras R (2018) Hydrocarbons and oil spills. In: Beiras R (ed) Marine pollution sources, fate and effects of pollutants in coastal ecosystems. Elsevier, Amsterdam, pp 1–385

Canadian Council of Ministers of the Environment (CCME) (1999) Canadian sediment quality guidelines for the protection of aquatic life: Polycyclic aromatic hydrocarbons (PAHs). http://ceqg-rcqe.ccme.ca/download/en/243. Last accessed 13 Feb 2021

Chapman P, Wade TL, Knap AH (2018) Environmental effects and management of oil spills on marine ecosystems. In: Islam MN, Jørgensen SJ (eds) Environmental management of marine ecosystems. CRC Press, Boca Raton, pp 119–146

Comisión Nacional del Límite Exterior de la Plataforma Continental (COPLA) (2017) El margen continental Argentino: entre los 35°S y los 55°S en el contexto del Artículo 76 de la Convención de las Naciones Unidas sobre el Derecho del Mar. CABA, Ministerio de Relaciones Exteriores y Culto, pp 1–455

Commendatore MG, Esteves JL (2007) An assessment of oil pollution in the coastal zone of Patagonia, Argentina. Environ Manage 40(5):814–821

Commendatore MG, Esteves JL, Colombo JC (2000) Hydrocarbons in coastal sediments of Patagonia, Argentina: levels and probable sources. Mare Pollut Bull 40(11):989–998

Commendatore MG, Esteves JL, Nievas ML (2008) Evaluación de sedimentos de la zona costera de Caleta Córdova. Etapa I: Intermareal superior. Informe técnico. Centro Nacional Patagónico-Ministerio de Ambiente y Control del Desarrollo Sustentable de la Provincia del Chubut, p 28

Commendatore MG, Nievas ML, Amin O, Esteves JL (2012) Sources and distribution of aliphatic and polyaromatic hydrocarbons in coastal sediments from the Ushuaia Bay (Tierra del Fuego, Patagonia, Argentina). Mar Environ Res 74:20–31

Commendatore MG, Franco MA, Gomes Costa P, Castro IB, Fillmann G, Bigatti G, Esteves JL, Nievas ML (2015) Butyltins, polyaromatic hydrocarbons, organochlorine pesticides, and polychlorinated biphenyls in sediments and bivalve mollusks in a mid-latitude environment from the Patagonian coastal zone. Environ Toxicol Chem 34(12):2750–2763

Coronato AM, Coronato F, Mazzoni E, Vázquez M (2008) The physical geography of Patagonia and Tierra del Fuego. Dev Quat Sci 11:13–55

Cuitiño JI, Varela AN, Ghiglione M, Richiano SM, Poire DG (2019) The Austral-Magallanes Basin (southern Patagonia): a synthesis of its stratigraphy and evolution. Lat Am J Sedimentol Basin Anal 26(2):155–166

de Haro C (2007) Conservación de los ambientes costeros de Santa Cruz, Argentina. Observaciones sobre el impacto de la actividad petrolera. http://www.marlimpio.delfinaustral.com/publicaciones/PETROLEO%20SANTA%20CRUZ%20Posgrado%20ECO%20ECO.pdf. Last accessed 20 Jan 2021

Decreto N° 701/14 de la provincia del Chubut. Acuerdo Transaccional entre la Provincia del Chubut y la firma Antares Naviera S.A. Rawson, 17 Junio de 2014. Boletín Oficial Año LVI - N° 1197. http://boletin.chubut.gov.ar/archivos/boletines/Junio%2018,%202014.pdf. Last accessed 18 June 2014

ENAP (2021) Nuestras Operaciones. Cuenca Austral. http://www.enap.com.ar/nuestras-operaciones-cuenca-austral.php. Last accessed 29 Jan 2021

Esteves JL, Commendatore MG (1993) Total aromatic hydrocarbons in water and sediment in a coastal zone of Patagonia, Argentina. Mar Pollut Bull 26(6):341–342

Esteves J, Commendatore M, Nievas M, Massara Paletto V, Amín O (2006) Hydrocarbon pollution in coastal sediments of Tierra del Fuego Islands, Patagonia Argentina. Mar Pollut Bull 52(5):582–590

European Union Commission (EUC) (2011) Commission regulation (EU) No 835/2011. Amending regulation (EC) No 1881/2006 as regards maximum levels for polycyclic aromatic hydrocarbons in foodstuffs. Off J Eur Union 5

Fazio A, Argüelles MB, Bertellotti M (2015) Change in southern right whale breathing behavior in response to gull attacks. Mar Biol 162(2):267–273

Filler DM, Kennicutt MCII, Snape I, Sweet ST, Klein AG (2014) Arctic and Antarctic spills. In: Fingas M (ed) Handbook of oil spill science and technology. Wiley, Hoboken, pp 497–512

Fundación Patagonia Natural (FPN) (2011) Empetrolamiento de pingüinos. Nuestra experiencia. http://www.patagonianatural.org/empetrolamiento-de-pinguinos/nuestra-experiencia. Last accessed 29 Jan 2021

García-Borboroglu P, Boersma PD, Ruoppolo V, Reyes L, Rebstock GA, Griot K, Heredia SR, Adornes AC, Da Silva RP (2006) Chronic oil pollution harms magellanic penguins in the Southwest Atlantic. Mar Pollut Bull 52(2):193–198

García Borboroglu P, Boersma P, Reyes L, Ruoppolo V (2008) Contaminación por hidrocarburos y su efecto sobre el pingüino de Magallanes. Puerto Madryn: Estado de conservación del mar patagónico. In: Foro para la conservación del mar patagónico y áreas de influencia (2008) Estado de conservación del mar patagónico y áreas de influencia. Puerto Madryn, Argentina, Edición del Foro. http://www.marpatagonico.org

Gill DA, Ritchie LA, Picou JS (2016) Sociocultural and psychosocial impacts of the Exxon Valdez oil spill: twenty-four years of research in Cordova, Alaska. Extr Ind Soc 3(4):1105–1116

Gil MN, Giarratano E, Barros V, Bortolus A, Codignotto JO, Schenke RD, Góngora GME, Lovrich G, Monti AJ, Pascual M (2019) Southern Argentina: the patagonian continental shelf. In: Sheppard C (ed) World seas: an environmental evaluation. Academic Press, Elsevier, London, pp 783–811

Góngora ME, González-Zevallos D, Pettovello A, Mendía L (2012) Characterization of the main fisheries in San Jorge Gulf, Patagonia, Argentina. Lat Am J Aquat Res 40(1):1–11

Guibert LM, Loviso CL, Marcos MS, Commendatore MG, Dionisi HM, Lozada M (2012) Alkane biodegradation genes from chronically polluted subantarctic coastal sediments and their shifts in response to oil exposure. Microb Ecol 64(3):605–616

Gundlach ER (2017) The great Patagonian oil spill (Metula), 41 years later. Int Oil Spill Conf Proc 2017(1):515–535

Hook SE (2020) Beyond thresholds: a holistic approach to impact assessment is needed to enable accurate predictions of environmental risk from oil spills. Integr Environ Assess and Manag 16 (6):813–830

Iantanos N, Pucci GN, Acuña AJ, Pucci OH (2008) Derrame de hidrocarburos en la playa de Caleta Córdova, ciudad de Comodoro Rivadavia, Argentina: su evolución. AIDIS Argentina, Ingeniería Sanitaria y Ambiental 100(4):113–117

Instituto Argentino del Gas y del Petróleo (IAPG) (2020) Argentina's Hydrocarbon Industry: 2019 Outlook, pp 1–18. http://www.aogexpo.com.ar/ENG/OverviewEN.pdf. Last accessed 29 Jan 2021

Jupp BP, Fowler SW, Dobretsov S, van der Wiele H, Al-Ghafri A (2017) Assessment of heavy metal and petroleum hydrocarbon contamination in the Sultanate of Oman with emphasis on harbours, marinas, terminals and ports. Mar Pollut Bull 121(1–2):260–273

Kellar NM, Speakman TR, Smith CR, Lane SM, Balmer BC, Trego ML, Catelani KN, Robbins MN, Allen CD, Wells RS (2017) Low reproductive success rates of common bottlenose dolphins *Tursiops truncatus* in the northern Gulf of Mexico following the Deepwater Horizon disaster (2010–2015). Endanger Species Res 33:143–158

Koutoudjian A (2011) Lineamientos para la incorporación de la problemática del Mar Argentino en la planificación territorial. Subsecretaría de Planificación Territorial de la Inversión Pública (ed), p 79

Krata P, Jachowski J (2021) Towards a modification of a regulatory framework aiming at bunker oil spill prevention from ships—a design aspect of bunker tanks vents location guided by CFD simulations. Reliab Eng Syst Saf 208:

Law N° 26,875/13 del Honorable Congreso de la Nación Argentina. Área Marina Protegida Namuncurá—Banco Burdwood—Creación, 5 Agosto de 2013. Boletín Oficial N° 32694. http://servicios.infoleg.gob.ar/infolegInternet/anexos/215000-219999/218102/norma.htm. Last accessed 7 Mar 2021

Law N° 27,490 del Honorable Congreso de la Nación Argentina. Ley N° 27,037 Modificación. Áreas Marinas Protegidas Namuncurá—Banco Burdwood II y Yaganes—Creación, 17 Diciembre de 2018. Boletín Oficial N° 34017. http://servicios.infoleg.gob.ar/infolegInternet/anexos/315000-319999/317651/norma.htm, Last accessed 7 Mar 2021

Marcos MS, Lozada M, Di Marzio WD, Dionisi HM (2012) Abundance, dynamics, and biogeographic distribution of seven polycyclic aromatic hydrocarbon dioxygenase gene variants in coastal sediments of Patagonia. App Environ Microbiol 78(5):1589–1592

Martínez-Gómez C, Fernández B, Valdés J, Campillo JA, Benedicto J, Sánchez F, Vethaak AD (2009) Evaluation of three-year monitoring with biomarkers in fish following the Prestige oil spill (N Spain). Chemosphere 74(5):613–620

Massara Paletto V, Commendatore MG, Esteves JL (2008) Hydrocarbon levels in sediments and bivalve mollusks from Bahía Nueva (Patagonia, Argentina): an assessment of probable origin and bioaccumulation factors. Mar Pollut Bull 56:2082–2105

Miloslavich P, Klein E, Díaz JM, Hernandez CE, Bigatti G, Campos L, Artigas F, Castillo J, Penchaszadeh PE, Neill PE (2011) Marine biodiversity in the Atlantic and Pacific coasts of South America: knowledge and gaps. PLoS ONE 6(1):

Ministerio de Economía, Subsecretaría de Programación Microeconómica (ME-SSPME) (2019) Informes de cadena de valor. Pesca—septiembre 2019, pp 1–26. https://www.senado.gob.ar/upload/32040.pdf. Last accessed 14 Feb 2021

Ministerio de Energía y Minería. Gobierno de Argentina (MEM 2017) Estudio para evaluar el potencial de inversión en exploración de hidrocarburos offshore. https://www.minem.gob.ar/prensa/26157/estudio-para-evaluar-el-potencial-de-inversion-en-exploracion-de-hidrocarburos-offshore. Last accessed on Jan 2021

Ministerio de Energía y Minería. Secretaría de Energía (MEM-SE) (2021) Producción de Petróleo y Gas (SESCO). http://datos.minem.gob.ar/dataset/produccion-de-petroleo-y-gas-tablas-dinamicas. Last accessed 29 Jan 2021

Monirith I, Ueno D, Takahashi S, Nakata H, Sudaryanto A, Subramanian A, Tanabe S (2003) Asia-Pacific mussel watch: monitoring contamination of persistent organochlorine compounds in coastal waters of Asian countries. Mar Pollut Bull 46(3):281–3001

National Research Council (NRC) (2003) Oil in the sea III: inputs, fate, and effects. The National Academic Press, pp 1–265

Nievas ML (2006) Diseño, simulación y optimización de un reactor biológico para el tratamiento de efluentes orgánicos provenientes de sentinas de buques, Ph.D. Thesis, Universidad Nacional del Sur, Bahía Blanca, Argentina

Nievas El Makte M, Sepúlveda MA (2016) Riesgo ambiental del buque CODEPECA 1 en la playa del muelle Almirante Storni. Informe técnico ST1400-02/16 del CESIMAR-CONICET, CCT-CONICET-CENPAT, Puerto Madryn, pp 1–28

Nievas ML, Commendatore M, Franco M, Esteves JL (2010) Hidrocarburos en plumas de pingüinos de Magallanes (*Spheniscus magellanicus*), generados por un derrame en el mar argentino. Informe Técnico CENPAT-CONICET, pp 1–12

O'Brian M (2006) Oil spills in ports. Ports Harbors 51(4):34–35

Pampa Azul (2017) Horizontes estratégicos para el Mar Argentino. Documento del Ministerio de Ciencia, Tecnología e Innovación Productiva, pp 1–358. https://www.pampazul.gob.ar/wp-content/uploads/2017/06/DOC_Pampa-azul_OK_baja1.pdf. Last accessed 14 Feb 2021

Pampa Azul (2021) Argentina cuenta con dos nuevas Áreas Marinas Protegidas. https://www.pampazul.gob.ar/argentina-cuenta-con-dos-nuevas-areas-marinas-protegidas/. Last accessed 14 Feb 2021

Penela-Arenaz M, Bellas J, Vázquez E (2009) Chapter five: effects of the Prestige oil spill on the biota of NW Spain: 5 years of learning. Adv Mar Biol 56:365–396

Petrotecnia (2002) Las empresas y el offshore en la Argentina. Petrotecnia 43(3):8–14

Petrotecnia (2004) Las terminales de hidrocarburos. Petrotecnia 45(2):16–37

Petrotecnia (2016) Novedades de la industria. Total comenzó la producción offshore Vega Pléyade. Petrotecnia 57(1):54

Poder Judicial de la Nación (PJN) 2020 Sentencia de la causa N° FCR 91000995/2009/TO1 caratulada "Avalos, Ricardo Raúl y otros s/Infracción Ley 24.051" Tribunal Oral en lo Criminal Federal de Comodoro Rivadavia, pp 1–98. https://opsur.org.ar/wp-content/uploads/2020/09/SENTENCIA.pdf. last accessed 29 Jan 2021

Prefectura Naval Argentina (PNA) (1998a). Designación de zonas de protección especial en el litoral Argentino. Ordenanza N 12/98 (DPMA), Tomo 6: Régimen para la Protección del Medio Ambiente. PNA, Buenos Aires

Prefectura Naval Argentina (PNA) (1998b). Rutas de los buques que transportan hidrocarburos y sustancias nocivas a granel en navegación marítima nacional Ordenanza N 13/98 (DPMA). Tomo 6: Régimen para la Protección del Medio Ambiente. PNA, Buenos Aires

Primost MA, Commendatore M, Torres PJ, Bigatti G (2018) PAHs contamination in edible gastropods from north Patagonian harbor areas. Mar Pollut Bull 135:828–831

Ruoppolo V, Callahan B, Heredia SR, Silva Filho RP, Poleschi C, Perancho J, Holcomb J (2007) Cabo Vírgenes mystery spill: challenges and lessons learned. Effects of oil on wildlife, 2007: Conference Proceedings, pp 161–169

Secretaría de Gobierno de Energía (SGE) (2019a) Síntesis de la evolución de reservas de hidrocarburos. Período 2009–2018

Secretaría de Gobierno de Energía (SGE) (2019b) Concurso Público Internacional Costa Afuera N °1. Adjudicación. Resolución N° RESOL-2019-276-APN-SGE#MHA

Sepúlveda MA, del Brio F, Revuelta F, Polifroni R, Olivera NL, Nievas El Makte ML (2016) Hydrocarbon spill at Puerto Madryn port, Patagonia Argentina: site characterization and biostimulation field trial. Latin American Biodeterioration and Biodegradation Symposium (LABS9) Biodeterioration & Biodegradation Society, Río de Janeiro, Brasil

Sericano JL (2000) The Mussel Watch approach and its applicability to global chemical contamination monitoring programmes. Int J Environ Pollut 13(1–6):340–350

Stokes DL, Boersma PD, de Casenave JL, García-Borboroglu P (2014) Conservation of migratory Magellanic penguins requires marine zoning. Biol Conserv 170:151–161

Sturla Lompré J, Nievas M, Franco M, Grossid V, Ferrando A, Militon C, Gilbert F, Cuny P, Stora G, Sepúlveda M, Esteves J, Commendatore M (2018) Fate of petroleum hydrocarbons in bioturbated pristine sediments from Caleta Valdés (Patagonia Argentina): An ex situ bioassay. Ecotoxicol Environ Safety 162:673–682

Sylwan CA, Droeven C, Iñigo J, Mussel F, Padva D (2011) Cuenca del Golfo San Jorge. VIII Congreso de Exploración y Desarrollo de Hidrocarburos. In: Simposio Cuencas Argentinas: visión actual. Instituto Argentino de Petróleo y Gas (IAPG), Mar del Plata, pp 139–183

TERMAP (2020) Port rules—Edition 2020, pp 1–77. http://www.termap.com.ar/TermapWeb/Paginas_e/Download/Port%20Rules%20Ed.2016.pdf. Last accessed 29 Jan 2021

Tiempo Fueguino.com.ar (2019) El derrame de petróleo en San Sebastián no afectó a la costa. https://www.tiempofueguino.com/el-derrame-de-petroleo-en-san-sebastian-no-afecto-a-la-costa/. Last accessed 21 Feb 2021

Van Wyk D, Genini L, Francos JP (2019) Optimización del desarrollo de un campo maduro con pozos horizontales con simulación numérica. Caso de estudio campo Caleta Córdova, Petrotecnia 3:126–133

Wang Z, Fingas M, Landriault M, Sigouin L, Feng Y, Mullin J (1997) Using systematic and comparative analytical data to identify the source of an unknown oil on contaminated birds. J Chromatogr A 775(1–2):251–265

Wang Z, Fingas M, Owens EH, Sigouin L, Brown CE (2001) Long-term fate and persistence of the spilled Metula oil in a marine salt marsh environment: degradation of petroleum biomarkers. J Chromatogr A 926:275–290

YPF (2011) Preliminary Prospectus Supplement dated March 14, 2011. U.S. Securities and Exchange Commission (SEC). https://www.sec.gov/Archives/edgar/data/904851/000120864611000119/c104424.htm. Last accessed 29 Jan 2021

Zhang B, Matchinski EJ, Chen B, Ye X, Jing L, Lee K (2019) Marine oil spills—Oil pollution, sources and effects. In: Sheppard C (ed) World seas: an environmental evaluation. Academic Press, Elsevier, London, pp 783–811

Chapter 16
Dumping of Toxic Waste into the Oceans

Donat-P. Häder

Abstract Human development is characterized by discarding items after the end of their useful lifetime rather than recycling them which stresses the finite resources of our planet. A growing avalanche of single-use plastic articles accumulates in the landscape and in the oceans. Hazardous chemical wastes are pumped by pipeline into the sea, carried by ships and dumped into the ocean or are incinerated off-shore. Pharmaceuticals, drugs and household wastes are washed into coastal ecosystems stressing the biota. In addition to household and industrial waste, soil and rubble from building activities are disposed of at sea. Especially after wars, weapons and ammunition including chemical, bacteriological and nuclear warfare agents are dumped into the oceans. Eventually, hazardous chemicals leak out of the containers threatening marine life. Radioactive waste from nuclear fission reactors and nuclear weapon production constitutes a major problem because of the long half life of the radionuclides generated during production. Therefore, many countries have resorted to dumping these at sea. In addition, catastrophic accidents on nuclear powered vessels and submarines as well as nuclear reactors release radioactive material to the environment which can reach the sea. After the conquest of the space around our planet, the number of objects orbiting the Earth has increased exponentially. Decommissioned spacecraft and damaged objects amount to more than 500,000 pieces, which fall out of orbit in a controlled or uncontrolled manner. Larger objects such as space stations and rockets are steered to the Southern Pacific Ocean where a "spacecraft cemetery" holds the remains of numerous spacecraft.

Keywords Aquatic ecosystems · Toxic chemical wastes · Weapons and ammunition · Radioactive waste · Spacecraft

D.-P. Häder (✉)
Deptartment of Biology, Friedrich-Alexander Universität, D-91054 Erlangen, Germany
e-mail: donat@dphaeder.de

© The Author(s), under exclusive license to Springer Nature Switzerland AG 2021
D.-P. Häder et al. (eds.), *Anthropogenic Pollution of Aquatic Ecosystems*,
https://doi.org/10.1007/978-3-030-75602-4_16

16.1 Introduction

Our technological development is not characterized by recycling. Many used items are deposited and wasted stressing the limited resources of our finite planet. The way of living of the fast growing human population would need more than one planet to provide the requested energy, materials, water and food (Foley 2010). Numerous studies indicate new technologies to reduce the resource intensity in order to warrant continuous use of the limited resources (Moore and Rees 2013).

One example is the agricultural production as a major driver exceeding planetary boundaries (Campbell et al. 2017). Exploiting the dwindling water resources for irrigation stresses the availability of water for human consumption, e.g., the agriculture in the US Midwest is based on the usage of million-year-old fossil water, but the wells need to be dug deeper every year. Another problem is the unrestricted use of fossil fuel. Even though new deposits of oil, gas and coal are detected, the supply is limited and will end sooner or later (Capellán-Pérez et al. 2014).

Rather than recycling used articles, many products in industry, household and traffic are being discarded at the end of their useful lifetime. Typical examples are single-use articles such as plastic bags, which end up littering the landscape and the oceans (Jambeck et al. 2015) for economical reasons even though recycling of solid plastic waste is fairly easy and straightforward (Ragaert et al. 2017) (cf. Chap 17 Microplastic pollution in the oceans, this volume). An estimated amount of 19–23 million metric tons has entered the aquatic ecosystems as of 2019 which is about 11% of all plastic waste. 14 million tons had sedimented to the ocean floor (Borrelle et al. 2020).

The dumping of unused items can be traced back to Neolithic communities who discarded the shells of eaten mollusks in large mounds often containing artifacts such as shards of broken pottery and even buried skeletons (Mayer et al. 2010). These dumps can be hundreds of meters long and several meters high. They are found in Europe, Africa, Asia, Australia and North and South America (Gaspar et al. 2011). The oldest mounds, which contain stone tools, were found in Terra Amata near Nice, France, which are dated to the middle Pleistocene about 400,000 years ago (Villa 1983). Archeologists use intentionally or accidently dumped artifacts to reconstruct historical cultures. Scientists use radiocarbon dating to determine the age of these artifacts (Bowman 1990) recently aided by dendrochronology (Dean 1997).

After humans had detected that dumping toxic materials in the environment can be dangerous for the biota and human health another option evolved: dumping at sea. This has been done for household wastes which were washed into rivers and the sea. The Cloaca Maxima in Rome, Italy, is an excellent example for removing wastes from a large population (Hopkins 2007). It took much longer to solve the problem of human waste removal by sewers in London to combat the cholera epidemics between 1831 and 1866 (Halliday 2001). China dumps sludge from municipal wastewater treatment plants into coastal waters (Zhang et al. 2013).

Analysis of several sludge samples showed high concentrations of heavy metals such as Cr, Zn, Cu, Pf, As, Hg and Cd.

Side products from chemical reactions cannot always be utilized. Therefore, chemical industry has to discard them as waste. In order not to spoil the soil for a long time, one of the strategies is to dump chemical wastes at sea, e.g., diluted sulfuric acid is a byproduct of the production of titanium dioxide and dyes and may include heavy metals and halogenated hydrocarbons. Hazardous electronic wastes as well as radioactive nuclear wastes are being dumped into the oceans by several countries (Ringius 1997; Rajput and Gupta 2019). In coastal areas, industrial and household wastes, rubble from construction and snow from road cleaning are being dumped into the sea (Kang et al. 2020; Kjellman and Swahn 2017). Outdated chemical weapons and ammunition and even used spacecraft, satellites, rockets and space stations are deposited in the sea (De Lucia and Iavicoli 2018; Lastumäki et al. 2020).

16.2 Chemical Wastes

During the 1960 and 1970s, large-scale exploitation of oil and gas was initiated in the North Sea which became possible by using advanced drilling equipment and technology (Craig et al. 2018; Sühring et al. 2020). These methods involved large amounts of hazardous chemicals and little thought was given to the environmental risks. The produced oil was stored in reservoirs so that the simultaneously drilled water could separate from the crude oil. This water contains traces of oil and chemicals, used by the installation, and is discharged into the sea. Only starting in 1979, British companies voluntarily agreed to regularly report their use and discharge of chemicals (Cefas-OCNS website 2018).

A recent review reports 276 hazardous chemicals released from sea-based sources including antifouling biocides, aquaculture, medicinal products and warfare agents and summarizes concentration and toxicity data (Tornero and Hanke 2016). In addition to legally or illegally releasing (mostly liquid) chemicals into the sea, highly toxic substances are being burned on ships and the remain dumped into the sea (Zerbe 2019), e.g., in the USA, organochlorines and other organohalogens have been disposed of by at-sea incineration until a program for regulating such methods under the Ocean Dumping Law and other international regulations throughout the world ended this practice (Kamlet 1981). Rivers transport large amounts of toxic materials into the sea. Dumping of diluted sulfuric acid in the North Sea has been stopped in 1988 after protests by Greenpeace (Take 2000).

The river Elbe was estimated to carry 180000 tons of nitrogen, 12000 tons of phosphorus, 1750 tons of zinc, 15 tons of mercury and 200 kg of PCB into the North Sea in 1986 (Scharmann 1994). Other chemicals reach the North Sea by dumping of diluted sulfuric acid, sludge from waste water treatment plants or dredged material from soil excavation.

Up to 1989, these wastes have been pumped into the sea by pipeline or transported by ship which dumped up to 1200 tons per day into the open sea, e.g., near the island of Helgoland, German bight (Neumüller 1985). Ships, especially tankers, have cleaned their tanks and dumped the excess heavy oil at sea (Kirchner and Kleemola-Juntunen 2018). Despite the large dilution in the ocean, concentrations of toxic chemicals can be so high that they become hazardous for the environment and health of marine organisms. Pathological conditions were found in cod and haddock near dump sites of sewage sludge (McVicar et al. 1988). Field bioassays have been developed for early detection of chronic impacts on marine organisms such as using killifish (*Fundulus grandis*) to monitor hazardous chemicals including industrial and municipal wastes (Pequegnat and Wastler 1980). In the Netherlands, five sites had been monitored from 1986 to 1988 to document pathologies in fish (*Limanda limanda*, common dab) (Vethaak and Meer 1991). One of the sites was an offshore dumping ground, another received large amounts of diluted sulfuric acid from titanium dioxide waste and three other sites served as references. The analysis found epidermal hyperplasia, papilloma, lymphocytic liver nodules and lesions as well as infections with the microsporidian parasite *Glugea* in fish from the affected sites.

16.3 Weapons and Ammunition

The coasts of the Baltic Sea and some nearby islands are good places to collect amber. However, there are frequent accidents because lumps of phosphorus look almost identical to amber being translucent and waxen (Frank et al. 2008). When phosphorus becomes dry and in contact with oxygen, it can ignite spontaneously and burns within seconds through cloths, skin and flesh at 1300 °C. In addition, tissue damage occurs because of the corrosive effects of phosphoric acid. About 13000 tons of chemical warfare material have been dumped at several sites in the Baltic Sea after World War 2 (WW2) (Baršienė et al. 2014). East of the island of Bornholm in the southern Baltic Sea, 32000 tons of chemical war materials were dumped after WW2 including shells and bombs as well as containers with toxic chemicals (Missiaen et al. 2010). Analysis of the chemical warfare agents identified Adamsite (diphenylaminechlorarsine), Clark (diphenylcyanoarsine), chlorobenzene, Yperite (sulfur mustard gas, first used by the Germans during WW1 in the battle of Ypres, July 1915), Tabun (nerve agent) and arsine oil, some of which came from deposits stockpiled during WW1 (Sanderson et al. 2010). This material leaks out of the decaying containers, accumulates in the sediment and is carried over long distances. Another large dumping area is the Gotland Basin in the Baltic Sea (Kaffka 1996; Glasby 1997). Figure 16.1 shows the dumpsites of chemical weapons in the oceans around the world. Chemical weapons can be located at their dumpsite by high resolution seismic and magnetic monitoring, but often the material is spread by wind and waves and bottom fishing activity (Missiaen and Feller 2008). There are numerous other sites of chemical weapons in the sea. Today, chemical weapons together with nuclear and bacteriological weapons are classified as weapons of mass destruction by the United Nations (Andrulewicz 2007). A Medline search returned 5863 articles on the subject showing that hundreds of thousands to tons of chemical munitions had been dumped after WW2 with large concentrations around Europe, Russia, Japan and the USA (Greenberg et al. 2016). At the end of WW2, more than 300,000 tonnes of chemical weapon munitions were left especially in Germany. In order to dispose of this material, bombed ships and burned hulks were collected in northern European waters, filled with the wastes and disposed of in the sea (Stock 1996). A large fraction of the German stocks (130,000–160,000 tonnes, containing 41,000–48,000 tones of chemical warfare) were dumped in the Skagerrak, 25 miles southeast of Arendal at a depth of 600–700 m (Tørnes et al. 2020). Some of the containers are leaking releasing the nerve agent Tabun into the water. The most common chemical warfare agents are sulfur mustard, Lewisite (chlorine containing organoarsenic compound) and several nerve agents followed by organoarsenicals, blood and choking agents. With time, they are released from their original casings into the water posing health risks for marine life including microbiota, invertebrates and vertebrates. Large amounts of TNT (2,4,6–trinitrotoluene) have also been discarded in the Baltic Sea. The enzymes 7–ethoxyresorufin–*O*–deethylase (EROD) and 7–methoxyresorufin–*O*–deethylase (MROD) in the livers from dab (*Limanda limanda*), plaice (*Pleuronectes platessa*) and flounder (*Platichthys flesus*) were found to be inhibited by the explosives leaking out of their containers (Koske et al. 2020). A catalogue of dumped conventional warfare in the Baltic Sea can be found in a recent review (Miętkiewicz 2020).

In order to detect the effects of chemical warfare material dumped into the water, Baltic mussels (*Mytilus tossulus*) were caged and deposited in two hotspot sites in the southern Adriatic Sea north of Sicily for 2.5 months at a depth of 35 and 65 m,

Fig. 16.1 Dumpsites of chemical weapons in the oceans. Courtesy https://nonproliferation.org/wp-content/uploads/2009/08/cw_dumped_at_sea_580x280.png, permission by Google Maps/Google Earth

respectively (Lastumäki et al. 2020). A similar study has been reported recently (Strehse and Maser 2020). Several biochemical and histochemical biomarkers, bioenergetic and genotoxic indicators showed the effects of released toxic material. DNA damage, organ lesions and high concentrations of arsenic and mercury have been found in two benthic fish species (Blackbelly rosefish *Helicolenus dactylopterus* and European congrid eel *Conger conger*) attributed to dumping of obsolete chemical warfare agents. The materials came from clean-ups of bombed harbors, stockpiles and factories. In addition, histological lesions and genotoxicity in gills were found (Della Torre et al. 2010).

Shells with gases were loaded on Russian barges for half a year in 1954 which were subsequently sunk in the White Sea near Severodvinsk. In addition, further chemical weapons such as mines and shells were also deposited in 1956 first also near Severodvinsk and later near Spitzbergen. These deposits and the released toxins are being discussed as the source for an ecological catastrophe in the summer of 1990 in the White Sea (Yufit et al. 1996). Also, the Americans, British and Canadians had dump programs for postwar leftover weapons and ammunition (Souchen 2017).

In addition to ammunition, firearms, military ships and even tanks have been deposited in the ocean including conventional ammunition, rockets and anti-artillery shells (Miętkiewicz 2020). Some obsolete military materials are being used for an applaudable environmental purpose: A 1200 square miles area off the coast of Alabama is used by the Army Corps of Engineers for the construction of artificial reefs by sinking car bodies, barges, dry docks, tanks, small planes and boat hulls (Minton and Heath 1998).

16.4 Household and Industrial Waste, Soil and Rubble from Building Activities

Waste from daily life has been discarded in the sea by all coast living cultures. In the 1980s, New York City solved the problem of mounting wastes in a cheap way. About 80% of the collected material from Manhattan was hauled to the coast, loaded onto wooden flat-bottom boats which were towed into the ocean, and the waste was dumped into the Atlantic Ocean (Walsh 1991). However, the beachfront communities in New York and New Jersey were enraged since the waste was frequently washed up to the beaches, and they started a civil suit which successfully ended the waste disposal practice in 1895. The procedure was replaced by land fill and municipal recycling. In the following, marshes were used as deposits. Even in the recent past, the practice of dumping household waste into the sea continues especially in developing nations as found along the Caribbean cost of Panama (Garrity and Levings 1993) and Indonesia (Willoughby et al. 1997). The accumulation and fragmentation of plastic debris is covered in Chap. 3 (Effects of pollution on fish) and Chap. 17 (Microplastic pollution in the oceans) (Barnes et al. 2009).

A more recent example is the accumulation of marine debris on Volunteer Beach on East Falkland (Malvinas) beginning of this century (Otley and Ingham 2003). Quantification found 77 ± 25 items per km and month with a weight of 117.3 ± 12 kg. The most common items were cotton fabric, plastic packing sheets and tape, string and broken plastic pieces attributed to fishing vessels.

Industry is another major source of waste disposal in the ocean (Duedall et al. 1983; Ketchum 2013). In addition to terrestrial runoff carrying water-borne waste products from cultivated and pastoral land, industrial liquid wastes have been discharged by pipelines into the oceans (Wood et al. 1993). The effluent has a density similar to that of fresh water, rises to the surface and is diluted there. The East China Sea is overburdened by the wastes from the growing population in the Yangtze River basin, including nitrogen, phosphate, hydrocarbons, heavy metals and organic material (Daoji and Daler 2004). However, the dam construction in the river will decrease the sediment loads. Arsenic and mercury were found in 31 sediment samples from the Str. Anna Trough, Kara Sea, Arctic Ocean resulting from deposition of industrial effluents and emission as well as dumping of military wastes (Siegel et al. 2001).

For a long time, the disposal of waste in the oceans was considered a plausible solution which poses no damage to marine life especially if the debris is contained in steel drums or other containers (Bascom 1974). Public attitude has changed in recent years: while in the past, the value of the marine environment was not highly esteemed, the general public has become more aware of environmental threats and conservation efforts (Easman et al. 2018). Also, the economic costs of household waste and marine anthropogenic litter are moving into the focus of administrators (Newman et al. 2015). By now, the litter in the oceans has piled up to form "garbage islands" (Fig. 16.2).

Fig. 16.2 Garbage island in the Pacific Ocean. Courtesy https://images.app.goo.gl/gStvyfSR8xr36wkm6

Jettison of excavated stone, soil, rubble and other material from building activities was common in the US oceans until the Marine Protection, Research and Sanctuary Act was passed in 1988 (Moore 1992). By further amendments to this act, all municipalities have agreed to prevent discharging industrial wastes and sewage sludge into the ocean by June 20, 1992. Comparison of bathymetric charts from 1845 to 1973 indicated that ocean dumping of solid waste in the Hudson shelf channel was a significant geologic process since about 850 million cubic meters of wastes had been dumped over 85 years; 318 million cubic meters of this material are assumed to be of anthropogenic origin (Williams 1979). Some of this material, such as rock, sand and excavation debris originated from the excavations for subway construction in Manhattan (Burroughs 1988). This material formed a mound of about 15 m on an area of about 90 square kilometers. About 85000 cubic meters of dredged and excavated material were dumped of the cost of Canada by split-hull barges (Mosher et al. 1997). The material can be monitored by using side-scan mosaicing based on the refection of the debris. Some urban communities near the coast disposed the collected snow during winter in rivers or directly into the sea (Oliver et al. 1974). This material can contain chloride, nickel and lead and poses a threat for fish and other aquatic organisms.

Another rapidly increasing source of solid debris is electro and electronic waste (e-waste) including smart phones, tablets and computers as well as washing machines, refrigerators, hair dryers and vacuum cleaners. In 2019, the world production of e-waste was 53.6 million tons which amounts to 7.3 kg per person with an increase of 21% over the last five years (Lingner 2020). Asia had a share of 24.9 million tons, America 12.1 million tons, Europe 12 million tons and Africa 2.9 million tons. Only 17.4% of this waste is recycled and most is dumped in terrestrial reservoirs or into the sea (Walters and Loureiro 2020). This e-waste contains toxic substances such as mercury, chlorofluorocarbons and brominated flame retardants. In addition, precious substances such as gold, copper, platinum and rare Earth materials with an estimated value of 57 billion US $ are wasted. Solid wastes from hospitals such as disposable items (gloves, syringes, needles, vials, etc.) are usually decontaminated and either incinerated or disposed of in municipal dump sites. But there is also illegal dumping of hospital and healthcare wastes mixed with municipal and industrial solid wastes, as found, e.g., on the Turkish Black Sea Coast (Berkun et al. 2005).

16.5 Terrestrial Runoff of Fertilizers and Pesticides from Agriculture

Excessive usage of fertilizers and pesticides reaches the oceans as terrestrial runoff and has resulted in high concentrations in coastal waters and even open ocean waters which changes the aquatic biogeochemistry (Doney 2010) such as the oceanic phosphorus cycle (Paytan and McLaughlin 2007). The massive influx of minerals fuels large phytoplankton blooms in vulnerable habitats changing the species composition of the biota (Beman et al. 2005). Over the past half a century dumping from sewage treatment plants, industry and farming has carried about 20 million tonnes of nitrogen and about 2 million tonnes of phosphorus into the Baltic Sea (Conley 2012). A large share of this material stems from the northwest Russian coast mainly from intensive agriculture (Alekseev and Smirnova 2016). This inadvertent fertilization results in large algal blooms, often cyanobacteria. When the organisms die and decay, they consume most of the available oxygen, so that the waters become death zones for fish and other animals. By this effect, an estimated 60000 square kilometers is converted into hypoxic dead zones every year. The spreading of dead zones has been recorded in many of the world oceans (Dybas 2005).

Terrestrial runoff also carries large concentrations of organochlorine pesticides into the oceans (Ma et al. 2015). Especially, estuarine ecosystems have been found to be strongly affected by pesticide runoff (Fulton et al. 1999). High concentrations of polychlorinated biphenyls and organochlorines pesticides have been detected in Arctic rivers, lakes and the Arctic Ocean which reach the target areas by snow deposition and melting (Cabrerizo et al. 2019). The herbicide atrazine is applied extensively in the US Midwest and is transported by the Mississippi River into the Louisiana estuaries where it poses a stress for photosynthesis and growth in phytoplankton populations (Starr et al. 2017). Pesticides entering waterways are often lethal to aquatic organisms and can cause mutations, e.g., in crustaceans (Major et al. 2018). Organochlorine pesticides have also been found in tissues of green sea turtles (*Chelonia mydas*) (Yaghmour et al. 2020). Finally, pesticides enter the food web where they are bioaccumulated and may end up in invertebrates, fish and mammals for human consumption (Hui 2018).

16.6 Radioactive Waste

The advent of nuclear power has been accompanied with the burning question of how to dispose of the radioactive waste which develops by contamination of various liquid and solid materials with the radionuclides produced during the fission reaction in nuclear power plants (Efremenkov 1989; Ramana 2017). Other sources of radioactive materials are medicine and research as well as nuclear weapons and nuclear powered vessels and submarines (Hakami 2016). These materials pose a serious threat for the biota and human life for some hundred thousand years unless they can be recycled (Magill et al. 2003). Several options for safe storage are being explored. In Korea, deep boreholes are being considered to bury the waste (Jeon et al. 2019). Early disposal options included near surface disposal (mounds, trenches and boreholes) and deep underground disposal (natural cavities, deep boreholes, well injection and excavated cavities) (Rahman et al. 2016). Germany is searching for underground cavities in solid rock formations for the existing nuclear waste which is still an urging and as of today unsolved problem even though the political leaders of the country have decided to phase-out from nuclear power (Kruse 2017). Deep boreholes and abandoned salt mines are being discussed as possible final disposals for high-level radioactive waste but no final decision has been made because of scientific concern and public protests (Bracke et al. 2019; von Berlepsch and Haverkamp 2016). Also, China and Sweden are investigating geological disposal of radioactive waste (Wang et al. 2018).

As an alternative to disposing radioactive waste in terrestrial deposits, many governments have decided to dump nuclear material into the ocean (Lavine 2020). The first dumping of radioactive waste was performed in 1946 about 80 km of the Californian coast (Rima et al. 1971). In October 1993, Russia had dumped 900 tons of nuclear waste into the Sea of Japan, but it was known that Russia had used this practice for the preceding three decades (Eaton 2020). This deposition of low-level nuclear waste into important Japanese fishing areas, made public by Greenpeace, raised global criticism and finally resulted in the London Convention in November 1993 prohibiting dumping of low-level nuclear waste into the oceans (Harrison 2016). Another dumping site for radioactive waste used by the former Soviet Union is the Stepovogo Fjord, eastern coast of Novaya Zemlya, which is an extension of the Ural Mountain into the Kara Strait. Release of radioactive material might affect the biota as investigated in species of marine organisms in 2012 (Heldal et al. 2018). Demersal fish such as navaga (*Eleginus nawaga*) and shorthorn sculpin (*Myoxocephalus scorpius*) were the dominant members of the fish fauna. However, there is no commercial fishing around the Stepovogo Fjord and very little activity in the Kara Sea because of the low commercial value of the fish species in this area. At this site also the Russian nuclear submarine K-27 was dumped (Gwynn et al. 2016). An investigation by a joint Norwegian-Russian expedition found no leakage from the submarine reactor. However, the concentrations of ^{137}Cs and ^{90}Sr were elevated, which was attributed to the dumping of other solid radioactive waste in this area. Also, the research Cruise 67 of the R/V Akademik Mstislav Keldysh did not find any hazard from dumped radioactive waste in the Barents Sea (Politova et al. 2018).

In addition to reducing radioactive residues, it is of vital importance to package the wastes in a safe way (Yadav and Kumar 2019). Early methods used concrete blocks which have the disadvantages of being porous and the possibility of changing the chemistry within the concrete (Fernández et al. 2018). A safer way might be surrounding the material with a solid lead borate glass matrix (Sawada et al. 2020).

In addition to intentionally dumped radioactive wastes into the oceans, there have been a number of accidental releases of radioactive material into aquatic ecosystems. The Three Mile Island power plant is located on an island in the river Susquehanna near Harrisburg in a densely populated region in the eastern USA. It was the site of the first of a series of nuclear power plant accidents. On March 28, 1979, a partial meltdown and a fire led to the release of radioactive water vapor into the atmosphere (Zaretsky 2018; Sills 2019). Despite of the accident and public opposition, the reactor was repaired and operated for another 40 years before it was shut down (Halden 2017).

On April 26, 1986, reactor 4 of the nuclear power plant near Chernobyl, Ukraine, experienced a meltdown resulting from operator errors, an inadequately evaluated test procedure and unsafe reactor design (Bevelacqua 2016). The accident scattered dangerous radioactive material around the reactor and the emitted clouds resulted in an increase in radiation over most of Europe (Kashparov et al. 2019). 135000 people within a radius of 30 km around the nuclear power plant had been evacuated. After four month between 31 and 54 people had died from radiation sickness and burns (Braithwaite 2019).

A third cataclysmic accident occurred in the Fukushima Daiichi Nuclear Power Plant on March 11, 2011, which resulted from a severe destructive earthquake followed by a massive tsunami (Geilhorn and Iwata-Weickgenannt 2016). The cores of three out of four reactors experienced a complete meltdown, followed by hydrogen explosions. Massive amounts of radionuclides were washed into the ocean when workers tried to cool the cores with seawater (Song 2018). Furthermore, radionuclides were dispersed by the atmosphere and deposited on the sea surface (Suh et al. 2017). In the years following the disaster, large numbers of marine products were analyzed to quantify the radioactive contamination (^{134}Cs, ^{137}Cs and ^{90}Sr). However, only a few desmersal and pelagic fishes were found to have a contamination which exceeded the Japanese regulatory limit of 100 Bq kg^{-1} wet weight (Wada et al. 2016).

16.7 Discarding Spacecraft at Sea

Ever since humans have started sending unmanned or manned missions into space, the number of spacecraft orbiting our planet has skyrocketed—excuse the pun. In June 2019, 2244 satellites were counted circling the Earth (Fischetti 2019). Thus, the question arises what to do with outdated spacecraft at the end of their lifetime. In fact, the lower Earth orbits have become so crowded that collisions between payloads and rockets have occurred. Currently, about 50,000 pieces of space debris from wrecked spacecraft increase growing concern about future collisions (Fig. 16.3). About 13,000 objects larger than 5 cm are constantly monitored by the American Space Surveillance Systems and other space agencies. If this problem is not solved, space may not be usable any more (Lechtenberg 2019). Solutions to deal with space debris include collective, laser-based, ion-beam shepherd-based, tether-based, sail-based, satellite-based, unconventional and dynamical systems-based methods (Mark and Kamath 2019). In most cases, the final solution is to remove the debris from its orbit and steer it toward the Earth surface. Many smaller low-Earth orbit satellites which are decommissioned burn up safely in the atmosphere (Shoemaker et al. no year). But large spacecraft and internal fuel tanks require a targeted reentry. For this purpose, NASA has developed the On-orbit Servicing, Assembly and Manufacturing (OSAM-1) mission, formerly known as Restore-L (Coll et al. 2020).

The Russion Mir station was launched on February 19, 1986, following the Saljut 7 station which reached its orbit four years earlier (Badhwar et al. 2002).

Fig. 16.3 Computer-generated image of objects in Earth orbit (95% being orbital debris) in the low-Earth orbit (about 2000 km), but not scaled to Earth. Courtesy NASA, public domain http://orbitaldebris.jsc.nasa.gov/photogallery/beehives.html#leo

It circled the Earth 86,325 times at a hight between 385 and 393 km and housed a number of Russian and international astronauts (DeLucas 1996). In face of the high costs, it was decided to end the mission, since the ISS had already been launched and the hope for financial support from Western countries did not materialize. Therefore, the flight of the Mir station was stopped on March 23, 2001, by a backward thrust which caused a controlled reentry. Most of the mass of 124,340 kg burned up in the atmosphere, but more than 1500 fragments with a total weight of about 40 tons fell into the Pacific Ocean south of the Fidji Islands (44°12′S/150°W).

Point Nemo is a small island (48°52′31.75″S, 123°23′33.07″W) in the South Pacific gyre characterized as a "Pole of Inaccessibility" with the longest distances from all continents with 3978 km from the Brazilian Itajai and 2778 km from Auckland, New Zealand (Garcia-Castellanos and Lombardo 2007). Russian, European and Japanese space agencies have used this lonely area to dump their space debris including parts of the Russian Mir station and numerous decommissioned satellites making it a spacecraft cemetery (Bell 2016; Kerr et al. 2017).

In 1979, the NASA Skylab crashed from its orbit (Caroff et al. 1979). Parts burned up in the atmosphere, others fell into the ocean and some parts smashed into

Australia. Nobody was hurt, but a community issued a fine notice to NASA for unauthorized waste disposal (Johnson 2007).

The Chinese space station Tiangong-1 (Heaven Palace) was launched in September 2011 (Kuo 2018). It was designed as a test bed for robot technologies but was decommissioned in March 2016. Because of the drag in the upper atmosphere, the station lost about 4 km height every day from its orbit. However, the exact location for the final drop (between 43° North and 43° South) could not be predicted. The ground risk assessment was performed by various simulations such as the object reentry survival analysis tool (ORSAT) and the NASA debris assessment software (DAS), spacecraft atmospheric reentry and aerothermal break-up (SCARAB) and the ESA debris risk assessment and mitigation analysis (DRAMA) (Choi et al. 2017). During its uncontrolled crash, the defunct Tiangong-1 space station reentered the atmosphere and mostly burned up, but about 10% of the 85,000 kg spacecraft landed northwest of Tahiti on April 1, 2018 (Pardini and Anselmo 2019).

Typically, 200–400 artificial objects return to Earth every year posing a risk by broken pieces of spacecraft. An additional risk is the radioactive debris. In January 1978, the Russian reconnaissance Cosmos 954 experienced a malfunction of its nuclear reactor over the Northwest Territories of Canada and fell into the suburbs but distant from densely populated areas. There were no casualties but numerous radioactive fragments hit the ground over a length of more than 1000 km (Klinkrad 2006).

16.8 Conclusions

Marine ecosystems provide many ecologically and economically important services. They deliver a large share of food for the fast growing human population. In addition, they sequester large quantities of the anthropogenically emitted carbon dioxide and absorb most of the heat generated through the emission of greenhouse gases mitigating climate change.

The aquatic biota is under a number of stress factors including increasing temperatures, ocean acidification and solar UV radiation. In addition, large amounts of anthropogenically generated waste products are dumped into the oceans or reach them by terrestrial runoff and riverine transport. These include household, industrial and agricultural waste products, solid waste such as materials from building activities, electro and electronic waste, weapons and ammunition, radioactive material and spacecraft debris.

Some of these disposals are already prohibited by international conventions but others need to be regulated in order to warrant the future health of aquatic ecosystems and a stable source of food for a growing human population.

References

Alekseev M, Smirnova E (2016) Waste water of north-west Russia as a threat to the Baltic. J Environ Eng Sci 11(3):67–78

Andrulewicz E (2007) Chemical weapons dumped in the Baltic Sea. In: Assessment of the fate and effects of toxic agents on water resources. Springer, pp 299–319

Badhwar G, Atwell W, Reitz G, Beaujean R, Heinrich W (2002) Radiation measurements on the MIR Orbital Station. Radiat Meas 35(5):393–422

Barnes DK, Galgani F, Thompson RC, Barlaz M (2009) Accumulation and fragmentation of plastic debris in global environments. Philos Transact Royal Soc London B 364(1526):1985–1998

Baršienė J, Butrimavičienė L, Grygiel W, Lang T, Michailovas A, Jackūnas T (2014) Environmental genotoxicity and cytotoxicity in flounder (*Platichthys flesus*), herring (*Clupea harengus*) and Atlantic cod (*Gadus morhua*) from chemical munitions dumping zones in the southern Baltic Sea. Marine Environ Res 96:56–67

Bascom W (1974) The disposal of waste in the ocean. Sci Am 231(2):16–25

Bell R (2016) Space Junk. Science Scope 39(8):70

Beman JM, Arrigo KR, Matson PA (2005) Agricultural runoff fuels large phytoplankton blooms in vulnerable areas of the ocean. Nature 434(7030):211–214

Berkun M, Aras E, Nemlioglu S (2005) Disposal of solid waste in Istanbul and along the Black Sea coast of Turkey. Waste Manag 25(8):847–855

Bevelacqua JJ (2016) The three Mile Island, Chernobyl, and Fukushima Daiichi accidents and their radiological impacts. Int Nucl Saf J 5(1):21–79

Borrelle SB, Ringma J, Law KL, Monnahan CC, Lebreton L, McGivern A, Murphy E, Jambeck J, Leonard GH, Hilleary MA (2020) Predicted growth in plastic waste exceeds efforts to mitigate plastic pollution. Science 369(6510):1515–1518

Bowman S (1990) Radiocarbon Dating, vol 1. University of California Press

Bracke G, Kudla W, Rosenzweig T (2019) Status of deep borehole disposal of high-level radioactive waste in Germany. Energies 12(13):2580

Braithwaite R (2019) Chernobyl: a 'Normal' Accident? Chernobyl (TV miniseries) Craig Mazin, writer. Johan Renck, director. Distributed by HBO (US) and Sky (UK), 2019. Survival 61 (5):149–158

Burroughs R (1988) Ocean dumping: information and policy development in the USA. Marine Policy 12(2):96–104

Cabrerizo A, Muir DC, Teixeira C, Lamoureux SF, Lafreniere MJ (2019) Snow deposition and melting as drivers of polychlorinated biphenyls and organochlorine pesticides in Arctic rivers, lakes, and ocean. Environ Sci Technol 53(24):14377–14386

Campbell BM, Beare DJ, Bennett EM, Hall-Spencer JM, Ingram JS, Jaramillo F, Ortiz R, Ramankutty N, Sayer JA, Shindell D (2017) Agriculture production as a major driver of the Earth system exceeding planetary boundaries. Ecol Soc 22(4)

Capellán-Pérez I, Mediavilla M, de Castro C, Carpintero Ó, Miguel LJ (2014) Fossil fuel depletion and socio-economic scenarios: an integrated approach. Energy 77:641–666

Caroff J, Trebol J, Legoff I, Sculfort J (1979) The reentry of SKYLAB: reentry tracking operations support. STIN 81:21100

Cefas-OCNS website (2018) Offshore-chemical-notification-scheme. https://www.cefas.co.uk/cefas-data-hub/. Accessed Aug 2018

Choi E-J, Cho S, Lee D-J, Kim S, Jo JH (2017) A study on re-entry predictions of uncontrolled space objects for space situational awareness. J Astron Space Sci 34(4):289–302

Coll GT, Webster G, Pankiewicz O, Schlee K, Aranyos T, Nufer B, Fothergill J, Tamasy G, Kandula M, Felt A (2020) Satellite servicing projects division restore-L propellant transfer subsystem progress 2020. In: AIAA propulsion and energy 2020 forum, p 3795

Conley DJ (2012) Save the Baltic Sea. Nature 486(7404):463–464

Craig J, Gerali F, MacAulauy F, Sorkhabi R (2018) History of the European oil and gas industry. Geological Society of London, Special Publications, 465:1–24, 21. https://doi.org/10.1144/SP465.23

Daoji L, Daler D (2004) Ocean pollution from land-based sources: East China Sea, China. Ambio 33(1/2):107–113

De Lucia V, Iavicoli V (2018) From outer space to ocean depths: the spacecraft cemetery and the protection of the marine environment in areas beyond national jurisdiction. Cal W Int'l LJ 49:345

Dean JS (1997) Dendrochronology. In: Chronometric dating in archaeology. Springer, pp 31–64

Della Torre C, Petochi T, Corsi I, Dinardo MM, Baroni D, Alcaro L, Focardi S, Tursi A, Marino G, Frigeri A (2010) DNA damage, severe organ lesions and high muscle levels of As and Hg in two benthic fish species from a chemical warfare agent dumping site in the Mediterranean Sea. Sci Total Environ 408(9):2136–2145

DeLucas LJ (1996) International space station. Acta Astronaut 38(4–8):613–619

Doney SC (2010) The growing human footprint on coastal and open-ocean biogeochemistry. Science 328(5985):1512–1516

Duedall I, Ketchum B, Park P, Kester D (1983) Global inputs, characteristics, and fates of ocean-dumped industrial and sewage wastes: an overview. Wastes Ocean 1:3–45

Dybas CL (2005) Dead zones spreading in world oceans. Bioscience 55(7):552–557

Easman ES, Abernethy KE, Godley BJ (2018) Assessing public awareness of marine environmental threats and conservation efforts. Marine Policy 87:234–240

Eaton JH (2020) Kicking the habit: Russia's addiction to nuclear waste dumping at sea. Denver J Int Law Policy 23(2):3

Efremenkov V (1989) Radioactive waste management at nuclear power plants. IAEA Bull 31 (4):37–42

Fernández R, González-Santamaría D, Angulo M, Torres E, Ruiz AI, Turrero MJ, Cuevas J (2018) Geochemical conditions for the formation of Mg silicates phases in bentonite and implications for radioactive waste disposal. Appl Geochem 93:1–9

Fischetti M (2019) Satellite surge. Sci Am 322(5):70

Foley J (2010) Boundaries for healthy planet. Sci Am 302(4):54–57

Frank M, Schmucker U, Nowotny T, Ekkernkamp A, Hinz P (2008) Not all that glistens is gold: civilian white phosphorus burn injuries. Am J Emerg Med 26(8):974. e973–974. e975

Fulton MH, Moore DW, Wirth EF, Chandler GT, Key PB, Daugomah JW, Strozier ED, Devane J, Clark JR, Lewis MA (1999) Assessment of risk reduction strategies for the management of agricultural nonpoint source pesticide runoff in estuarine ecosystems. Toxicol Ind Health 15(1–2):201–214

Garcia-Castellanos D, Lombardo U (2007) Poles of inaccessibility: a calculation algorithm for the remotest places on earth. Scottish Geogr J 123(3):227–233

Garrity SD, Levings SC (1993) Marine debris along the Caribbean coast of Panama. Mar Pollut Bull 26(6):317–324

Gaspar MD, Klokler DM, DeBlasis P (2011) Traditional fishing, mollusk gathering, and the shell mound builders of Santa Catarina, Brazil. J Ethnobiol 31(2):188–212

Geilhorn B, Iwata-Weickgenannt K (2016) Fukushima and the arts: negotiating nuclear disaster. Routledge

Glasby G (1997) Disposal of chemical weapons in the Baltic Sea. Sci Total Environ 206(2–3):267–273

Greenberg M, Sexton K, Vearrier D (2016) Sea-dumped chemical weapons: environmental risk, occupational hazard. Clin Toxicol 54(2):79–91

Gwynn JP, Nikitin A, Shershakov V, Heldal HE, Lind B, Teien H-C, Lind OC, Sidhu RS, Bakke G, Kazennov A (2016) Main results of the 2012 joint Norwegian-Russian expedition to the dumping sites of the nuclear submarine K-27 and solid radioactive waste in Stepovogo Fjord, Novaya Zemlya. J Environ Radioact 151:417–426

Hakami BA (2016) Radioactive waste: the problem and its management. Technology 7(1):89–96

Halden G (2017) Three Mile Island: the meltdown crisis and nuclear power in American popular culture. Taylor & Francis, Milton

Halliday S (2001) Death and miasma in Victorian London: an obstinate belief. BMJ 323 (7327):1469–1471

Harrison J (2016) 2. Marine pollution. C. Ocean dumping. Yearb Int Environ Law 27:163–168

Heldal HE, Bogstad B, Dolgov AV, Gwynn JP, Teien H-C (2018) Observations of biota in Stepovogo Fjord, Novaya Zemlya, a former dumping site for radioactive waste. Polar Biol 41 (1):115–124

Hopkins JN (2007) The Cloaca Maxima and the monumental manipulation of water in archaic Rome. Waters Rome 4:1–15

Hui YH (2018) Foodborne disease handbook: volume IV: seafood and environmental toxins. CRC Press

Jambeck JR, Geyer R, Wilcox C, Siegler TR, Perryman M, Andrady A, Narayan R, Law KL (2015) Plastic waste inputs from land into the ocean. Science 347(6223):768–771

Jeon B, Choi S, Lee S, Jeon S (2019) A conceptual study for deep borehole disposal of high level radioactive waste in Korea. Tunn Undergr Space 29(2):75–88

Johnson MP (2007) Skylab: the human side of a scientific mission. University of North Texas

Kaffka AV (1996) Sea-dumped chemical weapons: aspects, problems and solutions, vol 7. Springer Science & Business Media

Kamlet KS (1981) Ocean disposal of organochlorine wastes by at-sea incineration. In: Ocean dumping of industrial wastes. Springer, pp 295–314

Kang P, Zhang H, Duan H (2020) Characterizing the implications of waste dumping surrounding the Yangtze River economic belt in China. J Hazard Mater 383:121207

Kashparov V, Salbu B, Levchuk S, Protsak V, Maloshtan I, Simonucci C, Courbet C, Nguyen HL, Sanzharova N, Zabrotsky V (2019) Environmental behaviour of radioactive particles from Chernobyl. J Environ Radioact 208:

Kerr E, Macdonald M, Voigt P (2017) Taxonomy and analysis of issues facing post mission disposal concept, vol IAC-17-A6.5.10, pp 1–14

Ketchum B (2013) Ocean dumping of industrial wastes, vol 12. Springer Science & Business Media

Kirchner S, Kleemola-Juntunen P (2018) Dumping and oil pollution: regulatory approaches for vessel operations in an ice-free Central Arctic Ocean. Rev Eur Comp Int Environ Law 27 (1):28–34

Kjellman M, Swahn J (2017) IWW transport of recycling volumes

Klinkrad H (2006) Space debris: models and risk analysis. Springer Verlag, Berlin

Koske D, Goldenstein NI, Rosenberge T, Machulik U, Hanel R, Kammann U (2020) Dumped munitions: new insights into the metabolization of 2,4,6-trinitrotoluene in Baltic flatfish. Marine Environ Res 160:104992

Kruse H (2017) Has the exit from nuclear energy failed? Why nuclear power should not be surrendered. inis.iaea.org

Kuo L (2018) Tiangong-1 crash: Chinese space station comes down in Pacific Ocean

Lastumäki A, Turja R, Brenner M, Vanninen P, Niemikoski H, Butrimavičienė L, Stankevičiūtė M, Lehtonen KK (2020) Biological effects of dumped chemical weapons in the Baltic Sea: a multi-biomarker study using caged mussels at the Bornholm main dumping site. Marine Environ Res 161:105036

Lavine SD (2020) Russian dumping in the Sea of Japan. Denver J Int Law Policy 24(2):7

Lechtenberg T (2019) An operational algorithm for evaluating satellite collision consequence. Paper presented at the 2019 AAS/AIAA Astrodynamics Specialist Conference

Lingner M (2020) Pfand für Elektrogeräte. PC Magazin 2020(11):18–19

Ma Y, Xie Z, Halsall C, Möller A, Yang H, Zhong G, Cai M, Ebinghaus R (2015) The spatial distribution of organochlorine pesticides and halogenated flame retardants in the surface sediments of an Arctic fjord: the influence of ocean currents vs. glacial runoff. Chemosphere 119:953–960

Magill J, Berthou V, Haas D, Galy J, Schenkel R, Wiese J, Heusener G, Tommasi J, Youinou G (2003) Impact limits of partitioning and transmutation scenarios on the radiotoxicity of actinides in radioactive waste. Nucl Energy Lond 42(5):263–278

Major KM, Weston DP, Lydy MJ, Wellborn GA, Poynton HC (2018) Unintentional exposure to terrestrial pesticides drives widespread and predictable evolution of resistance in freshwater crustaceans. Evol Appl 11(5):748–761

Mark CP, Kamath S (2019) Review of active space debris removal methods. Space Policy 47:194–206

Mayer DEBY, Gümüş BA, İslamoğlu Y (2010) Fossil hunting in the Neolithic: shells from the Taurus Mountains at Çatalhöyük, Turkey. Geoarchaeology 25(3):375–392

McVicar A, Bruno D, Fraser C (1988) Fish diseases in the North Sea in relation to sewage sludge dumping. Mar Pollut Bull 19(4):169–173

Miętkiewicz R (2020) Dumped conventional warfare (munition) catalog of the Baltic Sea. Marine Environ Res 161:105057

Minton RV, Heath SR (1998) Alabama's artificial reef program: building oases in the desert. Gulf Mexico Sci 16(1):14

Missiaen T, Feller P (2008) Very-high-resolution seismic and magnetic investigations of a chemical munition dumpsite in the Baltic Sea. J Appl Geophys 65(3–4):142–154

Missiaen T, Söderström M, Popescu I, Vanninen P (2010) Evaluation of a chemical munition dumpsite in the Baltic Sea based on geophysical and chemical investigations. Sci Total Environ 408(17):3536–3553

Moore SJ (1992) Troubles in the high seas: a new era in the regulation of US ocean dumping. Envtl L 22:913

Moore J, Rees WE (2013) Getting to one-planet living. In: State of the World 2013. Springer, pp 39–50

Mosher D, Currie R, Sullivan D (1997) Monitoring of ocean disposal using side–scan mosaicing. Lead Edge 16(11):1667–1670

Neumüller O-A (ed) (1985) Römpps Chemie-Lexikon, vol 4, 8th edn. Franckh'sche Verlagshandlung, Stuttgart

Newman S, Watkins E, Farmer A, Ten Brink P, Schweitzer J-P (2015) The economics of marine litter. Marine anthropogenic litter. Springer, Cham, pp 367–394

Oliver BG, Milne JB, LaBarre N (1974) Chloride and lead in urban snow. Journal (Water Pollution Control Federation) 46(4):766–771

Otley H, Ingham R (2003) Marine debris surveys at Volunteer Beach, Falkland Islands, during the summer of 2001/02. Mar Pollut Bull 46(12):1534–1539

Pardini C, Anselmo L (2019) Uncontrolled re-entries of spacecraft and rocket bodies: a statistical overview over the last decade. J Space Saf Eng 6(1):30–47

Paytan A, McLaughlin K (2007) The oceanic phosphorus cycle. Chem Rev 107(2):563–576

Pequegnat W, Wastler T (1980) Field bioassays for early detection of chronic impacts of chemical wastes upon marine organisms. Helgoländer Meeresuntersuchungen 33(1–4):531

Politova N, Novigatsky A, Kozina N, Terpugova S (2018) Multidisciplinary research in the Barents Sea on cruise 67 of the R/V Akademik Mstislav Keldysh. Oceanology 58(3):499–501

Ragaert K, Delva L, Van Geem K (2017) Mechanical and chemical recycling of solid plastic waste. Waste Manag 69:24–58

Rahman ROA, Guskov A, Kozak MW, Hung Y-T (2016) Recent evaluation of early radioactive disposal practice. In: Natural resources and control processes. Springer, pp 371–400

Rajput K, Gupta A (2019) Re-cycle of E-waste in concrete by partial replacement of coarse aggregate. Eng Herit J (GWK) 3(1):5–8

Ramana M (2017) An enduring problem: radioactive waste from nuclear energy [Point of view]. Proc IEEE 105(3):415–418

Rima DR, Chase EB, Myers BM (1971) Subsurface waste disposal by means of wells: a selective annotated bibliography, vol 2020. US Government Printing Office

Ringius L (1997) Environmental NGOs and regime change: the case of ocean dumping of radioactive waste. Eur J Int Relat 3(1):61–104

Sanderson H, Fauser P, Thomsen M, Vanninen P, Soderstrom M, Savin Y, Khalikov I, Hirvonen A, Niiranen S, Missiaen T (2010) Environmental hazards of sea-dumped chemical weapons. ACS Publications
Sawada K, Enokida Y, Tsukada T (2020) Oxidation and vitrification of aluminum with lead borate glass for low level radioactive waste treatment. J Nucl Sci Technol 57(6):1–7
Scharmann L (1994) Meer und Küste-eine neue Raumordnungskategorie in Deutschland? Zum Erfordernis "integrierter" Entwicklungskonzepte am Beispiel der Meeresraumbelastung im deutschen Festlandssockel der Nordsee. Europa Reg 2(1):14–20
Shoemaker MA, Vavrina M, Gaylor DE, Mcintosh R, Volle M, Jacobsohn J (no year) OSAM-1 decomissioning orbit design, vol (Preprint) AAS 20–460
Siegel FR, Kravitz JH, Galasso JJ (2001) Arsenic and mercury contamination in 31 cores taken in 1965, St. Anna Trough, Kara Sea, Arctic Ocean. Environ Geol 40(4–5):528–542
Sills DL (2019) Accident at Three Mile Island: the human dimensions. Routledge, London
Song JH (2018) An assessment on the environmental contamination caused by the Fukushima accident. J Environ Manage 206:846–852
Souchen A (2017) "Under Fathoms of Salt Water:" Canada's Ammunition Dumping Program, 1944–1947. Can Mil Hist 26(2):3
Starr AV, Bargu S, Maiti K, DeLaune RD (2017) The effect of atrazine on Louisiana Gulf Coast estuarine phytoplankton. Arch Env Contamin Toxicol 72(2):178–188
Stock TD (1996) Sea-dumped chemical weapons and the chemical weapons convention. In: Sea-dumped chemical weapons: aspects, problems and solutions. Springer, pp 49–66
Strehse JS, Maser E (2020) Marine bivalves as bioindicators for environmental pollutants with focus on dumped munitions in the sea: a review. Marine Environ Res 158: 105006
Suh K-S, Kim S, Min B-I (2017) Atmospheric dispersion and sea surface deposition of radionuclides by the Fukushima nuclear accident. J Coast Res 79:85–88
Sühring R, Cousins A, Gregory L, Moran C, Papachlimitzou A, Phillips C, Rowles R, Supple S, Wilczynska M, Birchenough S (2020) The past, present, and future of the regulation of offshore chemicals in the North Sea—a United Kingdom perspective. ICES J Mar Sci 77 (3):1157–1166
Take I (2000) The better half of world society. Civilizing world politics society and community beyond the State Lanham, (Rowman & Littlefield), pp 199–214
Tornero V, Hanke G (2016) Chemical contaminants entering the marine environment from sea-based sources: a review with a focus on European seas. Mar Pollut Bull 112(1):17–38
Tørnes JA, Vik T, Kjellstrøm TT (2020) Leakage rate of the nerve agent tabun from sea-dumped munition. Marine Environ Res 161:161
Vethaak D, Meer JVD (1991) Fish disease monitoring in the Dutch part of the North Sea in relation to the dumping of waste from titanium dioxide production. Chem Ecol 5(3):149–170
Villa P (1983) Terra Amata and the Middle Pleistocene archaeological record of southern France. University of California Publications in Anthropology, Berkeley, p 13
von Berlepsch T, Haverkamp B (2016) Salt as a host rock for the geological repository for nuclear waste. Elements 12(4):257–262
Wada T, Fujita T, Nemoto Y, Shimamura S, Mizuno T, Sohtome T, Kamiyama K, Narita K, Watanabe M, Hatta N (2016) Effects of the nuclear disaster on marine products in Fukushima: an update after five years. J Environ Radioact 164:312–324
Walsh DC (1991) The history of waste landfilling in New York City. Groundwater 29(4):591–630
Walters R, Loureiro MAF (2020) Waste crime and the global transference of hazardous substances: a southern green perspective. Crit Criminol 1–18
Wang J, Chen L, Su R, Zhao X (2018) The Beishan underground research laboratory for geological disposal of high-level radioactive waste in China: planning, site selection, site characterization and in situ tests. J Rock Mech Geotech Eng 10(3):411–435
Williams SJ (1979) Geologic effects of ocean dumping on the New York Bight Inner Shelf. US Coastal Engineering Research Center
Willoughby N, Sangkoyo H, Lakaseru BO (1997) Beach litter: an increasing and changing problem for Indonesia. Mar Pollut Bull 34(6):469–478

Wood IR, Bell RG, Wilkinson DL (1993) Ocean disposal of wastewater. World Scientific, Singapore

Yadav D, Kumar P (2019) Phytoremediation of hazardous radioactive wastes. In: Assessment and management of radioactive and electronic wastes. IntechOpen, London

Yaghmour F, Samara F, Alam I (2020) Analysis of polychlorinated biphenyls, polycyclic aromatic hydrocarbons and organochlorine pesticides in the tissues of green sea turtles, *Chelonia mydas*, (Linnaeus, 1758) from the eastern coast of the United Arab Emirates. Mar Pollut Bull 160:111574

Yufit SS, Miskevich IV, Shtemberg ON (1996) Chemical weapons dumping and White Sea contamination. In: Sea-dumped chemical weapons: aspects, problems and solutions. Springer, pp 157–166

Zaretsky N (2018) Radiation nation: Three Mile Island and the political transformation of the 1970s. Columbia University Press, New York

Zerbe S (2019) Marine Lebensräume in Nord-und Ostsee. In: Renaturierung von Ökosystemen im Spannungsfeld von Mensch und Umwelt. Springer, pp 297–314

Zhang C, Chen H, Yu Y-X, Wang L-J, Han J-B, Tao P (2013) Pollution characteristics of heavy metals in sludge from wastewater treatment plants and sludge disposal in Chinese coastal areas. Huan jing ke xue=Huanjing kexue 34(4):1345–1350

Chapter 17
Microplastics as Pollutants in the Marine Environment

Anthony Andrady and Liping Zhu

Abstract While plastics are a ubiquitous material in our life, their use also brings serious environmental concerns. This chapter focuses on impacts associated with post-consumer plastic waste, especially microplastics, that find their way into the ocean environment. It introduces the sources of plastic debris in ocean and identifies the major categories of plastics involved and their characteristics. A great majority of the plastics debris in the ocean is found to originate on land and washed into the water due to tidal movements or wave action and the use of plastics at sea, especially in fishing gear, that also generates marine debris. There are no reliable mechanisms that remove plastic litter from the ocean in any practical timescale, resulting in their accumulation in the bottom sediment. The top ten items of beach litter provide a good indication of the classes of plastics that dominate marine debris. These are mostly made from five classes of plastics, including polyethylene (PE), polypropylene (PP), polystyrene (PS), polyester (PET) and cellulose acetate fibers. The chapter also includes a discussion of the key concern of persistent organic pollutants (POPs) being sorbed by microplastics. The ingestion of microplastics carrying the POPs and their potential bioavailability to the ingesting organisms is a serious concern. The scientific updates on the hazards of entanglement and ingestion of plastics, to marine animals, including vertebrate species as well as planktonic species are included. Given the body of evidence that shows plastic pollution in all ocean basins, proven efficient sorption of POPs by microplastics and their ingestion by a wide range of marine organisms, this emerging pollutant in the ocean deserves close scientific attention.

Keywords Microplastics · Pollution · Ingestion

A. Andrady (✉)
Department of Chemical and Biomolecular Engineering, North Carolina State University, Raleigh, NC 27607, USA

L. Zhu
Center for Advanced Low-Dimension Materials, State Key Laboratory for Modification of Chemical Fibers and Polymer Materials, Donghua University, Shanghai 201620, China

© The Author(s), under exclusive license to Springer Nature Switzerland AG 2021
D.-P. Häder et al. (eds.), *Anthropogenic Pollution of Aquatic Ecosystems*,
https://doi.org/10.1007/978-3-030-75602-4_17

17.1 Introduction

Plastics are a recent addition to the conventional mix of materials used in consumer goods, such as wood, paper, metal or glass, dating back only to the early 1950s when they first became available in commercial quantities (Gilbert 2017). Compared to other materials, however, plastics have enjoyed an unprecedented rate of growth, rapidly gaining market share from competing materials in a wide range of applications. Plastics were such a successful material that at the time of their introduction, they were regarded as the utopian 'wonder material' with an endless array of potential applications, supporting this rapid growth in their production (Ball 2020). It is the low-density, toughness, transparency, moisture resistance and above all the low cost of plastics compared to metal and glass that contributed to their dramatic success as a material (Thompson et al. 2009; Andrady and Neal 2009). Packaging, in particular, relied on the lightweight, break resistance, transparency and heat-seal-ability of polyolefin plastics leading the way for their subsequent use in food service applications as well. Single-use food containers, for instance, are popularly made from foamed expanded polystyrene (EPS) because of its good thermal insulation characteristics and lightweight. In the building industry, versatile poly(vinyl chloride) (PVC) is widely used: It is extruded into water pipes, window frame elements and siding panels as a rigid durable replacement for wood or metal (Petrović and Hamer 2018). But the same plastic, PVC, can also be compounded differently and calendared into flexible soft plasticized sheets to serve as membrane roofing or flooring panels. PVC remains the leading plastic used in the building products industry today. With such a unique mix of properties, it is not surprising that plastics continue to enjoy an exponential growth in production volume over the years. Enjoying an annual growth rate of about 8.5% global plastic resin production has consistently increased and presently stands at 390 MMT (million metric tons) a year.

Plastics have lived up to its promise and delivered far more than ever expected of the material, and today, they are used in diverse applications. In the transportation sector, plastics feature prominently in watercraft, aircraft and especially in automobiles (Girijappa et al. 2020), where the use of plastics can save energy. The Boeing 787 dream-liner made of 50% plastic, for instance, saves nearly 80% in fuel costs compared to comparable predecessors! Plastics are also indispensable in numerous medical devices as well as in the design of life-saving artificial organs (Padsalgikar 2017). Implants in the human body such as replacement hearts, knee or hip implants as well as blood bags, sterile tubing and medical textiles are routinely made of plastics. Plastics also contribute to alternative energy production, both in wind power systems (with plastics used in wind turbine blades) and solar photovoltaic (PV) installations (plastics making up most of the packaging modules of PV cells). In addition, plastics also conserve residential energy use *via* insulation such as with polyurethane foam. Over the years, synthetic textile fibers have proven to be an excellent replacement for the more expensive natural fibers that are tedious to harvest and process. Of all fibers, globally produced today over 70% is plastic (amounting to 79 MMT in 2018). Single-use plastics in food service and food packaging provided a level of convenience hitherto not enjoyed by the consumer. Plastics by now have become an integral part of modern lifestyle.

What was never fully realized, however, was the serious environmental costs associated with using plastics in such large volumes. By 1970s, the fraction of recalcitrant plastic waste in the municipal solid waste (MSW) stream as well as in urban litter had significantly increased worldwide (Kole et al. 2017). Popularity of single-use plastics in the food and beverage sector further exacerbated the situation (Giacovelli 2018). An early warning sign was the accumulation of unsightly plastic litter on public and even remote beaches (Merrell 1980). The most disconcerting aspect of urban plastic waste, however, was the finding back in the 1970s that the burgeoning plastics waste on land leaked into the coastal ocean environment (Carpenter and Smith 1972; Carpenter et al. 1972). It is generally believed that a majority of marine litter originates on land. Beach litter is likely to be washed into the water due to tidal movements or wave action but most of the debris likely reach the ocean via rivers as well as storm water runoff and sewer overflow during heavy rains. Inadequate management of MSW streams rich in plastic, and illegal dumping of plastic waste in the oceans also makes a significant contribution. While a great majority of the plastics debris in the ocean appears to originate on land, uses of plastics at sea also generate debris. Plastics from fishing gear (e.g., Nylons from gill nets) that are denser than seawater and routinely lost at sea, generally accumulate in the bottom sediment and are not counted in either beach surveys nor net-sampled floating stock. Numerous commercial plastic fishing gear (nets, traps and crab pots) made of polyolefin or nylons is lost each season (Adey et al. 2008). Abandoned netting discarded at sea and crab pots lost in a large number annually in commercial fishing will continue to 'ghost fish' for very long periods of time, damaging the fishery (Dixon and Dixon 1981; Gregory and Andrady 2005). In the North Pacific Gyre area, a vortex that suffers especially high levels of plastic pollution (with reported abundance of 69.5 kg/km^2) (Lebreton et al. 2018) about 46% of the plastic debris was found to be derived from fishing gear.

Early reports of ecological impacts of large plastic debris surfaced from the 1970s onwards (Laist 1997). As incidents of entanglement of marine animals in plastics debris (Allen et al. 2012; Hanni and Pyle 2000; Boren et al. 2006) and credible accounts of their ingestion by fish (de Vries et al. 2020), birds (Fry et al. 1987) and turtles (Bjorndal et al. 1994) drew public attention to the issue, and solutions were actively sought. Collecting plastic litter in general and especially marine plastic debris for recycling was considered a viable mitigation strategy for a while, but the cost of low-grade recycled resins was too high to be competitive in the market place. Recycling was too expensive for routine use with marine plastics debris unless supported by public funds. Even the recycling of general post-consumer plastic waste in MSW stream is still abysmally low in the US being limited mainly to polyester beverage bottles and high-density polyethylene (HDPE) milk jugs, for the most part. An estimated 7300 MMT of plastics have been produced since the second World War until 2015 in the world, and a majority of these is still believed to be in the environment (Geyer et al. 2017).

Presently, all ocean basins are known to be contaminated with plastics (Law et al. 2014), including even the polar regions (Waller et al. 2017; Eriksson et al. 2013), and the five ocean gyres (especially the North Pacific gyre) that accumulate plastic debris in high concentration (Eriksen et al. 2014). All beaches, the surface water, deep sea water column and the benthic sediment, worldwide, now contain plastic debris (Cózar et al. 2014; Woodall et al.; Peng et al. 2018; Browne et al. 2011; Cole et al. 2011; Van Cauwenberghe et al. 2013). That plastics will continue to accumulate in the oceans in the future because of production trends coupled with their inherent durability was realized only recently (Thompson et al. 2004). An estimated average of 8 MMT of plastics waste is annually introduced into the ocean from land-based sources (Jambeck et al. 2015). Regardless of their origin, plastics that enter the ocean accumulate in the sediment in the long term, there being no credible mechanism for their removal as plastics do not biodegrade in a practical timescale. The common sources of marine plastic debris in the ocean are illustrated in Fig. 17.1 (GAO-19-653 2019).

Fig. 17.1 Overview of the sources and types of plastic debris in the ocean. (Marine Debris Report. GAO-19-653 2019). *S* indicates source of debris, and *T* indicates types of plastic debris items of concern

17.2 Classes of Plastics

Only a handful of the 20 or so families of commercial plastics are represented in the beach plastic litter or in floating plastic debris in the ocean. The floating stock of plastic fragments in the World's oceans now estimated at 268, 940 MT by Cózar et al. (2014) and 93,000–236,000 MT by van Sebille et al. (2015), predominantly include polyethylene (PE), polypropylene (PP) and polystyrene (PS) (Suaria et al. 2020). About half of the 359 MMT of plastic produced globally in 2018 also constituted just these three classes of plastics, with approximately 100 MMT of PE, 56 MMT of PP and 16 MMT of PS. Two other plastics found in the ocean environment are cellulose acetate fibers from cigarette filters and polyester (PET) from carbonated beverage bottles.

Sampling plastics on beaches can provide a good indication of which classes of these might be found as debris in surface water (Kershaw 2016). In their annual beach cleaning worldwide, coordinated by the Ocean Conservancy (Washington DC), the following top ten items of beach debris are found consistently each year (Ocean Conservancy 2019) (see Table 17.1). Most items (except perhaps for cigarette filters) are made of commodity thermoplastics typically used in high volume in consumer goods, and their specific gravity is also included in the table. These values are for the base resin, and those <1.02 will float in seawater. But, a variety of additives may be added to the base resin during manufacture of plastic products to obtain the requisite properties. These can change the specific gravity of the plastic depending on the volume fraction of the additives used. In some instances, a positively buoyant plastic such as PE or PP may become denser than seawater because an additive with a relatively high density such as a metal oxide is incorporated into it at a high-volume fraction. Specific gravity is a significant metric in that only the plastics that float in seawater can degrade abiotically by exposure to

Table 17.1 Categories of plastic debris collected in international beach cleanup (35,890 coastal km. yielding 10.5 million kg of plastic waste) 2019 data

Debris item	Count (1000s)	Polymer type	Specific gravity
Cigarette filter	5716	Cellulose acetate	1.24
Food wrappers	3728	PE/PP	0.91–0.97/0.90–0.92
Straws/stirrers	3668	PE/PP	
Plastic cutlery	1968	PS	1.04–1.08
Plastic beverage bottles	1755	PET	−1.34–1.39
Plastic bottle caps	1390	PP	
Plastic grocery bags	964	PE	0.01–0.04
Other plastic bags	939	PE, PP, PVC	1.37–1.39 (PVC)
Plastic lids	729	PE, PP	
Plastic cups and plates	657	PS	

Courtesy Ocean Conservancy (2019)

Key: *PE* high-density or low-density polyethylene; *PP* polypropylene; *PET* poly(ethylene terephthalate); *PS* polystyrene; *EPS* expanded polystyrene; *PVC* poly(vinyl chloride)

solar UVR. Initiation of degradation can only occur at a significant rate if solar UVR is available to the plastic litter (Andrady 2011; Bond et al. 2018). However, the duration for which these buoyant plastics float on surface water is limited because surface fouling of the debris by algae and other marine species invariably increases its density slowly sinking the item (Zardus et al. 2008; Ye and Andrady 1991). Any photo-initiation of degradation must therefore occur only during the time period the plastic debris floats in water.

Even buoyant plastics such as PE or PP that initially float in sea water invariably sink in the water column to reach the sediment environment (Ye and Andrady 1991; Fazey and Ryan 2016). With larger debris, this is the result of fouling where colonies of marine organisms establish on the surface of debris increasing its density above that of seawater. Sinking debris will reach a depth below the photic zone inhibiting further photo-initiation. It may resurface again temporarily if predators remove the foulant, reducing its density, a phenomenon reported by early researchers (Ye and Andrady 1991). Very small fragments or fibers of plastics can agglomerate with other denser degree because of marine exopolymer present in seawater (Zardus et al. 2008), to similarly sink in seawater as a component of marine snow.

Plastics generally fall into two broad categories: thermoplastics that can be melted repeatedly into different shapes and therefore are recyclable as materials and thermosets that are crosslinked plastics that do not dissolve in solvents or melt on heating. While plastics commonly found in the marine environment, such as PE, PP, PS, nylon fishing gear and cellulose acetate cigarette filters, are all thermoplastics, some thermoset polymers such as polyurethane (PU) is used as foam in floats or insulation in watercraft and in marine coatings. While less frequent and perhaps more difficult to identify, pieces of thermoset PU foam and fragments of organic surface coatings are included in plastic debris in the ocean. Characteristics of the main types of plastics found in the marine environment are given in Table 17.2.

Table 17.2 Common characteristics of plastics typically found in the marine environment

	Density (g/cm^3)	Crystallinity (%)	Melting temp (°C)	Extensibility (%)
LDPE	0.910 to 0.925	50 to 60	98 to 115	100 to 650
HDPE	0.941 to 0.965	~90	125 to 132	600 to 1350
LLDPE	0.91 to 0.94	35 to 60	100 to 125	100 to 950
PP	0.90	30 to 50	175	100 to 600
PS (GPPS)	1.04	–	180 to 260	2 to 3
PS (HIPS)	1.04	–	180 to 270	15 to 65
PVC	1.4	10 to 30	100 to 260	15
PET	1.38	30 to 40	260	20
Nylon 6	1.14	–	262	25 to 50

The structures of typical thermoplastics found as marine debris are given below.

Low-density polyethylene [LDPE]	–[CH_2–CH_2–]–
High-density polyethylene [HDPE]	Same as above with different chain branching
Linear low-density polyethylene [LLDPE]	Same as above with different chain branching
Polypropylene [PP]	–[CH_2–$CH(CH_3)$–]–
Polystyrene [GPPS] HIPS: copolymer of PS with 5–10% poly (butadiene)	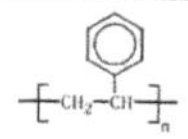
Poly(vinyl chloride) [PVC]	–[CH_2–CHCl–]–
Poly(ethylene terephthalate) [PET]	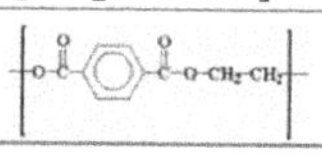
Nylon 6 [polycaprolactam]	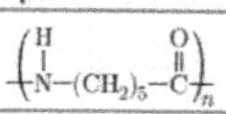

17.3 Fate of Plastics in the Ocean

Typically, beach litter exposed to solar ultraviolet radiation (UVR) undergoes extensive weathering degradation (Bond et al. 2018; Welden and Cowie 2017; Andrady 2011) reducing their mechanical integrity. Because of the high-surface temperatures achieved by plastic debris on sandy surfaces, plastic litter exposed to solar radiation on beaches undergoes particularly rapid degradation relative to those exposed floating in water (Andrady 2011). Weathering degradation of common plastics is an oxidative reaction sequence that is also accompanied by the scission of long polymer chains, reducing the average molecular weight of the plastic. Therefore, a consequence of extensive weathering of plastics is their reduction in mechanical strength often including surface embrittlement that ultimately breaks the material down into small fragments (ter Halle et al. 2016; Andrady 2017). The chemistry of the oxidative degradation process that cause chain scission of the plastic molecules leading to the loss of mechanical integrity are well known for common varieties of plastics found in marine debris such as PE, PP, PS and PET. With plastic macro-debris where the material is thick in cross section, the oxidative process starts at the surfaces exposed to air or water (Bracco et al. 2018). Because of the diffusion-controlled nature of the reaction (Audouin et al. 1994), only a thin surface layer of the plastic is often degraded extensively. After weakened or embrittled, the plastic litter poses a dramatically reduced entanglement threat to larger species, however, the smaller fragments produced on weathering may present an ingestion hazard to a range of smaller marine organisms.

17.4 Weathering and Fragmentation of Polymers

Generally, plastics exposed to the outdoor environment undergo oxidative fragmentation due to the action of several stressors acting in concert (Da Costa et al. 2018). Most important among them is solar UVR, the impact of which is exacerbated at higher sample temperatures as predicted by the Arrhenius relationship. With some plastics, biotic factors, hydrolysis and even atmospheric pollutants can play a secondary role in their deterioration on extended exposure outdoors (Andrady 2011). The combined effect of these factors in a given environment (such as the dry sediment on a beach) is referred to as natural weathering. It is a process that involves a series of changes that can vary from initial discoloration, reduction in mechanical properties, surface cracking and finally embrittlement where the plastic is friable and is reduced to small fragments on handling. It is a slow process; PE or PP film without any UV stabilizer, exposed outdoors in a test site in Florida, USA, for instance, would take close to a year to become embrittled by natural weathering (Andrady et al. 1993).

To observe the same photodamage within a shorter period of time, laboratory-accelerated weathering might be used. Typically, the material is exposed to a source of radiation that is spectrally similar to solar radiation but at a higher intensity compared to outdoor solar radiation (at air mass=1), and the samples are maintained at a higher temperature. The higher fluence rate of photons relative to outdoor insolation and higher temperatures accelerates the photodegradation allowing changes in the plastic to be observed in a much shorter timescale. A particularly useful measurement of changes in mechanical integrity of plastics is uniaxial tensile extensibility of a standard dumbbell-shaped sample. It is a measure that is particularly sensitive to changes in the average molecular weight of the polymer due to chain scission that accompanies oxidation. As the mechanical properties of plastics rely on the intact long-chain structure of the material, large changes in tensile extensibility are anticipated even at a low extent of chain scission. Other tests used to monitor oxidation-related changes in polymers exposed to either natural or accelerated weathering include infra-red spectroscopy, X-ray diffraction and molecular weight measurement (by gel permeation chromatography).

Interestingly, the photodegradation of common plastics is drastically retarded when the outdoor exposure is carried out with samples floating in seawater as opposed to in air (Gregory and Andrady 2005; Biber et al. 2019; Tang et al. 2019; Pegram and Andrady 1989; Andrady 1990). While the phenomenon has been observed by several researchers, the exact reason for it was not fully understood. Air (e.g., beach) and surface water exposures differ in three important respects: (a) the oxygen concentration in air is much higher than that dissolved in seawater; (b) the sample temperature of plastics exposed to sunlight in air are much higher than those exposed floating in seawater; and (c) fouling that covers the surface of plastics to shield them from sunlight occurs only in seawater exposures. One or more of these factors can contribute to the retardation of degradation in seawater relative to air observed for several classes of common plastics (Pegram and Andrady 1989; Gregory and Andrady 2005). However, the observation suggests that plastic debris directly disposed into seawater will have longer durability compared to those disposed on beaches. For instance, netting discarded at sea into water will degrade much slower, and being more durable would pose a much higher entanglement threat for a longer duration compared to one discarded on beach (and subsequently washed into the water) where it will undergo faster degradation. Much slower is the degradation of plastics in the water column below the photic zone as there is no facile mechanism of initiation of degradation. Slowest degradation in the marine environment takes place in the bottom sediment that is low in oxygen and is colder than midwater column. Figure 17.2 presents tensile data for a clear low-density polyethylene sheet (a 0.04 cm thick), exposed outdoors in air and while floating in seawater in Beaufort, NC, USA for a period of 12 months (Andrady 1990).

Under extensive oxidative degradation, the amorphous parts of semi-crystalline polymers are preferentially photodegraded (Bracco et al. 2018), weakening the plastic. Further, oxidation and reduction in average molecular weight invariably leads to fragmentation of the embrittled surface layers of the plastic. While plastic litter exposed to solar radiation is well known to fragment into micro- and nanoplastics yielding secondary microplastics (Weinstein et al. 2016; ter Halle et al. 2016), neither the mechanism nor the kinetics of the process are well understood. Microcracks that develop on the surface of a plastic fragment can propagate into its

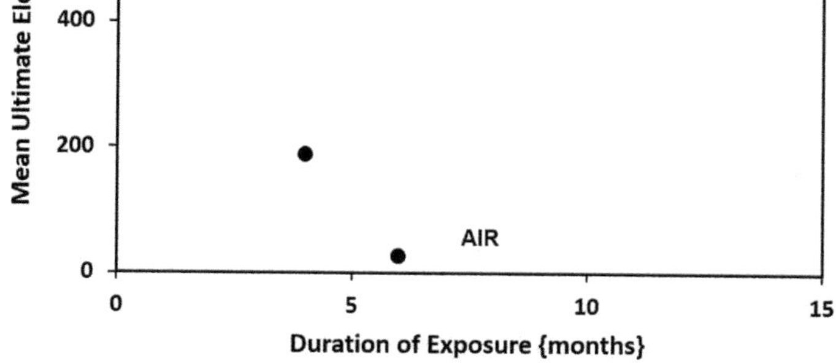

Fig. 17.2 Changes in tensile extensibility with duration of exposure for a LDPE film exposed in air or sea water in Beaufort, NC. *Courtesy* Andrady (1990)

interior, especially under mechanical stress, fragmenting it into two or more daughter fragments. This type of size reduction is commonly observed in weathered, embrittled plastic products in litter. The microplastic fragments formed can continue to photodegrade when they are exposed to UVR to produce nano-scale particles. Formation of nanoplastics from weathering degradation of plastics has been demonstrated in the laboratory (Gigault et al. 2016; Lambert and Wagner 2016) but has never been reported in any field studies.

However, a second mode of fragmentation by surface ablation has also been proposed (Andrady 2017) and leads to the generation of large numbers of small micro- or nanoplastics. This is a consequence of fragmentation being localized in a highly degraded surface layer resulting from the attenuation of UVR and diffusion-control of oxidation. Figure 17.3 is an artist's impression of this mode of fragmentation by surface ablation. For instance, a single square cm of PE laminate exposed to solar-simulated radiation in 20 mL of water generated 10^6 micro-fragments per mL on 112 days of laboratory-accelerated weathering exposure that was an order of magnitude higher than for the control. Six other plastics tested also yielded about the same concentration at the same duration of weathering, and of the plastics tested polylactic acid (PLA) yielded the highest particle concentration. Whether the strategy of using photodegradable plastics that undergo rapid fragmentation in the marine environment is ecologically beneficial is an open question (Napper and Thompson 2019). The ecological impacts of their use would depend on how small the fragments generated are and if the fragments subsequently biodegrade in a reasonable timescale.

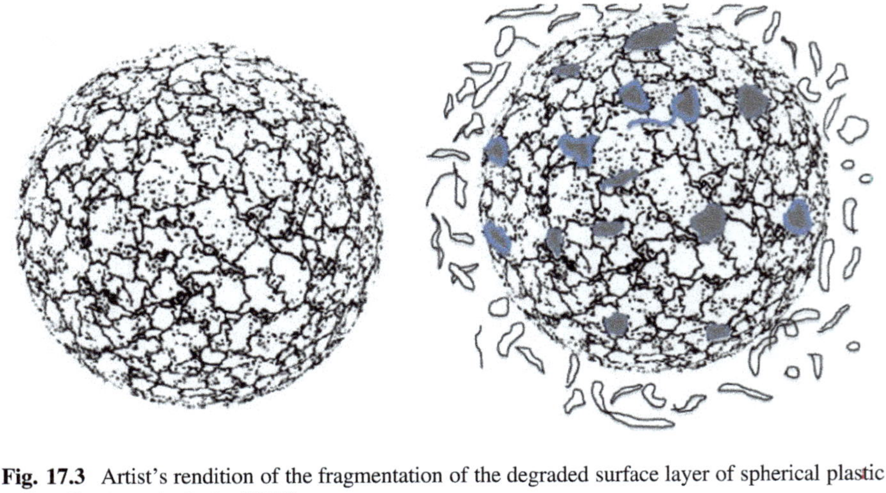

Fig. 17.3 Artist's rendition of the fragmentation of the degraded surface layer of spherical plastic pellet. *Courtesy* Andrady (2017)

17.5 Microplastics in the Ocean

The category of plastic litter loosely referred to as 'microplastics' (MPs) is the major public concern related to ocean pollution. Early work generally identified all plastic fragments <5 mm, a size class that can be ingested by a range of marine animals, as 'microplastics' (Auta et al. 2017; Galloway 2015). As defined, MPs include virgin resin pellets used in plastic processing operations, that were a significant part of beach debris in early studies (their abundance in coastal litter has decreased in the more recent reports). Strictly speaking, the term 'microplastic,' is a misnomer in that what is categorized in the class need not necessarily have micro-scale dimensions. Yet, the term is popularly used in research literature as a matter of convenience. MPs appear to be abundant in the ocean environment: The floating stock of MPs alone in the world's oceans was estimated to be 98–236 TMT (thousand metric tons) (van Sebille et al. 2015) and is estimated to reach 10^4 to 10^5 TMT by 2100 (Everaert et al. 2018). Microplastics pose the risk of being ingested by a wide range of animals.

The broad size range of MPs includes particles that are several mm in size to submicron particles not discernible by the naked eye. It might be practical to subdivide the range into several narrower size classes that also separate out true micro-scale particles. Of the alternative classifications proposed, the following scheme (Fig. 17.4) by the GESAMP G40 WG4 committee on microplastics (GESAMP 2016) is particularly useful.

This subdivision of the range recognizes plastic fragments of size d, where 1 μm < d < 1 mm as the 'true' 'microplastics' and those smaller than 1 μm as nanoplastics (NPs). However, the definition still does not properly take into account the shape factor (that can modify adverse effects from their ingestion), and how foam fragments as well as microfibers, the leading category of MPs present in seawater, should be classified is not clear.

MPs are also divided into two classes based on how they are generated: the primary and secondary MPs (Cole et al. 2011; Eerkes-Medrano et al. 2015; Andrady 2011). The primary MPs are manufactured and used as micro- or nano-scale particles, and their functionality depends on their small size. For

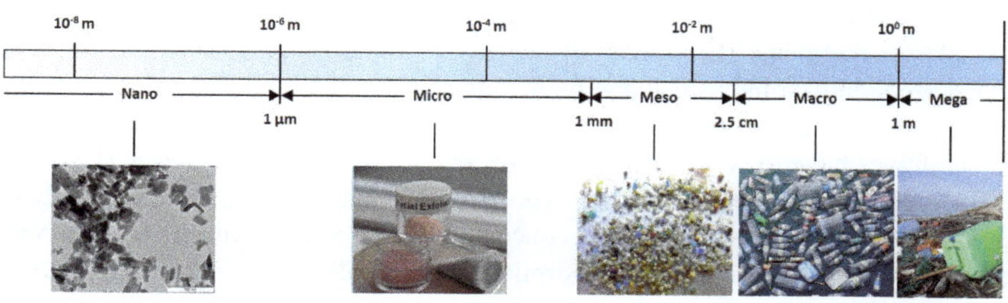

Fig. 17.4 Classification of plastics fragments by size. Based on the GESAMP Report #90 (2016). *Source* Fate and Effects of microplastics in the marine environment Part 1

instance, microbeads manufactured as spheres a few hundred microns in diameter are used as an exfoliant additive in cosmetic products (Fendall and Sewell 2009; Gouin et al. 2015), such as facial scrubs, sunscreens and toothpaste. A recent survey (Napper et al. 2015) of facial creams found most microbeads to be 164–327 μ in diameter, made of PE and used at 10^2 to 10^4 beads per mL in cosmetic formulations. These find their way to the ocean along with waste water (Leslie 2015) and might be ingested by small marine organisms. Microbeads with a modal size of <100 μm can be readily ingested even by zooplankton (Fendall and Sewell 2009). Other applications of microbeads include nano-powders used as fillers in composite materials and as 'plastic' blasting media used to remove paint. A larger fraction of MPs in the ocean, however, is secondary microplastics generated by the degradation and fragmentation of larger plastic debris in the marine environment, mostly due to natural weathering (Andrady 2011; Corcoran et al. 2009; Thompson et al. 2004; Song et al. 2017). The chemical reactions involved in the photooxidation sequence of plastics include the incorporation of polar functionalities into the polymer surface, reduction in average molecular weight and increase in the fractional crystallinity of the plastic. These changes weaken the degraded plastics facilitating their fracture under mechanical stress such as wave action leading to smaller fragments (Barnes et al. 2009; Moore 2008; Browne et al. 2007; Andrady 2011). Another common example of secondary MPs is the textile fiber fragments (or microfibers) formed during laundering of fabric that end up in wastewater. They are prevalent in the marine environment worldwide and are mostly made of polyester thermoplastic (Browne 2015). The weathering and fragmentation process is critical in the generation of a majority of microplastics.

17.6 Concentration of Persistent Organic Pollutants in Seawater by Microplastics

A particular concern on ingestion of MPs and NPs by marine organisms is that they may carry persistent organic pollutants (POPs) that might be bioavailable to it. There are two sources of potentially toxic organic compounds in plastic litter in the ocean (Fred-Ahmadu et al. 2020). As already mentioned, additives such as plasticizers and flame retardants are used at significant levels in plastics and can end up in the plastic fragments (Oehlmann et al. 2009). Some of these such as phthalate plasticizers and brominated flame retardants can be toxic and endocrine disruptor chemicals. However, their release into seawater is unlikely to be significant because the additives have to be hydrophobic to be able to dissolve in the plastic (Liu et al. 2016). However, the plastics being hydrophobic can absorb and concentrate POPs dissolved in seawater at very low concentrations (Teuten et al. 2009) in a very efficient manner. The equilibrium sorption of molecules dissolved in sea water is quantified by means of the partition coefficient $K_{P/W}$. It is the ratio of the concentrations of a solute between two solvent phases in contact with the compound as with MPs exposed to seawater. When a plastic material is exposed to low concentrations of a solute [such as PCB] dissolved in seawater at an equilibrium concentration S_W, it is picked up by the plastic and reach an equilibrium concentration S_P of the solute in the plastic. The value of the equilibrium partition coefficient $K_{P/W}$ depends on the units selected.

$$K_{P/W} = (S_P/S_W)$$

where $[K_{P/W}]$ is the partition coefficient for the system (Speight 2018). The concentrations might be expressed in units of S_P [mol/kg plastic or mg/kg plastic] and S_w [mol/L water or mg/L water]. On initial exposure of the plastic to seawater $(S'_P/S'_W) < K_{P/W}$, the sorption of the solute by the plastic (where S' is the non-equilibrium solubility) will continue until $(S'_P/S'_W) = (S_P/S_W) = K_{P/W}$ is reached at a given temperature. The description assumes that S'_W remains relatively constant (as with solutes dissolved in seawater) over the process. Note that the value of $K_{P/W}$ does not indicate how rapidly such equilibrium might be reached which in practice would depend on other factors such as temperature and the crystallinity of the polymer. The simple equation above assumes that the solute is evenly distributed in the entire volume of each phase. The sorption capacity of plastics found in marine litter for several pollutant species in seawater is available (Rios et al. 2007; Lee et al. 2014).

At a molecular level, the value of $K_{P/W}$ invariably depends on the free volume available in the polymer, and its phase properties. Common plastics such as PE and PP are semi-crystalline polymers with substantial levels of crystallinity in addition to amorphous material. The crystalline phase comprises of closely arranged ordered polymer chains with very little free volume and is not available to dissolve any solute (Guo et al. 2012). Solutes are almost exclusively located in the amorphous fraction. For instance, the value of $[K_{P/W}]_{PE}$ was shown to vary with the percent crystallinity of the PE (Xu et al. 2018; Guo et al. 2012). With plastics such as PS, PVC and PET, all substantially present in marine debris, even the non-crystalline phase will be glassy because their glass transition temperatures, T_g, are well above ambient temperature. Temperature-dependent sorption properties of amorphous polymers are generally much lower at $T < T_g$ as opposed to $T > T_g$. Data on specific plastic/seawater partition coefficient are scarce in the literature but can be approximated using octanol-water partition coefficients available for a wide range of organic compounds.

$$\log\left[K_{P/W}\right] = \alpha \log\left[K_{O/W}\right] + \beta$$

where $[K_{O/W}]$ is the octanol/water partition coefficient for the compound. The coefficients α and β are reported in the literature for different polymers of interest (Lohmann 2012). Kinetics of sorption (or desorption) of contaminants into microplastics have been described by a pseudo first-order reaction (Skrip et al. 2013).

$$\log(q_m - q_t) = \log q_m - k_1/2.303\,\mathrm{t}$$

where q_m and q_t are the amount of contaminant sorbed at equilibrium (mg/g) and that sorbed at time t, respectively, and k_1 is the rate constant (Ho et al. 2000).

During weathering, hydrophilic polar functional groups are progressively incorporated into the MPs, especially in its surface layers (Kaczmarek et al. 2002), as a result of oxidation. This has an impact on the kinetics of sorption and the equilibrium value of $K_{P/W}$ in seawater (Goedecke et al. 2017; Albertsson et al. 1987) because of the reduced hydrophobic character of the polymer resulting from weathering (Ho et al. 2000; Hüffer and Hofmann 2016).

Field samples of mesoplastics are analyzed for contaminants by extracting them with methylene chloride followed by fractionation of the extract on a chromatographic column. Fractions are analyzed using a mass spectrometric technique such as gas chromatography-ion-trap mass spectrometers (GC-ITMS) (Hirai et al. 2011). Early research by Endo et al. (2005) focused on the contamination of plastic pre-production pellets collected in the field, by polychlorinated biphenyls (PCBs), nonyl phenol (NP) and dichlorodiphenyltrichloroethane (DDT) from sea water. Data for geographic variation of these contaminants on pellets collected from beaches (Takada 2006) and from seawater (Rios et al. 2010) across the world was reported. These studies generally show two categories of contaminants associated with field samples: intentionally added chemicals or additives and those pollutants that partition into the MPs from seawater. The concentration of the latter in MPs of PE and PP collected from the field and in laboratory partition studies ranges from 1 to 10^3 ng/g with a high piece to piece variability (Hirai et al. 2011). Typical published values collected by Zhang et al. (2015) are provided in Table 17.3. Bisphenol A (BPA), NP and polybrominated diphenyl ethers (PBDs) are either additives or derived from the polymer itself but the others have been picked up and concentrated from seawater.

More recently, microplastics have been shown to absorb toxic metal species (Yu et al. 2021; Turner et al. 2020; Santos-Echeandía et al. 2020) and pharmaceuticals present in waste water streams. These include the popular antibiotics in the waste water stream and has two potential significant impacts: (a) they can deliver relatively high doses of the antibiotic to small organisms such as zooplanktons, with adverse effects (Yuan et al. 2020; Mohsen et al. 2019) and (b) encourage the growth of antibiotic-resistant organisms (Yang et al. 2019; Zhang et al. 2020) within foulant colonies supported by the microplastics. Antibiotic-resistant bacteria have been observed to be associated with fouled microplastics, and their dispersal by currents as well as via ingestion by animals is a serious concern. The clinical management of diseases such as tuberculosis, *Staphyllococcus* skin infections and Gonorrhea, for instance, is already complicated by the relevant bacteria being antibiotic resistant. The UV-thermal degradation of common plastics increases their sorption of both metal ions and POPs.

Table 17.3 Comparison of concentration reported for plastic resin pellets in previous studies

Sampling location	PCBs ($ng\ g^{-1}$)	PAHs ($ng\ g^{-1}$)	DDTs ($ng\ g^{-1}$)	HCHs ($ng\ g^{-1}$)	Reference
Japan	3.97 to 117				Mato et al. (2001)
Japan	28 to 2300				Endo et al. (2005)
Japan	5 to 892				Takada et al. (2006)
Mexico	27 to 980	39 to 1200			Rios et al. (2007)
Japan	169 to 453				Ogata et al. (2009)
Porto, Portugal	307	100			Frias et al. (2010)
Lisboa, Portugal	273	300			Frias et al. (2010)
San Diego, US	3.8 to 4.2	18 to 210	0.56 to 64		van Franeker et al. (2011)
Remote islands in the Pacific, Atlantic and Indian Oceans and the Caribbean Sea	0.1 to 9.9		0.8 to 4.1	0.6 to 1.7	Heskett et al. (2012)
Portugal-coast	10.5 to 307				Mizukawa et al. (2013)
Dalian, China	34.7 to 213.7	136.3 to 1586.9	1.15 to 101.54	nd-1.90	Zhang et al. (2015)
Qinghuangdao, China	21.5 to 232.2	397.6 to 2384.2	1.46 to 126.95	nd-1.49	Zhang et al. (2015)

Courtesy Zhang et al. (2015)
Note: *PCBs* polychlorinated biphenyls; *PAHs* polycyclic aromatic hydrocarbons; *DDTs* dichlorodiphenyltrichloroethanes; *HCHs* hexachlorocyclohexanes; *nd* not detected

17.7 Ecological Impacts of Plastics in the Ocean

17.7.1 Entanglement Hazard Posed by Macroplastic Debris

Early concerns on large pieces of plastics debris such as six-pack rings (Hanni and Pyle 2000; Boren et al. 2006) or discarded net fragments (Allen et al. 2012; Adimey et al. 2014) in the ocean primarily addressed the entanglement risk these pose to marine animals. Entanglement in plastics debris is a serious problem that affects diverse marine species; in 2015, Kühn et al. reviewed the relevant research and determined that 103 species (or 25.4% of the species) of seabirds, 51 species (or 41.5% of the species) of marine mammals, 2 species (or 3.2% of the species) of sea snakes and 92 species (or 0.06% of the species) of invertebrates suffered entanglement in marine plastic debris. With marine birds, turtles and mammals this rate of entanglement of ~30% of their species estimated in 2015 had increased from about 20.5% reported in a 1997 study (Laist 1997). Some of this increase might be attributed to the larger number of species considered in the latter study, but that entanglement is more common or is more accurately reported, is also evident from the data. Population-level adverse impacts of entanglement in plastic netting has been reported in the case of the Hawaiian monk seal (*Monachus schauinslandi*) (Lowry et al. 2011; Boland and Donohue 2003). The legal requirement to use photodegradable plastics in six-pack rings (in the US) has alleviated the entanglement hazard posed by this product but that due to fishing gear still persists. Fragments of derelict fishing gear such as gill nets or trawl sections abandoned in water are particularly hazardous because they continue to target fish and other species over long periods of time (Carr and Harris 1997; Boren et al. 2006).

Behavior of some species appears to increase their chances of getting entangled in plastic debris. Feeding behavior motivated by fish already caught in floating debris (Cliff et al. 2002) or species using the debris as a reef (Tschernij and Larsson 2003), for instance, might attract predator fish or marine mammals to the debris. Marine mammals, especially their young, enjoy playful curiosity that attracts them to the unusual, often colorful, net debris increasing their risk of entanglement (McIntosh et al. 2015). Some marine birds use seaweed in nest construction and might be attracted to monofilament or netting that looks similar, increasing their chance of entanglement (Jagiello et al. 2019). In three of the six North American Gannet populations, about 75% of the nests were found to contain fishing gear debris (Bond et al. 2012). Entanglement results in severely restricting movements and foraging activity of animals. In some cases, as with a six-pack ring or a net fragment looped around its body, the growth of the animal is severely restricted and may result in a slow death.

17.7.2 Ingestion of Plastic Debris by Marine Organisms

Ingestion of plastics by marine fauna is a major concern that can have a particularly adverse effect on the marine animals *via* a combination of physical and chemical effects. Physical effects of ingestion include the potential occlusion of gut passage and the triggering of satiation signals that interfere with the regular feeding behavior of animals. Once in the digestive tract MPs can obstruct the passage of food, for instance, in fish (Jovanović 2017), turtles (Santos et al. 2020) or seabirds (Pierce et al. 2004). Far more serious, however, are the chemical effects that may potentially introduce toxic pollutants into the organism.

Observations of birds (Ryan 1987) and turtles (Carr 1987) ingesting MPs were reported from late 1980s. The MPs in marine plastic litter are within a size range that can be readily ingested by a wide variety of marine organisms including zooplankton (Lattin et al. 2004), marine birds (Fry et al. 1987), fish (de Vries et al. 2020), turtles (Bjorndal et al. 1994) and other species (Baird and Hooker 2000). Among the species that ingest MPs are commercial fish and seafood species (de Vries et al. 2020) where ingestion not only negatively affects their health but also transmit pathogens (Wang et al. 2020) which is a particular concern with edible species. Ingestion of MPs by marine birds and sea turtles in particular has been highlighted in the literature. Why they ingest MPs (or even macroplastics) is not entirely clear. With birds or turtle species, however, it is likely that they seek out MPs mistaking them to be food, either because of their shape or because they are covered with foulants. The full ecological impact of MP ingestion, however, is not entirely clear because most of the reported studies are limited to short-term observations, excluding especially the intergenerational effects. Also, most laboratory studies tend to use concentrations of MPs that are well over what might be reasonably encountered in the marine environment. But, understanding the potential adverse impacts of ingestion is important because given the expected growth in plastic litter and potentially much higher levels of MP and NPs in future years (Critchell and Hoogenboom 2018).

It is convenient to consider ingestion-related distress to a marine organism in several steps: (a) ingestion of MP into the digestive tract of the organism; (b) potential leaching or removal of sorbed POPs and foulants; and (c) sorption of the released chemicals from the gut at the organ or cellular level. The various steps of the process are shown in Fig. 17.5. As discussed in Sect. 17.6, hydrophobic plastic materials readily absorb and concentrate toxic pollutants dissolved at low concentrations in sea water. Sorbed chemicals as well as additives in MPs can be toxic to ingesting organisms if the chemicals are bioavailable to them (Karapanagioti et al. 2011; Teuten et al. 2007). This in turn depends on the residence time of the MPs in the gut of the ingesting species. Being hydrophobic compounds, pollutant species that are soluble in the plastic matrix will not generally leach out readily unless the MPs come into contact with high-lipid food contents in the hydrophilic gut environment. Additives such as brominated flame retardants or phthalate plasticizers, however, are endocrine disruptor chemicals (Chen et al. 2019) that can cause adverse reactions at relatively low doses, especially in organisms with a small body size. If the released chemicals are in fact bioavailable to the animal, they can reach various organs resulting in dose-dependent toxicity (Teuten et al. 2007). Bioavailability is determined by the type of plastic involved and the species in question as well as on the complex, possibly countering mechanisms, that occur in gut environment (Engler 2012; Koelmans 2015).

A phenomenon of interest with ingestion (but not with entanglement) is bio-magnification where the effect of a pollutant in the marine ecosystem is magnified via the predation of species with ingested MPs by other organisms in the trophic chain. Predators who ingest 'contaminated' prey in large numbers receive a relatively high dose of the chemical toxicants along with the meal. While such

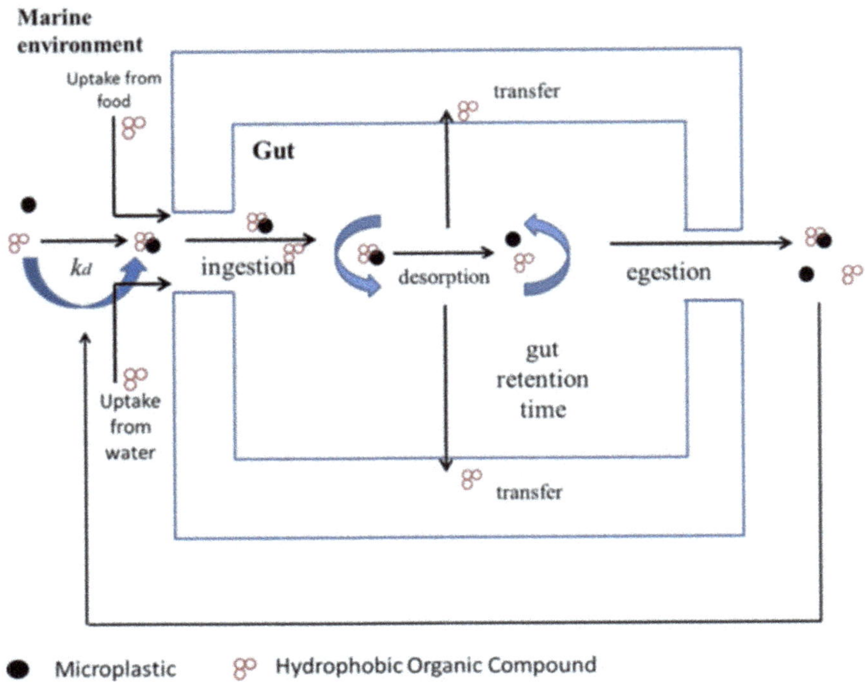

Fig. 17.5 Graphical illustration of mechanisms that contribute to bioavailability of POPs in ingested MPs. *Courtesy* Bakir et al. (2016)

transfer of MPs across trophic boundaries is known and has been demonstrated in the laboratory, the magnitude of the transfer coefficients for a marine trophic network has not been reliably assessed. Though experimental indications of the transfer of POPs to the ingesting organisms are reported, they often do not always unequivocally point to MPs as the source of transfer (Ryan et al. 1988; Tanaka et al. 2013). Food, water and sediment are all important sources of POPs as well, and the current understanding is that in general, the ingestion of MPs does not provide an additional significant pathway for transfer of POPs into the biota (Bakir et al. 2016) in competition with other sources.

NPs differ from MPs in several characteristics: (a) their very high specific surface area makes them highly reactive rendering them highly toxic (Oberdörster et al. 2005; Mattsson et al. 2018) and (b) they are able to permeate through tissue (gut wall) and translocate throughout the body (Stapleton 2019; Kashiwada 2006; von Moos et al. 2012) because of their minute dimensions. Particles smaller than about 50 nm can enter individual cells by endocytosis as monitored using lysosomal membrane stability as a biomarker (Avio et al. 2015; Jeong et al. 2017). This was shown in laboratory studies on the blue mussels (Browne et al. 2008) and shore crab (Farrell and Nelson 2013). Small enough NPs can even pass through the blood–brain barrier potentially delivering the POPs to the brain (Mattsson et al. 2017). A particularly good discussion of the impacts of ingestion of MPs from cellular to ecosystem level was provided by Galloway et al. (2017).

17.7.3 Ingestion of MPs by Marine Vertebrates

Comprehensive global data for ingestion of macroplastics or MPs by different marine animals are understandably difficult to compile. But knowing what taxa are affected and how the relevant statistics change in time is useful in assessing the seriousness of the problem. Kühn and van Franeker (2020) have compiled such data for reported cases of ingestion first in 2015 and then again in 2020. Their data can be compared with an even earlier dataset by Laist (1997) to locate any evident trends. Data reported by Kühn and van Franekerac (2020) for species that were also reported in the earlier years (1997 and 2015) are compared in Fig. 17.6, which suggests the taxa affected by ingestion of plastics have increased in the time period considered. Percentages of species with reported ingestion increased for all major groups and in case of marine turtles are a 100! With ingestion, the distress to the animals is variable, depending on the amount ingested, the geometry of fragments and the toxicity of POPs. Also, the higher percentages may also be due to higher levels of reporting because of the attention now being paid to the problem.

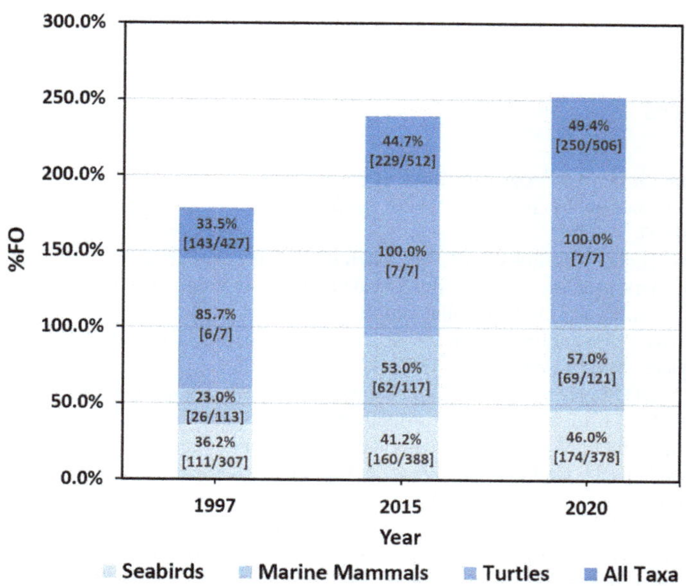

Fig. 17.6 Percentage of species of seabirds, marine mammals and turtles ingesting MPs. Data based on Kühn et al. (2015, 2020) and Laist (1997). The bar chart indicates data only for those species in each category reported in all 3 years

17.7.4 Ingestion by Planktonic Species

Ingestion of MPs by zooplankton species is a serious concern because they occupy the key trophic level between the base population of autotrophs and the higher trophic levels. If plankton populations were to drop or if their foraging behavior affected by ingestion of MPs, the impacts on productivity will be felt throughout the ocean food pyramid. Not only are the zooplankton a key food source for commercially important marine fish species, but they are also critically needed to maintain the nutrient and carbon cycling in the marine ecosystem. There is little doubt that zooplankton ingest MPs and possibly NPs as these particles are in the same size range as their staple prey (Galloway et al. 2017). A few field observations of ingestion have been reported (Katija et al. 2017; Desforges et al. 2015), but a majority of the research on both holoplanktonic as well as meroplanktonic species (Cole and Galloway 2015; Choi et al. 2018) are laboratory studies, often involving unrealistically high concentrations of MPs. A majority of these have been carried out on the order *Calonoida* or the copepod zooplankton, because they dominate the plankton population at many sites and support the commercial fish species. Botterell et al. (2019) have compiled the studies on ingestion of MPs by 28 taxanomic orders of zooplankton and found 45% of the ingestion studies suggested serious impacts such as reduced fecundity, a decreased feeding behavior [therefore limiting growth] and effects on the lifespan. But, with variable concentrations, geometries and plastic types used in different studies, it is difficult to draw any quantitative conclusions from the review.

References

Adey JM, Smith IP, Atkinson RJA, Tuck ID, Taylor AC (2008) 'Ghost fishing' of target and non-target species by Norway lobster *Nephrops norvegicus* creels. Mar Ecol Prog Ser 366:119–127

Adimey NM, Hudak CA, Powell JR, Bassos-Hull K, Foley A, Farmer NA, White L, Minch K (2014) Fishery gear interactions from stranded bottlenose dolphins, Florida manatees and sea turtles in Florida, U.S.A. Mar Pollut Bull 81:103–115

Albertsson A-C, Andersson SO, Karlsson S (1987) The mechanism of biodegradation of polyethylene. Polym Degrad Stab 18:73–87

Allen R, Jarvis D, Sayer S, Mills C (2012) Entanglement of grey seals *Halichoerus grypus* at a haul out site in Cornwall, UK. Mar Pollut Bull 64:2815–2819

Andrady AL (1990) Weathering of polyethylene (LDPE) and enhanced photodegradable polyethylene in the marine environment. J Appl Polym Sci 39:363–370

Andrady AL (2011) Microplastics in the marine environment. Mar Pollut Bull 62:1596–1605

Andrady AL (2017) The plastic in microplastics: a review. Mar Pollut Bull 119:12–22

Andrady AL, Neal MA (2009) Applications and societal benefits of plastics. Philosophical transactions of the Royal Society of London. Series B, Biol Sci 364:1977–1984

Andrady AL, Pegram JE, Song Y (1993) Studies on enhanced degradable plastics. II. Weathering of enhanced photodegradable polyethylenes under marine and freshwater floating exposure. J Environ Polym Degrad 1:117–126

Audouin L, Langlois V, Verdu J, de Bruijn JCM (1994) Role of oxygen diffusion in polymer ageing: kinetic and mechanical aspects. J Mater Sci 29:569–583

Auta HS, Emenike CU, Fauziah SH (2017) Distribution and importance of microplastics in the marine environment: a review of the sources, fate, effects, and potential solutions. Environ Int 102:165–176

Avio CG, Gorbi S, Milan M, Benedetti M, Fattorini D, d'Errico G, Pauletto M, Bargelloni L, Regoli F (2015) Pollutants bioavailability and toxicological risk from microplastics to marine mussels. Environ Pollut 198:211–222

Baird RW, Hooker SK (2000) Ingestion of plastic and unusual prey by a juvenile harbour porpoise. Mar Pollut Bull 40:719–720

Bakir A, O'Connor IA, Rowland SJ, Hendriks AJ, Thompson RC (2016) Relative importance of microplastics as a pathway for the transfer of hydrophobic organic chemicals to marine life. Environ Pollut 219:56–65

Ball P (2020) The plastic legacy. Nat Mater 19:938

Barnes DKA, Galgani F, Thompson RC, Barlaz M (2009) Accumulation and fragmentation of plastic debris in global environments. Philos Trans Royal Soc B Biol Sci 364:1985–1998

Biber NFA, Foggo A, Thompson RC (2019) Characterising the deterioration of different plastics in air and seawater. Mar Pollut Bull 141:595–602

Bjorndal KA, Bolten AB, Lagueux CJ (1994) Ingestion of marine debris by juvenile sea turtles in coastal Florida habitats. Mar Pollut Bull 28:154–158

Boland RC, Donohue MJ (2003) Marine debris accumulation in the nearshore marine habitat of the endangered Hawaiian monk seal, *Monachus schauinslandi* 1999–2001. Mar Pollut Bull 46:1385–1394

Bond AL, Montevecchi WA, Guse N, Regular PM, Garthe S, Rail J-F (2012) Prevalence and composition of fishing gear debris in the nests of northern gannets (*Morus bassanus*) are related to fishing effort. Mar Pollut Bull 64:907–911

Bond T, Ferrandiz-Mas V, Felipe-Sotelo M, van Sebille E (2018) The occurrence and degradation of aquatic plastic litter based on polymer physicochemical properties: a review. Crit Rev Environ Sci Technol 48:685–722

Boren LJ, Morrissey M, Muller CG, Gemmell NJ (2006) Entanglement of New Zealand fur seals in man-made debris at Kaikoura, New Zealand. Mar Pollut Bull 52:442–446

Botterell ZLR, Beaumont N, Dorrington T, Steinke M, Thompson RC, Lindeque PK (2019) Bioavailability and effects of microplastics on marine zooplankton: a review. Environ Pollut 245:98–110

Bracco P, Costa L, Luda MP, Billingham N (2018) A review of experimental studies of the role of free-radicals in polyethylene oxidation. Polym Degrad Stab 155:67–83

Browne MA (2015) Sources and pathways of microplastics to habitats. In: Bergmann M, Gutow L, Klages M (eds) Marine anthropogenic litter. Springer International Publishing, Cham, pp 229–244

Browne MA, Galloway T, Thompson R (2007) Microplastic—an emerging contaminant of potential concern? Integr Environ Assess Manag 3:559–561

Browne MA, Dissanayake A, Galloway TS, Lowe DM, Thompson RC (2008) Ingested microscopic plastic translocates to the circulatory system of the mussel, *Mytilus edulis* (L.). Environ Sci Technol 42:5026–5031

Browne MA, Crump P, Niven SJ, Teuten E, Tonkin A, Galloway T, Thompson R (2011) Accumulation of microplastic on shorelines woldwide: sources and sinks. Environ Sci Technol 45:9175–9179

Carpenter EJ, Smith KL (1972) Plastics on the Sargasso Sea surface. Science 175:1240

Carpenter EJ, Anderson SJ, Harvey GR, Miklas HP, Peck BB (1972) Polystyrene spherules in coastal waters. Science 178:749

Carr A (1987) Impact of nondegradable marine debris on the ecology and survival outlook of sea turtles. Mar Pollut Bull 18:352–356

Carr HA, Harris J (1997) Ghost-fishing gear: have fishing practices during the past few years reduced the impact? In: Coe JM, Rogers DB (eds) Marine Debris: sources, impacts, and solutions. Springer, New York, New York, NY, pp 141–151

Chen Q, Allgeier A, Yin D, Hollert H (2019) Leaching of endocrine disrupting chemicals from marine microplastics and mesoplastics under common life stress conditions. Environ Int 130:

Choi JS, Jung Y-J, Hong N-H, Hong SH, Park J-W (2018) Toxicological effects of irregularly shaped and spherical microplastics in a marine teleost, the sheepshead minnow (*Cyprinodon variegatus*). Mar Pollut Bull 129:231–240

Cliff G, Dudley SFJ, Ryan PG, Singleton N (2002) Large sharks and plastic debris in KwaZulu-Natal, South Africa. Mar Freshw Res 53:575–581

Cole M, Galloway TS (2015) Ingestion of nanoplastics and microplastics by Pacific oyster larvae. Environ Sci Technol 49:14625–14632

Cole M, Lindeque P, Halsband C, Galloway TS (2011) Microplastics as contaminants in the marine environment: a review. Mar Pollut Bull 62:2588–2597

Corcoran PL, Biesinger MC, Grifi M (2009) Plastics and beaches: a degrading relationship. Mar Pollut Bull 58:80–84

Cózar A et al (2014) Plastic debris in the open ocean. Proc Natl Acad Sci 111:10239

Critchell K, Hoogenboom MO (2018) Effects of microplastic exposure on the body condition and behaviour of planktivorous reef fish (*Acanthochromis polyacanthus*). PLoS ONE 13:

Da Costa JP, Nunes AR, Santos PSM, Girão AV, Duarte AC, Rocha-Santos T (2018) Degradation of polyethylene microplastics in seawater: insights into the environmental degradation of polymers. J Environ Sci Health Part A 53:866–875

de Vries AN, Govoni D, Árnason SH, Carlsson P (2020) Microplastic ingestion by fish: body size, condition factor and gut fullness are not related to the amount of plastics consumed. Mar Pollut Bull 151:110827

Desforges J-PW, Galbraith M, Ross PS (2015) Ingestion of microplastics by zooplankton in the Northeast Pacific Ocean. Arch Environ Contam Toxicol 69:320–330

Dixon TR, Dixon TJ (1981) Marine litter surveillance. Mar Pollut Bull 12:289–295

Eerkes-Medrano D, Thompson RC, Aldridge DC (2015) Microplastics in freshwater systems: a review of the emerging threats, identification of knowledge gaps and prioritisation of research needs. Water Res 75:63–82

Endo S, Takizawa R, Okuda K, Takada H, Chiba K, Kanehiro H, Ogi H, Yamashita R, Date T (2005) Concentration of polychlorinated biphenyls (PCBs) in beached resin pellets: variability among individual particles and regional differences. Mar Pollut Bull 50:1103–1114

Engler RE (2012) The complex interaction between marine debris and toxic chemicals in the ocean. Environ Sci Technol 46:12302–12315

Eriksen M (2014) Plastic pollution in the world's oceans: more than 5 trillion plastic pieces weighing over 250,000 tons afloat at sea. PLoS ONE 9(12):e111913.

Eriksson C, Burton H, Fitch S, Schulz M, van den Hoff J (2013) Daily accumulation rates of marine debris on sub-Antarctic island beaches. Mar Pollut Bull 66:199–208

Everaert G, Van Cauwenberghe L, De Rijcke M, Koelmans AA, Mees J, Vandegehuchte M, Janssen CR (2018) Risk assessment of microplastics in the ocean: modelling approach and first conclusions. Environ Pollut 242:1930–1938

Farrell P, Nelson K (2013) Trophic level transfer of microplastic: *Mytilus edulis* (L.) to *Carcinus maenas* (L.). Environ Pollut 177:1–3

Fazey FMC, Ryan PG (2016) Biofouling on buoyant marine plastics: an experimental study into the effect of size on surface longevity. Environ Pollut 210:354–360

Fendall LS, Sewell MA (2009) Contributing to marine pollution by washing your face: microplastics in facial cleansers. Mar Pollut Bull 58:1225–1228

Fred-Ahmadu OH, Bhagwat G, Oluyoye I, Benson NU, Ayejuyo OO, Palanisami T (2020) Interaction of chemical contaminants with microplastics: principles and perspectives. Sci Total Environ 706:

Frias JPGL, Sobral P, Ferreira AM (2010) Organic pollutants in microplastics from two beaches of the Portuguese coast. Mar Pollut Bull 60:1988–1992

Fry DM, Fefer SI, Sileo L (1987) Ingestion of plastic debris by Laysan Albatrosses and Wedge-tailed Shearwaters in the Hawaiian Islands. Mar Pollut Bull 18:339–343

Galloway TS (2015) Micro- and nano-plastics and human Health. In: Bergmann M, Gutow L, Klages M (eds) Marine Anthropogenic Litter. Springer International Publishing, Cham, pp 343–366

Galloway TS, Cole M, Lewis C (2017) Interactions of microplastic debris throughout the marine ecosystem. Nat Ecol Evol 1:0116

GAO-19-653, United States Government Accountability Office (2019) Marine Debris: interagency committee members are taking action, but additional steps could enhance the federal response, Washington, DC

GESAMP, International Maritime Organization (2016) Sources, fate and effects of microplastics in the marine environment: part two of a global assessment, London, p 220

Geyer R, Jambeck JR, Law KL (2017) Production, use, and fate of all plastics ever made. Sci Advan 3(7):e1700782

Giacovelli C, United Nations Environment Programme (2018) Single use plastics: a roadmap for sustainability

Gigault J, Pedrono B, Maxit B, Ter Halle A (2016) Marine plastic litter: the unanalyzed nano-fraction. Environ Sci Nano 3:346–350

Gilbert M (2017) Chapter 1—plastics materials: introduction and historical development. In: Gilbert M (ed) Brydson's plastics materials, 8th edn. Butterworth-Heinemann, Cambridge, MA, pp 1–18

Girijappa YGT, Ayyappan V, Puttegowda M, Rangappa SM, Parameswaranpillai J, Siengchin S (2020) Plastics in automotive applications. In: Reference module in materials science and materials engineering. Elsevier

Goedecke C, Mülow-Stollin U, Hering S, Richter J, Piechotta C, Paul A, Braun U (2017) A first pilot study on the sorption of environmental pollutants on various microplastic materials. J Environ Anal Chem 4:1–8

Gouin T et al (2015) Use of micro-plastic beads in cosmetic products in Europe and their estimated emissions to the North Sea environment. SOFW J 3:1–33

Gregory MR, Andrady AL (2005) Plastics in the marine environment. In: Andrady AL (ed) Plastics and the environment. Wiley, Hoboken, pp 379–401

Guo X, Wang X, Zhou X, Kong X, Tao S, Xing B (2012) Sorption of four hydrophobic organic compounds by three chemically distinct polymers: role of chemical and physical composition. Environ Sci Technol 46:7252–7259

Hanni KD, Pyle P (2000) Entanglement of pinnipeds in synthetic materials at South-east Farallon Island, California, 1976–1998. Mar Pollut Bull 40:1076–1081

Heskett M, Takada H, Yamashita R, Yuyama M, Ito M, Geok YB … Mermoz J (2012) Measurement of persistent organic pollutants (POPs) in plastic resin pellets from remote islands: toward establishment of background concentrations for International Pellet Watch. Mar Pollut Bull 64:445–448

Hirai H, Takada H, Ogata Y, Yamashita R, Mizukawa K, Saha M … Ward MW (2011) Organic micropollutants in marine plastics debris from the open ocean and remote and urban beaches. Mar Pollut Bull 62:1683–1692

Ho YS, Ng JCY, McKay G (2000) Kinetics of pollutant sorption by biosorbents: review. Sep Purif Methods 29:189–232

Hüffer T, Hofmann T (2016) Sorption of non-polar organic compounds by micro-sized plastic particles in aqueous solution. Environ Pollut 214:194–201

Jagiello Z, Dylewski Ł, Tobolka M, Aguirre JI (2019) Life in a polluted world: a global review of anthropogenic materials in bird nests. Environ Pollut 251:717–722

Jambeck JR, Geyer R, Wilcox C, Siegler TR, Perryman M, Andrady A, Narayan R, Law KL (2015) Plastic waste inputs from land into the ocean. Science 347:768

Jeong C-B et al (2017) Adverse effects of microplastics and oxidative stress-induced MAPK/Nrf2 pathway-mediated defense mechanisms in the marine copepod *Paracyclopina nana*. Sci Rep 7:41323

Jovanović B (2017) Ingestion of microplastics by fish and its potential consequences from a physical perspective. Integr Environ Assess Manag 13:510–515

Kaczmarek H, Kowalonek J, Szalla A, Sionkowska A (2002) Surface modification of thin polymeric films by air-plasma or UV-irradiation. Surf Sci 507–510:883–888

Karapanagioti HK, Endo S, Ogata Y, Takada H (2011) Diffuse pollution by persistent organic pollutants as measured in plastic pellets sampled from various beaches in Greece. Mar Pollut Bull 62:312–317

Kashiwada S (2006) Distribution of nanoparticles in the see-through medaka (*Oryzias latipes*). Environ Health Perspect 114:1697–1702

Katija K, Choy CA, Sherlock RE, Sherman AD, Robison BH (2017) From the surface to the seafloor: How giant larvaceans transport microplastics into the deep sea. Sci Advan 3(8): e1700715

Kershaw PJ, United Nations Environment Programme (UNEP) (2016) Marine Plastic Debris and microplastics—global lessons and research to inspire action and guide policy change, Nairobi, p 252

Koelmans AA (2015) Modeling the role of microplastics in bioaccumulation of organic chemicals to marine aquatic organisms: a critical review. In: Bergmann M, Gutow L, Klages M (eds) Marine anthropogenic litter. Springer International Publishing, Cham, pp 309–324

Kole PJ, Löhr AJ, Van Belleghem FGAJ, Ragas AMJ (2017) Wear and tear of tyres: a stealthy source of microplastics in the environment. Int J Environ Res Public Health 14:1265

Kühn S, van Franeker JA (2020) Quantitative overview of marine debris ingested by marine megafauna. Mar Pollut Bull 151:

Kühn S, Bravo Rebolledo EL, van Franeker JA (2015) Deleterious effects of litter on marine life. In: Bergmann M, Gutow L, Klages M (eds) Marine anthropogenic litter. Springer International Publishing, Cham, pp 75–116

Laist DW (1997) Impacts of marine debris: Entanglement of marine life in marine debris including a comprehensive list of species with entanglement and ingestion records. In: Coe JM, Rogers DB (eds) Marine Debris: sources, impacts, and solutions. Springer, New York, pp 99–139

Lambert S, Wagner M (2016) Characterisation of nanoplastics during the degradation of polystyrene. Chemosphere 145:265–268

Lattin GL, Moore CJ, Zellers AF, Moore SL, Weisberg SB (2004) A comparison of neustonic plastic and zooplankton at different depths near the southern California shore. Mar Pollut Bull 49:291–294

Law KL, Morét-Ferguson SE, Goodwin DS, Zettler ER, DeForce E, Kukulka T, Proskurowski G (2014) Distribution of surface plastic debris in the Eastern Pacific Ocean from an 11-year data set. Environ Sci Technol 48:4732–4738

Lebreton L, Slat B, Ferrari F, et al (2018) Evidence that the Great Pacific Garbage Patch is rapidly accumulating plastic. Sci Rep 8:4666

Lee H, Shim WJ, Kwon J-H (2014) Sorption capacity of plastic debris for hydrophobic organic chemicals. Sci Total Environ 470–471:1545–1552

Leslie HA, United Nations Environment Programme (UNEP) (2015) Plastics in cosmetics. Amsterdam

Liu L, Fokkink R, Koelmans AA (2016) Sorption of polycyclic aromatic hydrocarbons to polystyrene nanoplastic. Environ Toxicol Chem 35:1650–1655

Lohmann R (2012) Critical review of low-density polyethylene's partitioning and diffusion coefficients for trace organic contaminants and implications for its use as a passive sampler. Environ Sci Technol 46:606–618

Lowry LF, Laist DW, Gilmartin WG, Antonelis GA (2011) Recovery of the Hawaiian monk seal (*Monachus schauinslandi*): a review of conservation efforts, 1972 to 2010, and thoughts for the future. Aquatic Mammals 37:397–419

Mato Y, Isobe T, Takada H, Kanehiro H, Ohtake C, Kaminuma T (2001) Plastic resin pellets as a transport medium for toxic chemicals in the marine environment. Environ Sci Technol 35:318–324

Mattsson K, Johnson EV, Malmendal A, Linse S, Hansson L-A, Cedervall T (2017) Brain damage and behavioural disorders in fish induced by plastic nanoparticles delivered through the food chain. Sci Rep 7:11452

Mattsson K, Jocic S, Doverbratt I, Hansson L-A (2018) Chapter 13—nanoplastics in the aquatic environment. In: Zeng EY (ed) Microplastic contamination in aquatic environments. Elsevier, Cambridge, MA, pp 379–399

McIntosh RR, Kirkwood R, Sutherland DR, Dann P (2015) Drivers and annual estimates of marine wildlife entanglement rates: a long-term case study with Australian fur seals. Mar Pollut Bull 101:716–725

Merrell TR (1980) Accumulation of plastic litter on beaches of Amchitka Island, Alaska. Mar Environ Res 3:171–184

Mizukawa K, Takada H, Ito M, Geok YB, Hosoda J, Yamashita R … Ferreira CA (2013) Monitoring of a wide range of organic micropollutants on the Portuguese coast using plastic resin pellets. Mar Pollut Bull 70:296–302

Mohsen M, Wang Q, Zhang L, Sun L, Lin C, Yang H (2019) Heavy metals in sediment, microplastic and sea cucumber *Apostichopus japonicus* from farms in China. Mar Pollut Bull 143:42–49

Moore CJ (2008) Synthetic polymers in the marine environment: a rapidly increasing, long-term threat. Environ Res 108:131–139

Napper IE, Thompson RC (2019) Environmental deterioration of biodegradable, oxo-biodegradable, compostable, and conventional plastic carrier bags in the sea, soil, and open-air over a 3-year period. Environ Sci Technol 53:4775–4783

Napper IE, Bakir A, Rowland SJ, Thompson RC (2015) Characterisation, quantity and sorptive properties of microplastics extracted from cosmetics. Mar Pollut Bull 99:178–185

Oberdörster G, Oberdörster E, Oberdörster J (2005) Nanotoxicology: an emerging discipline evolving from studies of ultrafine particles. Environ Health Perspect 113:823–839

Ocean Conservancy, International Coastal Cleanup (2019) 2018 report: building a clean swell. Washington, DC

Oehlmann J et al (2009) A critical analysis of the biological impacts of plasticizers on wildlife. Philosophical transactions of the Royal Society of London. Ser B Biol Sci 364:2047–2062

Ogata Y, Takada H, Mizukawa K, Hirai H, Iwasa S, Endo S … Thompson RC (2009) International pellet watch: global monitoring of persistent organic pollutants (POPs) in coastal waters. 1. Initial phase data on PCBs, DDTs, and HCHs. Mar Pollut Bull 58:1437–1446

Padsalgikar AD (2017) 4—Biological properties of plastics. In: Padsalgikar AD (ed) Plastics in medical devices for cardiovascular applications. William Andrew Publishing, Cambridge, MA, pp 83–102

Pegram JE, Andrady AL (1989) Outdoor weathering of selected polymeric materials under marine exposure conditions. Polym Degrad Stab 26:333–345

Peng X, Chen M, Chen S, Dasgupta S, Xu H, Ta K, Du M, Li J, Guo Z, Bai S (2018) Microplastics contaminate the deepest part of the world's ocean. Geochem Perspect Lett 9:1–5

Petrović E, Hamer L (2018) Improving the healthiness of sustainable construction: example of polyvinyl chloride (PVC). Buildings 8:28

Pierce K, Harris RJ, Larned LS, Pokras M (2004) Obstruction and starvation associated with plastic ingestion in a Northern Gannet *Morus bassanus* and a Greater Shearwater *Puffinus gravis*. Mar Ornithol 32:187–189

Rios LM, Moore C, Jones PR (2007) Persistent organic pollutants carried by synthetic polymers in the ocean environment. Mar Pollut Bull 54:1230–1237

Rios LM, Jones PR, Moore C, Narayan UV (2010) Quantitation of persistent organic pollutants adsorbed on plastic debris from the Northern Pacific Gyre's "eastern garbage patch". J Environ Monit 12:2226–2236

Ryan PG (1987) The incidence and characteristics of plastic particles ingested by seabirds. Mar Environ Res 23:175–206

Ryan PG, Connell AD, Gardner BD (1988) Plastic ingestion and PCBs in seabirds: Is there a relationship? Mar Pollut Bull 19:174–176

Santos RG, Andrades R, Demetrio GR, Kuwai GM, Sobral MF, Vieira JdS, Machovsky-Capuska GE (2020) Exploring plastic-induced satiety in foraging green turtles. Environ Pollut 265:

Santos-Echeandía J, Rivera-Hernández JR, Rodrigues JP, Moltó V (2020) Interaction of mercury with beached plastics with special attention to zonation, degradation status and polymer type. Mar Chem 222:

Skrip S, Das GK, Chatterjee SG (2013) Analytical expressions for the adsorbate breakthrough curve from a fixed bed of adsorbent with first-order and second-order kinetic models. Indian Chem Eng 55:87–103

Song YK, Hong SH, Jang M, Han GM, Jung SW, Shim WJ (2017) Combined effects of UV exposure duration and mechanical abrasion on microplastic fragmentation by polymer type. Environ Sci Technol 51:4368–4376

Speight JG (2018) Chapter 9—molecular interactions, partitioning, and thermodynamics. In: Speight JG (ed) Reaction mechanisms in environmental engineering. Butterworth-Heinemann, Cambridge, pp 307–336

Stapleton PA (2019) Toxicological considerations of nano-sized plastics. AIMS Environ Sci 6:367–378

Suaria G, Perold V, Lee JR, Lebouard F, Aliani S, Ryan PG (2020) Floating macro- and microplastics around the Southern Ocean: results from the Antarctic Circumnavigation Expedition. Environ Int 136:105494

Takada H (2006) Call for pellets! International Pellet watch global monitoring of POPs using beached plastic resin pellets. Mar Pollut Bull 52:1547–1548

Tanaka K, Takada H, Yamashita R, Mizukawa K, Fukuwaka M-a, Watanuki Y (2013) Accumulation of plastic-derived chemicals in tissues of seabirds ingesting marine plastics. Mar Pollut Bull 69:219–222

Tang C-C, Chen H-I, Brimblecombe P, Lee C-L (2019) Morphology and chemical properties of polypropylene pellets degraded in simulated terrestrial and marine environments. Mar Pollut Bull 149:

ter Halle A, Ladirat L, Gendre X (2016) Understanding the fragmentation pattern of marine plastic debris. Environ Sci Technol 50:5668–5675

Teuten EL, Saquing JM, Knappe DRU, Barlaz MA, Jonsson S, Björn A, … Takada H (2007) Potential for plastics to transport hydrophobic contaminants. Environ Sci Technol 41:7759–7764

Teuten EL et al (2009) Transport and release of chemicals from plastics to the environment and to wildlife. Philos Trans Royal Soc B Biol Sci 364:2027–2045

Thompson RC, Olsen Y, Mitchell RP (2004) Lost at sea: where is all the plastic? Science 304:838

Thompson RC, Moore CJ, vom Saal FS, Swan SH (2009) Plastics, the environment and human health: current consensus and future trends. Philos Trans Royal Soc B Biol Sci 364:2153–2166

Tschernij V, Larsson PO (2003) Ghost fishing by lost cod gill nets in the Baltic Sea. Fish Res 64:151–162

Turner A, Holmes L, Thompson RC, Fisher AS (2020) Metals and marine microplastics: adsorption from the environment versus addition during manufacture, exemplified with lead. Water Res 173:115577

Van Cauwenberghe L, Vanreusel A, Mees J, Janssen CR (2013) Microplastic pollution in deep-sea sediments. Environ Pollut 182:495–499

Van Franeker JA, Blaize C, Danielsen J, Fairclough K, Gollan J, Guse N … Turner M (2011) Monitoring plastic ingestion by the northern fulmar *Fulmarus glacialis* in the North Sea. Environ Pollut 159:2609–2615

van Sebille E, Wilcox C, Lebreton L (2015) A global inventory of small floating plastic debris. Environ Res Lett 10:124006 von Moos N, Burkhardt-Holm P, Köhler A (2012) Uptake and effects of microplastics on cells and tissue of the blue mussel *Mytilus edulis* L. after an experimental exposure. Environ Sci Technol 46:11327–11335

Waller CL, Griffiths HJ, Waluda CM, Thorpe SE, Loaiza I, Moreno B, Pacherres CO, Hughes KA (2017) Microplastics in the Antarctic marine system: an emerging area of research. Sci Total Environ 598:220–227

Wang W, Ge J, Yu X (2020) Bioavailability and toxicity of microplastics to fish species: a review. Ecotoxicol Environ Saf 189:109913

Weinstein JE, Crocker BK, Gray AD (2016) From macroplastic to microplastic: degradation of high-density polyethylene, polypropylene, and polystyrene in a salt marsh habitat. Environ Toxicol Chem 35:1632–1640

Welden NA, Cowie PR (2017) Degradation of common polymer ropes in a sublittoral marine environment. Mar Pollut Bull 118:248–253

Woodall LC, Sanchez-Vidal A, Canals M, Paterson GLJ, Coppock R, Sleight V, Calafat A, Rogers AD, Narayanaswamy BE, Thompson RC (2014) The deep sea is a major sink for microplastic debris. Royal Soc Open Sci 1:140317

Xu B, Liu F, Brookes PC, Xu J (2018) The sorption kinetics and isotherms of sulfamethoxazole with polyethylene microplastics. Mar Pollut Bull 131:191–196

Yang Y, Liu G, Song W, Ye C, Lin H, Li Z, Liu W (2019) Plastics in the marine environment are reservoirs for antibiotic and metal resistance genes. Environ Int 123:79–86

Ye S, Andrady AL (1991) Fouling of floating plastic debris under Biscayne Bay exposure conditions. Mar Pollut Bull 22:608–613

Yu H, Zhang Z, Zhang Y, Fan P, Xi B, Tan W (2021) Metal type and aggregate microenvironment govern the response sequence of speciation transformation of different heavy metals to microplastics in soil. Sci Total Environ 752:141254

Yuan W, Zhou Y, Chen Y, Liu X, Wang J (2020) Toxicological effects of microplastics and heavy metals on the *Daphnia magna*. Sci Total Environ 746:

Zardus JD, Nedved BT, Huang Y, Tran C, Hadfield MG (2008) Microbial biofilms facilitate adhesion in biofouling invertebrates. Biol Bull 214:91–98

Zhang W, Ma X, Zhang Z, Wang Y, Wang J, Wang J, Ma D (2015) Persistent organic pollutants carried on plastic resin pellets from two beaches in China. Mar Pollut Bull 99:28–34

Zhang Y, Lu J, Wu J, Wang J, Luo Y (2020) Potential risks of microplastics combined with superbugs: Enrichment of antibiotic resistant bacteria on the surface of microplastics in mariculture system. Ecotoxicol Environ Saf 187:109852

Chapter 18
Effects of Ocean Acidification on Marine Primary Producers and Related Ecological Processes Under Multiple Stressors

Peng Jin and Kunshan Gao

Abstract Although the individual effects of ocean acidification (OA), warming, solar UV radiation, deoxygenation and heavy metal pollution on marine producers are well-studied, their interactive effects are still unclear, strongly limiting our ability to project the ecological consequences of ocean climate changes. This chapter aims to provide an overview of our understanding the eco-physiological effects of OA and its combination with warming, solar UV radiation, deoxygenation and heavy metals. While OA is known to enhance photorespiration in both diatoms and green macroalgae, it enhances growth of coastal diatoms and other macroalgae that are adapted to fluctuating diel pH changes and then potentially enhances its contribution to carbon sequestration in coastal waters. OA is supposed to decrease pelagic primary productivity under multiple stressors (e.g., in combination with ultraviolet radiation, deoxygenation, warming), especially in oligotrophic waters, due to insufficient repairing or improvising processes that require both macro- and trace nutrients for syntheses of required proteins. Under influences of OA, macroalgal communities would shift toward non-calcifying species; diatoms become less abundance in phytoplankton assemblages. OA decreases calcification in algal calcifiers and exposes them to more harmful UV radiation, leading to a further decline of photosynthesis. Therefore, both the magnitude and direction of response of microalgae and macroalgae to OA largely depend on the levels of other environmental drivers (e.g., warming, deoxygenation). OA also exerts tremendous impacts on marine food webs. Total fatty acids and the ratio of long-chain polyunsaturated to saturated fatty acids of microalgae decrease, while some toxic secondary metabolites (such as phenolic compounds) accumulate under OA conditions, indicating a decline of food quality. This decline of food quality in primary

P. Jin
School of Environmental Science and Engineering, Guangzhou University, Guangzhou 510006, China

K. Gao (✉)
State Key Laboratory of Marine Environmental Science and College of Ocean and Earth Sciences, Xiamen University, Xiamen 361102, China
e-mail: ksgao@xmu.edu.cn

© The Author(s), under exclusive license to Springer Nature Switzerland AG 2021
D.-P. Häder et al. (eds.), *Anthropogenic Pollution of Aquatic Ecosystems*,
https://doi.org/10.1007/978-3-030-75602-4_18

producers can be transferred to secondary producers and negatively affect them (e.g., decrease in growth and reproduction). Taken together, OA can influence the biochemical compositions and contents in primary producers and their transfer to higher trophic levels and marine food webs is likely to be destabilized.

Keywords Ocean acidification · Deoxygenation · Ocean warming · UV radiation · Heavy metals · Primary producers

18.1 Introduction

As a consequence of anthropogenic activities, including combustion of fossil fuels, tropical deforestation and land use changes, the atmospheric CO_2 concentration has increased by 47% from 280 before the Industrial Revolution to 411 ppm in 2020. The rising of this greenhouse gas and others has led to global and ocean warming over the past century. Meanwhile, a large quantity of CO_2 dissolved in the oceans results in a continuous decline of pH and the alterations of marine carbonate chemistry, which is known as ocean acidification (OA) (Caldeira and Wickett 2003). Furthermore, the upper mixed layer (UML) becomes shallower due to the enhanced stratification caused by ocean warming, so the upward transport of nutrients from deeper waters into UML decreases. The shoaling UML has been supposed to expose organisms within this layer to high levels of solar radiation due to a shorter pathlength. Another consequence of ocean warming is that the oxygen solubility in seawater is decreasing along with increased heterotrophic O_2 consumption, resulting in global deoxygenation and anoxic zone expansion (Keeling 2010; Gilly et al. 2013; Schmidtko et al. 2017; Breitburg et al. 2018).

The pH of the surface ocean water has dropped by $\sim$0.1 (about 30% increase in H^+ ions) since pre-industrial times, a rate tenfold higher than that in the past 300 million years (Hönisch et al. 2012). It is projected that, by the end of this century, proton concentrations in the surface ocean will be more than twice that of the pre-industrial period, leading to a drop in pH by 0.4 (H^+ concentrations increase by 150%) (IPCC 2014; Gattuso et al. 2015). Carbonate chemistry changes associated with OA include increases in concentrations of CO_2 and bicarbonate and decreases in levels of carbonate ions and pH. It is worth noting that the pH of coastal waters will drop by 0.45, being about 12% faster than open oceans (Cai et al. 2011). These changes occur on top of the strong spatiotemporal variability, which is especially true for coastal regions due to multiple drivers, such as upwelling (Capone and Hutchins 2013), tidal cycles (Dai et al. 2009; Jiang et al. 2011; Wang et al. 2014), anthropogenic nutrient inputs, aquaculture activities and changes in ecosystem structure and biological metabolism (Duarte et al. 2013; Waldbusser and Salisbury 2014). These changes are endangering marine organisms, ecosystems and associated services (IGBP et al. 2013).

The rising atmospheric CO_2 is tightly coupled with the global average temperature increase (IPCC 2014). Since the ocean plays a major role as the main (93%)

reservoir to trap additional heat associated with global warming, average global sea surface temperature (SST) increased with a rate of ~0.13 °C per decade since the beginning of the twentieth century (IPCC 2014). The temperatures of the last three decades have been warmer than any period since first instrumental records were obtained on a routine basis around 1880. There are, however, pronounced regional differences in SST patterns with some areas with negative anomalies, including the Southern Ocean and the South Pacific (Grist et al. 2016). Despite these variabilities, it is projected that the SST is likely to continue to rise in the twenty-first century with mean global sea surface warming varying between 1 °C and more than 3 °C depending on the emission scenarios considered, with the greatest warming in the tropics and subtropics (Collins et al. 2013).

Increased use of chlorofluorocarbons (CFCs) as refrigerants and propellants led to an alarming depletion of stratospheric ozone in the 1980s (Molina and Rowland 1974) and to a corresponding increase in ultraviolet B irradiance (UV-B) reaching the Earth's surface (Madronich et al. 1998). Although the ozone depletion has been curbed by implementation of the Montreal Protocol, the ozone layer will unlikely recover to 1980 levels within the coming decades (Weatherhead and Andersen 2006). Because of other factors, such as rising temperature, emissions of trace gases (Osso et al. 2011) and changes in the atmospheric circulation pattern, may lead to the increase of the UV-B level at low latitudes (Williamson et al. 2014). In addition to the increased UV radiation (UVR) on the Earth's surface, marine organisms in the UML are exposed to higher levels of UVR due to the intensified stratification (shortened mixing paths) in a warming ocean. The responses of marine organisms to UVR depend on both the dose and intensity of harmful radiation to which an individual organism is exposed. Daytime UVR dose differs by latitude, altitude and the optical characteristics of the water body. For example, in the oligotrophic ocean, characterized by deep penetration of UVR (Tedetti et al. 2007), UVR can penetrate as deep as 80 m. However, in the coastal waters, where the concentrations of chromophoric dissolved organic matter (CDOM) are high, much of the UVR is absorbed, leading to low levels of UVR in the water column. In the coastal waters of the South China Sea, UV-A and UV-B can only reach 3–3.5 and 1.7–2.3 m, respectively, due to the high concentrations of suspended particles (Gao et al. 2007; Li and Gao 2012). Irrespective of pelagic and coastal waters that give rise to contrasting differences in UVR penetration, impacts of UVR on marine organisms have been extensively documented in some comprehensive reviews and meta-analysis studies (Häder et al. 2015; Jin et al. 2017; Gao et al. 2019 and references therein).

Ocean warming and intensified stratification (due to reduced specific gravity of seawater with increased temperature) of the upper ocean have been considered responsible for decreasing dissolved oxygen concentration in seawater (ocean deoxygenation) (Keeling 2010). The global oceanic oxygen content has decreased by more than two percent over the past 60 years, with large variations in oxygen loss in different ocean basins and at different depths (Schmidtko et al. 2017). At the same time, algal blooms increased in both scale and frequency and near-shore eutrophication fueled by runoffs have exacerbated deoxygenation and expanded the oxygen-depleted zones, covering more than 245,000 km^2 over the past half-century (Diaz and Rosenberg 2008; Breitburg et al. 2018). Modeling studies predicted that the global ocean oxygen inventory will decline by 1–7% over the next century, with declines continuing for a thousand years or more into the future (Bopp et al. 2002; Schmittner et al. 2008; Oschlies et al. 2008; Frolicher et al. 2009). It has been estimated that warming by 1 °C can reduce O_2 in the oceans by 2%, which is believed to have a wide range of biological and ecological consequences (Breitburg et al. 2018 and references therein).

Heavy metals, including cadmium (Cd) and copper (Cu), from natural and anthropogenic sources, are the most common types of coastal contaminants. They enter aquatic systems via various ways, including agricultural and stormwater runoff, industrial effluents, sewage treatment discharge and fossil fuel combustion (Davis et al. 2001; Flint and Davis 2007; Pan and Wang 2012). Unlike other pollutants which can be degraded to less harmful components by chemical or biological processes, these heavy metals are considered as nondegradable pollutants. They can accumulate to high levels in sediments, as a sink, or be released from sediments, acting as a source back to overlying water via natural or anthropogenic disturbances (Pan and Wang 2012). It has been widely reported that heavy metals are major environmental stressors for marine organisms (Chakraborty et al. 2010; Patra et al. 2019). More importantly, toxic heavy metals can be taken up by marine organisms, entering the food chain and being potentially transferred to higher trophic levels and then affecting human health (Wang 2002). In the context of ocean acidification, the decrease in concentration of OH^- and CO_3^{2-} ions can affect the solubility, adsorption, toxicity and rates of redox processes of metals in seawater (Millero et al. 2009; Stockdale et al. 2016). The metals that form strong complexes with carbonate ions are supposed to be most strongly affected by a decrease in pH, resulting in an increase in their free ionic forms in the future with progressive OA (Millero et al. 2009). How OA and heavy metals would interactively affect various marine organisms has gained increasing research interests over the last decade (e.g., Moreira et al. 2016; Nardi et al. 2018; Cao et al. 2019; Dong et al. 2019; Zhang et al. 2020).

The wide range of biological and ecological effects of OA has been extensively examined. However, most studies have been carried out in laboratory-based perturbation experiments focusing on single species exposed to different pH scenarios. Few studies have been focusing on the impacts at community or ecosystems levels involving more than one environmental driver (Riebesell and Gattuso 2015; Hurd et al. 2018; Doney et al. 2020). It has been demonstrated that OA can interact with other environmental drivers to affect the responses of marine organisms directly or indirectly; however, it is challenging to perform such experiments under influences of multiple stressors as there are some bottlenecks in technologies and methods (Boyd et al. 2018). This chapter summarizes the effects of OA alone and in combination with temperature, oxygen, UV and heavy metals on marine primary producers, as well as their combined potential impacts on marine food webs and corresponding ecological and biogeochemical processes.

18.2 Individual Effects of OA

Marine photosynthetic organisms contribute nearly half of the global primary production, playing an important role in global biogeochemical cycles (Field et al. 1998) and carbon remediation as blue carbon (Ortega et al. 2019; Mcleod et al. 2011). Diatoms, which are a relevant group contributing largely to oceans primary production, can be found from the euphotic layer to the deep sea (2000–4000 m depth) (Agustí et al. 2015). They dominate phytoplankton in abundance in cold waters and act as the main driver of the marine biological carbon pump (Smetacek 1985). Their silicification significantly diminishes under OA conditions both in laboratory mono-species cultures and natural communities (Milligan et al. 2004; Petrou et al. 2019). The effect of OA on silicification differs among taxa, with several species having significantly reduced silica incorporation at elevated CO_2 levels projected for the end of this century (Petrou et al. 2019). Nevertheless, the biogenic silicon content (BSi) may also increase per cell under combined conditions of OA and stratification (Xu et al. 2014). Furthermore, the reduction of BSi incorporation in diatoms can affect the predation rate of diatoms by zooplankton as well as the number of fecal pellets (Liu et al. 2016a, b). Taken together, all of these findings suggest that OA may influence the silicon cycle by both altering the composition of the diatom assemblages and reducing cell ballasting, which probably alter vertical flux of these elements to the deep ocean. Besides silicon, carbon is also sensitive to OA. In the oligotrophic South China Sea, OA can decrease photosynthetic carbon fixation, by reducing the abundance of diatoms (Gao et al. 2012). Under low sunlight levels, OA can promote the growth of three diatoms, while it inhibits their growth under high sunlight conditions (Gao et al. 2012). Mitochondrial respiration indeed increased under OA condition by ~34% in *Phaeodactylum tricornutum* (Wu et al. 2010) and by 35% in *Thalassiosira pseudonana* (Yang and Gao 2012). Increased acidity of seawater under OA conditions could disturb cell surface (Flynn et al. 2012) or even intracellular pH stability, so that microalgae may need to allocate extra energy to transport ions against the acid–base perturbation. The net effects of OA on marine primary producers may depend on the balance between those positive and negative effects, which can be modulated by other environmental drivers (Fig. 18.1). When quantifying the responses of diatoms to OA, their adaptation history should also be considered. For instance, coastal and pelagic diatom species of *Thalassiosira* showed contrasting responses to OA, with the growth of the coastal *Thalassiosira weissflogii* being more resilient than the pelagic *Turbonilla oceanica*, the latter showing reduced net photosynthesis (Li et al. 2016). These results highlight the importance of taking into account the natural environmental variabilities when examining the effects of OA (Vargas et al. 2017). Regarding to biochemical contents and compositions, OA has been shown to increase the content of toxic phenolic compounds in the coccolithophorid *Emiliania huxleyi*, diatoms, and natural phytoplankton assemblages (Jin et al. 2015). A common response to OA is an increase in cellular carbon-to-nitrogen ratios (see Riebesell and Tortell 2012 for a detailed review), indicating a decrease in

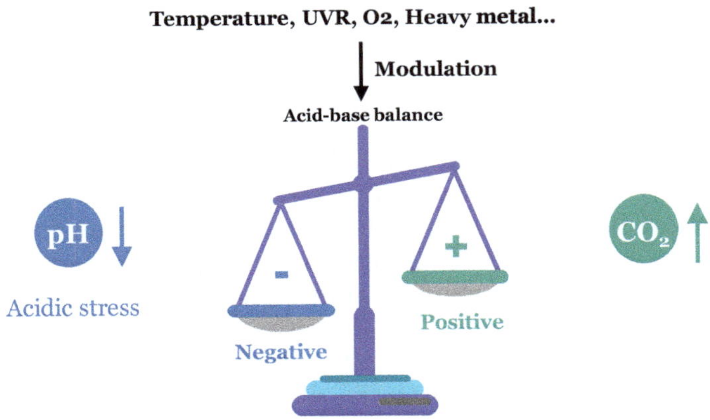

Fig. 18.1 Conceptual diagram showing the positive and negative effects of OA on marine primary producers. The net effects of OA on marine primary producers may depend on the balance between those positive and negative effects, which can be modulated by other environmental drivers, such as temperature, UV radiation, O_2 and heavy metals. For details see text

food quality. A diverse set of responses is also seen in diatoms, which in some cases show opposite trends even in closely related species (Burkhardt et al. 1999). By synthesizing sediment-trap data from in situ mesocosm studies in different oceans, highly variable impacts of OA on carbon-to-nitrogen ratios, reaching up to a 20% increase or decrease (Taucher et al. 2021). These changes are driven by pCO_2 effects on a variety of plankton taxa and corresponding shifts in food web structures (Taucher et al. 2021).

Due to the short generation span, high population densities and standing genetic variations of phytoplankton, they have a high potential to promote swift evolutionary responses to OA (Reusch and Boyd 2013; Collins et al. 2020). Indeed, there is growing evidence of adaptation by phytoplankton through evolutionary responses to OA (Lohbeck et al. 2012; Jin et al. 2013a, b; Hutchins et al. 2015; Li et al. 2017). For example, the calcification of the coccolithophorid *E. huxleyi* was partly restored after a long-term adaptation to OA condition for ~500 generations (Lohbeck et al. 2012). In another coccolithophorid, *Gephyrocapsa oceanica* (non-calcifying strain), the long-term (~670 generations) OA selected cells showed increases in growth, photosynthetic carbon fixation, cellular particulate organic carbon (POC) or nitrogen (PON) contents (Jin et al. 2013a, b). However, the long-term OA selection (1000 generations) reduced the growth and calcification of the calcifying strain of *G. oceanica* (Tong et al. 2018). For diatoms, photosynthesis, respiration, and growth of *P. tricornutum* decreased after growing under OA conditions for 1800 generations (Li et al. 2017). In the cyanobacterium *Trichodesmium*, nitrogen fixation and growth rates significantly increased in the cells under OA selection for 4.5 years (Hutchins et al. 2015), which is in contrast with results from a short-term study (Hong et al. 2017). While long-term evolutionary experiments can logistically be difficult to manipulate multiple drivers

(Brennan and Collins 2015), it is obvious that physiological and molecular changes during long-term adaptation often disagree with those obtained from short-term (less than 10–30 generations) incubations. In the context of OA, which is supposed to proceed for centuries, long-term adaption studies are indispensable; however, responses of short-term exposures to pH drop and CO_2 rise are also ecologically relevant in view of the dynamic physical conditions that drive changes in carbonate chemistry.

In terms of macroalgae, it has been extensively reported that OA can stimulate their photosynthesis, photochemical efficiency and growth in various species (e.g., *Pyropia yezoensis*: Gao et al. 1991; *Ulva* spp.: Huan et al. 2016; *Gracilaria* spp.: Gao et al. 1993; *Lomentaria articulate*: Kübler et al. 1999; *Halimeda opuntia, Udotea luna, Jania adhaerens, Neogoniolithon strictum, Sargassum fluitans, Canistrocarpus cervicornis, Laurencia intricate*: Zweng et al. 2018; *Plocamium angustum*: Cornwall and Hurd 2019). However, for most of the macroalgae species without CO_2 concentration mechanisms (CCMs), their growth or photosynthesis did not benefit from OA (Cornwall et al. 2012; Kram et al. 2016; Britton et al. 2019; Loos et al. 2019; Cornwall and Hurd 2019). In addition, lowered pH under OA may reduce growth due to the acidic stress during the night period; thus, the overall effects of OA on macroalgae depend on the magnitude of the positive effect of elevated CO_2 at daytime and the negative effect of declined pH during night time. It is worth noting that those positive and negative effects would also be altered by light levels, nutrient availability, and/or species-specific physiology. Nevertheless, macroalgae experiences diel and tidal cycle-induced pH variations, they are supposed to possess the capability to tolerate certain levels of pH changes, though their microscopic stages of their life cycles may be more sensitive to OA (see review by Ji and Gao 2020 and references therein). As documented in a comprehensive meta-analysis, calcifying microorganisms seem to be more sensitive to OA (Nagelkerken and Connell 2015); this is also true for calcifying macroalgae. A common response of calcifying macroalgae to OA is a reduction of calcification rate, as the lower pH and saturation rate of calcium carbonate either lead to dissolution of $CaCO_3$ or make it difficult for the algae to calcify, since they require energy to cope with the acidic stress (Gao et al. 1993; Hall-Spencer et al. 2008; Gao and Zheng 2010; Hofmann et al. 2013; McCoy et al. 2020). Additionally, more energy is required for the carbonate ion efflux to support calcification (Raven and Crawfurd 2012). Thereby, the photosynthetic efficiency might be inhibited by reduced calcification (Martin and Gattuso 2009; Hofmann et al. 2012; Cornwall et al. 2013). There were also some exceptions that OA did not decrease calcification, even increased it in some cases (Ries et al. 2009; Price et al. 2011; Scherner et al. 2016). The possible explanations are that some coralline algae and calcareous green algae can develop a protective external layer or create a micro-environment favorable for calcification (Zweng et al. 2018; Scherner et al. 2016). At a community level, the macroalgae community structure may change under OA, as different taxa or species showed different responses (Johnson et al. 2014; Celis-Plá et al. 2017). It is expected that the macroalgal community would shift toward non-calcifying species, and this was confirmed by some of the previous studies

(Connell and Russell 2010). The decreasing abundance of calcifying macroalgae was also found near natural CO_2 seeps, where Hall-Spencer et al. (2008) found a significant decline of coralline algal abundance along the pH gradients. On the other hand, large macroalgae may be more tolerant to OA since the diffusion boundary layer at their thallus surface plays a buffering role (Hurd 2015), and increased water current speed can reduce the buffering diffusion boundary layer, which can stimulate photosynthesis (Gao 1992) and hypothetically make the macroalgae more sensitive to OA.

18.3 Combined Effects of OA with Warming

OA and warming are concurrent major global environmental problems caused by human CO_2 emissions. Quantifying how they interactively influence primary producers is crucial to understand the marine service and functions in the context of climate change. Different species showed diverse responses to OA and warming in laboratory studies. For instance, OA and warming enhanced the growth of the picophytoplankton *Synechococcus*, but did not show any effect on another picophytoplankton *Prochlorococcus* (Fu et al. 2007). In diatoms, OA and a 4 °C increase in temperature enhanced the growth of *Skeletonema marinoi* (Kremp et al. 2012), but did not affect the growth of *Thalassiosira* or *Chaetoceros* (Hyun et al. 2014). In spite of these results, a handful studies have examined the long-term evolutionary responses of phytoplankton to OA (see above) or warming (Padfield et al. 2016; Chakravarti et al. 2017; Schaum et al. 2017, 2018; O'Donnell et al. 2018, 2019; Jin and Agustí 2018; Jin et al. 2020) alone, studies investigating the long-term evolutionary responses of phytoplankton to the combination of OA and warming are very limited so far. In a long-term experimental study ($\sim$400 days) on a coccolithophore *E. huxleyi* by Schlüter et al. (2014), particulate inorganic (PIC) and organic (POC) carbon contents were restored to values under present-day ocean conditions, owing to adaptive evolution and were higher by up to 100% and 50% under combined OA and warming treatment, respectively, than in non-adapted controls. Both OA and warming can promote the growth of *E. huxleyi*, but the magnitude of the growth adaptation due to warming in that experiment was roughly three times larger than the adaptation due to ocean acidification (Schlüter et al. 2014). It has also been found that the maximum persistence temperature (T_{max}) of the coccolithophore *E. huxleyi* significantly decreased after long-term selection under ocean acidification conditions (Listmann et al. 2016), reflecting a narrowed thermal window under influence of OA. In addition, such an effect alters under increased growth temperature levels, where the changed thermal window was only observed in a high-temperature selection line (Listmann et al. 2016). In addition to those single species studies, a quantitative meta-analysis focusing on multispecies at community level showed that primary productivity by temperate non-calcifying plankton increases with OA and warming, whereas tropical plankton decreases productivity because of acidification (Nagelkerken and Connell 2015). Species diversity and abundances in tropical waters as well as temperate species decline under the influence of OA, with shifts favoring novel community compositions dominated by microorganisms and non-calcifiers (Nagelkerken and Connell 2015).

For macroalgae, OA and warming act synergistically to promote the growth of the red alga *Plocamium cartilagineum* (Kram et al. 2016) and increase the respiration rate of *Sargassum fusiforme* (Zou et al. 2011). The reproduction of the green tide alga *Ulva rigida* is stimulated by warming with a further increase when combined with OA (Gao et al. 2017a, b). However, these two drivers have been shown to decrease the biomass production in the brown algae *Sargassum muticum* and *Cystoseira tamariscifolia* (Olabarria et al. 2013) and inhibit early stage development of the giant kelp *Macrocystis pyrifera*, lowering germination rates and increasing mortality of the kelp's spores (Gaitán-Espitia et al. 2014). On the other hand, no interactions of OA and warming were detected during meiospore development in the brown macroalgae *Macrocystis*, *Pyrifera* and *Undaria pinnatifida* (Leal et al. 2017). For calcifying macroalgae, the combination of OA and warming inhibited the growth and calcification of coral reef macroalgae (Sinutok et al. 2011; Martin and Hall-Spencer 2017; Graba-Landry et al. 2018). On the contrary, warming can act antagonistically to impair the negative effects of OA on a crustose coralline alga by promoting its growth and photochemical efficiency (Kim et al. 2020).

Results from the combined effects of OA and warming on single species of microalgae and macroalgae strongly suggest that OA and warming could increase the potential for an overall simplification of ecosystem structure and function, with reduced energy flow along trophic levels. For instance, a significant direct negative effect of OA on dimethylsulfide (DMS) production has been reported about temperate phytoplankton communities (Nagelkerken and Connell 2015). DMS is a driver of the food web structure, acting as an anti-grazing defensive compound as well as providing a chemical cue to attract predators to prey phytoplankton forage (Hay and Kubanek, 2002; Wolfe et al. 1997; Nevitt et al. 1995). Changes in oceanic DMS release under OA conditions can, therefore, alter the trophic interactions in the ocean (Pohnert et al. 2007) and the global temperature changes because it contributes to cloud formation as a climate gas (Six et al. 2013; Vallina and Simó 2007). Besides DMS, the interactions of OA and warming on the composition of dissolved organic carbon (DOC) and particulate organic carbon (POC) would also have the potential to affect marine ecological processes. While no effect of OA on the composition of DOC was observed (Zark et al. 2015), it can enhance the production of POC (Czerny et al. 2013), exhibiting a positive effect on the stock of blue carbon. How OA and warming will interactively influence the production of POC and composition of DOC has yet to be determined. Since some DOC resists bacterial lysis and survives for hundreds or even thousands of years, playing a vital role in the carbon sink (Jiao et al. 2010), it is particularly important to elucidate the combined effects of OA and warming on the carbon sink/source processes in marine organisms and/or ecosystems.

18.4 Combined Effects of OA with UV Radiation

UV radiation affects the photochemical and photo-biological processes in seawater (Bais et al. 2018). Despite its critical importance, most laboratory and deck culture experiments do not include ultraviolet radiation (UVR, 280–400 nm), limiting our ability to extrapolate the reported effects of OA from laboratory studies to natural conditions (Gao et al. 2012). A comprehensive meta-analysis has quantified the responses of aquatic photosynthetic organisms to UV-B radiation, showing that the overwhelming majority of the experiments demonstrated that increased UV-B exposures decreased the organisms' performance (Jin et al. 2017). Only ~3% of the experiments showed improved performance with increased UV-B and just ~6% of the experiments showed reduced performance with reduced UV-B (Jin et al. 2017). On the other hand, low and moderate levels of UV-A promote photosynthetic carbon fixation of phytoplankton (Gao et al. 2007) and macroalgae (Xu and Gao 2010). UVR can affect the photosynthetic organisms at various levels, with the mortality rate being the most sensitive response to elevated UV-B, followed by changes in cellular and molecular traits, with small-celled microalgae being more sensitive than large-celled microalgae (Wu et al. 2015; Jin et al. 2017). Since OA has tremendous impacts on primary producers, it is not surprising that OA would interact with UVR. For example, the combination of OA and UVR acts synergistically to reduce the calcification of algae (Gao et al. 2009; Gao and Zheng 2010; Xu and Gao 2015). Similarly, the diatom *T. pseudonana* is more sensitive to UVR when grown under OA conditions (Sobrino et al. 2008). OA also exacerbates the harmful effects of UVR on PSII function in the diatom *T. weissflogii* through reducing the PsbD removal rate and the ratio of RbcL to PsbA (Gao et al. 2018). On the contrary, it was also found that the coccolithophorid *G. oceanica* grown under OA conditions exhibited less UV-B-induced inhibition, implying an antagonistic effect of OA and UV-B (Jin et al. 2013b). Regarding macroalgae, the responses of the green algae *Ulva linza* to the combination of OA and UVR seem to be life-stage specific (Ma et al. 2019). Exposures to UVR increased PSII activity in juvenile thalli, but the addition of OA did not impose any influence, while a combination of OA and UVR further decreased the activity of PSII of adult thalli compared to UVR alone (Ma et al. 2019).

18.5 Combined Effects of OA with Deoxygenation

Marine organisms need O_2 in metabolic processes, and decreased O_2 availability affects their physiology. A prolonged and extreme hypoxia stress can even lead to the death of these organisms. There are large variabilities of the half lethal concentration (LC50) of dissolved oxygen among different species. In typical hypoxic zones, the oxygen content is below 2 mg L^{-1}, and hypoxia is usually coupled with high pCO_2 and low pH. It was reported that hypoxic areas are increasing at a rate of ~5.5% per year in coastal waters due to the interaction between eutrophication and warming-induced deoxygenation (Vaquer-Sunyer and Duarte 2008). The eutrophication could increase the susceptibility of coastal waters to ocean acidification with an additional drop in pH of 0.05 units as predicted for future OA at the end of this century (Cai et al. 2011). In addition, the decrease in oxygen concentration could promote denitrification, reducing the concentration of nitrate ions and influencing the ocean *N* cycle, primary productivity and marine biological carbon pump efficiency (Hutchins et al. 2019).

Lower oxygen levels can be expected to have effects on photoautotrophic organisms, since the ratio of oxygenase to carboxylase activity of the enzyme ribulose-1,5-bisphosphate carboxylase/oxygenase (Rubisco) depends on the selectivity of CO_2 over O_2. Declined oxygen concentrations may favor the carboxylase activity of Rubisco and inhibit the photorespiration in algae (Gao and Campbell 2014). However, little effect of low oxygenation on physiology and growth of algae have been reported, although levels above air equilibrium can be harmful (Black et al. 1976; Kitaya et al. 2008; Raso et al. 2012; Haas et al. 2014). In the seagrass, *Zostera marina* photosynthesis was stimulated by declining oxygen concentrations from 231 to 8 μmol L^{-1} at sub-saturating but not at saturating light levels (Kim et al. 2018). At the same time, the respiration rates dropped fivefold at the lower oxygen level. The possible mechanism for the lack or disappearance of low oxygen effect could involve the presence or active levels of CCMs in most algae, which maintain a sufficiently high CO_2:O_2 ratio at the active site of Rubisco to minimize oxygenase activity and photorespiration (Giordano et al. 2005). However, there have been no studies on the interaction between deoxygenation and OA to date. It appears that net photosynthesis of diatoms and *Ulva* spp. decreased with increased respiration index [$\log_{10}(pO_2/pCO_2)$] (Gao and Campbell 2014), implying a trend of photosynthetic enhancement under lowered O_2 levels. In the study of Kim et al. (2018), low oxygen caused a ~threefold down-regulation of γ-carbonic anhydrase (γ-CA) genes, which was highly unexpected. A recent study by Sun et al. (personal communication) showed that lowered O_2 availability enhanced CCMs activity in a diatom and primary production of a coastal phytoplankton assemblage, though to a smaller extent under elevated CO_2 concentrations, reflecting an opposing effect of OA and deoxygenation.

The ratio of pO_2 and pCO_2 in seawater is decreasing with time as a consequence of global changes, and this decrease is amplified in deep waters (Brewer and Peltzer 2009). In the upwelling systems, the physical mixing process causes the conveyance of deep seawater to the surface layer also affecting this ratio in the upper layer. The change of the ratio of pO_2 and pCO_2 in seawater would have tremendous ecological impacts. The declining of this ratio in the upwelling system along the California coast leads to a huge economic loss in local shellfish farming as well as a decrease in the available food source for zooplankton (Doney et al. 2012). However, as documented above, the combined effects of OA and deoxygenation on primary producers and their ecological consequences have been little examined; more studies toward this aspect are needed.

18.6 Combined Effects of OA with Heavy Metals

There are growing research interests in examining the combined effects of OA with heavy metals on marine organisms. However, most of the studies focused on animals (Götze et al. 2014; Moreira et al. 2016; Nardi et al. 2018; Cao et al. 2018, 2019); few studies have been reported on primary producers. OA can enhance the tolerance of the marine diatom *P. tricornutum* to heavy metal Cd, thus benefit their survival under Cd exposure, suggesting an antagonistic interaction between OA and heavy metals (Dong et al. 2020). This is probably due to the enhanced resistance to Cd under OA conditions (Liu et al. 2016a, b). Similarly, OA appeared to alleviate the impact of Cd toxicity on *P. tricornutum* not only in indoor experiments but also in outdoor mesocosm experiments (Zhang et al. 2020). A transcriptomics approach was applied to elucidate the underlying mechanisms, suggesting that the genes involved in Cd efflux and phytochelatin production were up-regulated and genes involved in Cd influx were down-regulated in the cells under OA conditions (Zhang et al. 2020). OA stimulated the antioxidant enzyme activities of SOD, CAT and APX, which were previously shown to alleviate oxidative stress induced by Cd exposures in diatoms (Mu et al. 2018; Dong et al. 2020).

In terms of macroalgae, elevated Zn concentrations up to 100 $\mu g\ L^{-1}$ result in a decrease in growth, photosynthesis and respiration of the red alga *Pyropia yezoensis*, OA alleviates the damages caused by Zn exposures, reflecting an antagonistic interaction between OA and Zn as well (Ma et al. 2019). In the green tide alga *Ulva prolifera*, the inhibition of Cu on the relative growth rate and net photosynthetic rate was reduced at 1000 μatm pCO_2 but was magnified at 1400 μatm pCO_2, indicating that a modest increase of pCO_2 can alleviate the toxicity of Cu to *U. prolifera* while a further increase of OA may exacerbate it (Gao et al. 2017a, b). This study highlights the fact that the interaction levels between OA and heavy metals might be OA scenario-dependent. For seagrasses, de los Santos et al. (2019) found that the toxicity and bioconcentration of Cu in *Zostera noltei* were not affected by OA, though complex OA and Cu interactions were observed in its photosynthetic responses.

The combined effects of OA and heavy metals on primary producers may have large impacts on ecological processes. For instance, a significant reduction of Cd transfer from diatoms to the scallop *Argopecten irradians* was observed under OA conditions (Zhang et al. 2020), which could have broader impacts on the biogeochemical cycle of Cd in the marine ecosystem. No significant effect of OA on the dietary transfer of Cu from seagrass to amphipods was observed (de los Santos et al. 2019) though. While it seems that OA can alleviate the impact of heavy metals to algae, the interactions and involved mechanisms need to be further investigated.

18.7 OA Effects Under Multiple Drivers

As reviewed above, along with the ongoing OA, ocean warming, deoxygenation, shoaling of UML and subsequently increasing UV exposure to cells within this layer and heavy metal pollutions are supposed to compound the effects of OA (Boyd et al. 2018; Gao et al. 2020; Ji and Gao 2020). Understanding the combined effect of these environmental changes is one of the most important research priorities (Riebesell and Gattuso 2015; Boyd et al. 2018).

Although studies on single drivers still dominate the relevant literature, an increasing number of studies consider the effects of OA in combination with one or more drivers. Jin et al (2019) investigated the combined effects of OA and warming, light intensity, light fluctuating frequency and UV on the coccolithophorid *G. oceanica* and found that the combination of OA and warming acts synergistically with decreasing fluctuating light frequency increases its photochemical efficiency and photosynthetic carbon fixation rate. In another coccolithophorid species, *E. huxleyi*, while OA and warming reduced the ratio of calcification to photosynthesis, UVR increased this ratio, particularly for cells under OA conditions, indicating an antagonistic interaction between UVR and other environmental drivers (Tong et al. 2019). Regarding the combined effects of OA, warming and nutrient, Li et al. (2018) found that neither OA nor warming affected the growth of *T. pseudonana* under nitrate replete conditions, but they both inhibited the growth under nitrate limited conditions. The study by Brennan and Collins (2015) looking at the growth responses of the green alga *Chlamydomonas reinhardtii* to eight environmental drivers (e.g., CO_2, temperature, pH, light) showed that the growth decreased in a predictable way with an increasing number of environmental drivers. In terms of the long-term evolutionary responses, even in multi-driver environments, only a few dominant drivers explain most of the evolutionary changes in growth (Brennan et al. 2017). Many of the functional traits (cell size, chlorophyll content) often revert to ancestral values during adaptation in multi-driver environments (Brennan et al. 2017).

18.8 OA Effects on Marine Food Webs

As been discussed previously in this chapter, OA and its interactions with other marine environmental changes have shown a wide range of effects on primary producers. However, the studies examining the effects of OA on marine food webs are still rare.

A decline of food quality in primary producers under OA conditions has been reported in many previous studies (e.g., Rossoll et al. 2012; Bermúdez et al. 2015; Jin et al. 2015). Since approximately 10–20% of essential biomolecules in primary producers are incorporated into new biomass at the next trophic level, it is expected that the decline in food quality in primary producers under OA conditions will have the potential to affect the zooplankton which prey on them. This hypothesis was tested by Rossoll et al. (2012), in which the declines in both total fatty acids and the ratio of long-chain polyunsaturated to saturated fatty acids constrained growth and reproduction of the copepod *Acartia tonsa* feeding on the diatom *T. pseudonana* grown under OA conditions. Similarly, a decreased proportion of polyunsaturated fatty acids (PUFA) in the natural phytoplankton assemblage was mirrored by a reduction in the relative PUFA content of the dominant copepod *Calanus finmarchicus* (Bermúdez et al. 2016). On the other hand, OA has sometimes been shown to increase food availability and thus helps alleviate the negative effects of acidification on invertebrates. In contrast, no significant changes in fatty acids were found in phytoplankton, and fatty acid profiles of mesozooplankton were not altered under elevated CO_2 in a mesocosm study in eutrophicated coastal water (Wang et al. 2019).

In addition to lipids and fatty acids, the changes of other secondary metabolites in primary producers can also be transferred to consumers. Phytoplankton species (diatoms and coccolithophores) increased their phenolic compounds (toxic) when grown under OA conditions, and the accumulation of phenolics in phytoplankton led to higher phenolics contents in copepods that fed with the phytoplankton (Jin et al. 2015) (Fig. 18.2). This suggests a food chain effect of OA from primary producers to consumers. Food chain effects of OA were also observed in the brown alga *Saccharina japonica*, in which OA increased the accumulation of iodide and the amounts of iodine released into seawater, while abalone fed on the kelp exhibited increased levels of iodine (Xu et al. 2019) (Fig. 18.2).

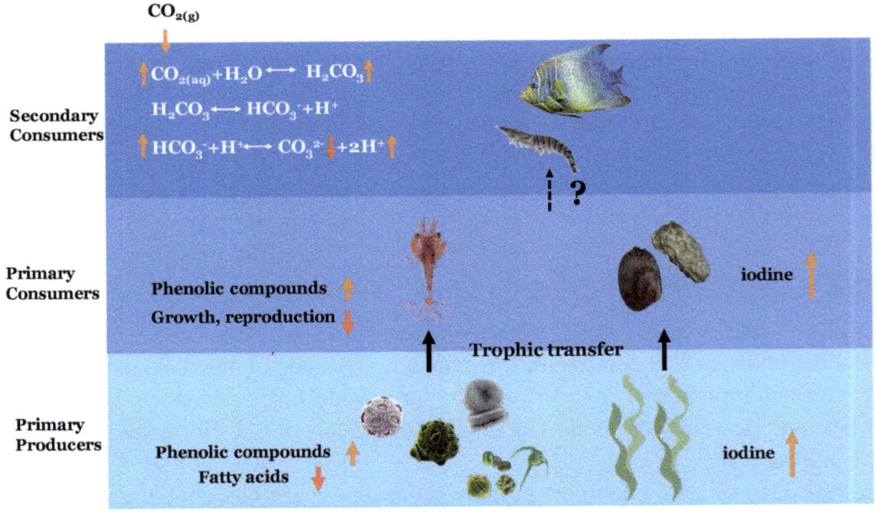

Fig. 18.2 Conceptual diagram showing the food chain effects of ocean acidification (OA). OA increased the contents of phenolic compounds and iodine in the primary producers (phytoplankton and kelp), and these changes lead to accumulation in the secondary producers that feed on them. Whether these food chain effects can scale up in multiple trophic levels (e.g., primary to secondary consumers) needs further investigation. The orange arrows indicate increased, and red ones indicate decreased levels of the compounds, growth or reproduction

In an oligotrophic plankton community in the subtropical North Atlantic Ocean, the toxic microalga *Vicicitus globosus* increased its abundance under OA conditions, strongly inhibiting the development of the zooplankton community, thereby disrupting the trophic transfer of organic matter from primary producers (Riebesell et al. 2018). This food chain alteration prolonged the residence of particulate matter in the water column and caused a strong decline in export flux. In another mesocosm study carried out in a Swedish fjord, enhanced primary production under OA increased the survival rate of Atlantic herring (*Clupea harengus*) larvae (Sswat et al. 2018).

In a mesocosm experiment mimicking Australian marine benthic systems, including all of the major groups of organisms that might be expected: cyanobacteria, microalgae, macrophytes, copepods, predatory invertebrates, ascidians, sponges, detritivores, molluscs, macro-crustaceans and fish, Nagelkerken et al. (2020) explored how would the marine food web respond to a future combination of OA and warming. Although limited impacts of OA were observed, the primary producer and secondary consumer biomass and productivity increased, while the biomass and productivity significantly decreased among primary consumers, under a combination of OA and warming conditions, suggesting the biomass and productivity were greatly reorganized among trophic levels (Nagelkerken et al. 2020). They also pointed out that such trophic imbalance is unlikely to be stable in the long term. The worse is that this trophic imbalance represents a transitory state, and it is likely to collapse the system in such a way that primary producers dominate and secondary consumers (e.g., fish) are largely lost. Since secondary marine consumers, such as fish, are an important nutritional source for human beings, the

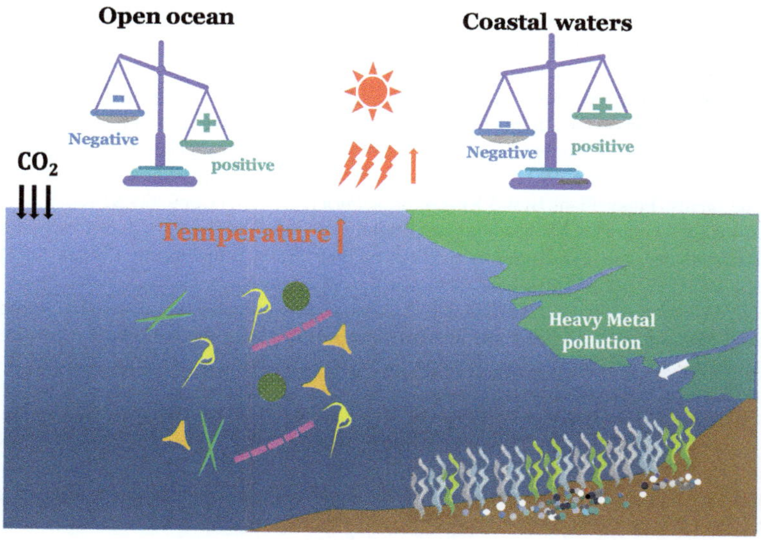

Fig. 18.3 This conceptual diagram shows the effects of ocean acidification on marine primary producers under multiple stressors. For details see text

consequences inferred are worrying. Taken together, OA can influence the biochemical compositions and contents in primary producers and their transfer to higher trophic levels, and marine food webs are likely destabilized.

To summarize, in general OA showed positive effects by enhancing growth of coastal diatoms and macroalgae that are adapted to fluctuating diel pH changes and then potentially enhanced its contribution in carbon sequestration in coastal waters (Fig. 18.3). However, OA is supposed to decrease pelagic primary productivity under multiple stressors (e.g., in combination with ultraviolet radiation, deoxygenation, warming), especially in oligotrophic waters. Therefore, both the magnitude and direction of the response of microalgae and macroalgae to OA largely depend on the levels of other environmental drivers (e.g., warming, deoxygenation). Furthermore, OA can influence the biochemical compositions and contents in primary producers and their transfer to higher trophic levels and marine food webs is likely to be destabilized.

Acknowledgements This work was funded by the National Natural Science Foundation (41720104005, 41806141, 41721005). The authors are grateful to the laboratory engineers Xianglan Zeng and Wenyan Zhao for their technical supports.

References

Agustí S, González-Gordillo JI, Vaqué D, Estrada M, Cerezo MI, Salazar G, Gasol JM, Duarte CM (2015) Ubiquitous healthy diatoms in the deep sea confirm deep carbon injection by the biological pump. Nat Commun 6:1–8. https://doi.org/10.1038/ncomms8608

Bermúdez R, Feng Y, Roleda MY, Tatters AO, Hutchins DA, Thomas L, Boyd PW, Hurd CL, Ulf R, Monika W (2015) Long-term conditioning to elevated pCO_2 and warming influences the fatty and amino acid composition of the diatom *Cylindrotheca fusiformis*. PLoS ONE 10(5): https://doi.org/10.1371/journal.pone.0123945

Bermúdez JR, Winder M, Stuhr A, Almén AK, Engström-Öst J, Riebesell U (2016) Effect of ocean acidification on the structure and fatty acid composition of a natural plankton community in the Baltic Sea. Biogeosciences 13(24):6625–6635. https://doi.org/10.5194/bg-13-6625-2016

Black C, Burris J, Everson R (1976) Influence of oxygen concentration on photosynthesis in marine plants. Funct Plant Biol 3:81–86. https://doi.org/10.1071/pp9760081

Bopp L, Le QC, Heimann M, Manning AC, Mofray P (2002) Climate-induced oxygen fluxes: implications for the contemporary carbon budget. Glob Biogeochem Cycles 16(2):1022. https://doi.org/10.1029/2001GB001445

Boyd PW, Collins S, Dupont S, Fabricius K, Gattuso J, Havenhand J, Hutchins DA, Riebesell U, Rintoul MS, Vichi M (2018) Experimental strategies to assess the biological ramifications of multiple drivers of global ocean change—a review. Global Change Biol 24(6):2239–2261. https://doi.org/10.1111/gcb.14102

Breitburg D, Levin LA, Oschlies A, Grégoire M, Zhang J (2018) Declining oxygen in the global ocean and coastal waters. Science 359(6371):eaam7240. https://doi.org/10.1126/science.aam7240

Brennan G, Collins S (2015) Growth responses of a green alga to multiple environmental drivers. Nat Clim Change 5:892–897. https://doi.org/10.1038/nclimate2682

Brennan GL, Colegrave N, Collins S (2017) Evolutionary consequences of multidriver environmental change in an aquatic primary producer. Proc Natl Acad Sci USA 114:9930–9935. https://doi.org/10.1073/pnas.1703375114

Brewer PG, Peltzer ET (2009) Oceans: limits to marine life. Science 324(5925):347–348. https://doi.org/10.1126/science.1170756

Burkhardt S, Zondervan I, Riebesell U (1999) Effect of CO_2 concentration on C:N: P ratio in marine phytoplankton: a species comparison. Limnol Oceanogr 44:683–690. https://doi.org/10.4319/lo.1999.44.3.0683

Cai WJ, Hu XP, Huang WJ, Murrell MC, Lehrter JC, Lohrenz SE, Chou WC, Zhai WD, Hongbaugh JT, Wang YC, Zhao PS, Guo XH, Gundersen K, Dai MH, Gong GC (2011) Acidification of subsurface coastal waters enhanced by eutrophication. Nat Geosci 4(11):766–770. https://doi.org/10.1038/ngeo1297

Caldeira K, Wickett ME (2003) Anthropogenic carbon and ocean pH. Nature 425(6956):365. https://doi.org/10.1038/425365a

Cao R, Liu Y, Wang Q, Dong Z, Yang D, Hui L, Wen R, Qu Y, Zhao J (2018) Seawater acidification aggravated cadmium toxicity in the oyster *Crassostrea gigas*: metal bioaccumulation, subcellular distribution and multiple physiological responses. Sci Total Environ 642:809–823. https://doi.org/10.1016/j.scitotenv.2018.06.126

Cao RW, Zhang TY, Li X, Zhao YT, Wang Q, Yang DL, Qu Y, Liu H, Dong ZJ, Zhao JM (2019) Seawater acidification increases copper toxicity: a multi-biomarker approach with a key marine invertebrate, the Pacific Oyster *Crassostrea gigas*. Aquat Toxicol 210:167–178. https://doi.org/10.1016/j.aquatox.2019.03.002

Capone DG, Hutchins DA (2013) Microbial biogeochemistry of coastal upwelling regimes in a changing ocean. Nat Geosci 6(9):711–717. https://doi.org/10.1038/ngeo1916

Celis-Plá PSM, Martínez B, Korbee N,Hall-Spencer JM, Figueroa FL (2017) Photoprotective responses in a brown macroalgae *Cystoseira tamariscifolia* to increases in CO_2 and temperature. Mar Environ Res 130(sep):157–165. https://doi.org/10.1016/j.marenvres.2017.07.015

Chakraborty P, Raghunadh BPV, Acharyya T, Bandyopadhyay D (2010) Stress and toxicity of biologically important transition metals (Co, Ni, Cu and Zn) on phytoplankton in a tropical freshwater system: an investigation with pigment analysis by HPLC. Chemosphere 80(5):548–553. https://doi.org/10.1016/j.chemosphere.2010.04.039

Chakravarti LJ, Beltran VH, Van Oppen MJ (2017) Rapid thermal adaptation in photosymbionts of reef-building corals. Global Change Biol 23:4675–4688. https://doi.org/10.1111/gcb.13702

Collins M, Knutti R, Arblaster J Dufresne J-L, Fichefet T, Friedlingstein P, Gao X, Gutowski WJ, Johns T, Krinner G, Shongwe M, Tebaldi C, Weaver AJ, Wehner MF, Allen MR, Andrews T, Beyerle U, Bitz CM, Bony S, Booth BBB (2013) Long-term climate change: projections, commitments and irreversibility. In: Stocker T (ed) Climate change 2013: the physical science basis (chapter: 12). Cambridge University Press, pp 1029–1136

Collins S, Boyd PW, Doblin MA (2020) Evolution, microbes, and changing ocean conditions. Annu Rev Mar Sci 12:181–208. https://doi.org/10.1146/annurev-marine-010318-095311

Connell SD, Russell BD (2010) The direct effects of increasing CO_2 and temperature on non-calcifying organisms: increasing the potential for phase shifts in kelp forests. Proc R Soc B 277:1409–1415

Cornwall CE, Hurd C (2019) Variability in the benefits of ocean acidification to photosynthetic rates of macroalgae without CO_2-concentrating mechanisms. Mar Freshwater Res 71(3):275–280. https://doi.org/10.1071/MF19134

Cornwall CE, Hepburn CD, Pritchard DW (2012) Carbon-use strategies in macroalgae: differential responses to lowered pH and implications for ocean acidification. J Phycol 48:137–144. https://doi.org/10.1111/J.1529-8817.2011.01085.X

Cornwall CE, Hepburn CD, Pilditch CA, Hurd CL (2013) Concentration boundary layers around complex assemblages of macroalgae: implications for the effects of ocean acidification on understory coralline algae. Limnol Oceanogr 58(1):121–130. https://doi.org/10.4319/lo.2013.58.1.0121

Czerny J, Schulz KG, Boxhammer T, Bellerby RGJ, Engel A, Krug S, Ludwig A, Nachtigall K, Nondal G, Niehoff B (2013) Implications of elevated CO_2 on pelagic carbon fluxes in an Arctic mesocosm study—an elemental mass balance approach. Biogeosciences 10(5):3109–3125. https://doi.org/10.5194/bg-10-3109-2013

Dai M, Lu Z, Zhai W (2009) Diurnal variations of surface seawater pCO_2 in contrasting coastal environments. Limnol Oceanogr 54:735–745. https://doi.org/10.4319/lo.2009.54.3.0735

Davis AP, Shokouhian M, Ni SB (2001) Loading estimates of lead, copper, cadmium and zinc in urban runoff from specific sources. Chemosphere 44:997–1009. https://doi.org/10.1016/s0045-6535(00)00561-0

de los Santos CB, Arenas F, Neuparth T (2019) Interaction of short-term copper pollution and ocean acidification in seagrass ecosystems: toxicity, bioconcentration and dietary transfer. Mar Pollut Bull 142:155–163. https://doi.org/10.1016/j.marpolbul.2019.03.034

Diaz RJ, Rosenberg R (2008) Spreading dead zones and consequences for marine ecosystems. Science 321(5891):926–929. https://doi.org/10.1126/science.1156401

Doney SC, Ruckelshaus M, Duffy JE, Barry JP, Chan F, English CA, Galindo HM, Grebmeier JM, Hollowed AB, Knowlton N (2012) Climate change impacts on marine ecosystems. Annu Rev Mar Sci 4(1):11–37. https://doi.org/10.1146/annurev-marine-041911-111611

Doney SC, Busch DS, Cooley SR, Kroeker KJ (2020) The impacts of ocean acidification on marine ecosystems and reliant human communities. Annu Rev Env Resour 45:83–112. https://doi.org/10.1146/annurev-environ-012320-083019

Dong F, Zhu X, Qian W, Wang P, Wang J (2019) Combined effects of CO_2-driven ocean acidification and Cd stress in the marine environment: enhanced tolerance of *Phaeodactylum tricornutum* to Cd exposure. Mar Pollut Bull 150:110594 https://doi.org/10.1016/j.marpolbul.2019.110594

Dong F, Zhu X, Qian W, Wang P, Wang J (2020) Combined effects of CO_2-driven ocean acidification and Cd stress in the marine environment: enhanced tolerance of *Phaeodactylum tricornutum* to Cd exposure. Mar Pollut Bull 150: https://doi.org/10.1016/j.marpolbul.2019.110594

Duarte CM, Hendriks IE, Moore TS, Ylva (2013) Is ocean acidification an open-ocean syndrome? Understanding anthropogenic impacts on seawater pH. Estuar Coast 36(2):221–236. https://doi.org/10.1007/s12237-013-9594-3

Field CB, Behrenfeld MJ, Randerson JT, Falkowski P (1998) Primary production of the biosphere: integrating terrestrial and oceanic components. Science 281(5374):237–240. https://doi.org/10.1126/science.281.5374.237

Flint KR, Davis AP (2007) Pollutant mass flushing characterization of highway stormwater run off from an ultra-urban area. J Environ Eng 133:616–626. https://doi.org/10.1061/(ASCE)0733-9372(2007)133:6(616)

Flynn KJ, Blackford JC, Baird ME, Raven JA, Clark DR, Beardall J, Browniee C, Fabian H, Wheeler GL (2012) Changes in pH at the exterior surface of plankton with ocean acidification. Nat Clim Change 2(10):510–513. https://doi.org/10.1038/nclimate1696

Frolicher TL, Joos F, Plattner GK, Steinacher M, Doney SC (2009) Natural variability and anthropogenic trends in oceanic oxygen in a coupled carbon cycle-climate model ensemble. Glob Biogeochem Cycles 23(1) n/a-n/a. https://doi.org/10.1029/2008gb003316

Fu FX, Warner ME, Zhang YH, Feng Y, Hotchins DA (2007) Effects of increased temperature and CO_2 on photosynthesis, growth, and elemental ratios in marine *Synechococcus* and *Prochlorococcus* (*Cyanobacteria*). J Phycol 43(3):485–496. https://doi.org/10.1111/j.1529-8817.2007.00355.x

Gaitán-Espitia JD, Hancock JR, Padilla-Gamiño JL, Rivest EB, Blanchette CA, Reed DC, Hofmann GE (2014) Interactive effects of elevated temperature and pCO_2 on early-life-history stages of the giant kelp *Macrocystis pyrifera*. J Exp Mar Biol Ecol 457:51–58. https://doi.org/10.1016/j.jembe.2014.03.018

Gao K (1992) Effects of seawater current speed on the photosynthetic oxygen evolution of *Sargassum thunbergii* (Phaeophyta). Jap J Phycol 39:291–293. https://doi.org/10.1111/1440-1835.ep17141356

Gao KS, Campbell D (2014) Photophysiological responses of marine diatoms to elevated CO_2 and decreased pH: a review. Funct Plant Biol 41(5):449–459. https://doi.org/10.1071/FP13247

Gao KS, Zheng YQ (2010) Combined effects of ocean acidification and solar UV radiation on photosynthesis, growth, pigmentation and calcification of the coralline alga *Corallina sessilis* (Rhodophyta). Global Change Biol 16(8):2388–2398. https://doi.org/10.1111/j.1365-2486.2009.02113.x

Gao KS, Aruga Y, Asada K, Ishihara T, AkanoMasataka T (1991) Enhanced growth of the red alga *Porphyra yezoensis* Ueda in high CO_2 concentrations. J Appl Phycol 3(4):355–362. https://doi.org/10.1007/BF00026098

Gao KS, Aruga Y, Asada K, Kiyohara M (1993) Influence of enhanced CO_2 on growth and photosynthesis of the red algae *Gracilaria sp.* and *G chilensis*. J Appl Phycol 5(6):563–571. https://doi.org/10.1007/bf02184635

Gao KS, Wu YP, Li G, Wu HY, Villafañe VE, Helbling EW (2007) Solar UV-radiation drives CO_2-fixation in marine phytoplankton: a double-edged sword. Plant Physiol 144:54–59. https://doi.org/10.1104/pp.107.098491

Gao KS, Ruan ZX, Villafañe BVE, Gattuso JP, Helbling EW (2009) Ocean acidification exacerbates the effect of UV radiation on the calcifying phytoplankter *Emiliania huxleyi*. Limnol Oceanogr 54(6):1855–1862. https://doi.org/10.4319/lo.2009.54.6.1855

Gao KS, Xu JT, Gao G, Li Y, Riebesell U (2012) Rising CO_2 and increased light exposure synergistically reduce marine primary productivity. Nat Clim Change 2:519–523. https://doi.org/10.1038/nclimate1507

Gao G, Clare AS, Rose C, Caldwell GS (2017a) Eutrophication and warming-driven green tides (*Ulva rigida*) are predicted to increase under future climate change scenarios. Mar Pollut Bull 114:439–447. https://doi.org/10.1016/j.marpolbul.2016.10.003

Gao G, Jin P, Liu N, Li F, Tong S, Hutchins DA, Gao K (2017b) The acclimation process of phytoplankton biomass, carbon fixation and respiration to the combined effects of elevated temperature and pCO_2 in the northern South China Sea. Mar Pollut Bull 118(1–2):213–220. https://doi.org/10.1016/j.marpolbul.2017.02.063

Gao G, Xu Z, Qi S, Wu H (2018) Increased CO_2 exacerbates the stress of ultraviolet radiation on photosystem II function in the diatom *Thalassiosira weissflogii*. Environ Exp Bot 156:96–105. https://doi.org/10.1016/j.envexpbot.2018.08.031

Gao K, Beardall J, Häder DP, Hall-Spencero JM, Gao G, Hutchins DA (2019) Effects of ocean acidification on marine photosynthetic organisms under the concurrent influences of warming, UV Radiation, and Deoxygenation. Front Mar Sci 6:322. https://doi.org/10.3389/fmars.2019.00322

Gao KS, Gao G, Wang Y, Dupont S (2020) Impacts of ocean acidification under multiple stressors on typical organisms and ecological processes. Mar Life Sci Technol 1–13. https://doi.org/10.1007/s42995-020-00048-w

Gattuso JP, Magnan A, Billé R, Cheung WWL, Turley C (2015) Contrasting futures for ocean and society from different anthropogenic CO_2 emissions scenarios. Science 349(6243):aac4722. https://doi.org/10.1126/science.aac4722

Gilly WF, Beman JM, Litvin SY, Robison BH (2013) Oceanographic and biological effects of shoaling of the oxygen minimum zone. Annu Rev Mar Sci 5:393–420. https://doi.org/10.1146/annurev-marine-120710-100849

Giordano M, Beardall J, Raven JA (2005) CO_2 concentrating mechanisms in algae: mechanisms, environmental modulation and evolution. Ann Rev Plant Biol 56:99–131. https://doi.org/10.1146/annurevarplant56032604.144052

Götze S, Matoo OB, Beniash E, Saborowski R, Sokolova IM (2014) Interactive effects of CO_2 and trace metals on the proteasome activity and cellular stress response of marine bivalves *Crassostrea virginica* and *Mercenaria mercenaria*. Aquat Toxicol 149:65–82. https://doi.org/10.1016/j.aquatox.2014.01.027

Graba-Landry A, Hoey AS, Matley JK, Sheppard-Brennand H, Poore AGB, Byrne M, Dworjanyn SA (2018) Ocean warming has greater and more consistent negative effects than ocean acidification on the growth and health of subtropical macroalgae. Mar Ecol Prog Ser 595 (May14):55–69

Grist JP, Josey SA, Jacobs ZL, Marsh R, Sebille EV (2016) Extreme air-sea interaction over the North Atlantic subpolar gyre during the winter of 2013–2014 and its sub-surface legacy. Clim Dynam 46(11):4027–4045. https://doi.org/10.1007/s00382-015-2819-3

Haas AF, Smith JE, Thompson M, Deheyn DD (2014) Effects of reduced dissolved oxygen concentrations on physiology and fluorescence of hermatypic corals and benthic algae. PeerJ 2: https://doi.org/10.7717/peerj235

Häder D-P, Williamson CE, Wängberg S-A, Rautio M, Rose KC, Gao K, Helbling EW, Sinha RP, Worrest R (2015) Effects of UV radiation on aquatic ecosystems and interactions with other environmental factors. Photochem Photobiol 14(1):108–126. https://doi.org/10.1039/c4pp90035a

Hay ME, Kubanek J (2002) Community and ecosystem level consequences of chemical cues in the plankton. J Chem Ecol 28(10):2001–2016. https://doi.org/10.1023/A:1020797827806

Hall-Spencer JM, Rodolfo-Metalpa R, Martin S, Ransome E, Fine M, Turner SM, Rowley SJ, Tedesco D, Buia MC (2008) Volcanic carbon dioxide vents show ecosystem effects of ocean acidification. Nature 454:96–99. https://doi.org/10.1038/nature07051

Hofmann LC, Yildiz G, Hanelt D, Bischof K (2012) Physiological responses of the calcifying rhodophyte, *Corallina officinalis* (L), to future CO_2 levels. Mar Biol 159:783–792. https://doi.org/10.1007/s00227-011-1854-9

Hofmann LC, Straub S, Bischof K (2013) Elevated CO_2 levels affect the activity of nitrate reductase and carbonic anhydrase in the calcifying rhodophyte *Corallina officinalis*. J Exp Bot 64(4):899–908. https://doi.org/10.1093/jxb/ers369

Hong H, Shen R, Zhang F, Wen ZF, Chang SW, Lin WF, Kranz SA, Luo YW, Kao S, Morel FMM, Shi DL (2017) The complex effects of ocean acidification on the prominent N_2-fixing cyanobacterium *Trichodesmium*. Science 356(6337):527. https://doi.org/10.1126/science.aal2981

Hönisch B, Ridgwell A, Schmidt DN, Thomas E, Gibbs SJ, Sluijs A, Zeebe R, Kump L, Martindale RC, Greene SE (2012) The geological record of ocean acidification. Science 335 (6072):1058–1063. https://doi.org/10.1126/science.1208277

Huan L, Gu W, Gao S, Wang G (2016) Photosynthetic activity and proteomic analysis highlights the utilization of atmospheric CO_2 by *Ulva prolifera* (Chlorophyta) for rapid growth. J Phycol 52:1103–1113. https://doi.org/10.1111/jpy.12469

Hurd CL, Andrew L, Bronte T, Boyd PW (2018) Current understanding and challenges for oceans in a higher-CO_2 world. Nat Clim Change 8(8):686–694. https://doi.org/10.1038/s41558-018-0211-0

Hutchins DA, Walworth NG, Webb EA, Saito MA, Moran D, Mcllvin MR, Gale J, Fu FX (2015) Irreversibly increased nitrogen fixation in *Trichodesmium* experimentally adapted to elevated carbon dioxide. Nat Commun 6:8155. https://doi.org/10.1038/ncomms9155

Hutchins DA, Jansson JK, Remais J, Rich VI, Trivedi P (2019) Climate change microbiology-problems and perspectives. Nat Rev Microbiol 17(6):391–396. https://doi.org/10.1038/s41579-019-0178-5

Hyun B, Choi KH, Jang PG, Jang MC, Moon CH, Shin K (2014) Effects of increased CO_2 and temperature on the growth of four diatom species (*Chaetoceros debilis*, *Chaetoceros didymus*, *Skeletonema costatum* and *Thalassiosira nordenskioeldii*) in laboratory experiments. J Environ Sci Int 23(6):1003–1012. https://doi.org/10.5322/JESI.2014.23.6.1003

IGBP, IOC, SCOR (2013) Ocean acidification summary for policymakers-third symposium on the ocean in a high-CO_2 world. International geosphere-biosphere programme, Stockholm

IPCC (2014) Synthesis report. Contribution of working groups I. II and III to the fifth assessment report of the intergovernmental panel on climate change, 151

Jiang ZP, Huang JR, Dai MH, Kao SJ, Hydes DJ, Chou WC, Jan S (2011) Short-term dynamics of oxygen and carbon in productive nearshore shallow seawater systems off Taiwan: observations and modelling. Limnol Oceanogr 56(5):1832–1849

Jiao NZ, Herndl GJ, Hansell DA, Benner R, Kattner G, Wilhelm SW, Kirchman DL, Weinbauer MG, Luo T, Chen F (2010) Microbial production of recalcitrant dissolved organic matter: long-term carbon storage in the global ocean. Nat Rev Microbiol 8:593–599. https://doi.org/10.1038/nrmicro2386

Jin P, Agustí S (2018) Fast adaptation of tropical diatoms to increased warming with trade-offs. Sci Rep 8(1):17771. https://doi.org/10.1038/s41598-018-36091-y

Jin P, Gao K, Villafañe VE, Campbell DA, Helbling EW (2013a) Ocean acidification alters the photosynthetic responses of a *coccolithophorid* to fluctuating UV and visible radiation. Plant Physiol 162(4):2084–2094. https://doi.org/10.1104/pp.113.219543

Jin P, Gao KS, Beardall J (2013b) Evolutionary responses of a coccolithophorid *Gephyrocapsa oceanica* to ocean acidification. Evolution 67(7):1869–1878. https://doi.org/10.1111/evo.12112

Jin P, Wang TF, Liu NN, Dupont S, Beardall J, Boyd PW, Riebesell U, Gao K (2015) Ocean acidification increases the accumulation of toxic phenolic compounds across trophic levels. Nat Commun 6:8714. https://doi.org/10.1038/ncomms9714

Jin P, Duarte CM, Agusti S (2017) Contrasting responses of marine and freshwater photosynthetic organisms to UVB radiation: a meta-analysis. Front Mar Sci 4:45. https://doi.org/10.3389/fmars.2017.00045

Ji Y, Gao K (2020) Effects of climate change factors on marine macroalgae: a review. Adv Mar Biol. https://doi.org/10.1016/bs.amb.2020.11.001

Jin P, Liu N, Gao K (2019) Physiological responses of a coccolithophore to multiple environmental drivers. Mar Pollut Bull 146(Sep.):225–235. https://doi.org/10.1016/j.marpolbul.2019.06.032

Jin P, Gonzàlez G, Agustí S (2020) Long-term exposure to increasing temperature can offset predicted losses in marine food quality (fatty acids) caused by ocean warming. Evol Appl 13 (9):2497–2506. https://doi.org/10.1111/eva.13059

Johnson MD, Price NN, Smith JE (2014) Contrasting effects of ocean acidification on tropical fleshy and calcareous algae. Peer J 2: https://doi.org/10.7717/peerj.411

Keeling RF (2010) Ocean deoxygenation in a warming world. Annu Rev Mar Sci 2(1):199. https://doi.org/10.1146/annurev.marine.010908.163855

Kim M, Brodersen KE, Szabó M, Larkum AWD, Raven JA, Ralph PJ, Pernice M (2018) Low oxygen affects photophysiology and the level of expression of two carbon metabolism genes in the seagrass *Zostera muelleri*. Photosynth Res 136(2):147–160. https://doi.org/10.1007/s11120-017-0452-1

Kim JH, Kim N, Moon H, Lee S, Jeong SY, Diaz-Pulido G, Edward MS, Kang JH, Kang EJ, Oh HJ, Hwang JD, Kim IN (2020) Global warming offsets the ecophysiological stress of ocean acidification on temperate *crustose coralline algae*. Mar Pollut Bull 157: https://doi.org/10.1016/j.marpolbul.2020.111324

Kitaya Y, Xiao L, Masuda A, Wzawa T, Tsuda M, Omasa K (2008) Effects of temperature, photosynthetic photon flux density, photoperiod and O_2 and CO_2 concentrations on growth rates of the symbiotic dinoflagellate, *Amphidinium* sp. J Appl Phycol 20(5):737–742. https://doi.org/10.1007/s10811-008-9331-7

Kram SL, Price NN, Donham EM, Johnson MD (2016) Variable responses of temperate calcified and fleshy macroalgae to elevated pCO_2 and warming. ICES J Mar Sci 73(3):693–703. https://doi.org/10.1093/icesjms/fsv168

Kremp A, Godhe A, Egardt J, Dupont S, Suikkanen S, Casabianca S, Penna A (2012) Intraspecific variability in the response of bloom-forming marine microalgae to changed climate conditions. Ecol Evol 2(6):1195–1207. https://doi.org/10.1002/ece3.245

Kübler JE, Johnston AM, Raven JA (1999) The effects of reduced and elevated CO_2 and O_2 on the seaweed *Lomentaria articulata*. Plant, Cell Environ 22(10):1303–1310. https://doi.org/10.1046/j.1365-3040.1999.00492.x

Leal PP, Hurd CL, Fernández PA, Roleda MY (2017) Meiospore development of the kelps *Macrocystis pyrifera* and *Undaria pinnatifida* under ocean acidification and ocean warming: independent effects are more important than their interaction. Mar Biol 164:7. https://doi.org/10.1007/s00227-016-3039-z

Li G, Gao KS (2012) Variation in UV irradiance related to stratospheric ozone levels affects photosynthetic carbon fixation of winter phytoplankton assemblages from surface coastal water of the South China Sea. Mar Biol Res 8(7):670–676. https://doi.org/10.1080/17451000.2011.653368

Li FT, Wu YP, Hutchins DA, Fu FX, Gao KS (2016) Physiological responses of coastal and oceanic diatoms to diurnal fluctuations in seawater carbonate chemistry under two CO_2 concentrations. Biogeosciences 13(22):6247–6259. https://doi.org/10.5194/bg-2016-281

Li FT, Beardall J, Collins S, Gao KS (2017) Decreased photosynthesis and growth with reduced respiration in the model diatom *Phaeodactylum tricornutum* grown under elevated CO_2 over 1800 generations. Global Change Biol 23:127–137. https://doi.org/10.1111/gcb.13501

Li FT, Beardall J, Gao KS (2018) Diatom performance in a future ocean: interactions between nitrogen limitation, temperature, and CO_2-induced seawater acidification. ICES J Mar Sci 75 (4):1451–1464. https://doi.org/10.1093/icesjms/fsx239

Listmann L, Leroch M, Schlüter L, Thomas MK, Reusch TBH (2016) Swift thermal reaction norm evolution in a key marine phytoplankton species. Evol Appl 9(9):1156–1164. https://doi.org/10.1111/eva.12362

Liu FJ, Li SX, Zheng FY, Huang XG (2016) Risk assessment of excessive CO_2 emission on diatom heavy metal consumption. Sci Total Environ 566–567(oct.1):1349–1354. https://doi.org/10.1016/j.scitotenv.2016.05.196

Liu HB, Chen MR, Zhu F, Harrison PJ (2016b) Effect of diatom silica content on copepod grazing, growth and reproduction. Front Mar Sci 3:1–7. https://doi.org/10.3389/fmars.2016.00089

Lohbeck KT, Riebesell U, Reusch TB (2012) Adaptive evolution of a key phytoplankton species to ocean acidification. Nat Geosci 5:346–351. https://doi.org/10.1038/ngeo1441

Loos LMVD, Schmid M, Leal PP, McGraw CM, Britton D, Revill AT, Virtue P, Nichols PD, Hurd CL (2019) Responses of macroalgae to CO_2 enrichment cannot be inferred solely from their inorganic carbon uptake strategy. Ecol Evol 9(1):125–140. https://doi.org/10.1002/ece3.4679

Ma J, Wang W, Qu L, Liu XY, Wang ZQ, Qiao S, Wu HL, Gao G, Xu JT (2019) Differential photosynthetic response of a green tide alga *Ulva linza* to ultraviolet radiation, under short- and long-term ocean acidification regimes. Photochem Photobiol 95:990–998. https://doi.org/10.1111/php.13083

Madronich S, McKenzie RL, Björn LO, Caldwell MM (1998) Changes in biologically active ultraviolet radiation reaching the Earth's surface. Photochem Photobiol B 46:5–19. https://doi.org/10.1016/S1011-1344(98)00182-1

Martin S, Gattuso JP (2009) Response of Mediterranean coralline algae to ocean acidification and elevated temperature. Global Change Biol 15(8):2089–2100. https://doi.org/10.1111/j.1365-2486.2009.01874.x

Martin S, Hall-Spencer JM (2017) Effects of ocean warming and acidification on Rhodolith/Maërl beds. In: Riosmena-Rodríguez, Nelson W, Aguirre J (eds) Rhodolith/Maërl beds: a global perspective. Springer, Switzerland, pp 55–85

McCoy SJ, Santillán-Sarmiento A, Brown MT, Widdicombe S, Wheeler GL (2020) Photosynthetic responses of turf-forming red macroalgae to high CO_2 conditions. J Phycol 56:85–96. https://doi.org/10.1111/jpy.12922-19-026

Mcleod E, Chmura GL, Bouillon S, Salm R, Björk M, Duarte CM, Lovelock CE, Schlesinger WH, Silliman BR (2011) A blueprint for blue carbon: toward an improved understanding of the role of vegetated coastal habitats in sequestering CO_2. Front Ecol Environ 9(10):552–560. https://doi.org/10.1890/110004

Millero FJ, Woosley R, DiTrolio B, Waters J (2009) Effect of ocean acidification on the speciation of metals in seawater. Oceanography 22(4):72–85. https://doi.org/10.5670/oceanog.2009.98

Milligan AJ, Varela DE, Brzezinski MA, Morel FMM (2004) Dynamics of silicon metabolism and silicon isotopic discrimination in a marine diatom as a function of pCO_2. Limnol Oceanogr 49:322–329. https://doi.org/10.2307/3597843

Molina MJ, Rowland FS (1974) Stratospheric sink for chlorofluoromethanes: chlorine atom-catalysed destruction of ozone. Nature 249(5460):810–812. https://doi.org/10.1038/249810a0

Moreira A, Figueira E, Soares AMVM, Freitas R (2016) The effects of arsenic and seawater acidification on antioxidant and biomineralization responses in two closely related *Crassostrea* species. Sci Total Environ 545–546:569–581. https://doi.org/10.1016/j.scitotenv.2015.12.029

Nagelkerken I, Connell SD (2015) Global alteration of ocean ecosystem functioning due to increasing human CO_2 emissions. P Natl Acad Sci USA 112:13272–13277. https://doi.org/10.1073/pnas.1510856112

Nagelkerken I, Goldenberg SU, Ferreira CM, Hadayet Ullah, Connell SD (2020) Trophic pyramids reorganize when food web architecture fails to adjust to ocean change. Science 369 (6505):829–832. https://doi.org/10.1126/science.aax0621

Nardi A, Benedetti M, Fattorini D, Regoli F (2018) Oxidative and interactive challenge of cadmium and ocean acidification on the smooth scallop *Flexopecten glaber*. Aquat Toxicol 196:53–60. https://doi.org/10.1016/j.aquatox.2018.01.008

Nevitt GA, Veit RR, Kareiva P (1995) Dimethyl sulfide as a foraging cue for Antarctic procellariform seabirds. Nature 376(6542):680–682. https://doi.org/10.1038/376680ao

O'Donnell DR, Hamman CR, Johnson EC, Klausmeier CA, Litchman E (2018) Rapid thermal adaptation in a marine diatom reveals constraints and trade-offs. Global Change Biol 24 (10):4554–4565

O'Donnell DR, Du ZY, Litchman E (2019) Experimental evolution of phytoplankton fatty acid thermal reaction norms. Evol Appl 12:1201–1211. https://doi.org/10.1111/eva.12798

Olabarria C, Arenas F, Viejo RM, Gestoso I, Vaz-Pinto F, Incera M, Rubal M, Cacabelos E, Veiga P, Sobrino C (2013) Response of macroalgal assemblages from rockpools to climate change effects of persistent increase in temperature and CO_2. Oikos 122(7):1065–1079. https://doi.org/10.1111/j.1600-0706.2012.20825.x

Ortega A, Geraldi NR, Alam I, Kamau AA, Duarte CM (2019) Important contribution of macroalgae to oceanic carbon sequestration. Nat Geosci 12(9):748–754. https://doi.org/10.1038/s41561-019-0421-8

Oschlies A, Schulz KG, Riebesell U, Schmittner A (2008) Simulated 21st century's increase in oceanic suboxia by CO_2-enhanced biotic carbon export. Glob Biogeochem Cycles 22 GB4008. https://doi.org/10.1029/2007gb003147

Osso A, Sola Y, Bech J (2011) Evidence for the influence of the North Atlantic Oscillation on the total ozone column at northern low latitudes and midlatitudes during winter and summer seasons. J Geophys Res Atmos 116(D24):1–12. https://doi.org/10.1029/2011jd016539

Padfield D, Yvon-Durocher G, Buckling A, Jennings S, Yvon-Durocher G (2016) Rapid evolution of metabolic traits explains thermal adaptation in phytoplankton. Ecol Lett 19:133–142. https://doi.org/10.1111/ele.12545

Pan K, Wang WX (2012) Trace metal contamination in estuarine and coastal environments in China. Sci Total Environ 421–422(Apr.1):3–16. https://doi.org/10.1016/j.scitotenv.2011.03.013

Patra P, Mohandass C, Chakraborty P (2019) Snapshot of environmental condition in different tropical estuarine systems by using *S. cucullata* (an edible oyster) as bioindicator. Environ Sci Pollut Res 26(11):11342–11354. https://doi.org/10.1007/s11356-019-04564-w

Petrou K, Baker KG, Nielsen DA, Hancock AM, Schulz KG, Davidson AT (2019) Acidification diminishes diatom silica production in the Southern Ocean. Nat Clim Change 9:781–786. https://doi.org/10.1038/s41558-019-0557-y

Pohnert G, Steinke M, Tollrian R (2007) Chemical cues, defence metabolites and the shaping of pelagic interspecific interactions. Trends Ecol Evol 22(4):198–204. https://doi.org/10.1016/j.tree.2007.01.005

Price NN, Hamilton SL, Tootell JS, Smith JE (2011) Species-specific consequences of ocean acidification for the calcareous tropical green algae *Halimeda*. Mar Ecol Prog Ser 440:67–78. https://doi.org/10.3354/meps09309

Raso S, Genugten B, Vermuë M (2012) Effect of oxygen concentration on the growth of *Nannochloropsis* sp at low light intensity. J Appl Phycol 24(4):863–871. https://doi.org/10.1007/s10811-011-9706-z

Raven JA, Crawfurd K (2012) Environmental controls on *coccolithophore* calcification. Mar Ecol Prog Ser 470:137–166. https://doi.org/10.3354/meps09993

Reusch TB, Boyd PW (2013) Experimental evolution meets marine phytoplankton. Evolution 67 (7):1849–1859. https://doi.org/10.1111/evo.12035

Riebesell U, Gattuso JP (2015) Lessons learned from ocean acidification research. Nat Clim Change 5:12–14. https://doi.org/10.1038/nclimate2456

Riebesell U, Tortell PD (2012) Effects of ocean acidification on pelagic organisms and ecosystems. In: Gattuso JP, Hanson L (eds) Ocean acidification. Oxford University Press, Oxford, pp 99–121

Riebesell U, Aberle-Malzahn N, Achterberg EP, Algueró-Muiz M, Taucher J (2018) Toxic algal bloom induced by ocean acidification disrupts the pelagic food web. Nat Clim Change 8 (12):1082–1086. https://doi.org/10.1038/s41558-018-0344-1

Ries JB, Cohen AL, McCorkle DC (2009) Marine calcifiers exhibit mixed responses to CO_2-induced ocean acidification. Geology 37(12):1131–1134. https://doi.org/10.1130/G30210A.1

Rossoll D, Bermúdez R, Hauss H, Schulz KG, Riebesell U, Sommer U, Winder M, Thrush S (2012) Ocean acidification-induced food quality deterioration constrains trophic transfer. PLoS ONE 7(4): https://doi.org/10.1371/journal.pone.0034737

Schaum CE, Barton S, Bestion E, Buckling A, Garcia-Carreras B, Lopez P, Lowe C, Pawar S, Smirnoff N, Trimmer M, Yvon-Durocher G (2017) Adaptation of phytoplankton to a decade of experimental warming linked to increased photosynthesis. Nat Ecol Evol 1(4):94. https://doi.org/10.1038/s41559-017-0094

Schaum CE, Buckling A, Smirnoff N, Studholme DJ, Yvon-Durocher G (2018) Environmental fluctuations accelerate molecular evolution of thermal tolerance in a marine diatom. Nat Commun 9(1):1719. https://doi.org/10.1038/s41467-018-03906-5

Scherner F, Pereira CM, Duarte G, Horta PA, Pereira SMB (2016) Effects of ocean acidification and temperature increases on the photosynthesis of tropical reef calcified macroalgae. PLoS ONE 11(5): https://doi.org/10.1371/journal.pone.0154844

Schlüter L, Lohbeck KT, Gutowska MA, Gröger JP, Riebesell U, Reusch TBH (2014) Adaptation of a globally important coccolithophore to ocean warming and acidification. Nat Clim Change 4(11):1024–1030. https://doi.org/10.1038/nclimate2379

Schmidtko S, Stramma L, Visbeck M (2017) Decline in global oceanic oxygen content during the past five decades. Nature 542(7641):335–339. https://doi.org/10.1038/nature21399

Schmittner A, Oschlies A, Matthews HD, Galbraith ED (2008) Future changes in climate, ocean circulation, ecosystems, and biogeochemical cycling simulated for a business-as-usual CO_2 emission scenario until year 4000 AD. Glob Biogeochem Cycles 22(1). https://doi.org/10.1029/2007gb002953

Sinutok S, Hill R, Doblin MA, Wuhrer R, Ralph PJ (2011) Warmer more acidic conditions cause decreased productivity and calcification in subtropical coral reef sediment-dwelling calcifiers. Limnol Oceanogr 56(4):1200–1212. https://doi.org/10.4319/lo.2011.56.4.1200

Six KD, Kloster S, Llynia T, Archer SD, Zhang K, Maier-Reimer E (2013) Global warming amplified by reduced sulphur fluxes as a result of ocean acidification. Nat Clim Chang 3 (11):975–978. https://doi.org/10.1038/NCLIMATE1981

Smetacek VS (1985) Role of sinking in diatom life-history cycles: ecological, evolutionary and geological significance. Mar Biol 184(3):239–251. https://doi.org/10.1007/BF00392493

Sobrino C, Ward ML, Neale PJ (2008) Acclimation to elevated carbon dioxide and ultraviolet radiation in the diatom *Thalassiosira pseudonana*: effects on growth, photosynthesis, and spectral sensitivity of photoinhibition. Limnol Oceanogr 53(2):494–505. https://doi.org/10.4319/lo.2008.53.2.0494

Sswat M, Stiasny MH, Taucher J, Maria AM, Bach LT, Jutfelt F, Riebesell U, Clemmesen C (2018) Food web changes under ocean acidification promote herring larvae survival. Nat Ecol Evol 2(5):836–840. https://doi.org/10.1038/s41559-018-0514-6

Stockdale A, Tipping E, Lofts S, Mortimer RJG (2016) Effect of ocean acidification on organic and inorganic speciation of trace metals. Environ Sci Technol 50(4):1906–1913. https://doi.org/10.1021/acs.est.5b05624

Taucher J, Boxhammer T, Bach LT, Paul AJ, Schartau M, Stange P, Riebesell U (2021) Changing carbon-to-nitrogen ratios of organic-matter export under ocean acidification. Nat Clim Change 11:52–57. https://doi.org/10.1038/s41558-020-00915-5

Tedetti M, Sempéré R, Vasilkov A, Charrière B, Nérini D, Miller WL, Kawamura K, Raimbault P (2007) High penetration of ultraviolet radiation in the South East Pacific waters. Geophys Res Lett 34(12):L12610. https://doi.org/10.1029/2007GL029823

Tong S, Gao K, Hutchins DA (2018) Adaptive evolution in the coccolithophore *Gephyrocapsa oceanica* following 1000 generations of selection under elevated CO_2. Glob Change Biol 24:3055–3064. https://doi.org/10.1111/gcb.14065

Tong S, Hutchins DA, Gao K (2019) Physiological and biochemical responses of Emiliania huxleyi to ocean acidification and warming are modulated by UV radiation. Biogeosci Discuss 16(2):561–572. https://doi.org/10.5194/bg-2017-269

Vallina SM, Simó R (2007) Strong relationship between DMS and the solar radiation dose over the global surface ocean. Science 315(5811):506–508. https://doi.org/10.1126/science.1133680

Vaquer-Sunyer R, Duarte CM (2008) Thresholds of hypoxia for marine biodiversity. Proc Natl A Sci India B 105(40):15452–15457. https://doi.org/10.1073/pnas.0803833105

Vargas CA, Lagos NA, Lardies MA, Duarte C, Manríquez PH, Aguilera VM, Broitman B, Widdicombe S, Dupont S (2017) Species-specific responses to ocean acidification should account for local adaptation and adaptive plasticity. Nat Ecol Evol 1(4):0084. https://doi.org/10.1038/s41559-017-0084

Waldbusser GG, Salisbury JE (2014) Ocean acidification in the coastal zone from an organism's perspective: multiple system parameters, frequency domains, and habitats. Annu Rev Mar Sci 6:221–247. https://doi.org/10.1146/annurev-marine-121211-172238

Wang WX (2002) Interactions of trace metals and different marine food chains. Mar Ecol Prog Ser 243(Nov):295–309. https://doi.org/10.3354/meps243295

Wang GZ, Jing WP, Wang SL, Xu Y, Wang ZY (2014) Coastal acidification induced by tidal-driven submarine groundwater discharge in a coastal coral reef system. Environ Sci Technol 48(22):13069–13075. https://doi.org/10.1021/es5026867

Wang T, Jin P, Wells ML, Trick CG, Gao K (2019) Insensitivities of a subtropical productive coastal plankton community and trophic transfer to ocean acidification: results from a microcosm study. Mar Pollut Bull 141:462–471. https://doi.org/10.1016/j.marpolbul.2019.03.002

Weatherhead EC, Andersen SB (2006) The search for signs of recovery of the ozone layer. Nature 441(7089):39–45. https://doi.org/10.1038/nature04746

Williamson CE, Zepp RG, Lucas RM, Madronich S, Austion AT, Ballaré CL, Norval M, Sulzberger B, Bais AF, McKenzie RL, Robinson SA, Häder D-P, Paul ND, Bornman JF (2014) Solar ultraviolet radiation in a changing climate. Nat Clim Change 4(6):434–441. https://doi.org/10.1038/nclimate2225

Wolfe GV, Steinke M, Kirst GO (1997) Grazing-activated chemical defence in a unicellular marine alga. Nature 387(6636):894–897. https://doi.org/10.1038/43168

Wu Y, Gao K, Riebesell U (2010) CO_2-induced seawater acidification affects physiological performance of the marine diatom *Phaeodactylum tricornutum*. Biogeosciences 7(9):2915–2923. https://doi.org/10.5194/bg-7-2915-2010

Wu YP, Li ZZ, Du WJ, Gao KS (2015) Physiological response of marine centric diatoms to ultraviolet radiation, with special reference to cell size. J Photochem Photobiol B 153:1–6. https://doi.org/10.1016/j.jphotobiol.2015.08.035

Xu JT, Gao KS (2010) UV-A enhanced growth and UV-B induced positive effects in the recovery of photochemical yield in *Gracilaria lemaneiformis* (Rhodophyta). J Photochem Photobiol B 100(3):117–122. https://doi.org/10.1016/j.jphotobiol.2010.05.010

Xu K, Gao KS (2015) Solar UV irradiances modulate effects of ocean acidification on the coccolithophorid *Emiliania huxleyi*. Photochem Photobiol 91(1):92–101. https://doi.org/10.1111/php.12363

Xu JT, Gao KS, Li YH, Hutchins DA (2014) Physiological and biochemical responses of diatoms to projected ocean changes. Mar Ecol Prog Ser 515(nov.18):73–81. https://doi.org/10.3354/meps11026

Xu D, Brennan GL, Xu L, Zhang XW, Fan X, Han WT, Mock T, McMinn A, Hutchins DA, Ye N (2019) Ocean acidification increases iodine accumulation in kelp-based coastal food webs. Global Change Biol 25:629–639. https://doi.org/10.1111/gcb.14467

Yang GY, Gao KS (2012) Physiological responses of the marine diatom *Thalassiosira pseudonana* to increased pCO_2 and seawater acidity. Mar Environ Res 79(4):142–151. https://doi.org/10.1016/j.marenvres.2012.06.002

Zark M, Riebesell U, Dittmar T (2015) Effects of ocean acidification on marine dissolved organic matter are not detectable over the succession of phytoplankton blooms. Sci Adv 1(9):e1500531 https://doi.org/10.1126/sciadv.1500531

Zhang XS, Xu D, Huang SJ, Wang SH, Han WT, Liang CW, Zhang Y, Fan X, Zhang XW, Wang YT, Wang W, Egan S, Saha M, Li F, Naihao Ye (2020) The effect of elevated pCO_2 on cadmium resistance of a globally important diatom. J Hazard Mater 396:122749 https://doi.org/10.1016/j.jhazmat.2020.122749

Zou DH, Gao KS, Luo HJ (2011) Short- and long-term effects of elevated CO_2 on photosynthesis and respiration in the marine macroalga *Hizikia fusiformis* (Sargassaceae, Phaeophyta) grown at low and high N supplies. J Phycol 47:87–97

Zweng RC, Koch MS, Bowes G (2018) The role of irradiance and C-use strategies in tropical macroalgae photosynthetic response to ocean acidification. Sci Rep 8(1):9479. https://doi.org/10.1038/s41598-018-27333-0

GPSR Compliance
The European Union's (EU) General Product Safety Regulation (GPSR) is a set of rules that requires consumer products to be safe and our obligations to ensure this.

If you have any concerns about our products, you can contact us on

ProductSafety@springernature.com

In case Publisher is established outside the EU, the EU authorized representative is:

Springer Nature Customer Service Center GmbH
Europaplatz 3
69115 Heidelberg, Germany

www.ingramcontent.com/pod-product-compliance
Ingram Content Group UK Ltd.
Pitfield, Milton Keynes, MK11 3LW, UK
UKHW020712071026
15590UKWH00010B/479

* 9 7 8 3 0 3 0 7 5 6 0 4 8 *